70 Springer Series in Solid-State Sciences

Edited by Peter Fulde

Springer Series in Solid-State Sciences

Editors: M. Cardona   P. Fulde   K. von Klitzing   H.-J. Queisser

Volumes 1–39 are listed on the back inside cover

Jürgen Hafner

From Hamiltonians to Phase Diagrams

The Electronic and Statistical-Mechanical Theory
of sp-Bonded Metals and Alloys

With 147 Figures

Springer-Verlag Berlin Heidelberg New York
London Paris Tokyo

Professor Dr. Jürgen Hafner

Institut für Theoretische Physik, Techn. Universität Wien, Karlsplatz 13,
A-1040 Wien, Austria

Series Editors:

Professor Dr., Dr. h. c. Manuel Cardona
Professor Dr., Dr. h. c. Peter Fulde
Professor Dr. Klaus von Klitzing
Professor Dr. Hans-Joachim Queisser

Max-Planck-Institut für Festkörperforschung, Heisenbergstrasse 1
D-7000 Stuttgart 80, Fed. Rep. of Germany

ISBN-13:978-3-642-83060-0   e-ISBN-13:978-3-642-83058-7
DOI: 10.1007/978-3-642-83058-7

Library of Congress Cataloging-in-Publication Data. Hafner, J. (Jürgen), 1945-. From Hamiltonians to phase diagrams. (Springer series in solid-state sciences ; 70). Bibliography: p. Includes index. 1. Metal-metal bonds. 2. Alloys. 3. Phase transformations (Statistical physics) 4. Phase diagrams. I. Title. II. Series. QD461.H16 1987 530.4'1 87-9846

© Springer-Verlag Berlin Heidelberg 1987
Softcover reprint of the hardcover 1st edition 1987

2153/3150-543210

Preface

The development of the modern theory of metals and alloys has coincided with great advances in quantum-mechanical many-body theory, in electronic structure calculations, in theories of lattice dynamics and of the configurational thermodynamics of crystals, in liquid-state theory, and in the theory of phase transformations. For a long time all these different fields expanded quite independently, but now their overlap has become sufficiently large that they are beginning to form the basis of a comprehensive first-principles theory of the cohesive, structural, and thermodynamical properties of metals and alloys in the crystalline as well as in the liquid state. Today, we can set out from the quantum-mechanical many-body Hamiltonian of the system of electrons and ions, and, following the path laid out by generations of theoreticians, we can progress far enough to calculate a pressure-temperature phase diagram of a metal or a composition-temperature phase diagram of a binary alloy by methods which are essentially rigorous and from first principles.

This book was written with the intention of confronting the materials scientist, the metallurgist, the physical chemist, but also the experimental and theoretical condensed-matter physicist, with this new and exciting possibility. Of course there are limitations to such a vast undertaking as this. The selection of the theories and techniques to be discussed, as well as the way in which they are presented, are necessarily biased by personal inclination and personal expertise. The book begins with a discussion of the fundamental concepts that guide our way, starting with things that are as basic as the adiabatic approximation or the idea of the self-consistent field. To the condensed-matter theorist this is of course all familiar and he or she might just glance rapidly over the first chapter, but I hope that the non-specialist will find this introduction helpful. The introduction also sets the framework for the remaining chapters which should provide the reader with a working knowledge of the electronic theory of interatomic interactions, the theory of cohesion and structure, the thermodynamics of crystals, liquid-state theory, etc., as far as is necessary to reach the final goal. These chapters also present a necessarily very personal view of the state of the art in each field. The presentation is as self-contained as possible; no familiarity with advanced theoretical concepts beyond the level of basic quantum mechanics, statistical mechanics and solid-state theory is assumed.

Most of what is presented in this book resides upon the concept of effective interatomic potentials. To date, the most reliable technique for calculating these potentials in metals and alloys is based on reponse theory, which is valid only in systems whose ions scatter the electrons weakly and are describable by pseudopotentials. These are the sp-bonded or simple metals. For this reason most of the book is devoted to this particular class of metals and their alloys, with only short excursions into the theory of d-band or transition metals to sketch the essential differences between sp- and d-bonding and to draw the reader's attention to some interesting new developments. But here, as everywhere else in this book, I could not aim for encyclopaedic completeness.

The book reflects the long-time research interests of the author. Several chapters grew out of lectures given at the Technical University of Vienna and at the Laboratoire de Thermodynamique et Physico-Chimie Métallurgiques of the Ecole Normale Supérieure d'Electrochimie et d'Electrométallurgie de Grenoble. The ideas presented here have been shaped in many discussions with friends and colleagues all over the world. I do not attempt to name them all, as such a list could never be complete. However, I could not forget to mention Volker Heine and the late Heinz Bilz – it has always been a great privilege to work with them and I greatly enjoyed their hospitality at the Cavendish Laboratory of the University of Cambridge and at the Max-Planck-Institut für Festkörperforschung. I should also not forget to mention my friends Pierre Hicter, Ferencz Igloi, Sitaram Jaswal, Gerhard Kahl, Alain Pasturel, Gottfried Punz, and Werner Weber, whose cooperation I deeply appreciate. To all others who might have been cited but have not, I extend my apologies, and I sincerely hope that they will accept my gratitude.

I thank my colleagues at the Institute of Theoretical Physics under the direction of Prof. Otto Hittmair for creating a congenial athmosphere for my scientific work.

I thank Prof. Peter Fulde who suggested that I should write this book. I am grateful to Dr. H.K.V. Lotsch for his patience and to Dr. A.M. Lahee for her efficient cooperation and for making several excellent suggestions concerning the presentation.

Finally, I should not forget to thank my wife Anneliese and my sons Thomas and Michael. They certainly had to bear an absentminded husband and father over many evenings and weekends and they have tolerated it with good humour.

The book is dedicated to my parents.

Vienna, January 1987   *Jürgen Hafner*

Contents

1. Introduction

1.1 Why This Book Was Written and What It Contains

William Hume-Rothery's famous book "Electrons, Atoms, Metals, and Alloys" [1.1] is written in the form of a Galilean dialogue: a "young scientist" and an "older metallurgist" discuss the importance of quantum mechanical theories for re-shaping and re-orienting an old traditional field such as metallurgy. The "older metallurgist" articulates the practitioners scepticism against the introduction of rather abstract, highly mathematical theories into his familiar art. Nowadays, forty years after this book was first published, it is no longer necessary to advocate the importance of quantum mechanics in materials science. However, to many people the idea of establishing a correlation between the Hamiltonian H of the many-body system of electrons and ions and the pressure-temperature phase diagram of a metal or the composition-temperature phase diagram of a binary alloy may still appear as rather far-fetched. Nonetheless, this is precisely what we intend: we want to persuade the reader that condensed matter theory is now sufficiently advanced to allow us to set out with the Hamiltonian H,

$$H = T_i + V_{i-i} + T_e + V_{e-e} + V_{e-i} \tag{1.1}$$

consisting of the kinetic energy operators for the ions and electrons (T_i, T_e) and the potentials V_{i-i}, V_{e-e}, and V_{e-i} describing the forces acting between ions, between electrons, and finally between ions and electrons, and from this to end up with a phase diagram for a metal or alloy.

The foundations of the subject were laid in a famous paper published more then 50 years ago; in 1933 *Wigner* and *Seitz* [1.2] applied the Schrödinger equation to the metallic bond. Their treatment is based on the adiabatic decoupling of electrons and ions, and on the transformation of the complex many-electron problem into an effective one-electron form. For the monovalent metal sodium they were able to calculate the cohesive energy, lattice constant, and bulk modulus with an accuracy of better than 10 %. For a very long time their achievement remained unparalleled; it took several decades until similarly accurate results were obtained for polyvalent metals. The relative simplicity of the monovalent alkali metals stems from the fact that the electron-electron interaction can be neglected, giving an effective potential seen by the valence electrons that is essentially equal to the potential of the ion core. In a polyvalent metal, however, a valence electron experiences a strong electrostatic interaction with the other valence electrons in its vicinity. Thus the effective one-electron potential replacing $(V_{e-i} + V_{e-e})$ in (1.1) must be calculated self-consistently, since it depends on the wave functions of the electrons which are themselves solutions of the Schrödinger equation. Moreover, the quantum-mechanical nature of the interaction has to be taken into account, the Pauli principle keeps the electrons apart, thus lowering the Coulomb repulsion by an amount known as the exchange energy. These statistical correlations are described by the Hartree-Fock approximation. Furthermore there exist dynamical correlations even between electrons of anti-parallel spin which contribute to the correlation energy. *Brooks* [1.3] was the first to point out the importance of these effects in determining the cohesive properties of polyvalent simple metals. During the last twenty years a major breakthrough in condensed matter theory has occured with the discovery that these very complex exchange and correlation phenomena can be described quite accurately by adding a simple local exchange and correlation potential $V_{xc}(r)$ to the Hartree potential $V_H(r)$, which itself describes the average Coulomb interaction between the electrons. The resulting local density functional (LDF) equations [1.4, 5] describe the energetics of atoms, molecules and solids with a surprising accuracy [1.6–8]. Fifty-years after Wigner and Seitz, the cohesive properties can now be obtained to within 10 % of their experimental values for *all* metals (although of course, a spin-polarized version of the LDF is necessary to treat magnetic effects in the cohesive properties [1.9] and a relativistic version is necessary for the very heavy metals [1.10]).

For a solid, the iterative solution of the self-consistent field − or of the local-density functional equations, is greatly facilitated by the use of Bloch's theorem, which allows the problem of diagonalizing a $N\alpha \times N\alpha$ Hamiltonian matrix in real space (N stands for the number of particles, and α for the number of electronic states per particle) to be reduced to the repeated diagonalization of a $r\alpha \times r\alpha$ Hamiltonian matrix in reciprocal space (where r stands for the number of ions in the unit cell). This leads to a description of the bonding between the atoms in a band framework: the discrete energy levels of the free atoms are broadened to form bands of states as the atoms are brought together to build the solid. *Friedel* [1.11] was the first to elucidate the rather different nature of the simple- and transition-metal bands. In the former case the highly mobile *s*- and *p*-electrons are rather close to the nearly-free-electron limit, in the latter case the more localized *d*-electrons are better described in a tight-binding approximation. The knowledge of the energy band behaviour provides a microscopic picture of the metallic bonding in the crystalline elemental metals and in ordered stoichiometric intermetallic compounds.

The band picture breaks down if the system does not possess translational symmetry, since the wave vector of the electron is no longer a conserved quantity. This is the case e.g. for a substitutional alloy with a random distribution of the two atomic species. In principle, the ordering energy (i.e. the difference in the energies of the completely ordered and completely disordered states) can be expressed directly in terms of effective pair interactions between the first, second, third ... nearest neighbour atoms [1.12]. For the nearly-free-electron-like simple metals a successful scheme for calculating these interatomic potentials is based on the response of the homogeneous electron gas to the electron-ion potential. The response theory is applicable only to the nearly-free-electron (NFE) simple metals whose ion cores scatter the electrons only weakly and are describable by pseudopotentials [1.13–15]. For the tight-binding (TB) transition metals, we still do not have a method to derive interatomic potentials from first-principles that really works (although some important steps have been made in this direction just lately [1.16]). Only for the problem of ordering on a specified lattice has it been possible to derive a generalized TB-perturbation approach for the ordering potential [1.17, 18].

Substitutional disorder is not the only reason for the violation of translational symmetry. The translational invariance is also perturbed by the thermal disorder in vibrating crystals at $T \neq 0\,\mathrm{K}$, and it is completely destroyed upon melting. The topological disorder in melts and glasses creates serious difficulties in formulating an electronic theory of the cohesive and structural properties. In each case progress will hinge on a knowledge of the interatomic fordes in the metal or alloy. This is why the bulk of this review will deal with simple sp-bonded metals and alloys; only here, with the firm background of response theory and pseudopotentials, we can proceed to calculations of the thermodynamic properties of the hot vibrating crystal and of the melt. These calculations then form the basis of the construction of a phase diagram. However, we do include short excursions into the properties of transition metals and their alloys as well, just to emphasize the essential difference between the bonding in simple- and in transition-metal systems.

The material to be presented is organized as follows: in the remainder of this chapter we first recapitulate the thermodynamic considerations that are the basis of phase diagrams. This is followed by a brief discussion of the fundamental physical principles underlying the reduction of the many-body electron-ion Hamiltonian (1.1) to an effective ionic Hamiltonian expressible as a sum of volume-, pair-, triplet-, and multi-ion interactions; adiabatic decoupling of the ionic and electronic degrees of freedom; reduction of the many-electron Hamiltonian to an effective one-electron form using the self-consistent field, local-density-functional, and linear-response approaches; methods for solving this one-electron equation for periodic and aperiodic structures. The basic concepts behind the calculation of the thermodynamic functions of crystals (harmonic and anharmonic lattice dynamics, phonon statistics, order-parameter formulation of the configurational thermodynamics) and liquids (correlation functions, integral equations of liquid-state theory) are introduced. These discussions will provide the framework for the remaining chapters.

Chapter 2 describes the calculation of the effective interatomic forces in metals and alloys in the linear-response and pseudopotential framework. It contains a detailed discussion of the characteristic variations in the interatomic interactions in the sp-bonded metals across the periodic table, and of the changes in the interatomic potentials that occur upon alloying. In Chapter 3 we will discuss the phase stability of pure crystalline metals, including dynamic effects and pressure- and temperature-induced phase transitions. In Chapter 4 we describe methods to deduce the structure (i.e. the interatomic correlation functions) and the thermodynamic properties of liquid metals from the known interatomic force law. These provide the basis for the calculations of the pT phase diagrams of pure metals to be reviewed in Chapter 5.

Chapters 6 to 11 deal with binary alloys: phenomenological approaches to alloy phase formation (Chap. 6), and microscopic approaches to solid solutions (Chap. 7), stoichiometric intermetallic compounds (Chap. 8), and liquid alloys (Chap. 9) will be discussed. In Chapter 10 we present calculations of binary alloy phase diagrams and the book closes with an outlook on metastable phases (Chap. 11).

1.2 The Thermodynamic Origin of Phase Diagrams

A phase diagram is a graphical representation of the loci of the thermodynamic variables when equilibrium among the phases of a system is established under a given set of conditions. The phase diagrams most familiar to metallurgists, physicists and chemists are those for one-component systems where temperature and pressure are the axes, and those for two-component systems which plot the phase boundaries as a function of temperature and composition.

A spontaneously occuring phase transition is always accompanied by a decrease of the relevant thermodynamic potential. During a spontaneous $\alpha \rightarrow \beta$ transformation under isothermal-isobaric conditions it is the Gibbs free energy G that decreases, i.e.

$$G_\beta - G_\alpha = \Delta G \leq 0 \ . \tag{1.2}$$

The magnitude of the free-energy decrease at the transformation is regarded as the driving force for the phase change. Whereas the Gibbs free energy changes continuously through a transition ($\Delta G = 0$ at the equilibrium temperature, pressure, composition ...), the derivatives of G can have discontinuities at the transition. The lowest order of the derivatives for which discontinuities occur defines the order of the phase transition. Through the basic thermodynamic laws

$$\left(\frac{\partial G}{\partial T}\right)_{\mathrm{p}} = -S \ , \quad \left(\frac{\partial G}{\partial p}\right)_{\mathrm{T}} = V \tag{1.3}$$

we find that the entropy S and the volume V change abruptly at a first-order transition. Through the relations

$$\left(\frac{\partial^2 G}{\partial T^2}\right) = -\left(\frac{\partial S}{\partial T}\right)_{\mathrm{p}} = -\frac{1}{T}C_{\mathrm{p}} \tag{1.4a}$$

$$\left(\frac{\partial^2 G}{\partial p^2}\right) = \left(\frac{\partial V}{\partial p}\right)_{\mathrm{T}} = -V\chi_T \tag{1.4b}$$

$$\left(\frac{\partial^2 G}{\partial T \partial p}\right) = \left(\frac{\partial V}{\partial T}\right)_{\mathrm{p}} = V\alpha_{\mathrm{p}} \tag{1.4c}$$

we find that a second-order phase transition is accompanied by abrupt changes in the specific heat, the isothermal compressibility, and the isobaric expansion coefficient.

1.2.1 First-Order Transitions Without Compositional Change – The pT Phase Diagram

The basic free-energy equation for a temperature-induced phase transition at constant pressure is

$$G = H - TS$$
$$= H + T\left(\frac{\partial G}{\partial T}\right)_{\mathrm{p}} = H_0 - \int_0^T \left(\frac{1}{T}\int_0^T C_{\mathrm{p}}dT\right)dT \ , \tag{1.5}$$

and for a pressure-induced transformation at constant temperature we have

$$G = E - TS + pV$$
$$= F + pV = F - V\left(\frac{\partial F}{\partial V}\right)_{\mathrm{T}} \ . \tag{1.6}$$

Equation (1.6) shows that if suffices to calculate the free energy of the competing phases as a function of volume. The difference in the Gibbs free energy G at a given pressure p is then given by the difference between the

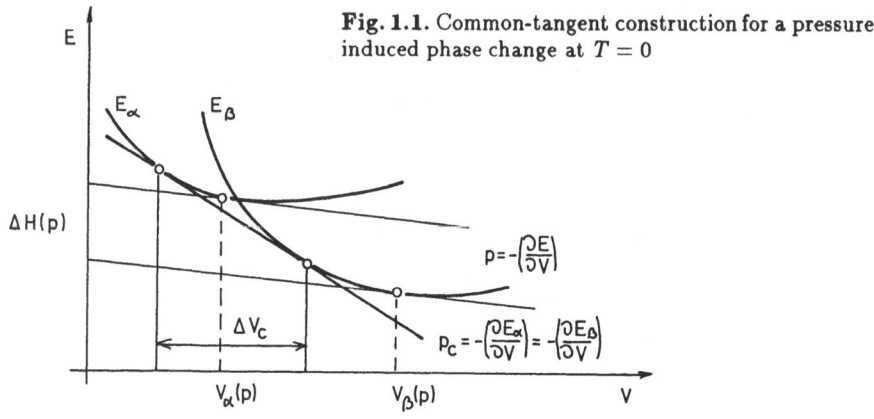

Fig. 1.1. Common-tangent construction for a pressure-induced phase change at $T = 0$

$$p = -\left(\frac{\partial E}{\partial V}\right)$$

$$p_c = -\left(\frac{\partial E_\alpha}{\partial V}\right) = -\left(\frac{\partial E_\beta}{\partial V}\right)$$

intersections of parallel tangents

$$p = -\left(\frac{\partial F_\alpha}{\partial V}\right)_{V=V_\alpha} = -\left(\frac{\partial F_\beta}{\partial V}\right)_{V=V_\beta} \tag{1.7}$$

with the ordinate axis. The transformation occurs at the pressure p_c at which the tangents to the $F_i(V)$ curves coincide (Fig. 1.1). The phase-transition line in the pT plane is constructed by repeating this construction for different temperatures.

1.2.2 First-Order Transitions with Compositional Change – The Alloy Phase Diagram

The phase transitions at constant pressure p and temperature T in a binary system of A and B may be discussed in a similar fashion. In Fig. 1.2 the segment 1–2 represents the change in free energy at a transition $\alpha \rightarrow \beta$ without a change in composition. Now we consider the formation of a small

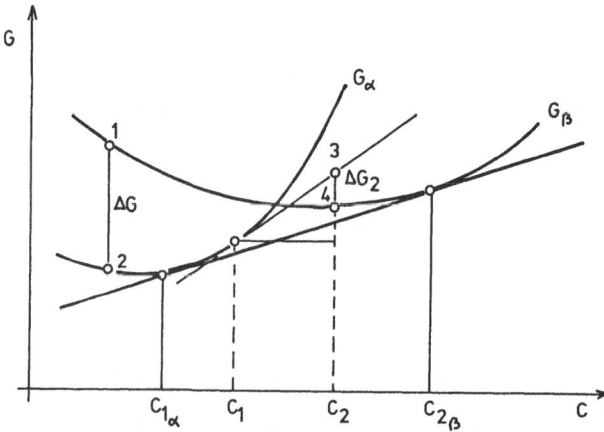

Fig. 1.2. Free energies of two competing phases α and β of a binary mixture as a function of composition

β particle with composition c_2 out of an α matrix with composition c_1. The segment 3–4 represents the thermodynamic driving force for the formation of the β particle, i.e.

$$\Delta G_2 = [G_\beta(c_2) - G_\alpha(c_1)] - (c_2 - c_1)\frac{dG_\alpha}{dc}\bigg|_{c=c_1}. \qquad (1.8)$$

Note that c_2 must be beyond the point of intersection of the two free-energy curves in order for the driving force to be negative. Equilibrium between the α and the β phases clearly requires thermodynamic driving forces for the formation of particles of equilibrium composition (c_1, c_2), to be equal, i.e. the limits of stability of the two competing phases are again given by the common tangent construction

$$\frac{\partial G_\alpha}{\partial c}\bigg|_{c=c_1} = \mu_\alpha(c_1) = \frac{\partial G_\alpha}{\partial c}\bigg|_{c=c_2} = \mu_\beta(c_2) = \frac{G_\alpha(c_1) - G_\beta(c_2)}{c_1 - c_2} \qquad (1.9)$$

requiring the equality of the chemical potentials μ_i at equilibrium (Fig. 1.2).

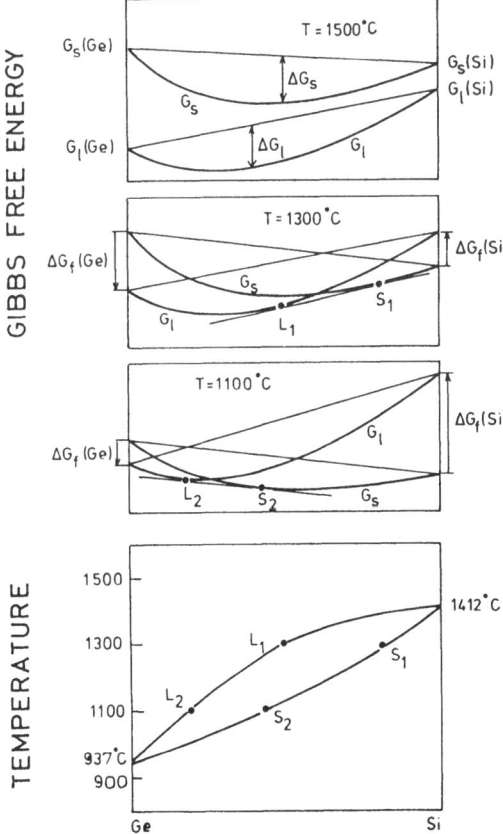

Fig. 1.3. Phase diagram of a binary alloy with unrestricted miscibility in both the solid and in the liquid states (an example would be the Ge-Si phase diagram). The upper three panels show the Gibbs free energies of the liquid and solid solutions as a function of composition at three different temperatures to illustrate the common tangent construction. ΔG_f is the heat of fusion of the pure metals, $\Delta G_{s(l)}$ the heat of formation of the solid (liquid) alloy

7

The complete alloy phase diagram follows from a series of such common-tangent constructions at different temperatures. This is illustrated in Fig. 1.3 for the example of the Ge-Si phase diagram. The upper three panels show the Gibbs free energies of the solid and liquid phases at three different temperatures. The common tangent divides the composition range into three sections; the Ge-rich liquid solution between Ge and c_l, the Si-rich solid solution between c_s and Si, and the two-phase region in between. The lowest panel shows the lens-shaped phase diagram resulting from these Gibbs free-energy curves that is typical for a system with unrestricted miscibility in the solid and the liquid states.

1.2.3 Second-Order Transitions

Except for spinodal decomposition and order-disorder transformations, all transitions discussed in this book are first order. Therefore the characteristics of a second-order transition will be discussed only very briefly. The order parameter (the density difference in the case of spinodal decomposition, the preferred occupation of a sublattice by one species of atoms in the case of long-range ordering of substitutional alloys) has a maximum value at $T = 0\,\mathrm{K}$. With increasing temperature, the increase in entropy that accompanies a decrease in ordering outweighs the loss in enthalpy, so that the order parameter decreases and finally vanishes at a temperature T_c. This critical temperature is characterized by a divergence of the second derivatives of G : specific heat, compressibility and thermal expansion coefficient (Fig. 1.4).

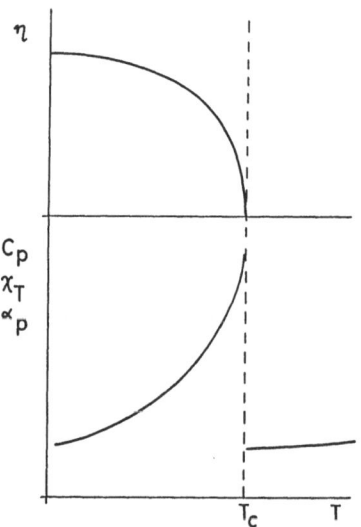

Fig. 1.4. Variation of the order parameter η, of the specific heat c_p, the compressibility χ_T, and the thermal expansion coefficient α_p at a second-order phase transition

1.3 Adiabatic Decoupling of the Ionic and Electronic Degrees of Freedom

In the canonical ensemble of statistical mechanics the Helmholtz free energy F corresponding to the Hamiltonian H is given by

$$F = -\beta^{-1} \ln Z_N \ , \tag{1.10}$$

where $\beta = (k_B T)^{-1}$ and Z_N is the canonical partition function

$$Z_N = \mathrm{Tr} \left\{ \exp\left(-\beta H\right) \right\} \ , \tag{1.11}$$

where the trace runs over both electronic and ionic states of the system. The Hamiltonian H is the electron-ion Hamiltonian of (1.1), which we now formulate somewhat more explicitly as

$$
\begin{aligned}
H &= \sum_i \frac{p_i^2}{2m} + \frac{e^2}{2} \sum_{i \neq j} \frac{1}{|r_i - r_j|} + \sum_{i,l} u(|r_i - R_l|) \\
&\quad + \sum_l \frac{P_l^2}{2M} + \frac{1}{2} \sum_{l \neq m} \phi(|R_l - R_m|) \\
&= T_e + V_{e-e} + V_{e-i} + T_i + V_{i-i} \ .
\end{aligned} \tag{1.12}
$$

Here $\{r_i\}$ and $\{R_l\}$ are the coordinates of the NZ electrons and N ions, $\{p_i\}$ and $\{P_l\}$ corresponding momenta and m and M the masses. The interaction of an electron with a single ion is expressed by $u(r)$ – for simplicity it is assumed that the electron-ion potential is local, in practice we will replace u by the bare ionic pseudopotential (see Sect. 2.1). The function $\phi(R)$ is the bare ion-ion interaction which is taken to be a central pairwise potential. For large separations $\phi(R)$ will be Coulombic, but for short distances it should incorporate contributions from the Born-Mayer repulsion, the van der Waals attraction, and from the polarizability of the ions.

In the adiabatic approximation of *Born* and *Oppenheimer* [1.19,20] the electronic and ionic degrees of freedom are separated

$$
\begin{aligned}
Z_N &= \mathrm{Tr}_{\{r_i, p_i\}} \mathrm{Tr}_{\{R_l, P_l\}} \{ \exp\left(-\beta H\right) \} \\
&= \mathrm{Tr}_{\{R_l, P_l\}} \{ \exp\left[-\beta(T_i + V_{i\text{-}i}) \right] \} \\
&\quad \times \mathrm{Tr}_{\{r_i, p_i\}} \{ \exp\left[-\beta(T_e + V_{e\text{-}e} + V_{e\text{-}i}) \right] \} \ .
\end{aligned} \tag{1.13}
$$

In the last step we have assumed that the ion-ion potential energy is independent of the electronic states. Physically this means that we assume that the electrons are always in their ground state corresponding to the instantaneous positions of the ions; the electrons follow any rearrangement of the ions adiabatically. This is justified because of the large differences in the electronic and ionic masses, and in the characteristic time scales for electronic and ionic motion.

The second term in (1.13) is just $\exp\left[-\beta F_e(\{R_l\})\right]$ where $F_e(\{R_l\})$ is the free energy of the system of electrons in the presence of the external field $V_{e\text{-}i}(\{R_l\})$. If $F_e(\{R_l\})$ can be calculated (albeit in some approximation), the electronic degrees of freedom no longer appear explicitly in the partition function and the problem is reduced to that of a system of pseudo-ions with an effective Hamiltonian given by

$$\tilde{H}_i = T_i + V_{i\text{-}i} + F_e(\{R_l\}) \ . \tag{1.14}$$

In practice, the electronic entropy contribution to F_e will usually be negligible, so that for most cases it is sufficient to deal only with the internal energy of the electrons $E_e(\{R_l\})$. Furthermore, for the sp-bonded metals the ion cores are small, so that for interionic distances of interest in normal metals and alloys we can simply set $\phi(R) = Z^2 e^2/R$.

In order to evaluate the thermodynamic properties, we must perform the trace over the ionic variables in

$$\begin{aligned}
Z_N &= \text{Tr}_{\{R_l, P_l\}}\{\exp\left(-\beta\tilde{H}_i\right)\} \\
&= \frac{1}{N! h^{3N}} \int d^3R_1 \ldots d^3R_N \int d^3P_1 \ldots d^3P_N \exp\left(-\beta\tilde{H}_i\right) \ . \tag{1.15}
\end{aligned}$$

The integral over the momenta can be carried out with the standard result

$$Z_N = (2\pi M/h^2)^{3N/2} Q_N \ , \tag{1.16}$$

where Q_N is the configurational partition function given by

$$Q_N = \frac{1}{N!} \int d^3R_1 \ldots d^3R_N \exp\left[-\beta\left(\frac{1}{2}\sum_{l\neq m}\frac{Z^2 e^2}{|R_l - R_m|} + E_e(\{R_l\})\right)\right] \ . \tag{1.17}$$

For an evaluation of (1.17), a sufficiently simple form of $E_e(\{R_l\})$ is required – the first step must be a reduction of the many-electron Hamiltonian $H_e = T_e + V_{e\text{-}e} + V_{e\text{-}i}$ into an effective one-electron form.

1.4 The One-Electron Approximation

The Hamiltonian of the many-electron system contains the interaction between all of the electrons,

$$H_e = \sum_i \left[-\frac{\hbar^2}{2m} \frac{\partial^2}{\partial \mathbf{r}_i^2} + u(\mathbf{r}_i) + \frac{e^2}{2} \sum_j \frac{1}{|\mathbf{r}_i - \mathbf{r}_j|} \right] , \tag{1.18}$$

and an exact solution of this quantum mechanical problem is far beyond reach. The simplest possibility to reduce (1.18) to a sum of one-electron terms is to write the many-electron wave function as a product of one-electron orbitals $\psi(\mathbf{r}_i)$ and to find the best possible product wave function by a variational calculation. In the variational calculation, one minimizes

$$\langle \psi_1(\mathbf{r}_1) \psi_2(\mathbf{r}_2) \ldots \psi_N(\mathbf{r}_N) | H_e | \psi_1(\mathbf{r}_1) \ldots \psi_N(\mathbf{r}_N) \rangle$$
$$- \sum_i E_i \langle \psi_i(\mathbf{r}_i) | \psi_i(\mathbf{r}_i) \rangle , \tag{1.19}$$

where the Lagrange multipliers E_i, have been introduced to satisfy the normalization requirement of the one-electron orbitals. This results in the Hartree equations for the best one-electron orbitals $\psi_i(\mathbf{r})$ [1.21]

$$\left[-\frac{\hbar^2}{2m} \frac{\partial^2}{\partial \mathbf{r}^2} + u(\mathbf{r}) + V_H(\mathbf{r}) \right] \psi_i(\mathbf{r}) = E_i \psi_i(\mathbf{r}) , \tag{1.20}$$

where the Hartree potential

$$V_H(\mathbf{r}) = e^2 \sum_{j \neq i} \int \frac{|\psi_j(\mathbf{r}')|^2}{|\mathbf{r} - \mathbf{r}'|} d^3 \mathbf{r}' \tag{1.21}$$

is the potential energy of the electron in the average electrostatic field produced by all other electrons. The total energy of the electron system is just

$$E_e = \sum_i E_i - \frac{e^2}{2} \iint \frac{\varrho(\mathbf{r}) \varrho(\mathbf{r}')}{|\mathbf{r} - \mathbf{r}'|} d^3 \mathbf{r} \, d^3 \mathbf{r}' \tag{1.22}$$

with the electron density $\varrho(\mathbf{r}) = \sum_i |\psi_i(\mathbf{r})|^2$. The second term in (1.22) is to compensate for the fact that the electron-electron Coulomb energy is counted twice in the sum of the one-electron eigenvalues.

If we now add the requirement that the total wave function be antisymmetric with respect to an interchange of any two electrons (this is accomplished by writing the total wave function in the form of a Slater determinant), it is easy to see that it gives an additional term in the energy (1.22) of

$$E_x = \frac{e^2}{2} \sum_{i \neq j} \iint \psi_i^*(\mathbf{r}) \psi_j^*(\mathbf{r}') \frac{1}{|\mathbf{r} - \mathbf{r}'|} \psi_i(\mathbf{r}') \psi_j(\mathbf{r}) d^3 \mathbf{r} \, d^3 \mathbf{r}' \tag{1.23}$$

for each pair of electron with parallel spin. Equation (1.23) describes the exchange energy, the inclusion of a corresponding term in (1.22) leads the Hartree-Fock equations. The exchange potential operating on the electron in the state $\psi_i(\mathbf{r})$ is not a scalar potential, but a nonlocal operator. This considerably complicates calculations. Moreover, there are also correlations between antiparallel spin which arise from the strong Coulomb repulsion between electrons. The correlation energy E_c may be defined as the difference between the Hartree-Fock energy E_{HF} and the correct value for the ground-state energy.

The long-range character of the electron-electron interaction leads to serious difficulties in the calculation of the correlation energy even for the homogeneous electron gas that cannot be handled even approximately along the lines of conventional perturbation theory. *Bohm* and *Pines* [1.22] showed that as a consequence of the long-range interaction of the electrons the many-electron system gives rise to collective oscillations – the plasma oscillations. If the long-range Coulomb correlations are treated as quantized plasma oscillations, the remaining short-range interactions may be treated by perturbation methods. This results in approximate formulae for the correlation energy which may be considered as an interpolation between the contributions from the long- and the short-range interactions, both limits being accurately calculated. According to *Nozières* and *Pines* [1.23] the correlation energy of the homogeneous electron gas at metallic densities is given by[1]

$$E_c = 0.031 \ln R_s - 0.115 \text{ Ry} , \qquad (1.24)$$

where R_s is the radius of a sphere which contains on average one electron

$$\frac{4\pi R_s^3}{3} = \frac{V}{ZN} . \qquad (1.25)$$

To date, the most accurate evaluations of the correlation energy stem from computer simulations [1.24]; various theories of the correlation energy have been reviewed by *Hedin* and *Lundquist* [1.25].

Expressing the kinetic and exchange energy in the same form, the ground state energy of the homogeneous electron gas is given by

$$\begin{aligned} E_{eg} &= E_{kin} + E_x + E_c \\ &= 2.21/R_s^2 - 0.916/R_s + (0.031 \ln R_s - 0.115) . \end{aligned} \qquad (1.26)$$

Equation (1.26) expresses the total energy of the homogeneous electron system as a *function* of the electron densities. According to the density-functional theorem of *Hohenberg, Kohn,* and *Sham* [1.4,5] the ground-state energy of any many-electron system is unique *functional* of the electron density and is minimal for the true density $\varrho(\mathbf{r})$. Thus

---

[1] Throughout the rest of this book we shall us atomic units defined by $\hbar = 1$, $e^2 = 2$, $m = 1/2$; length is measured in Bohr radii ($a_0 = \hbar^2/me^2$), energy in Rydbergs (ionization energy of the hydrogen atom: $e^2/2a_0$).

$$E_e = E_e(\varrho(\mathbf{r}))$$

$$= \sum_i E_i - \iint \frac{\varrho(\mathbf{r})\varrho(\mathbf{r'})}{|\mathbf{r} - \mathbf{r'}|} d^3\mathbf{r}\, d^3\mathbf{r'}$$

$$+ \int \varrho(\mathbf{r})[\varepsilon_{xc}(\varrho(\mathbf{r})) - \mu_{xc}(\varrho(\mathbf{r}))]d^3\mathbf{r} \ . \tag{1.27}$$

Many-particle effects are included here through an effective exchange and correlation energy $\varepsilon_{xc}(\varrho)$ and an exchange-correlation potential $\mu_{xc}(\varrho)$ [1.5] whose exact form is still unknown. The *local-density-functional (LDF) approximation* consists in approximating μ_{xc} by the simple functional

$$\mu_{xc}(\varrho(\mathbf{r})) = \frac{\delta(\varrho(\mathbf{r})\varepsilon_{xc}(\varrho(\mathbf{r})))}{\delta\varrho(\mathbf{r})} \ , \tag{1.28}$$

where $\varepsilon_{xc}(\varrho)$ is the exchange and correlation energy of a free electron gas of density ϱ. One could use for example the Nozières-Pines expression (1.26); other convenient parametrizations of the exchange-correlation functional have been reviewed by *Callaway* and *March* [1.26]. A one-electron equation can again be derived variationally; it has the form of the Hartree equations (1.20) with in addition a local exchange and correlation potential $\mu_{xc}(\varrho(\mathbf{r}))$. In fact, a similar general approach was given much earlier as an extension of the Thomas-Fermi theory by *Gombas* [1.27] and taken up again later by *Slater* [1.28], but a firm theoretical background was provided only by the Hohenberg-Kohn-Sham theorem.

The LDF form of the one-electron equations will be the basis for the treatment of many-electron effects throughout the rest of this book. For the solution of the atomic Hartree-Fock and LDF equations, standard programs are now available [1.29, 30].

1.5 Tightly-Bound and Nearly-Free Electrons; Potentials and Pseudopotentials

In describing the states of a molecule or a solid, the first step consists of writing the molecular or crystalline orbitals $|\psi\rangle$ as a linear combination of some basis states $|s_i\rangle$

$$|\psi\rangle = \sum_i u_i|s_i\rangle \ . \tag{1.29}$$

A variational procedure leads immediately to a set of equations for the expansion coefficients u_i,

$$\sum_i H_{ij}u_i - Eu_j = 0 \ , \tag{1.30}$$

where the H_{ij} are the matrix elements of the electronic Hamiltonian $H_e =$

$T_e + V_{e-e} + V_{e-i}$ (taken in its one-electron form) in the basis of the s_i's, $H_{ji} = \langle s_j | H_e | s_i \rangle$ and where we have assumed that the basis states are orthonormal, $\langle s_j | s_i \rangle = \delta_{ij}$. An obvious choice for the basis states would be to take the atomic orbitals of all the individual atoms – this is the linear combination of atomic orbitals (LCAO) method [1.31]. In that case the number of linear equations in (1.30) would be equal to the number of atoms multiplied with the number of electronic states per atom – an intractably large number for any solid. But of course Bloch's theorem tells us how this problem can be reduced to a manageable size. As a consequence of the translational periodicity of the perfect crystal, the u_i's must be of the form $u_{l\alpha}(\mathbf{k}) \sim u_\alpha \exp(\mathrm{i}\mathbf{k}\,\mathbf{R}_l)$, i.e. the trial orbitals must be of the form of a Bloch wave with wave vector $\mathbf{k}$. The wave vector is restricted to lie within the first Brillouin-zone with the allowed $\mathbf{k}$ values being determined by the periodic boundary conditions. The Bloch ansatz greatly reduces the dimension of the problem – the number of coupled equations is now just equal to the number of atoms in the unit cell times the number of orbitals per atom.

Assuming that the crystal potential $V(\mathbf{r}) = (V_{e-i} + V_{e-e})$ in H_e is the sum of atomic potentials $v(\mathbf{r} - \mathbf{R}_l)$, the LCAO band structure of a simple cubic lattice with s-orbitals may now be quickly found. If the matrix elements H_{ss} couple only to the nearest neighbours, we find [1.31] a solution of the secular equation belonging to (1.30) of the form

$$E(\mathbf{k}) = E_s + 2H_{ss}(\cos k_x a + \cos k_y a + \cos k_z a) \ , \tag{1.31}$$

where E_s is the atomic energy eigenvalue of the s-state and a the lattice constant. The calculation for the p-states proceeds in much the same way. Figure 1.5a shows the typical sinusoidal form of the s and p tight-binding bands.

How would these simple bands change if one could in some way eliminate the strong atomic potentials that give rise to the atomic states upon which the bands are based? The gaps between the bands would decrease and finally we would reach the limit of the free-electron bands shown in Fig. 1.5b. The lowest band here is given by the free-electron kinetic energy $E(\mathbf{k}) = k^2$ and the other bands are of the form $E(\mathbf{k}) = (\mathbf{k} + \mathbf{G})^2$; they are centred at other reciprocal lattice vectors $\mathbf{G}$. The band structure of a real metal is much closer to the free-electron than to the tight-binding limit. This would suggest that we should expand the crystalline orbitals in plane waves right away, resulting in a secular equation of the form

$$\det \left| [(\mathbf{k} + \mathbf{G})^2 - E]\delta_{GG'} + \langle \mathbf{k} + \mathbf{G} | V(\mathbf{r}) | \mathbf{k} + \mathbf{G}' \rangle \right| = 0 \ , \tag{1.32}$$

where the matrix elements of the crystal potential are taken between plane-wave states. In the simplest possible approximation, involving only two plane waves with wave vectors differing by a reciprocal lattice vector $\mathbf{G}$,

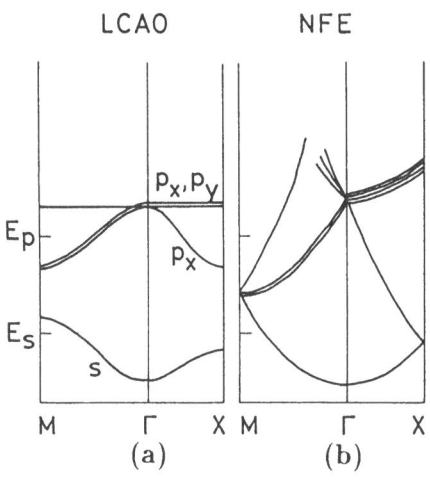

LCAO   NFE

E_p

p_x, p_y

p_x

E_s

s

M   Γ   X M   Γ   X
   (a)   (b)

Fig. 1.5a,b.   Conduction electron bands for a simple cubic structure calculated in the simplest LCAO approximation involving only s and p states (a), and calculated in the free-electron limit (b)

this results in the formation of gaps in the band of magnitude $2V(\boldsymbol{G}) = 2\langle -\boldsymbol{G}/2|V|\boldsymbol{G}/2\rangle$ at $\boldsymbol{k} = \pm\boldsymbol{G}/2$. Now the paradoxical fact is that the NFE method is entirely inapplicable to real atomic potentials, except for metallic hydrogen. For any other metal the number of terms in the secular equation (1.32) would be enormously large – of the order of 10^6 if we attempt to take the plane-wave expansion to convergence. Yet the experimentally measured band structure of metals and even semiconductors is not that far from the free-electron limit shown in our figure.

The reason for the difficulties with the plane-wave expansion is of course that the lowest eigenvalue does not converge to the lowest NFE-like conduction band, but to the tightly-bound band of $1s$ atomic orbitals. The next four eigenvalues correspond to the $2s$ and to the three $2p$-orbitals, and only the fifth eigenvalue of (1.32) would be representative of the lowest conduction band of a metal such as sodium, magnesium or aluminum. In fact the lowest NFE band in the metal has to be constructed starting from the fifth unperturbed free-electron band!

So we have to conclude that a plane-wave expansion is applicable only to *weak potentials* with *no bound states* (in our context the most useful definition of a strong potential is that it has a bound state).

Two different ways to overcome this problem were proposed as long ago as the thirties. One possibility consists of modifying the plane wave at small distances from the nucleus so that the new basis states are better suited for describing the rapid oscillations in the conduction-band states, which are a consequence of the orthogonality constraint following from Pauli's principle. In the second approach, one attempts to eliminate the core states and the strong potential responsible for binding them by a mathematical transformation of the Schrödinger equation.

An augmented plane wave (APW) [1.32, 33] is a plane wave outside a sphere with radius R_M inscribed in the Wigner-Seitz cell, and a sum over angular momentum eigenstates within that sphere,

$$\psi_{\boldsymbol{k}}^{\mathrm{APW}}(\boldsymbol{r}) = \langle r|\mathrm{APW}, \boldsymbol{k}\rangle$$

$$= \frac{1}{\sqrt{V}} \begin{cases} 4\pi \sum_{lm} iR_l(r, E)Y_{lm}(\theta_{\boldsymbol{k}}, \varphi_{\boldsymbol{k}})Y_{lm}(\theta, \varphi) \; ; & r \le R_M \\ \exp{(i\boldsymbol{k}\,\boldsymbol{r})} & r > R_M \end{cases} \tag{1.33}$$

where Y_{lm} is a spherical harmonic and $\theta_{\boldsymbol{k}}$, $\phi_{\boldsymbol{k}}$ stand for the directional coordinates of the wave vector $\boldsymbol{k}$. The point is that any APW satisfies the Schrödinger equation for a potential that is spherically symmetric for $r > R_M$, and we can expect a reasonable convergence for an APW expansion even outside R_M.

An orthogonalized plane wave (OPW) [1.34] consists of a plane wave plus a linear combination of core states,

$$\psi_{\boldsymbol{k}}^{\mathrm{OPW}}(\boldsymbol{r}) = \langle r|\mathrm{OPW}, \boldsymbol{k}\rangle$$

$$= \frac{1}{\sqrt{V}}\exp{(i\boldsymbol{k}\,\boldsymbol{r})} + \sum_{nlm,i} b_{nlm,i}(\boldsymbol{k})\psi_{nlm}(\boldsymbol{r} - \boldsymbol{R}_i) \;, \tag{1.34}$$

where the sum runs over all the core states of all the atoms. The $b_{nlm,i}$'s are chosen so that the OPW is automatically orthogonal to all the core states, thus

$$b_{nlm,i}(\boldsymbol{k}) = \frac{1}{\sqrt{V}} \int \psi_{nlm}^*(\boldsymbol{r} - \boldsymbol{R}_i)\exp{(i\boldsymbol{k}\,\boldsymbol{r})}d^3\boldsymbol{r} \;. \tag{1.35}$$

This results in a remarkably good approximation to the real wave function in the core region. The success of the OPW expansion relies upon the fact that (1.34) spans a subspace of the Hilbert space which is entirely orthogonal to that spanned by the core states and the lowest conduction-band state corresponds precisely to the lowest eigenvalue in the subspace defined by (1.34). Most of the techniques of electronic band-structure calculations developed in the last five decades are based on mixed-basis representations analogous to (1.33) and (1.34). On the other hand *Hellman* [1.35, 36] and *Gombas* [1.37] pointed out that a valence electron experiences not only the force field of the ion, but also undergoes a change in its kinetic energy arising from the Pauli principle which forces the valence orbitals to "orthogonalize" to those of the core. Hellmann emphasized that this change in the kinetic energy might be formally considered as a kind of repulsive potential. This repulsion largely cancels the strongly attractive crystal potential so that we are left with a weak effective *"pseudopotential"*. However, the significance of these ideas was hardly recognized until *Antoncik* [1.38] and *Phillips* and *Kleinman* [1.39, 40] showed that a very similar result may be obtained starting from the OPW expansion of the conduction-band orbitals. The OPW secular equation

$$\det|\langle \mathrm{OPW}, \boldsymbol{k} + \boldsymbol{G}|(H_e - E)|\mathrm{OPW}, \boldsymbol{k} + \boldsymbol{G}'\rangle| = 0 \tag{1.36}$$

may be rearranged in the form of a plane-wave secular equation (1.32)

$$\det\left|\left[(\boldsymbol{k}+\boldsymbol{G})^2 - E\right]\delta_{\boldsymbol{G}\,\boldsymbol{G}'} + \langle\boldsymbol{k}+\boldsymbol{G}|W^{\mathrm{OPW}}|\boldsymbol{k}+\boldsymbol{G}'\rangle\right| = 0 \ , \qquad (1.37)$$

where the pseudopotential W^{OPW} provides the coupling between the plane waves and the atoms. The explicit form of the pseudopotential is obtained by noting that the core orbitals ψ_{nlm} are eigenfunctions of the Hamiltonian H_{e} with eigenvalues E^{c}_{nlm}. It is easy to show from (1.35–37) that [1.13, 41]

$$\begin{aligned}
\langle\boldsymbol{k}+\boldsymbol{G}|W^{\mathrm{OPW}}|\boldsymbol{k}+\boldsymbol{G}'\rangle &= \langle\boldsymbol{k}+\boldsymbol{G}|V|\boldsymbol{k}+\boldsymbol{G}'\rangle \\
&+ \sum_{nlm,i}(E - E^{\mathrm{c}}_{nlm})b_{nlm,i}(\boldsymbol{k}+\boldsymbol{G})b_{nlm,i}(\boldsymbol{k}+\boldsymbol{G}')
\end{aligned} \qquad (1.38)$$

or written in a Dirac notation

$$W^{\mathrm{OPW}} = V + \sum_{nlm,i}(E - E^{\mathrm{c}}_{nlm,i})|\psi_{nlm,i}\rangle\langle\psi_{nlm,i}| \ . \qquad (1.39)$$

Closely related expressions for pseudopotentials can be derived by similar transformations of, e.g. the APW-secular-equation [1.41, 42].

So far (1.37) is just a reformulation of a secular equation, defining another technique for doing electronic band-structure calculations – we still have to solve the secular equation, calculate the electron density distribution, solve Poisson's equation for a new potential and to iterate to self-consistency.

The important point is to realize that (1.37) is equivalent to a new equation

$$(T + W)|\varphi\rangle = E|\varphi\rangle \ , \qquad (1.40)$$

where W is the operator defined in (1.37). Equation (1.39) is just the plane-wave secular equation belonging to this new "pseudo-Schrödinger" equation. Equation (1.40) is exactly the mathematical transformation of the original Schrödinger equation $(T + V)|\psi\rangle = E|\psi\rangle$ that we have been looking for. The exact orbitals $|\psi\rangle$ are obtained by projecting the "pseudoorbitals" $|\varphi\rangle$ onto the subspace orthogonal to the core states, the spectrum of eigenvalues belonging to (1.40) coincides with the conduction-band eigenvalues of the full Hamiltonian.

The cancellation of the strongly attractive electron-ion potential inside the core by the repulsive component in (1.39) is illustrated in Fig. 1.6 for the example of a free Si^{4+} ion. As the core states in (1.39) have definite angular momenta, the second term picks out and treats differently the different angular momentum components of the pseudoorbital φ. The s-component of φ is approximately constant inside the ion core as the oscillations in the radial part have been eliminated and an s-wave has no angular nodes. The s-component of the pseudopotential shows a nearly complete cancellation inside the core region.

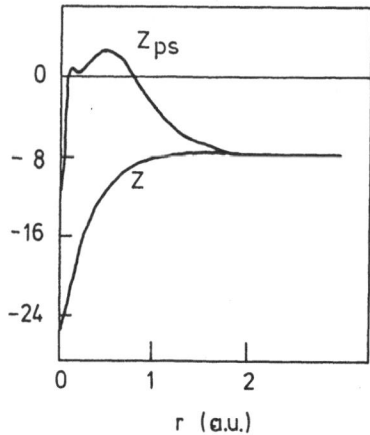

Fig. 1.6. The electron-ion potential $V(r)$ and the pseudopotential $W(r)$ (for s-states) of a Si^{4+} ion. Both potentials are expressed in the form $V(r) = Z(r)/r$, $W(r) = Z_{ps}(r)/r$. Note that both converge to the Coulomb potential $-8/r$ (in Ry) outside the core region. Inside the core, the pseudopotential is nearly zero. After [1.41]

Thus W is a weak potential (i.e. it has no bound states) and this suggests that we attempt a calculation of the total energy using a perturbation method.

1.6 Response Theory and Interatomic Interactions

We now explore an alternative way for calculating the electronic ground-state energy. We start from a homogeneous electron gas of density $\varrho_0 = NZ/V$ in the presence of a compensating positive background charge. The perturbation of this system created by the introduction of the ions is given by

$$\delta V(r) = W(r) - V_+(r)$$
$$= \sum_l w(r - R_l) - 2\varrho_0 \int \frac{d^3r'}{|r - r'|} \, , \tag{1.41}$$

where we have written the ionic pseudopotential $W(r)$ as a sum over individual ionic contributions $w(r)$, and $V_+(r)$ is the potential of the background charge (here we follow rather closely the presentation given in [1.43]). The free energy of the perturbed system can be calculated using the Hellmann-Feynman theorem

$$F = F_0 - \varrho_0^2 \int \frac{d^3r\, d^3r'}{|r - r'|} + \int_0^1 d\lambda \int d^3r \; \delta V(r)\varrho(r; \lambda) \; . \tag{1.42}$$

$\varrho(r; \lambda)$ is the electron density in an external potential $V_+ + \lambda \delta V$. If δV is weak − and this is the case if it is describable by pseudopotentials − we can calculate the screening charge density $\varrho(r; \lambda)$ via linear response theory

$$\varrho(r; \lambda) = \varrho_0 + \int d^3r' \Gamma(|r - r'|)\lambda \delta V(r') \tag{1.43}$$

with the density response function $\Gamma(|\mathbf{r} - \mathbf{r}'|)$ of the interacting electron gas [1.44]. [Strictly speaking $\varrho(\mathbf{r}; \lambda)$ is the density corresponding to the pseudoorbitals]. Inserting (1.43) in (1.42) leads to

$$
\begin{aligned}
F_e = F_0 &- \varrho_0^2 \int \frac{d^3\mathbf{r}\, d^3\mathbf{r}'}{|\mathbf{r} - \mathbf{r}'|} + \varrho_0 \int d^3\mathbf{r}\, \delta V(\mathbf{r}) \\
&+ \frac{1}{2} \iint d^3\mathbf{r}\, d^3\mathbf{r}'\, \delta V(\mathbf{r}) \Gamma(|\mathbf{r} - \mathbf{r}'|) \delta V(\mathbf{r}') \ .
\end{aligned}
\tag{1.44}
$$

Formally, (1.44) is a calculation of the free energy of the system of electrons to second order in the perturbing potential $\delta V(\mathbf{r})$. It reveals its full significance only if used in the adiabatic Hamiltonian (1.14). After a brief manipulation we find

$$
\begin{aligned}
\tilde{H}_i &= T_i + V_{i\text{-}i} + F_e(\{\mathbf{R}_l\}) \\
&= T_i + F_0(V) + \frac{1}{2} \sum_{l \neq m} \Phi(|\mathbf{R}_l - \mathbf{R}_m|; V) \ ,
\end{aligned}
\tag{1.45}
$$

i.e. the potential energy part of the effective ionic Hamiltonian consists of a volume energy and a sum over effective pair interactions [1.45,46]. The pair potential $\Phi(R; V)$ consists of the direct Coulomb repulsions [the self-interaction part of the ionic Coulomb interactions is cancelled by the negative of the self-interaction part of the background charge, i.e. the second term in (1.44)] and an indirect attraction between the pseudo-atoms (a pseudo-atom consists of the ion core and the screening electron cloud accumulated around it, each pseudo-atom is by itself electrically neutral) mediated by the conduction electrons

$$
\begin{aligned}
&\Phi(|\mathbf{R} - \mathbf{R}'|; V) \\
&= \frac{2Z^2}{R} + \iint d^3\mathbf{r}\, d^3\mathbf{r}'\, w(\mathbf{r} - \mathbf{R}) \Gamma(|\mathbf{r} - \mathbf{r}'|) w(\mathbf{r} - \mathbf{R}') \ .
\end{aligned}
\tag{1.46}
$$

The volume energy consists of various electron-gas and ionic self-interaction contributions; details will be given in Chap. 2. Thus, given a weak ionic pseudopotential it is relatively easy to bring the effective ionic Hamiltonian into the form of a volume energy plus a pair interaction term.

However, we have to note that these pair potentials are quite different from the classical pair potentials of a Lennard-Jones type conventionally used to describe the interatomic forces in rare-gas crystals. The effective pair interaction $\Phi(R; V)$ (1.46) is now volume dependent, because the electronic density response function $\Gamma(|\mathbf{r} - \mathbf{r}'|)$ varies with the electron density. Furthermore, the quantum nature of the electron-electron interaction has a pronounced influence on the form of the effective interatomic potential. The k-space density response function $\Gamma(|\mathbf{k}|)$ has a logarithmic singularity at $k = 2k_F$ where k_F is the Fermi momentum [1.46,47]. This singularity is a consequence of the fact that at $T = 0\,K$ all electron states with $|k| < k_F$ are occupied and all with $|k| > k_F$ are empty. This singularity in the response function determines the assymptotic behaviour of the pair potential at large interatomic separations. We have [1.48]

$$\Phi(R) \to |w(2k_F)|^2 \frac{\cos(2k_F R)}{(2k_F R)^3} \tag{1.47}$$

for large R. The pair potential at large R has the form of damped oscillations with the wavelength $\tilde{\lambda}_F = 2\pi/2k_F$, the amplitude of the oscillations is set by the matrix element of the pseudopotential with a momentum transfer equal to $2k_F$. The repulsive ion-ion interactions dominate at small interatomic distances, but they are strongly screened by the conduction electrons. At higher electron densities the screening is essentially of a Thomas-Fermi form. At intermediate distances (this would be distances corresponding to the separation of the first- and second-nearest neighbours in condensed phases), the form of the pair interaction is determined by a rather complex interplay of the pseudopotential and of the response function [1.49].

Equations (1.46, 47) have been derived in a linear response approximation. Higher-order response functions introduce three-, four-, and general many-body forces. This, together with the technical details of an actual calculation (which is more conveniently performed in reciprocal space) will be discussed in detail in Sect. 2. The important point is that the calculation via the response function formalism does not depend on the use of Bloch's theorem – thus it has equal validity for periodic and aperiodic structures.

1.7 The Statistical Mechanics of a Vibrating Lattice

The thermodynamic functions of a vibrating crystal at finite temperatures have to be calculated starting from the canonical distribution function $Z_N = \mathrm{Tr}\left\{\exp\left(-\beta \tilde{H}_i\right)\right\}$ with the effective ionic Hamiltonian given by (1.45). With $\tilde{H}_i$ given in the pair potential approximation, the calculation of the thermodynamic properties might be achieved in the harmonic approximation by expanding the potential energy part of (1.45) to second order in the displacements u_l of the ions from their equilibrium positions R_l. This results in a harmonic Hamiltonian which is quadratic in the dynamical variables

$$\tilde{H}_i \simeq \frac{1}{2}\sum_{l\alpha} \dot{u}_{l\alpha}^2 + \frac{1}{2}\sum_{\substack{ll' \\ \alpha\beta}} \Phi_{\alpha\beta}(ll') u_{l\alpha} u_{l'\beta} \tag{1.48}$$

with the force constants $\Phi_{\alpha\beta}(ll')$ given in terms of the second derivatives of the effective pair interaction (1.46), see e.g. in [1.50]. Assuming a harmonic time dependence of the vibrational displacements, $u_{l\alpha}(t) \sim u_{l\alpha} \exp{(i\omega t)}$, yields an eigenvalue problem of dimension $3N \times 3N$, in exact analogy to the LCAO eigenvalue problem (1.31) if we replace H_{ij} by $\Phi_{\alpha\beta}(ll')$ and E by $M\omega^2$. Again the Bloch theorem shows us that for a system with translational periodicity this may be reduced to the repeated diagonalization of the dynamical matrix $D_{\alpha\beta}(k)$,

$$D_{\alpha\beta}(k) = \sum_l \Phi_{\alpha\beta}(ll') \exp{\left[ik(R_l - R_{l'})\right]} \ , \tag{1.49}$$

i.e. of the Fourier transform of the force-constant matrix $\Phi_{\alpha\beta}(ll')$, with a wave vector k restricted to the first Brillouin zone. In the harmonic approximation, the vibrational states of the crystal are described as linear superpositions of normal modes of vibrations with squared frequencies $\omega_j^2(k)$, i.e. the eigenvalues of (1.49). Quantum-mechanically the normal modes may be considered as collective excitations of the crystal, i.e. as quasi-particles with zero spin which we call phonons. Then the thermodynamics of the vibrating crystal can be formulated in terms of the statistical mechanics of an ideal gas of non-interacting phonons [1.50].

The harmonic approximation breaks down when the atomic displacements become so large that the higher-order terms in the expansion of the potential energy can no longer be neglected. In that case the normal modes are no longer independent and the phonons start to interact. Again the self-consistent field method provides the simplest approach to this problem. We assume that each atom moves in the time-averaged force field of its neighbours, which has to be calculated self-consistently. This *self-consistent phonon approximation* [1.51] is the analogue of the Hartree approach for electrons.

The calculation of the anharmonic frequency spectrum of the normal modes is a formidable computational task. A simpler approach to the thermodynamic properties consists in using thermodynamic perturbation theory. The simplest, yet very often sufficiently accurate form of thermodynamic perturbation theory is a variational technique based on the Gibbs-Bogoljubov inequality [1.52]. This inequality states that the free energy F of a system is always lower than the free energy F_0 of a reference system with Hamiltonian H_0, plus the expectation value of the perturbation $(H - H_0)$ evaluated in this reference system, i.e.

$$F \leq F_0 + \langle H - H_0 \rangle \ . \tag{1.50}$$

In our case H is the effective Hamiltonian (1.45), H_0 can be chosen to be a system of Einstein or Debye oscillators. By minimizing the right hand side of (1.50) with respect to the parameters of the reference system (this would be the Einstein or Debye frequency), we determine an upper bound to the exact free energy of the system. The accuracy of the approach can be improved by using more refined reference systems and by adding further corrections.

1.8 Periodic, Aperiodic and Quasi-Periodic Structures

Depending on the external thermodynamic variables (and on the thermal history of the material) the ions can either form a periodic network (as in ideal crystals), an aperiodic structure (such as a liquid or a glass) or a quasi-periodic network (as in "quasi-crystals"). A "quasi-crystal" is a crystal with a point-group symmetry incompatible with translational symmetry. It has long-range orientational order, but only "quasi-periodic" – in the sense of a quasi-periodic function in mathematics [1.53] – translational order [1.54]. To appreciate the new aspects that arise in a theory of the nonperiodic state of matter (apart from the rather trivial fact that Bloch's theorem does not apply), it is helpful to think for a moment what structural information we can draw from experiment.

A diffraction experiment on a single crystal specifies the three-dimensional structure completely. From the positions of the Bragg reflections we get the reciprocal lattice (and hence the space-group symmetry of the crystal), and the positions of the individual atoms within the unit cell may be determined by analyzing the intensities of the Bragg reflections [1.55].

The diffraction patterns of a glass or a liquid consist only of diffuse rings. The intensity distribution of these rings is essentially proportional to the static structure factor $S(q)$, which is just the Fourier transform of the pair-correlation function $g(R)$ measuring the probability of finding a second atom at a distance R from a given atom [1.56]. Thus, in this case a diffraction experiment describes only a one-dimensional projection of the real three-dimensional structure. Figure 1.7 shows a schematic representation of the diffraction patterns and of the pair-correlation functions in crystals, quasi-crystals, and glasses or liquids.

The diffraction pattern of a quasi-crystal consists of a set of Bragg peaks that densely fill reciprocal space in a countable way. The intensity of these peaks depends on the distribution of the atoms within the quasi-periodically repeated three-dimensional units (of which the famous "Penrose-tiles" [1.54, 57] are just one possible realisation). In view of the infinitely many peaks, it is impossible to determine the distribution of the atoms uniquely. In fact, it has been pointed out that the structure factor, i.e. the thermodynamic expectation value of the squares of the Fourier components of the atomic density distribution $n(R)$, $S(q) = \langle |n(q)|^2 \rangle$ is insufficient for an absolute structure determination. One needs to know the phases as well as the magnitudes of the complex Fourier components of $n(R)$. The

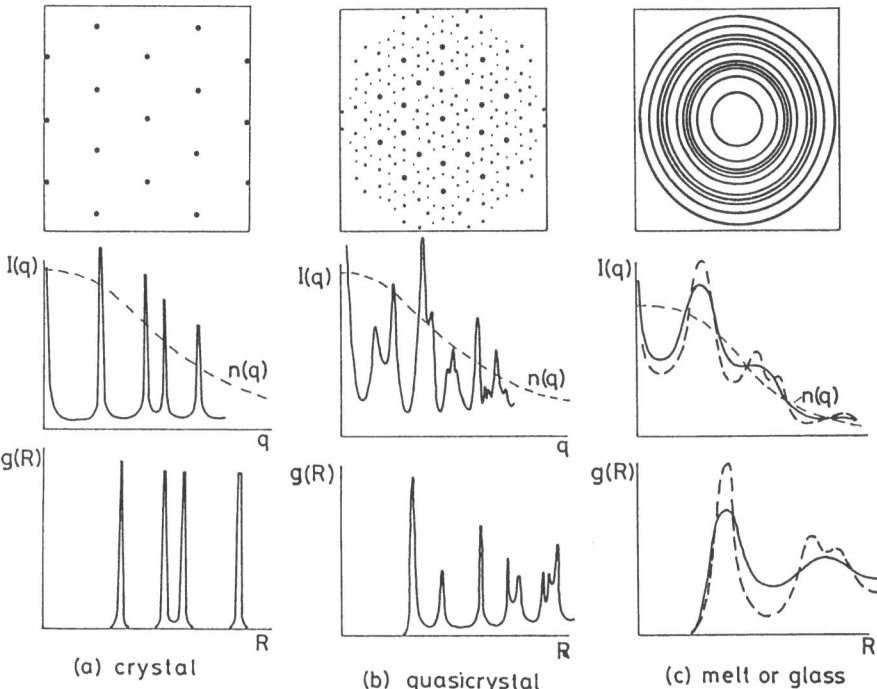

Fig. 1.7a–c. Schematic representation of the diffraction patterns, the angular average over the diffracted intensity $I(q) \propto N\varrho(q)S(q)$ [$\varrho(q)$ is the atomic scattering form factor, $S(q)$ is the static structure factor], and of the pair-correlation function $g(R)$ in a crystal (the diffraction pattern is shown along a hexagonal axis) (**a**), in a quasi-crystal (the diffraction pattern is shown along a pentagonal axis) (**b**), and in liquids or glasses (**c**). In (**c**), the solid line refers to a liquid, the broken lines to a glass, $I(q)$ and $g(R)$ include in all cases the effects of thermal disorder and finite instrumental resolution

only exception is the ideal crystal, where all the phases associated with the sharp Bragg peaks are equal [1.58].

Of course, these are the extreme cases. Nature actually offers much more variety. Take for example the case of a substitutionally disordered alloy with a regular mean lattice. In this case we observe sharp Bragg reflections from the mean lattice and a diffuse background (the "Laue monotonic scattering") arising from the random distribution of the two atomic species over the available atomic positions. Any kind of deviation from complete randomness creates a modulation of the diffuse background which contains the information on the short- or long-range partial order. Another example is the observation of structures that are quasi-periodic in a plane, but periodic in the direction perpendicular to that plane [1.59].

Any violation of translational periodicity creates two new challenges to the theorist. The first is the calculation of the eigenstates of a Hamiltonian without the use of Blochs theorem. This concerns not only the electronic band-structure problem, but also the calculation of the phonon spectrum which is the basis of the calculation of the vibrational contributions to the thermodynamic functions. The second is the necessity to calculate the actual atomic structure from the interatomic force law. For crystalline structures at $T = 0\,\mathrm{K}$, to investigate the structural stability of a material means to calculate the ground-state energy in various possible structures (but of course the answer found in that way can never be really exhaustive). At finite temperatures the thermal displacements of the atoms from their equilibrium positions have to be taken into aocount. Via the vibrational spectrum, the thermal averages of the atomic displacements depend on the forces acting between the atoms. These in turn depend on how far the atoms are actually displaced from their equilibrium positions. Thus both averages – those over the atomic displacements and those over the force constants – have to be calculated self-consistently (this is just the self-consistent phonon problem mentioned in the last section). The same problem arises of course when we consider the liquid state of matter: the free energy of the liquid depends on the correlation functions and on the interatomic potential. The equilibrium structure is the one that minimizes the free energy for a given interatomic potential. To cast this condition into the form of a mathematical relation between the interatomic potential $\Phi(R)$ and the pair correlation function $g(R)$ is one of the fundamental tasks of the statistical mechanics of liquids.

1.9 Elementary Excitations in Aperiodic Structures

We now turn to the problem of calculating the eigenvalues and the ground-state energy of a non-periodic system. In fact, as far as the electronic part of the total energy is concerned we have already shown that this problem can be circumvented: the linear-response expression (1.44) for the total electronic free energy does not depend on the use of Bloch's theorem and is equally applicable to periodic and aperiodic systems (but only to simple, and not to transition metals). However, the problem of diagonalizing a Hamiltonian in the absence of translational symmetry arises again if we attempt to calculate the vibrational excitations of a disordered system. We begin by discussing the simplest problem, that of substitutional disorder on a crystalline lattice.

1.9.1 Effective Medium Approximations

The information we are looking for (electronic or vibrational density of states, ground state energy, vibrational energy) may be expressed in terms of the Green's function $G = (E - H)^{-1}$. The simplest approach to the calculation of G consists in making a series expansion in powers of some supposedly small perturbation associated with the disorder. We therefore write

$$H = H_0 + \Delta H \ , \tag{1.51}$$

where the unperturbed Hamiltonian H_0 has perfect crystalline symmetry and ΔH contains all the effects of disorder. We attempt to keep the effects of disorder small by defining H_0 as the Hamiltonian $\langle H \rangle$ of a virtual crystal with the average parameters of the material – in this case the ensemble average of the perturbation vanishes, i.e. $\langle \Delta H \rangle = 0$. In the *virtual crystal approximation (VCA)* the effects of disorder are ignored, the spectrum of the disordered material is taken to be that of a regular crystal with the average parameters.

To obtain significant improvement over the VCA, it is not sufficient to evaluate only the first few terms of a Green's-function expansion. The eigenstates of the system are determined by the poles of the Green's function. Adding a finite number of terms leaves the analytic structure of the Green's function of the unperturbed system unchanged. The Green's function expansion has the form

$$G_{ij} = G_{ij}^0 + \sum_{k,k'} G_{ik}^0 \Delta H_{kk'} G_{k'j}^0 + \dots \tag{1.52}$$

where $G^0 = (E - H_0)^{-1}$ is the Green's function of the virtual crystal [(1.52) has been written in a basis of some local orbitals, i.e. in a site representation]. However, we note that the subset corresponding to multiple scattering of an excitation from the same site without excursion to other sites may be summed up in the form of a Dyson equation, since all G_{ii}^0 are equal, say to G_{00}^0. This series defines the scattering matrix T_i at the site i

$$\begin{aligned}
T_i &= \Delta H_{ii} + \Delta H_{ii} G_{ii}^0 + \Delta H_{ii} G_{ii}^0 \Delta H_{ii} G_{ii}^0 \Delta H_{ii} + \dots \\
&= \Delta H_{ii} + \Delta H_{ii} G_{ii}^0 T_i \\
&= \frac{\Delta H_{ii}}{1 - \Delta H_{ii} G_{00}^0} \ .
\end{aligned} \tag{1.53}$$

To proceed further in a simple way, we have to assume that the disorder is purely diagonal, i.e. $\Delta H_{ij} = \Delta H_i \delta_{ij}$. This means that we assume that the effects of disorder are seen mainly as variations of the local atomic energy levels (in the case of an LCAO-type electronic Hamiltonian) or as variations of the atomic mass (if we consider lattice vibrations). With this restriction, and using (1.53) in the expansion (1.52) we obtain the series

$$\begin{aligned}
G_{ij} = G_{ij}^0 &+ \sum_{k \neq i,j} G_{ik}^0 T_k G_{kj}^0 \\
&+ \sum_{\substack{k \neq i, k' \\ k' \neq j}} G_{ik}^0 T_k G_{kk'} T_{k'} G_{k'j}^0 + \dots \ ,
\end{aligned} \tag{1.54}$$

where now T_i is the random variable related through (1.53) to the local perturbation ΔH_i on site i. As we have summed up all single-site scattering events, we expect a much better answer from the *average T-matrix approximation (ATA)* [1.60] where each T_k is replaced by its ensemble average $\langle T \rangle$, than from the VCA. In the ATA the sum (1.54) may be calculated straightforwardly and we find that the effect is to shift the poles of the Green's function G^0 by a complex self-energy

$$\Sigma^{ATA}(E) = \langle T \rangle / (1 + \langle T \rangle G_{00}^0) \ . \tag{1.55}$$

$\mathrm{Re}\{\Sigma^{ATA}(E)\}$ measures the shift of the eigenvalues produced by the disorder, and $\mathrm{Im}\{\Sigma^{ATA}(E)\}$ determines their broadening (i.e. the lifetime of the excitations which is now finite). The ATA Hamiltonian

$$H^{ATA} = H_0 + \Sigma^{ATA}(E) \tag{1.56}$$

has full crystalline symmetry and can be handled using the conventional k-space techniques. In the ATA there are further scattering events arising from the departure of the local T-matrices T_i from their average $\langle T \rangle$.

The *coherent potential approximation (CPA)* describes the disordered system by an effective medium characterized by a Green's function G_{eff} with the full crystalline symmetry such that on average there are no further single-site scattering events. Clearly this requires

$$\langle T^{CPA} \rangle = 0 \ . \tag{1.57}$$

Equation (1.57) is an integral equation for the complex self-energy $\Sigma^{CPA}(E)$ [1.61–64]. One way of expressing this approximation is to say that the CPA regards the effective medium around an atom to be the disordered crystal itself, and calculates this medium self-consistently. The CPA is at present considered to be the best single-site approximation for treating electrons and phonons in disordered systems.

The problem with the CPA is that the inclusion of the effects of off-diagonal disorder makes the theory quite complicated and that it is still a single-site theory. The difficult step from a single-site to a cluster CPA complicates the theory even more. For topologically disordered materials the equivalent of the CPA is the effective medium approximation (EMA) [1.65]. Even an attempt to outline the basic principles of the EMA would go far beyond the level of this introductory review. The increasing complexity of these methods arises from the fact that one attempts to extend the k-space techniques based on the use of Bloch's theorem to problems where they are less and less appropriate. So the answer can only be, quoting *Heine* [1.66], to "throw out k-space".

1.9.2 The Recursion Method and Related Techniques

The information on the electronic and vibrational energies is contained in the density of states (electronic or vibrational) defined by

$$n(E) = \sum_i \delta(E - E_i) \ , \tag{1.58}$$

or in the local density of states defined by [1.11,67]

$$n(\mathbf{r}, E) = \sum_i |\psi_i(\mathbf{r})|^2 \delta(E - E_i) \tag{1.59}$$

in terms of the eigenvalues E_i and the eigenstates ψ_i of the Hamiltonian H. Equation (1.59) defines the local density of states in terms of the eigenstates of the system so that nothing has been gained as yet. The important point is that $n(\mathbf{r}, E)$ is related to the imaginary part of the Greens function $G(\mathbf{r}, \mathbf{r}'; E)$ through

$$n(\mathbf{r}, E) = -\frac{1}{\pi} \mathrm{Im} \left\{ G(\mathbf{r}, \mathbf{r}; E) \right\} \ . \tag{1.60}$$

In terms of the eigenfunctions ψ_i of the Hamiltonian, $G(\mathbf{r}, \mathbf{r}'; E)$ is given by

$$G(\mathbf{r}, \mathbf{r}'; E) = \sum_i \frac{\psi_i(\mathbf{r})\psi_i^*(\mathbf{r}')}{E - E_i + i\delta} \ . \tag{1.61}$$

We find that the information we want depends just on one diagonal matrix element of the Green's function, and this can be calculated separately in a very effective way. One possible way to achieve this is to proceed via the recursion technique. Using a representation of the eigenstates in terms of local atomic orbitals or atomic vibrations $\psi_i = \sum u_{il}\varphi_l$ [in analogy to (1.30)] and assuming the φ_l to be orthonormal, we find that $n(E)$ is the sum over the local densities of states $n_l(E)$ associated with one local orbital. Thus

$$n(E) = \sum_l n_l(E) \quad \text{with} \tag{1.62}$$

$$
\begin{aligned}
n_l(E) &= -\frac{1}{\pi} \mathrm{Im} \left\{ G_{ll}(E) \right\} \\
&= -\frac{1}{\pi} \mathrm{Im} \left\{ \sum_{i,j} \langle \varphi_l | \psi_i \rangle \langle \psi_i | (E - H + i\delta)^{-1} | \psi_j \rangle \langle \psi_j | \varphi_l \rangle \right\} \\
&= \sum_i |u_{il}|^2 \delta(E - E_i) \ , \tag{1.63}
\end{aligned}
$$

where we have used the fact that $\langle \varphi_l | \psi_i \rangle = u_{il}$ and that the ψ_i form a complete orthonormal set. The diagonal matrix element $G_{ll}(E)$ associated with the local state φ_l may be calculated by defining a new orthonormal basis set u_n via the recurrence relation [1.68]

$$b_{n+1}|u_{n+1}\rangle = H|u_n\rangle - a_n|u_n\rangle - b_n|u_{n-1}\rangle \qquad (1.64)$$

with $|u_0\rangle = |\varphi_l\rangle$. The numerical coefficients a_n and b_n are given by

$$a_n = \langle u_n|H|u_n\rangle \quad b_{n+1} = \langle u_n|H|u_{n+1}\rangle \ ,$$

and all other matrix elements of H vanish in the new basis, where H is therefore tridiagonal. Thus the calculation of the diagonal matrix element $G_{ll}(E)$ is straightforward and we find that $n_l(E)$ has a continued fraction representation

$$n_l(E) = -\frac{1}{\pi}\mathrm{Im}\,\{1/\{E - a_0 - b_1^2/[E - a_1 - b_2^2/(E - a_2 - \ldots)]\}\} \ .$$

$$(1.65)$$

Equations (1.62–65) sketch the recursive solution of the Schrödinger equation or of the dynamic eigenvalue problem in the harmonic approximation. The *recursion method* of *Haydock* et al. [1.68] is a very convenient way to calculate total energies or spectral properties of aperiodic systems. It has many relatives: the method of moments [1.69], the cluster Bethe-lattice method [1.70], or the equation-of-motion method [1.71], but the recursion method itself is perhaps the most efficient. None of these methods require any simplifying assumptions about the form of the Hamiltonian beyond the requirement that it is quadratic in the dynamical variables. They also avoid the single-site approximation, and thus take full account of the effects of the local environment.

1.10 Configurational Thermodynamics of Solids and Liquids

The cornerstone of a phase-diagram calculation is of course the determination of the thermodynamically stable equilibrium structure for a given set of thermodynamic variables (temperature, composition, pressure, ...). As long as we consider only pure crystalline metals, ordered intermetallic compounds, or ideal solid solutions (i.e. solid solutions where the distribution of the two atomic species over the sites of the mean lattice is completely random), the methods discussed so far are sufficient. We know how to calculate the ground-state energy of the system and how to add the vibrational contribution to the free energy. For the ideal solid solutions we have to add the ideal entropy of mixing, but beyond that there are no further configurational contributions to the free energy. So we can construct the free energy curves and derive the phase boundaries as described in Sect. 1.2.

However, the distribution of two different kinds of atoms over the sites of a lattice is generally not random; usually some sort of correlation is present. This means that between the two extremes of a random solid solution and a completely ordered superstructure there is a continuum of possible configurations and our task is to predict the one that minimizes the free energy. The situation is very similar in liquids were we have to find the configuration of minimum free energy within a continuum of possible liquid structures.

1.10.1 Order-Disorder Transitions

The first step towards obtaining a free-energy functional suitable for describing order-disorder transformations in solid solutions is to define an appropriate set of configurational coordinates by performing various averages over single-site, pair, and many-atom distributions. This leads to the important concepts of short-range order parameters (Warren-Cowley order parameters) [1.72], long-range order parameters (Bragg-Williams order parameters) [1.73], of concentration waves [1.74], and cluster-probabilities [1.12].

Given a set of configurational coordinates $\{\alpha\}$, the configurational partition function Q_N can be written as [1.12]

$$Q_N = \sum_{\{\alpha\}} g\{\alpha\} \exp\left(-\beta E\{\alpha\}\right) = \sum_{\{\alpha\}} \exp\left(-\beta F\{\alpha\}\right) , \qquad (1.66)$$

where $g\{\alpha\}$ is the statistical weight of the configuration defined by $\{\alpha\}$. The second equality in (1.66) defines the non-equilibrium free-energy function $F\{\alpha\}$,

$$F\{\alpha\} = E\{\alpha\} - TS\{\alpha\} \qquad (1.67)$$

with the configurational entropy $S\{\alpha\}$ given by

$$S\{\alpha\} = k_B \ln g\{\alpha\} . \qquad (1.68)$$

The equilibrium free energy is

$$F = -\beta^{-1} \ln Q_N . \qquad (1.69)$$

An important simplification is achieved when the sum in the partition function (1.66) is replaced by the maximum term – this means that fluctuations about the most probable state are neglected. Then to find the equilibrium free energy requires the minimization of the non-equilibrium free-energy function, i.e.

$$F \simeq \underset{\{\alpha\}}{\text{Min}} F\{\alpha\} \ . \tag{1.70}$$

The minimization in (1.70) is carried out with respect to the configurational coordinate (the order parameter). The lowest-order approximation based on single-site averages leads to the Bragg-Williams model. Other mean-field models, including the Landau theory, fit into the same framework.

1.10.2 From the Interatomic Force Law to the Structure of a Liquid

The relevant configurational coordinate for describing the structure of a liquid is the pair-correlation function. As we have already discussed in Sect. 1.8 this is the only structural information directly accessible to experiment. For systems with pair interactions only, the thermodynamic functions can be expressed in terms of $g(R)$ and the pair potential $\Phi(R)$. To establish a correlation between $\Phi(R)$ and $g(R)$ is the central task fo liquid-state theory.

An l-particle correlation function, $g^{(l)}(R_1, \ldots, R_l)$, is the thermodynamic expectation value of the l-particle density [1.75]

$$n^l g^{(l)}(R_1, \ldots, R_l) = \left\langle \sum_{i_1, \ldots, i_l} \delta(R_1 - R_{i_1}) \ldots \delta(R_l - R_{i_l}) \right\rangle$$

$$= \frac{N!}{(N-l)!} \frac{1}{Q_N} \int \ldots \int dR_{l+1} \ldots dR_N \exp\left[-\beta U(R_1, \ldots, R_N)\right] \tag{1.71}$$

where Q_N is the configurational partition function and $U(R_1, \ldots, R_N)$ is the potential energy of an N-particle system with the number denstiy $n = N/V$. Equation (1.71) shows that l- and $(l+1)$-body correlations are related through a recurrence relation of the form

$$g^{(l)}(R_1, \ldots, R_l) = \frac{n}{(N-l)} \int g^{(l+1)}(R_1, \ldots, R_{l+1}) dR_{l+1} \tag{1.72}$$

implying that the l-body correlation function may be determined if we know $g^{(l+1)}$ and so on. This is of course quite trivial: all distributions of a subset are already contained in the total distribution. If the few-body correlation functions are to be consistent with the total distribution, they are necessarily interdependent. Formally, the N-particle distribution function must be determined such as to minimize the total free energy of the system for a given interatomic interaction and a given set of thermodynamic parameters. The important point is that (1.72) establishes an infinite hierarchy of correlation functions which must be broken at some level if a closed set of relations for the lower-order correlation functions is to be obtained. Of course, this destroys the consistency with the higher-order correlation functions. The second part of (1.71) relates structure and interatomic forces through the potential energy. For the pair-correlation function this relation reads (we use now the explicit pair-potential representation of the potential energy)

$$n^2 g(|\mathbf{R}_1 - \mathbf{R}_2|) = \frac{N!}{(N-2)! Q_N} \iint d^3 \mathbf{R}_3 \dots d^3 \mathbf{R}_N$$

$$\times \exp\left[-\frac{\beta}{2} \sum_{k,l}' \Phi(|\mathbf{R}_k - \mathbf{R}_l|)\right] . \tag{1.73}$$

Taking the gradient of (1.73) with respect to $\mathbf{R}_1$ yields after a brief manipulation the Born-Green-Yvon equation [1.76]

$$\nabla_1 g(|\mathbf{R}_1 - \mathbf{R}_2|) = -\beta g(|\mathbf{R}_1 - \mathbf{R}_2|)\nabla_1 \Phi(|\mathbf{R}_1 - \mathbf{R}_2|)$$

$$- \beta n \int d^3 \mathbf{R}_3 \nabla_1 \Phi(|\mathbf{R}_1 - \mathbf{R}_3|)g^{(3)}(\mathbf{R}_1, \mathbf{R}_2, \mathbf{R}_3) . \tag{1.74}$$

Equation (1.74) is just the first member of an infinite hierarchy of integro-differential equations. The simplest (but unsatisfactory) closure that converts (1.74) into a closed integro-differential equation for $g(R)$ is the superposition approximation of *Kirkwood* [1.77]

$$g^{(3)}(\mathbf{R}_1, \mathbf{R}_2, \mathbf{R}_3) = g(|\mathbf{R}_1 - \mathbf{R}_2|)g(|\mathbf{R}_2 - \mathbf{R}_3|)g(|\mathbf{R}_3 - \mathbf{R}_1|) \tag{1.75}$$

resulting in the Born-Green equation. The search for more appropriate closure relations and the analytical or numerical solution of the resulting integral equations was, and continues to be, a central problem of liquid state theory. A variety of liquid-state integral equations and their application to metallic systems are discussed in Chaps. 4 and 9 and in Appendix D.

Any integro-differential equation has its equivalent variational problem. In the case of the integral equation of liquid-state theory, this variational condition requires the stationarity of some approximate expression for the non-equilibrium free energy with respect to variations of the correlation function [1.78]. So we find that the treatment of the configurational thermodynamics in both the solid and the liquid state is based on the same basic approximations.

1.10.3 Order-Parameter Approach to Freezing and Melting

In Sect. 1.2 we learnt how the phase boundaries which limit the range of stability of the competing phases may be derived from a comparison of the relevant thermodynamic potentials. In the subsequent sections we discussed the strategy for calculating these thermodynamic functions from first principles.

A successful calculation of a phase boundary requires of course that the free energies of the competing phases have all been determined with a maximum error that is significantly lower than the relevant heat of transformation. This criterion is particularly difficult to satisfy for the solid-liquid transition (melting and freezing), since the methods used for calculating the free energies of the crystal and of the melt are so widely different. In fact it turns out that on each side of the melting line, the absolute error in the calculated free energies is nearly of the order of the heat of fusion. This means that a successful calculation of the fluid-solid coexistence curve by the total free energy approach is possible only via a judicious combination of the methods used to describe the solid and the liquid. The absolute errors on both sides of the transition must have the same sign and nearly equal magnitude.

For second-order phase transitions (where the same problem is even more serious since the latent heat of transformation is zero) a better approach was proposed long ago by *Landau* [1.79,80]. The Gibbs free energy of one phase is obtained from a Taylor expansion of the Gibbs free energy of the second phase in powers of an order parameter which describes the change in symmetry at the transition. The coefficients of the expansion are determined by the microscopic bonding properties of the system.

The order parameter theories can also be extended to treat first-order phase transitions. For the solid-liquid transition the difficult point is to define an appropriate order parameter. In their pioneering work on freezing *Kirkwood* and *Monroe* [1.81] proposed to consider an expansion of the free energy in powers of the difference $\Delta n(\boldsymbol{R}) = n_s(\boldsymbol{R}) - n_l$ between the periodically fluctuating single-particle density $n_s(\boldsymbol{R})$ of the crystal and the constant density n_l of the homogeneous liquid. More recently, this approach was cast into the language of modern density-functional theory for classical systems [1.82–84]. The coefficients of the order-parameter expansion [i.e. the functional derivatives of the free energy with respect to $\Delta n(\boldsymbol{R})$] are found to be the correlation functions (more precisely the direct correlation functions) of the liquid. Today we can calculate the structural stability of a crystal, starting from a theory of the homogeneous melt [1.85]. This theory describes freezing as an instability of the homogeneous liquid with respect to periodic fluctuations of the density. The inverse way can also be followed: in that case we would expand the free energy of the liquid around that of the perfect crystalline solid in terms of a "disorder" parameter describing the gradual introduction of defects. Melting occurs if the crystal becomes saturated with defects. We will return to these methods in Chaps. 5 and 10.

With this very brief sketch of the modern density-functional theory of freezing and melting we close this general overview of the path that will take us from the Hamiltonian of the electron-ion system to the phase diagram, and of the many problems we shall have to confront on the way. The following chapters should provide the reader with a working knowledge of the methods that are currently used to solve these problems and of the state of the art that has been reached in each field. Again we try to keep the presentation rather concise and leave more technical aspects to a detailed discussion in the appendices.

2. Interatomic Forces in Metals and Alloys

Our first task will be the calculation of the ground-state energy of the system of electrons and ions as a function of the ionic coordinates $\{R_l\}$, starting from the full electron-ion Hamiltonian given by (1.1). Furthermore, we shall seek an expansion of $E(\{R_l\})$ in terms of a volume energy and sums over pair, triplet and higher-order interactions.

As shown in Chap. 1 a first important simplification can be made by noting that the characteristic frequencies of the moving electrons and ions are radically different. This allows us to eliminate the electronic degrees of freedom by taking in (1.1) the trace over the electronic variables. This is precisely the adiabatic Born-Oppenheimer approximation [2.1,2] which replaces (1.1) by an effective adiabatic Hamiltonian $\tilde{H}_i$,

$$\tilde{H}_i = T_i + V_{i-i} + E_e(\{R_l\}) \; , \tag{2.1}$$

where $E_e(\{R_l\})$ stands for the electronic ground-state energy at fixed ionic coordinates $\{R_l\}$ and where the bare ion-ion interaction may be considered with good accuracy as being purely Coulombic. To date the only reliable scheme for obtaining the desired decomposition of E_e from first principles is based on the response of the homogeneous electron gas to the electron-ion potential, which is valid only for systems whose ions scatter the electrons weakly and are describable by pseudopotentials. Therefore we will discuss these methods first. New techniques for deriving interatomic potentials by a generalized perturbation expansion within the tight-binding scheme [2.3–5] or by decomposing the bonding energy of transition metals into a sum of volume forces and pairwise interactions [2.6–8], will be discussed later.

2.1 Pseudopotentials

Pseudopotentials have been introduced with the intention of simplifying the calculation of the electronic structure by eliminating the necessity to include the tightly bound electrons forming the ionic core and the strongly attractive potential responsible for binding them. Like many important ideas in condensed matter theory, the concept of a weak pseudopotential replacing the strong ionic potential without modifying the eigenvalues of the conduction-electron states dates back to the thirties. The foundations were laid in a series of papers by *Hellmann* [2.9,10] and *Gombas* [2.11]. Hellmann shows that a conduction electron in the force field of the ion core feels not only the electrostatic potential of the ion, but in addition its kinetic energy is changed. This change in the kinetic energy has its origin in the Pauli principle which forbids two electrons of parallel spin to occupy the same orbital and forces the conduction electron to "orthogonalize" to the core electron wave functions.

Later, progress in the theory of pseudopotentials was achieved along two distinctly different lines of development.

2.1.1 The Operator Approach

The original problem adressed by pseudopotential theory is that of solving the Schrödinger equation

$$H|\psi_n\rangle = (T + V)|\psi_n\rangle = E_n|\psi_n\rangle \qquad (2.2)$$

for an electron moving in the effective potential V. The potential V includes Coulombic contributions from the nuclei and all the electrons, as well as exchange and correlation effects. In principle, V has to be determined self-consistently, but this problem will be dealt with in the framework of linear response theory. One starts from the fact that in simple metals the electron states separate naturally into two classes: "core" and "valence", and from the necessity for the valence orbitals to be orthogonal to the core orbitals. Hence it is desirable to introduce a new pseudo-Hamiltonian H_{ps} which has the same spectrum of valence eigenvalues, but produces no lower eigenvalues. It was first shown by *Phillips* and *Kleinman* [2.12] and by *Antoncik* [2.13] that this transformation may be achieved by defining a new nonlocal and energy-dependent operator

$$W(E) = V + \sum_c |\psi_c\rangle\langle\psi_c|(E - E_c) \ , \qquad (2.3)$$

where c stands for the set of quantum numbers characterizing a core state and for the ionic coordinates. If E_n is a valence-electron eigenvalue of H with eigenfunction ψ_n, then it follows by substitution of (2.3) in (2.2) that E_n is also an eigenvalue of the pseudo-Hamiltonian $H_{\mathrm{ps}} = T + W(E)$ with an eigenfunction φ_n (the "pseudoorbital") which satisfies

$$H_{\mathrm{ps}}|\varphi_n\rangle = (T + W(E))|\varphi_n\rangle = E_n|\varphi_n\rangle \ . \qquad (2.4)$$

The true valence orbital $|\psi_n\rangle$ and the pseudoorbital $|\varphi_n\rangle$ are related through the equation

$$|\psi_n\rangle = \left(1 - \sum_c |\psi_c\rangle\langle\psi_c|\right)|\varphi_n\rangle \ , \qquad (2.5)$$

i.e. the true orbital is just the projection of the pseudoorbital onto the subspace orthogonal to the core states. It follows at once from (2.5) that there is no unique correspondence between ψ and φ: we may add any linear combination of core orbitals to φ without changing ψ. Subsequent work by *Austin* et al. [2.14] has shown that the pseudopotential $W(E)$ is not a unique operator – there is a great variety of possible choices – any operator of the form

$$H_{ps} = T + V + \sum_c |\psi_c\rangle\langle F_c| \qquad (2.6)$$

(where F_c is a completely arbitrary wavefunction) has the properties of a pseudo-Hamiltonian specified by equations (2.4) and (2.5). *Cohen* and *Heine* [2.15] have proposed to exploit the arbitrariness in the choice of the pseudopotential by constructing an "optimized" pseudopotential giving the smoothest possible pseudoorbital. The smoothest orbital obeys the condition

$$\frac{\langle\nabla\varphi|\nabla\varphi\rangle}{\langle\varphi|\varphi\rangle} = -\frac{\langle\varphi|\Delta\varphi\rangle}{\langle\varphi|\varphi\rangle} \simeq \text{Min} \qquad (2.7)$$

(the smoothest orbital corresponds to the minimum kinetic energy – see [2.16]).

It is clear from (2.4) that the condition (2.7) is equivalent to $\langle\varphi|W(E)|\varphi\rangle$ $\sim$ Max, i.e. by optimizing the pseudopotential we ensure a maximum cancellation of the strongly attractive atomic potential V by the repulsive part of the pseudopotential [the second term in (2.6)]. This means that the optimized pseudopotential is weak and therefore amenable to the use of perturbation theory. A variety of such optimized pseudopotentials have been proposed in the literature [2.16–18]. Optimized pseudopotentials based on LDF theory and an expansion of the valence states in terms of linear combinations of plane waves and core orbitals are presented in detail in Appendix A. *Harrison* [2.19] and *Moriarty* [2.20–22] have proposed a generalization of the operator approach which is applicable to metals with empty or filled d-bands. The modified pseudopotential includes a "hybridization" term which has a resonant form, the width 2Γ of the resonance serving as the expansion parameter in a perturbative approach. *Heine* [2.23] had shown that such a term arises by combining a plane-wave and a tight-binding description of the band structure.

Let us note that the transformation (2.5) does not conserve the norm of the valence orbital: the electron density given by $|\varphi(\mathbf{r})|^2$ differs from the true electron density $|\psi(\mathbf{r})|^2$ in the region of the ionic cores. The integral difference between the two charge densities is called the orthogonalization hole. Evidently this is a very appropriate picture: because of the Pauli principle, the presence of an ion introduces a "hole" into the homogeneous distribution of conduction electrons. In this region a valence electron has a very high kinetic energy, hence it travels very rapidly through the core region and the probability of finding an electron there is diminished. The orthogonalization hole must of course be considered for a proper description of the scattering of valence electrons by the ions. For the more technical details of the construction of a pseudopotential we refer to Appendix A.1.

2.1.2 The Scattering Approach

Another method to define a correct pseudopotential proceeds by representing the pseudopotential and pseudoorbitals by simple analytic functions. These functions are selected so as to guarantee the desired properties, i.e. nodeless pseudoorbitals and weak pseudopotentials (of course they must be self-consistent in the sense that they are solutions of the same pseudo-Schrödinger equations). The parameters of this ansatz are adjusted to reproduce the spectral properties of the valence electrons as known from spectroscopic data or from an all-electron calculation.

The pseudopotential of Heine, Abarenkov and Animalu (HAA) is prototypic for this class of potentials [2.24, 25]. The potential consists of a Coulombic tail at large radius, discontinuously changing to an energy- and angular-momentum-dependent constant inside some (again possibly angular-momentum-dependent) core radius R_l, i.e.

$$w^{\mathrm{HAA}}(r) = \begin{cases} -A_l(E) & r < R_l \\ -\dfrac{2Z}{r} & r > R_l \end{cases}, \tag{2.8}$$

(atomic units will be used throughout). This approach is sufficiently flexible to allow the pseudopotential to satisfy other requirements thus making it more effective: it has been shown that the HAA pseudopotential may be optimized in the sense of criterion (2.7) by allowing the core radius to vary with angular momentum thereby avoiding a discontinuity [2.26–28]. *Dagens, Rasolt* and *Taylor* [2.29, 30] proposed a pseudopotential of the HAA type designed to yield within linear screening the correct charge density induced by the actual potential of a single ion in the homogeneous electron gas (which will contain nonlinear terms). The idea is again that by this requirement, higher order terms are folded back into lower order. The drawback is that this assumption is not systematically derivable from any accurate approximation for calculating total energies.

Extensions of the HAA-type pseudopotentials to d-band metals have been worked out by *Animalu* [2.31] and by *Dagens* [2.32]. The $l = 2$ angular momentum component of the pseudopotential inside the core radius is assumed to be of a resonant form, $A_2(E) \simeq U/(E - E_d)$, where U is a finite rank operator which acts only on the d-states and E_d is a d-band energy. Dagens points out that a perturbation scheme can be expected to work only as long as the d-band width is significantly smaller than $E_d - E_F$, where E_F is the Fermi level. Essentially this restricts the applicability of the scheme to d-bands that are either completely full (noble metals) or empty (the heavy alkaline earth metals).

Again all these pseudopotentials are not "norm-conserving". An identity similar to the Friedel-sum rule relates the energy derivative of the pseudoorbital φ, calculated at a distance R from the nucleus, to the integrated charge contained within a sphere of that radius [2.33, 34]:

$$-2\pi \left[(r\varphi)^2 \frac{d}{dE} \frac{d}{dr} \ln \varphi \right]_{r=R} = 4\pi \int_0^R \varphi^2 r^2 dr \qquad (2.9)$$

(in atomic units). The integrated difference between the charge densities described by $|\psi(r)|^2$ and $|\varphi(r)|^2$ is usually referred to as the depletion hole. The depletion hole charge must be treated in the same manner as the orthogonalization hole charge.

However, the problem with the orthogonalization or depletion hole is not an unavoidable consequence of replacing the all-electron potential by a pseudopotential. Recently it has been realized that "norm-conservation" can be imposed as a supplementary condition to be satisfied by the pseudoorbitals [2.34–40]. The important consequences of norm-conservation have been stressed by *Bachelet* et al. [2.40]. Note that the logarithmic derivative of the wave function is simply related to the scattering phase shift (see e.g. *Schiff* [2.41]). Then the identity (2.9) tells us that if two potentials V and W with eigenfunctions ψ and φ and the same eigenvalue spectrum have the same integrated charge within a sphere of radius R [and $\psi(R) = \varphi(R)$ by construction], then not only their scattering phase shifts (at $r = R$) but also the linear energy variation of their phase shifts (around E) are identical. Therefore the scattering properties of the full potential and the pseudopotential have the same energy variation to first order and thus, when the pseudopotential is transferred to other systems (from the free atom to the metal or to the alloy), norm-conservation serves to optimize the transferability. (See Appendix A.2 for a detailed discussion of norm-conserving pseudopotentials.) On the other hand one must bear in mind that in contrast to the case of Phillips-Kleinman pseudopotentials (2.3, 6), orthogonalization of the orbitals of norm-conserving pseudopotentials to the core orbitals does not yield a well-defined object. Thus norm-conserving pseudopotentials are not well suited to address problems such as Knight shifts etc.

Pseudopotentials of this type are usually constructed on the basis of a local density functional (LDF) description of exchange and correlation effects for a reference configuration (usually the free atom). As the LDF-theory is a statistical description, it applies to the all-electron situation. To obtain the ion-core pseudopotential which will be the basis of a perturbation-theoretic treatment, the atomic pseudopotential must first be unscreened by subtracting the valence-electron contribution to the local density functional [2.40]. Similar problems exist within the operator approach [2.21, 22, 42, 43]. See also Appendix A.

2.1.3 Model Potentials

Pseudopotentials derived from first principles are either energy dependent or non-local and these features complicate their application. Although such calculations are in fact feasible, simple local and energy-independent pseudopotentials (to distinguish, we shall call them model potentials) have found a wide-spread use. The use of local model potentials makes the explanation of the methodology much simpler and allows the discussion of trends through the periodic table. For actual quantitative calculations, however, it is desirable to retain both nonlocality and energy dependence.

An especially simple and useful form of a model potential is the empty-core (EC) potential proposed by *Ashcroft* [2.44]

$$w^{EC}(r) = \begin{cases} 0 & r < R_c \\ -\dfrac{2Z}{r} & r > R_c \end{cases} , \tag{2.10}$$

which has the Fourier transform

$$w^{EC}(q) = -\frac{8\pi Z}{V_a q^2} \cos(qR_c) . \tag{2.11}$$

Equation (2.11) shows why pseudopotential perturbation theory may be expected to work: the first zero of the pseudopotential matrix element occurs at $q_0 = \pi/2R_c$ and this happens to be very close to the smallest reciprocal lattice vectors of most metals. These are precisely the terms that enter a perturbation expansion of the total energy – we find them to be small and can hence expect the convergence of the perturbation series to be quite good. In a crystal at equilibrium a mutual cancellation of the electronic scattering occurs for all momentum transfers not equal to vectors of the reciprocal lattice. For a liquid or a glass there is no complete cancellation, but as indicated by the small values of the static structure factor at low q, the contribution of the large low-q matrix elements to the ground-state energy is still small.

It is unlikely that much can be learnt by considering the vast variety of parametrized model potentials that has appeared in the literature. Their usefulness decreases exponentially with the number of parameters and they will be referred to only if they serve to make a special point.

2.2 Response Theory

The general form of the desired expansion of the electronic ground state energy $E_e(\{R_l\})$ is given by (see e.g. [2.45])

$$E_e = E_0 + E_1 + E_2 + E_3 + \dots \qquad \text{with} \qquad (2.12)$$

$$E_n = V_a \sum_{q_1, \dots, q_n} \Gamma^{(n)}(q_1, \dots, q_n) w(q_1) \dots w(q_n)$$
$$\times S(q_1) \dots S(q_n) \delta(q_1 + \dots + q_n) \,, \qquad (2.13)$$

where V_a is the atomic volume, $w(q)$ is the Fourier transform of the pseudopotential of a single ion (the pseudopotential form factor), and $S(q) = N^{-1} \sum_j \exp(iqR_j)$ is the structure factor expressing the spatial arrangement of the ions (for simplicity we have assumed the pseudopotential to be local; for the corrections arising from nonlocality, see below). The δ-function assures the conservation of momentum and the "multipole functions" $\Gamma^{(n)}(q_1, \dots, q_n)$ are general characteristics of the electron gas, they depend only on the electron-electron interaction and are independent of the positions of the ions.

The calculation of the $\Gamma^{(n)}$ is a rather complex task of many-body theory but the lowest order terms can also be derived by conventional perturbation theory [2.46, 47], for details see Appendix B. For the first-order term one has immediately (ϱ is the electron density)

$$\Gamma^{(1)}(0) = \varrho \,. \qquad (2.14a)$$

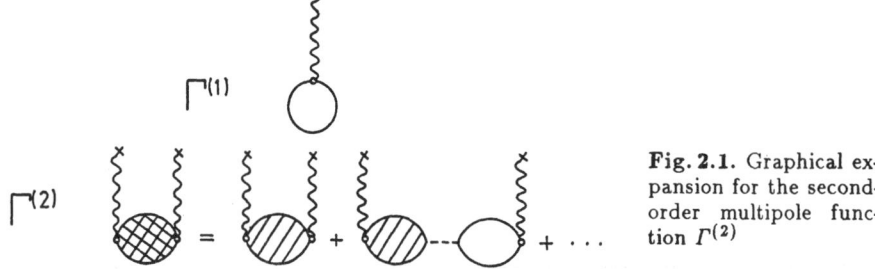

Fig. 2.1. Graphical expansion for the second-order multipole function $\Gamma^{(2)}$

In the graphical expansion for $\Gamma^{(2)}$ (Fig. 2.1) the wavy lines represent the interaction of an electron with the external potential, the dashed lines the electron-electron interaction and $\chi(q)$ is the irreducible block with respect to these lines, i.e. the polarization operator. The summation is carried out directly with the result

$$\Gamma^{(2)}(q, -q) = \tfrac{1}{2} \chi(q)/\varepsilon(q) \,, \qquad (2.14b)$$

where

$$\varepsilon(q) = 1 - \frac{8\pi}{q^2} \chi(q) \tag{2.15}$$

is the static dielectric function of the homogeneous electron gas. For a detailed discussion of the response theory of the electron gas the reader is referred to Appendix B. The polarizability is given by the Lindhard function [2.47]

$$\chi(q) = \varrho \sum_{k} \frac{n_k - n_{k+q}}{k^2 - (k+q)^2}$$
$$= -\frac{3}{2} \frac{\varrho}{E_F} \left[\frac{1}{2} + \frac{1}{4} \frac{1-\eta^2}{\eta} \ln \left(\left| \frac{1+\eta}{1-\eta} \right| \right) \right] \quad \eta = \frac{q}{2K_F}, \tag{2.16}$$

n_k is the Fermi function describing the occupation of the electron states. Equations (2.15) and (2.16) have been calculated in the random phase approximation (RPA), i.e. the electrons interact only through Coulomb forces; for exchange and correlation corrections, see below. In the same approximation, the higher-order multipole functions have the general structure

$$\Gamma^{(n)}(q_1, \ldots q_n) = \Lambda^{(n)}(q_1, \ldots q_n)/\varepsilon(q_1)\ldots\varepsilon(q_n) \quad n>2 . \tag{2.14c}$$

The explicit form of the third-order term has been derived by *Lloyd* and *Sholl* [2.48], see also *Brovman* et al. [2.49]; approximate forms for the fourth- and fifth-order expressions have been given by *Hammerberg* and *Ashcroft* [2.50] and *Kagan* et al. [2.51].

It is advantageous to take into account the condition of charge neutrality in an explicit way. A Fourier transformation of the potential energy (2.1), followed by a collecting together of the separately divergent $q = 0$ terms, then leads to the following form for the total energy E_g:

$$E_g = E_{es} + E_e , \tag{2.17}$$

where E_{es} is the electrostatic energy or Madelung energy, representing the energy of point ions of charge Z immersed in a uniform distribution of neutralizing electronic charges. E_e is again the electronic ground state energy from which have been subtracted the divergent parts corresponding to the interaction of the ions with the uniform distribution of the conduction electrons and to their self-interation (for the techniques for evaluating the Madelung sums, see [2.52] and Appendix C). In spite of the mutual cancellation of the Coulomb interactions at $q = 0$, the pseudopotential matrix element $w(q)$ retains a non-zero $q = 0$ component expressing the volume-averaged non-Coulombic interactions

$$\frac{b}{V_a} \equiv \lim_{q \to 0} \left[w(q) - \frac{8\pi Z}{V_a q^2} \right] . \tag{2.18}$$

Now it is easy to obtain explicit expressions for the various terms in the expansion (2.12),

$$E_0 = Z E_{eg}$$

$$E_1 = bZ/V_a$$

$$E_2 = \frac{V_a}{2} \sum_{q \neq 0} \frac{\chi(q)}{\varepsilon(q)} |w(q)|^2 |S(q)|^2$$

$$E_3 = V_a \sum_{q_1, \ldots, q_n} \frac{\Lambda^{(n)}(q_1, \ldots, q_n)}{\varepsilon(q_1) \ldots \varepsilon(q_n)} w(q_1) \ldots w(q_n)$$
$$\times S(q_1) \ldots S(q_n) \delta(q_1 + \ldots + q_n) . \tag{2.19}$$

The zeroth order term is identical to the valence Z times the ground state energy E_{eg} of the homogeneous electron gas. Remembering the definition of the static structure factor, eqs. (2.19) lead directly to the desired formulation of the effective Hamiltonian H_i (here $R_{ij} \equiv |R_i - R_j|$)

$$H_i = T_i + E_V + \frac{1}{2!N} \sum_{i \neq j} \Phi^{(2)}(R_{ij}, V)$$
$$+ \frac{1}{3!N} \sum_{i \neq j \neq k} \Phi^{(3)}(R_{ij}, R_{jk}, R_{ki}, V) + \ldots \tag{2.20}$$

where E_V is a large volume energy and $\Phi^{(p)}$ are potentials describing a p-body interaction. For example, consider the second-order contribution in the expansion given by (2.19). Defining

$$F(q) = \frac{V_a}{2} \frac{\chi(q)}{\varepsilon(q)} |w(q)|^2 , \tag{2.21}$$

the energy-wavenumber characteristic, we write

$$E_2 = \sum_{q \neq 0} |S(q)|^2 F(q)$$
$$= \frac{1}{2N} \sum_{i \neq j} \left\{ \frac{2}{N} \sum_q e^{iq(R_i - R_j)} F(q) \right\} + \frac{1}{N} \sum_q F(q) - F(0) \tag{2.22}$$

where the bracketed term defines $\Phi^{(2)}_{ind}(R_{ij}, V)$. We find that $F(q)$ is the

Fourier transform of an indirect, electron-mediated ion-ion interaction $\Phi_{\text{ind}}^{(2)}$, whereas the second and third terms in (2.22) contribute to the volume energy. The direct Coulomb repulsion between the ions may be treated in a similar way and combined with the indirect interaction to form an effective pair potential $\Phi^{(2)}$, the $q = 0$ term contributing to the volume energy. Thus, to second order in the pseudopotential, $\Phi^{(2)}$ and E_V are given by

$$\Phi_2^{(2)}(R, V) = \frac{2Z^2}{R} + \frac{2}{N} \sum_q e^{iqR} F(q) \quad \text{and} \tag{2.23}$$

$$E_V = Z E_{\text{eg}} + Zb/V_a + \frac{1}{N} \sum_q F(q) + \lim_{q \to 0} \left[\frac{8\pi Z^2}{V_a q^2} + F(q) \right] . \tag{2.24}$$

For a local pseudopotential *Finnis* [2.53] has shown that the limit may be evaluated using the compressibility sum rule of the electron gas [2.52, 54, 55] with the result

$$E_V = Z(E_{\text{eg}} - \tfrac{1}{2} V_a B_{\text{eg}}) + \tfrac{1}{2} \Phi_{\text{ind}}^{(2)}(R = 0) , \tag{2.25}$$

where we have in addition used the definition of the indirect ion-ion interaction (2.22). In this form the physical content of the various contributions is evident: if we take the pair summation $(2N)^{-1} \sum_{i \neq j} \Phi^{(2)}(R_{ij})$, we have per ion one half of the interaction ion-ion plus ion-screening cloud of the other ion. The volume energy E_V consists of the electrostatic interaction between the ion and its own screening cloud and two electron-gas terms (2.25). Let us note at once that most of the binding energy is found to be in the intra-atomic term, the two electron-gas terms contribute typically 15–25%, the pair summation 2–5%.

The third-order terms can be treated in a similar way. With the definition

$$F^{(3)}(q_1, q_2, q_3) = V_a \frac{\Lambda^{(3)}(q_1, q_2, q_3)}{\varepsilon(q_1)\varepsilon(q_2)\varepsilon(q_3)} w(q_1) w(q_2) w(q_3) , \tag{2.26}$$

the three body potential $\Phi_3^{(3)}$ is given by (we drop the argument V)

$$\Phi_3^{(3)}(R_{ij}, R_{jk}, R_{ki}) = \frac{3!}{N^2} \sum_{q_1, q_2} F^{(3)}(q_1, q_2, -q_1 - q_2)$$
$$\times \exp[iq_1(R_i - R_j)] \exp[iq_2(R_j - R_k)] , \tag{2.27}$$

the third-order contribution to the two-body potential is given by

$$\Phi_3^{(2)}(R) = \frac{3.2!}{N} \sum_{q_1, q_2} F^{(3)}(q_1, q_2, -q_1 - q_2) e^{iq_1 R}$$
$$- \frac{3.2!}{N} \sum_{q_1} F^{(3)}(q_1, -q_1, 0) e^{iq_1 R} \tag{2.28}$$

43

and to the volume energy by

$$E_{V,3} = \frac{1}{N^2} \sum_{q_1,q_2} F^{(3)}(q_1, q_2, -q_1 - q_2)$$

$$- \frac{3}{N} \sum_{q_1} F^{(3)}(q_1, -q_1, 0) - F^{(3)}(0,0,0) \ . \tag{2.29}$$

Hence we find that the volume energy E_V contains contributions from all orders of the expansion $(2.12, 13)$ and the p-body potential $\Phi^{(p)}$ describing the simultaneous indirect, electron-mediated interaction between p ions may be represented in the form of a series in powers $k \geq p$ of the electron-ion interaction

$$\Phi^{(p)}(\{R_i\}) = \sum_{k \geq p} \Phi^{(p)}_k(\{R_i\}) \ , \tag{2.30}$$

but − as we have already emphasized [see the discussion after (2.11)] − this is not an expansion in terms of a small parameter. For terms of a given order in $w(q)$, the separation of this term into pairwise and non-pairwise contributions is considerably less effective than the actual determination of that term. As a rule one finds that the two-body and three-body contributions to E_3 are of the same order of magnitude and considerably larger than E_3 itself. This means that it is not useful to try to improve the pair potential by going to the higher order terms of the expansion (2.30) without at the same time calculating the corresponding many-body interactions.

We also remark that the effective interaction between the ions depends not only on their separation: the multipole functions $\Gamma^{(n)}$ depend on the electron density (and hence on the volume) and so do the p-body potentials $\Phi^{(p)}$. The peculiar importance of the density dependence of the interatomic forces appears very clearly if we consider the relation between the dynamical compressibility − derived from the velocity of sound − and the static compressibility − calculated by differentiating the ground state energy twice with respect to the atomic volume. In performing this differentiation one must remember that in addition to its explicit volume dependence, the ground-state energy depends implicitly on the volume through the interatomic distances (or momentum transfers, depending on whether a real or a reciprocal space representation is used) and through its dependence on the electron density. Thus the total differential reads

$$\frac{d}{dV} = \frac{\partial}{\partial V}\bigg|_{R_j,\varrho} + \frac{R_j}{3V} \frac{\partial}{\partial R_j}\bigg|_{\varrho,V} - \frac{\varrho}{V} \frac{\partial}{\partial \varrho}\bigg|_{R_j,V} \ . \tag{2.31}$$

As a consequence, even though the interaction is described by second-order forces only, if the volume changes, not only the interparticle distance changes, but also their interaction potential changes, due to its dependence on the electron density. This introduces higher-order forces between the ions, even though they are derived from E_2 in the static approach [2.56,57]. In the dynamic approach we consider the expansion of sound waves at constant volume and consequently, to second order, only pair forces contribute to the dynamical compressibility. To obtain results consistent with the static compressibility, one must include some terms of third and fourth-order in the electron-ion interaction (note that the compressibility problem is closely related to the Cauchy problem for the elastic constants [2.49,53,58]).

Mathematically, the interrelation between the density dependence of the pair (or generally p-body forces) and the $(p + 1)$-body forces is expressed by a number of exact relations of the type known as Ward identities (*Luttinger* and *Ward* [2.59]) valid for the multipole functions of any Fermi liquid. Moreover, what applies to the compressibility, also applies to any property involving volume changes, e.g. internal pressure [2.53], vacancy formation volume [2.60] or relaxation around defects [2.61–64].

Before we turn to a discussion of the general characteristics of the interatomic interactions as defined in eqs. (2.20) through (2.30), we will briefly consider the corrections introduced by nonlocality and exchange and correlation between the conduction electrons.

2.2.1 Nonlocality

For the sake of simplicity our overview of the methodology has been based on local pseudopotentials. The nonlocal nature of the pseudopotential does not introduce any new physical aspects, only additional mathematical complications. In compensation for the increased numerical effort, calculations based on nonlocal pseudopotentials are considerably more accurate. If the form factor of the pseudopotential now depends on the initial momentum k, the first order contribution E_1 is now given by [2.16,2.52]

$$E_1 = \sum_k \langle k|w_{\mathrm{NC}}|k\rangle n_k \qquad (2.32)$$

instead of (2.14a, 18). n_k is the Fermi function describing the occupation of the electron states. The suffix NC designates the non-Coulombic part of the pseudopotential. The characteristic blocks in the expansion for E_2 (Fig. 2.1) still have the same form, but it is no longer possible to express them by the polarization operator alone. If one represents the first term of the sum depicted in Fig. 2.1 by $X_1(q)$ and the second by $X_2(q)$, then the sum over the polarization blocks yields

$$E_2 = \frac{V_a}{2} {\sum_q}' |S(q)|^2 \left(X_1(q) - \frac{8\pi}{q^2} \frac{1}{\varepsilon(q)} |X_2(q)|^2 \right) . \qquad (2.33)$$

In the local approximation X_1 and X_2 reduce to $X_1(q) = w(-q) X_2(q) =$

$|w(q)|^2 \chi(q)$ and of course we recover (2.15) and (2.19). Again keeping within the RPA, in the nonlocal case $X_1(q)$ and $X_2(q)$ are given by [2.16, 2.45]

$$X_1(q) = \int \frac{n_k - n_{k+q}}{E_k - E_{k+q}} |\langle k|w|k+q\rangle|^2 \frac{d^3k}{(2\pi)^3} \tag{2.34}$$

and

$$X_2(q) = \int \frac{n_k - n_{k+q}}{E_k - E_{k+q}} \langle k|w|k+q\rangle \frac{d^3k}{(2\pi)^3} \ . \tag{2.35}$$

One finds that the screened pseudopotential is now given by

$$\langle k|\tilde{w}|k+q\rangle = \langle k|w|k+q\rangle + w_{sc}(q) \tag{2.36}$$

with the local screening potential

$$w_{sc}(q) = \frac{8\pi}{q^2 \varepsilon(q)} X_2(q) \ . \tag{2.37}$$

Similar complications arise for the higher-order terms, but all the general results presented above are preserved. An expansion of the total energy to third order in a nonlocal pseudopotential has been worked out by *McLaren* [2.65], albeit with some simplifying assumptions; actual nonlocal calculations have only been performed to second order until now. The numerical calculation of the multi-dimensional integrals defining the analogues of X_1 and X_2 for the higher order terms appears indeed to be quite hopeless. For higher order calculations, nonlocal pseudopotentials have to be localized according to some appropriate description and of course this introduces some uncertainty (see e.g. [2.66]).

2.2.2 Screening Beyond the Random Phase Approximation

Effects of exchange and correlation between the conduction electrons may be taken into account through a replacement of the RPA-dielectric function (2.15, 16) by the effective dielectric function $\tilde{\varepsilon}(q)$

$$\tilde{\varepsilon}(q) = 1 - \frac{8\pi}{q^2}[1 - G(q)]\chi(q) \ , \tag{2.38}$$

where $G(q)$ is the exchange and correlation correction (often also called the local field) determining an effective electron-electron interaction of the form

$$v_{eff}(q) = [1 - G(q)]\frac{8\pi}{q^2} \ . \tag{2.39}$$

Unfortunately, there is not room here for a detailed review of the full array of local fields $G(q)$ to be found in the literature. The reason for the existence of so many different approximations arises from the fact that they require a careful treatment of electronic many-body effects, which is an exceedingly complex problem. The simplest possible approach is the LDF-form of the local field proposed by *Hedin* and *Lundquist* [2.67,68] and *Taylor* [2.69] (for more details see Appendix B):

$$G(q) = \gamma(q/k_F)^2 , \qquad \text{where} \tag{2.40}$$

$$\gamma = \frac{1}{4} - \frac{\pi\alpha}{24} \frac{d}{dR_s} \left[R_s^{-2} \frac{d}{dR_s} E_c(R_s) \right] \tag{2.41}$$

with $\alpha = (\frac{4}{9}\pi)^{1/3}$. R_s is the electron-sphere radius ($R_s = R_a Z^{-1/3}$, where R_a is the atomic radius) and E_c the electron-gas correlation energy, taken in some convenient parametrization [2.55, 70–72]. γ has been determined from the condition that the dynamic compressibility of the electron gas derived from the low-q limit of $\tilde{\varepsilon}(q)$ [2.54], should be equal to the static compressibility calculated by differentiating the ground-state energy, see also Appendix B.

This compressibility sum rule is also the most important guide through the vast literature on local fields. Besides this, there are other sum rules relating the static dielectric function to the correlation energy, the electron-electron pair correlation function etc. (see e.g. the recent review of *Ichimaru* [2.73]), but as the relative success of the LDF-$G(q)$ demonstrates [2.69], the compressibility sum rule is by far the most important. Analyzing the proposed local fields in the light of the sum rules shows that all the expressions published before 1970 are now outdated and superceded by more recent work. Among the remaining approaches, those of *Vashishta* and *Singwi* [2.74], *Toigo* and *Woodruff* [2.75], *Geldart* and *Taylor* [2.76], and *Ichimaru* and *Utsumi* [2.77] are worthy of attention.

In Fig. 2.2 the various local fields are shown for an electron density corresponding approximately to Al. We find that the main differences occur

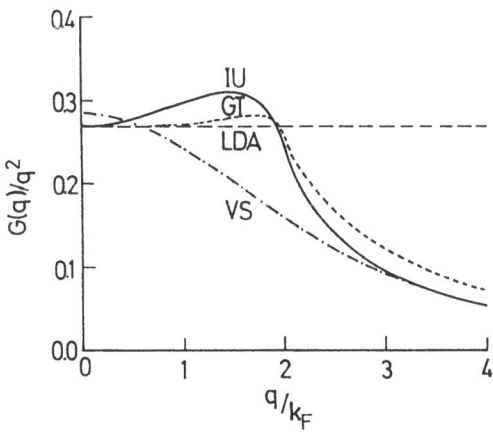

Fig. 2.2. Local fields $G(q)$ at an electron density of $R_s = 2.07$ calculated using local-density approximation (LDA) and the approximations of Geldart and Taylor (GT), Vashishta and Singwi (VS) and Ichimaru and Utsumi (IU) respectively

for $q \sim 2k_F$, but – as already noted by *Hedin* and *Lundquist* [2.67] – the large q behaviour of the local fields is not very important, because the polarizability drops off very rapidly near $q = 2k_F$. More important are the differences in the region $1.5k_F < q < 2k_F$. We will find that they have a rather important influence on the interatomic forces, especially for short distances.

The Lindhard form of the electron-gas polarizability is based on a sharp Fermi surface, hence an infinite mean-free-path of the electrons. An approximate generalization of the Lindhard function to include finite mean-free-path effect has been worked out by *Leavens* et al. [2.78]. This could well be important for liquid metals at rather elevated temperatures.

2.3 Effective Pair Potentials in Pure Metals

In the preceding sections we have seen that to construct a simple-metal pair potential both a pseudopotential and a dielectric function must be selected. Hence the first problem is to make a choice among the multitude of solutions proposed for both quantities. For the pseudopotential the choice will clearly be one of the optimized potentials for which convergence is forced to occur at second order. The only attractive feature of local model potentials is their simplicity – they are useful in analyzing trends. For the local fields, the sum rules provide the necessary guideline and reduce the choice to a few alternatives. In this section our aim will be to study the properties of the effective pair potentials: first we will review a few general properties of the pair potentials, than we will study the influence of different choices of $G(q)$ and $w(q)$ for a few examples, and finally we will review the trends in the interatomic potentials through the periodic table.

The general form of the interatomic potential is well known, it consists of a strongly repulsive core plus an oscillatory potential at intermediate and long range distances.[1] The asymptotic (large R) behaviour of $\Phi(R)$ is a consequence of the logarithmic singularity of the Lindhard polarizability (2.16) at $q = 2k_F$. At large distances $(R \to \infty)$ $\Phi(R)$ assumes the Friedel form (see [Ref. 2.52, p. 276]).

$$\Phi(R) \propto \tilde{w}(2k_F)^2 \frac{\cos{(2k_F R)}}{(2k_F R)^3} \ . \tag{2.42a}$$

Note that the amplitude of the oscillations is set by the magnitude of the screened pseudopotential at $q = 2k_F$. The higher-order multipole functions $\Gamma^{(n)}$ $(n > 2)$ have an even higher degree of non-analyticity than $\Gamma^{(2)}$. However, the definitions of the corresponding p-body forces involve additional

---
[1] The form (2.23) of the effective pair potential is just one possible formulation – other equivalent expressions appear frequently in the literature. See Appendix B.

integrations in q-space, so that the asymptotic form of the leading term in (2.30), $\Phi_p^{(p)}$ is still determined by the non-analyticity of the polarizability. *Harrison* [2.79] has shown that the asymptotic form of the p-body potential is given by

$$\Phi^{(p)}(\boldsymbol{R}_1, \ldots, \boldsymbol{R}_p) \propto \frac{\cos\left[k_{\mathrm{F}}(R_{12} + R_{23} + \ldots + R_{p1})\right]}{R_{12}R_{23}\ldots R_{p1}(R_{12} + R_{23} + \ldots + R_{p1})} \ , \quad (2.42\mathrm{b})$$

which reduces to (2.42a) for $p = 2$ (see also [2.80]).

The repulsive part of the potential is well known from Thomas-Fermi theory where it takes the form [Ref. 2.81, p. 87]

$$\Phi(R) = \frac{2Z^2}{R} M(i\kappa_{\mathrm{TF}}) \exp\left(-\kappa_{\mathrm{TF}} R\right) \qquad (2.43)$$

in which $M(q)$ is related through

$$w(q) = \frac{8\pi}{V_{\mathrm{a}} q^2} M(q) \qquad (2.44)$$

to the bare (unscreened) electron-ion pseudopotential $w(q)$. The Thomas-Fermi wavevector is given by $\kappa_{\mathrm{TF}} = (4k_{\mathrm{F}}/\pi)^{1/2}$. The exponential damping of $\Phi(R)$ stems mathematically from the pole in $\varepsilon_{\mathrm{TF}}(q)^{(-1)}$ at $q = \pm i\kappa_{\mathrm{TF}}$. From an analysis of the analytical properties of $\varepsilon(q)$, *Jacucci* and *Taylor* [2.82] have shown that a similar exponentially damped Coulomb potential term is expected in the RPA at all metallic densities. If one goes beyond the RPA, the expected behaviour will depend on the choice of the local field $G(q)$. *Shaw* and *Heine* [2.83] were the first to point out that generally the pair potential will be more attractive in the nearest-neighbour region if exchange and correlation are taken into account. Exchange effects keep the electrons apart and lessen the Coulomb repulsion. Generally, the screened Coulomb term will exist only at higher electron densities. *Jacucci* and *Taylor* [2.82] show that the limiting electron densities are $R_{\mathrm{s}} \sim 1.52$ for the Vashishta-Singwi $G(q)$ and $R_{\mathrm{s}} \sim 2.55$ for the LDF, Geldart-Taylor and Toigo-Woodruff $G(q)$; for the Ichimaru-Utsumi local field, the limiting R_{s} is even larger. Hence we expect the screened Coulomb repulsion to become more important in the sequence VS–LDF–GT–IU. Figure 2.3 shows pair potentials for Al and Na, calculated using the same local empty-core pseudopotential and different choices of $G(q)$. We find that the form of the pair potential at the nearest-neighbour distance depends quite critically on the interplay between the repulsive and the oscillatory terms; the oscillations extend under the repulsive core and may be exposed by a variation of its radius.

The choice of a pseudopotential appears to be less critical: in Fig. 2.4 we compare pair potentials for Al based on an orthogonalized plane wave

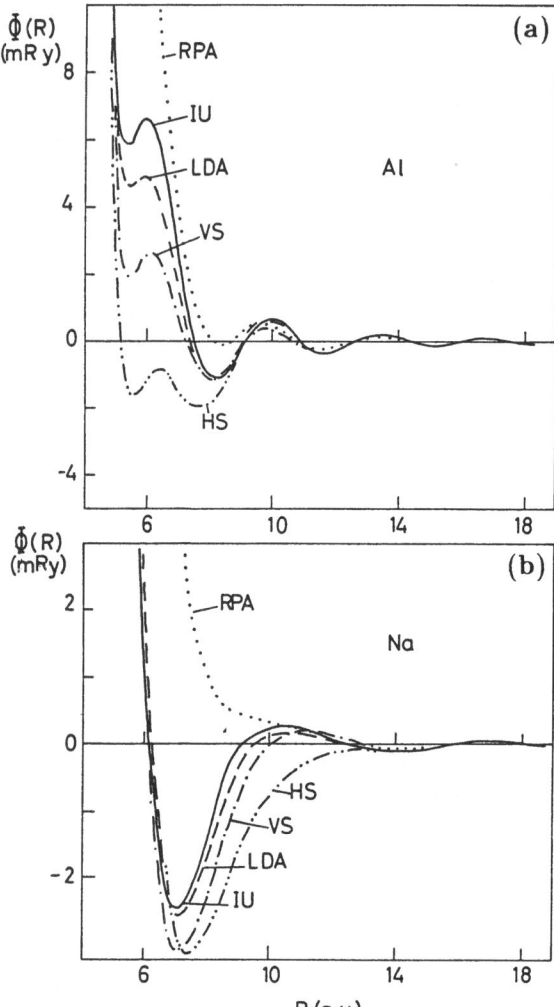

$\tilde{\Phi}(R)$
(mR y)

(a)

RPA

IU

LDA

Al

VS

HS

$\tilde{\Phi}(R)$
(mRy)

(b)

RPA

Na

HS

VS

LDA

IU

R (a.u.)

Fig. 2.3a,b. Pair potential for Al ($R_s = 2.07$, $R_c = 1.11$ a.u.) **(a)** and for N_a ($R_s = 3.93$, $R_c = 1.71$ a.u.) **(b)** calculated using the local empty-core pseudopotential and different choices for the local field (see Fig. 2.2), HS stands for the Hubbard-approximation discussed in Appendix B

(OPW) – derived pseudopotential [2.42] and on the Dagens-Rasolt-Taylor [2.29] pseudopotentials (note that both are optimized to enforce convergence of the perturbation series at low order) and again different choices of the local field. We think that it is fair to say that the two approaches seem to be reasonably convergent.

The comparison of the sodium and aluminum potentials in Fig. 2.3 demonstrates that there are important variations in the effective pair potential from one element to another – of course these variations are vital for an understanding of the structural properties. These important trends in the interatomic potentials have recently been analyzed by *Hafner* and *Heine* [2.84]. They start by noting that each element must be characterized by its valence Z, its atomic radius R_a and its pseudopotential. For simplic-

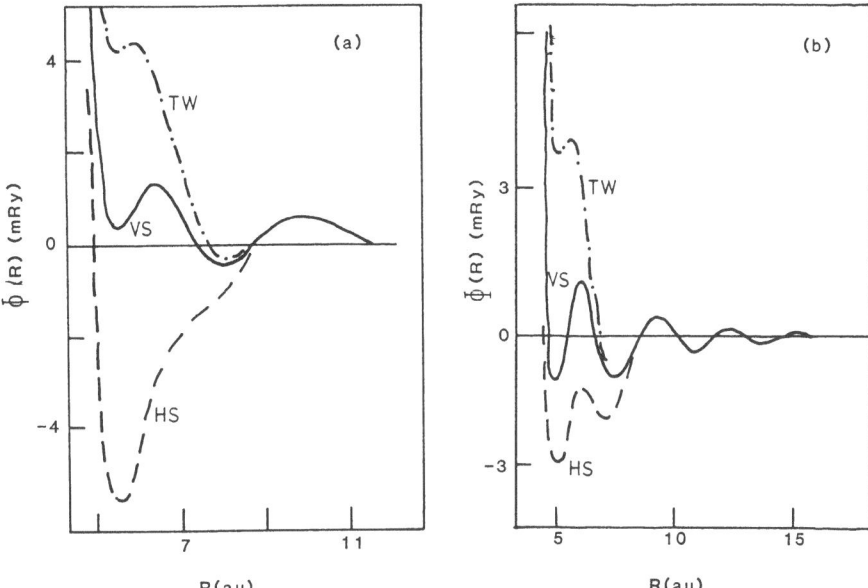

Fig. 2.4a,b. Pair potentials for Al ($R_s = 2.07$ a.u.) calculated using the optimized nonlocal pseudopotential [2.42] **(a)** and using the DRT effective pseudopotential [2.29] **(b)** and different choices for the local field (see Figs. 2.2 and 2.3)

ity, the pseudopotential is taken to be of the local empty-core form (2.10, 11) and for the local field the Ichimaru-Utsumi form has been adopted. Hafner and Heine study the variation of $\Phi(R)$ in the three-dimensional parameter manifold Z, R_a, and R_c (Note that all variables, even Z, can be treated as continuous variables. This computer-generated extension of the data base is extremely helpful to see the characteristic trends). The analysis is facilitated by two helpful transformations: the parameters are taken to be R_c, $R_s = R_a Z^{-1/3}$ and $y = R_a Z^{1/3}$ (y is the "orthogonal compelement" of the electron density). The pair potential $\Phi(R)$ is expressed in a reduced form Φ_{red}, defined by factorizing out the Coulomb term:

$$\Phi(R) = (Z^2/R)\Phi_{\mathrm{red}}(R) \; ; \tag{2.45a}$$

the distance is expressed by a scaled variable x,

$$x = \frac{2k_{\mathrm{F}} R}{2\pi} \tag{2.45b}$$

(which places the ideal Friedel oscillations at integer values of x).

The simplest feature is the variation of $\Phi(R)$ with y. It follows directly from (2.21, 23, 45a) that

$$\Phi_{\mathrm{red}}(R) = 2 + 32 \int\limits^{\infty} \frac{\chi(q)}{\varepsilon(q)} [M(q)]^2 \frac{\sin qR}{q^3} \, dq \tag{2.46}$$

51

so that it depends only on R_c and R_s. In (2.46), $M(q)$ is a normalized pseudopotential matrix element with the Coulomb part factored out [see(2.44)] so that for an empty core potential (2.10,11) we have simply $M(q) = \cos(qR_c)$. Thus y determines merely the variation of the amplitude but not the form of $\Phi(R)$ and is irrelevant to any structural argument.

The density dependence of $\Phi(R)$ at constant R_c/R_s is considered in Fig. 2.5. In the reduced distance x, the Friedel oscillations are constant in phase and almost constant in amplitude. The most important point is the expansion of the repulsive core as the electron density is increased – for the highest density it even covers the first minimum (note that in units of x, the screening length varies as $R_s^{-1/2}$).

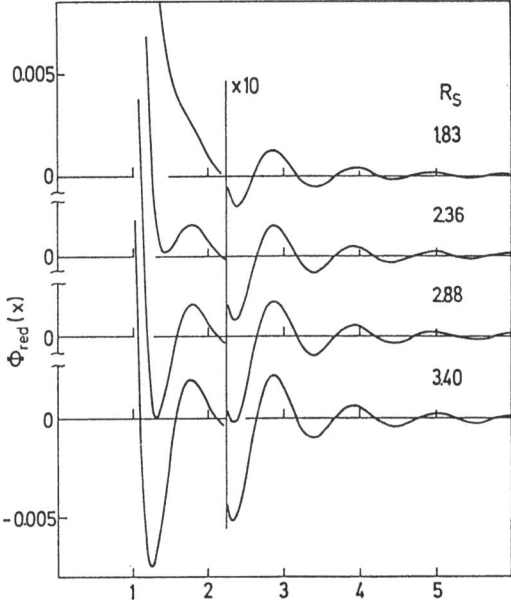

Fig. 2.5. Reduced interatomic potential $\Phi_{red}(x)$ as a function of R_s at constant $R_c/R_s = 0.50$ (after [2.84])

The variation of $\Phi(R)$ with R_c/R_s for a series of constant R_s values is shown in Fig. 2.6. One finds (i) a drastic change in the amplitude of the oscillations. According to (2.42) and (2.16) the amplitude goes as $w(2k_F)^2 \sim \cos^2(2k_F R_c) = \cos^2(2R_c/\alpha R_s)$. This means that for $R_c/R_s = 0.4092$ the first node of the pseudopotential occurs at $q = 2k_F$ ($q_0 = 2k_F$) and the leading term of the Friedel oscillations is suppressed by interference with the ionic pseudopotentials, (ii) the damping is accompanied by an increasing phase shift. This also influences the position R_{min} of the first minimum in $\Phi(R)$; R_{min} follows R_c. Hafner and Heine find that over a fairly wide range of R_c and R_s, R_{min} is in fact determined by R_c alone. In a more recent paper [2.85] the trends found in the numerical analysis are backed up by the study of simple analytical approximations to the pair potential (see also [2.86]).

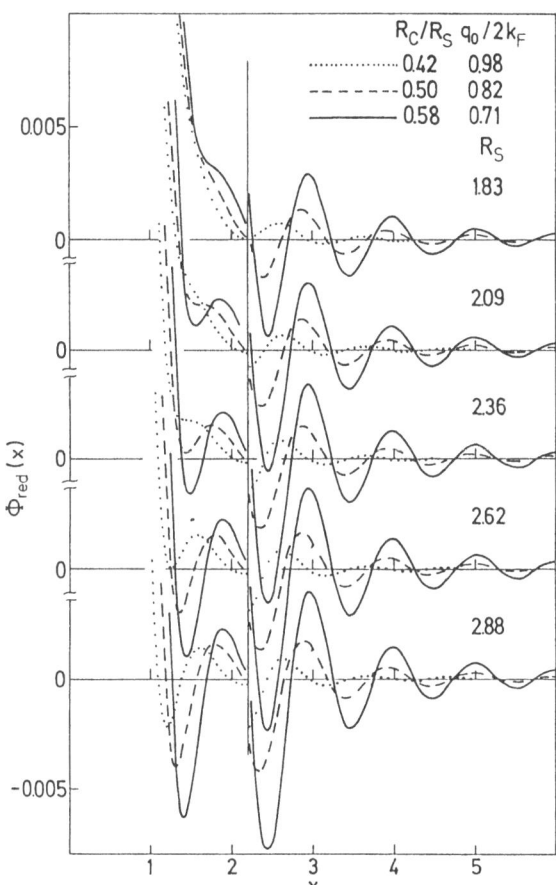

Fig. 2.6. Reduced interatomic potential $\Phi_{red}(x)$ as a function of R_c/R_s [(*full line*) $R_c/R_s = 0.58$, (- - -) $R_c/R_s = 0.50$, (···) $R_c/R_s = 0.42$] for different electron densities (after [2.84])

Hence the covering/uncovering of the minimum close to the nearest-neighbour distance is related to two essential mechanisms:

(i) the repulsive core moves over the oscillations with increasing electron density. To a good approximation, the diameter of the repulsive core is given by

$$D_r \simeq 2R_c + 2\lambda_{sc} \ , \tag{2.47}$$

where λ_{sc} is some screening length close to the Thomas-Fermi wavelength $\lambda_{TF} = \pi/\kappa_{TF}$. This is just the "scotch egg" behaviour already described by *Heine* and *Weaire* [Ref. 2.52, p. 420]

(ii) the amplitude effect: as R_c/R_s decreases the pseudopotential $w(2k_F)$ decreases and the Friedel oscillations become weaker. At sufficiently small R_c the first minimum can even be lost, turning $\Phi(R)$ into essentially a screened Coulomb form. With this analysis, a convincing explanation of the form of the interatomic potential at the nearest-neighbour distance and its variation with electron density and pseudopotential is available. Evidently this will be very helpful in analyzing structural trends.

2.4 Effective Pair Potentials in Binary Alloys

The generalization of the concept outlined above for pure metals to binary alloys is straightforward; instead of a single kind of ions we introduce now a mixture of two different ion types (described by their respective pseudopotentials) into the electron gas and we screen the ions with the response functions evaluated at the average density. All we have to do is to proceed to the replacement

$$S(\boldsymbol{q})w(\boldsymbol{q}) \rightarrow S_A(\boldsymbol{q})w_A(\boldsymbol{q}) + S_B(\boldsymbol{q})w_B(\boldsymbol{q}) \quad \text{or}$$
$$S(\boldsymbol{q})\overline{w}(\boldsymbol{q}) + D(\boldsymbol{q})\Delta w(\boldsymbol{q}) \ ,$$

where $w_A(q)$, $w_B(q)$ are the pseudopotential form factors of the individual components in the alloy and $\overline{w}$, Δw are their concentration average and difference

$$w(q) = c_A w_A(q) + c_B w_B(q) \ ,$$

$$\Delta w(q) = w_A(q) - w_B(q) \ . \tag{2.48}$$

The partial structure factors of the two atomic species are defined by [2.17,87]

$$S_i(\boldsymbol{q}) = N^{-1} \sum_l [\tfrac{1}{2} \pm g(\boldsymbol{R}_l)] e^{-i\boldsymbol{q}\boldsymbol{R}_l} \ , \quad i = A, B \ , \tag{2.49}$$

where the plus sign applies to $i = A$ and the minus to $i = B$ with

$$g(\boldsymbol{R}_l) = \begin{cases} +\tfrac{1}{2} & \text{if } \boldsymbol{R}_l \text{ is occupied by } A \\ -\tfrac{1}{2} & \text{if } \boldsymbol{R}_l \text{ is occupied by } B \end{cases} . \tag{2.50}$$

It follows that the average over the entire lattice is $\overline{g} = \tfrac{1}{2} - c$, where $c = c_B$ is the B-ion concentration. Further we define the structure factor $S(\boldsymbol{q})$ of the basic (mean) lattice and the difference structure factor $D(\boldsymbol{q})$

$$S(\boldsymbol{q}) = N^{-1} \sum_l e^{i\boldsymbol{q}\boldsymbol{R}_l}$$

$$D(\boldsymbol{q}) = N^{-1} \sum_l [\overline{g} - g(\boldsymbol{R}_l)] e^{i\boldsymbol{q}\boldsymbol{R}_l} \ . \tag{2.51}$$

To second order in the pseudopotential the relevant expressions for the ground-state energy in reciprocal space (2.19) are readily generalized to

$$E_1 = \frac{\overline{Z}}{V_a}(c_A b_A + c_B b_B)$$

$$
\begin{aligned}
E_2 &= \sum_{i,j=A,B} \sum_{q \neq 0} S_i(q) S_j^*(q) F_{ij}(q) \\
&= \sum_{q \neq 0} \{ |S(q)|^2 F_{NN}(q) + [S(q)D^*(q) + S^*(q)D(q)] F_{Nc}(q) \\
&\quad + |D(q)|^2 F_{cc}(q) \} ,
\end{aligned}
\tag{2.52}
$$

where $Z = c_A Z_A + c_B Z_B$ is the average valence and b_A, b_B stand for the $q = 0$ Fourier components of the non-Coulombic part of the pseudopotential, see (2.18). We have now a set of three partial energy-wavenumber characteristics

$$F_{ij}(q) = \frac{V_a}{2} \frac{\chi(q)}{\varepsilon(q)} w_i(q) w_j(q) , \tag{2.53}$$

expressing the indirect i-j interaction in Fourier space. Alternatively they can be replaced by the characteristics

$$F_{NN}(q) = \frac{V_a}{2} \frac{\chi(q)}{\varepsilon(q)} |\overline{w}(q)|^2 ,$$

$$F_{cc}(q) = \frac{V_a}{2} \frac{\chi(q)}{\varepsilon(q)} |\Delta w(q)|^2 ,$$

$$F_{Nc}(q) = \frac{V_a}{2} \frac{\chi(q)}{\varepsilon(q)} \overline{w}(q) \Delta w(q) , \tag{2.54}$$

describing respectively the indirect part of the mean interparticle potential, an ordering potential and a cross-term. (We could also say that F_{NN} couples to fluctuations in the number density and F_{cc} to concentration fluctuations, see Chap. 9.)

Finally, the real-space expressions for the ground-state energy (2.20–25) are replaced by

$$E = E_V + \frac{1}{2N} \sum_{\substack{i,j \\ =A,B}} \sum_{\substack{l(i) \neq \\ m(j)}} \Phi_{ij}(R_{lm}, V) \tag{2.55}$$

(the sum over $l(i)$ extends over the sites R_l occupied by i-ions) with the volume energy

$$E_V = \overline{Z}(E_{eg} - \tfrac{1}{2} V_a B_{eg}) + \tfrac{1}{2} \sum_{i=A,B} c_i \Phi_{ii}^{ind}(R = 0) , \tag{2.56}$$

55

and the pair potentials

$$\Phi_{ij}(R) = \frac{2Z_i Z_j}{R} + \frac{2}{N} \sum_q e^{i\mathbf{q}\mathbf{R}} F_{ij}(q) \ , \quad i,j = A, B \tag{2.57}$$

between the two atomic species [2.88]. Again the $\Phi_{ij}(R)$ may be replaced by a set of pair potentials expressing an average potential, an ordering potential and a cross term given respectively by

$$\Phi_{\mathrm{NN}}(R) = c_A^2 \Phi_{AA}(R) + c_B^2 \Phi_{BB}(R) + 2c_A c_B \Phi_{AB}(R) \ ,$$

$$\Phi_{\mathrm{cc}}(R) = c_A c_B [\Phi_{AA}(R) + \Phi_{BB}(R) - 2\Phi_{AB}(R)] \ ,$$

$$\begin{aligned}\Phi_{\mathrm{Nc}}(R) = 2c_A c_B [c_A \Phi_{AA}(R) \\ - c_B \Phi_{BB}(R) + (c_B - c_A)\Phi_{AB}(R)] \ . \end{aligned} \tag{2.58}$$

The formal problems being settled, the most important question is now the *transferability* of the pseudopotential, i.e. are we entitled to use the same pseudopotential in the pure metal A and for describing the component A in the A-B alloy? It is more or less usual to take the transferability of a pseudopotential for granted, but in principle the answer must be no. The pseudopotential is not a purely atomic, but rather a collective quantity; it describes the scattering of electrons by a particular ion type *in a given surrounding medium*. If this changes, the pseudopotential changes, too.

Let us for example follow the construction of a pseudopotential for a binary A-B alloy in the operator approach [2.17]. The Pauli principle now requires the valence electron wavefunction to be orthogonal to core states of type A or type B, depending on the occupancy of the particular site considered. This means, that the Phillips-Kleinman equation (2.3) is generalized [2.17] to

$$\begin{aligned}W(E) = V &+ \sum_{c(A)} |\psi_c^A\rangle\langle\psi_c^A|(E - E_c^A) \\ &+ \sum_{c(B)} |\psi_c^B\rangle\langle\psi_c^B|(E - E_c^B) \ , \end{aligned} \tag{2.59}$$

and a conduction band Bloch function will now be given by an expression of the type

$$|\varphi\rangle = |\psi\rangle + \sum_{c(A)} a_c^A |\psi_c^A\rangle + \sum_{c(B)} a_c^B |\psi_c^B\rangle \tag{2.60}$$

the coefficients a_c^A, a_c^B being determined by some appropriate criterion, e.g. that of minimum kinetic energy for the pseudo-valence orbital (2.7). Now it is clear that the degree of admixture of core orbitals ψ_c^A will vary with the concentration and with the type of the second component, hence the pseudopotential w_A will be different in the pure metal A and in the A-B alloy. Also the individual pseudopotentials for A and B ions (2.59) have to be evaluated at a different reference energy E. Note that the normalization of the pseudoorbital will also be affected (cf. Sect. 2.1) and the effective valence will also be concentration dependent.

This leaves us with two alternatives: either we have to construct pseudopotentials that are optimized specifically for a given concentration of the A-B alloy or we must attempt to optimize the transferability (e.g. by using the "norm-conserving" criterion, see Sect. 2.1.2). This second alternative – though evidently very attractive – has not as yet been attempted, although quite promising results have been obtained with the specifically optimized pseudopotentials.

If we take a simple alloy such as K-Rb, the change in the pair potentials on alloying will be rather weak and will depend essentially on the change in the average electron density. The resulting Φ_{K-K}, Φ_{K-Rb} and Φ_{Rb-Rb} are very similar (Fig. 2.7) and compatible with the formation of substitutional alloys both in the solid and in the liquid state. However, if we consider the formation of alloys from metals with large differences in electron density and/or electronegativity, the pair potentials reflect important chemical effects. These will be discussed briefly in the following section using very simple arguments.

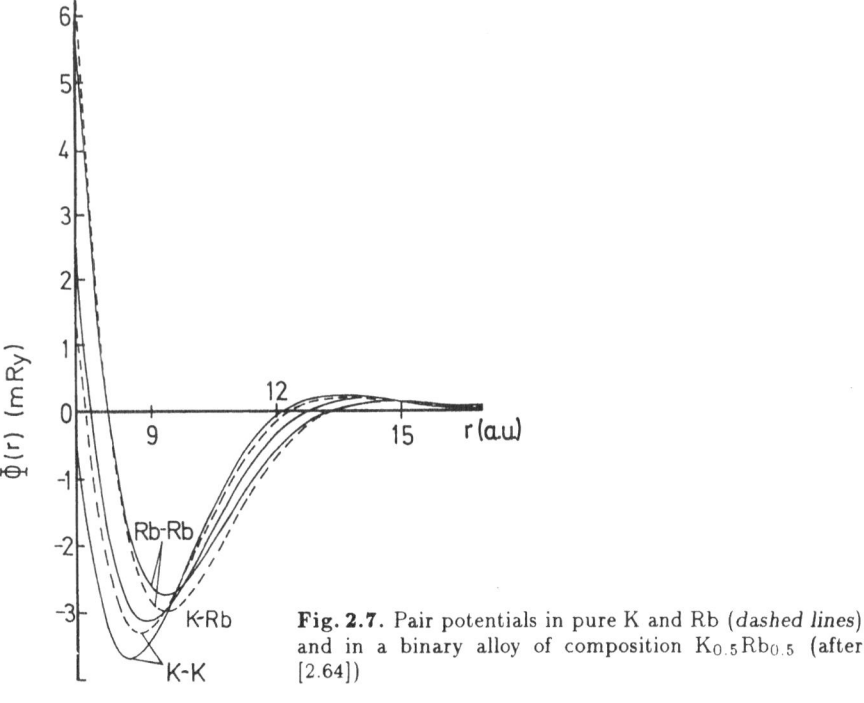

Fig. 2.7. Pair potentials in pure K and Rb (*dashed lines*) and in a binary alloy of composition $K_{0.5}Rb_{0.5}$ (after [2.64])

2.4.1 Chemical Compression

Let us imagine an A-B-alloy, A being a low-electron-density material and B having a considerably higher electron density. In our description of the alloy, we start from a homogeneous distribution of all valence electrons. In the next step we have to orthogonalize the valence orbitals to the core orbitals. Now the A ions "see" an increased electron density in the alloy, while at the B sites the density appears reduced. Thus the A ions have to form a bigger orthogonalization hole, whereas the hole at the B ions is reduced in the alloy. This is equivalent to a transfer of electrons from A to B, i.e. in the direction expected from the electronegativity difference. The ions together with their orthogonalization charges are then screened. As the effective charge of ion A has been increased, it induces a larger screening charge, and vice versa for the B ion (the entire "pseudo-atom" = ion + orthogonalization hole + screening charge has to remain electrically neutral). According to the electroneutrality principle, this equilibration of potential gradients occurs even on a local level: the additional screening charge around the electropositive ion is accumulated in the core region (see Fig. 2.8). This is accompanied

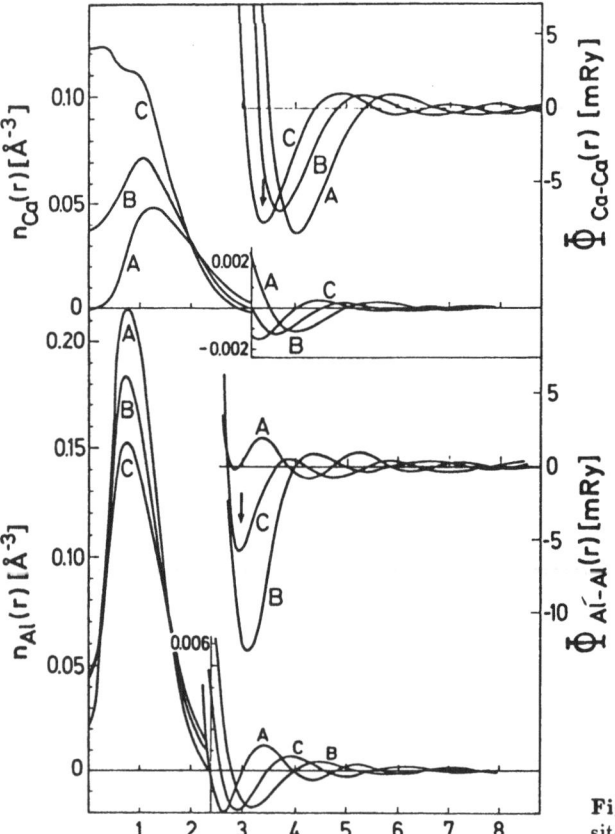

Fig. 2.8. Caption see opposite page

by an inward shift of the Friedel-oscillations [2.89]. Around the B ions the effet is just the reverse. The effective interatomic potentials follow the redistribution of the screening charge. For the electropositive ion the change in the repulsive part of $\Phi_{AA}(R)$ amounts to a compression of the A ion. This effect is wellknown in the metallurgical literature [2.90–92] under the name of "chemical compression". An early qualitative discussion of competing charge transfer mechanisms due to electronegativity and electroneutrality effects has been given by *Pauling* [2.93]. In a slightly different context, the importance of *two* charge transfer mechanisms has recently been stressed by *Varma* and *Schlüter* [2.94]. Note that although the actual amount of charge transferred in both directions nearly compensates, the effect on the chemical bond is quite different: the orthogonalization effect acts on charges that are strictly localized to the core region, whereas the screening affects the overlapping valence charges responsible for the bonding.

Figure 2.8 illustrates this effect for the example of Ca-Al alloys. This is a very striking example since the exceptionally high stability of the cubic Laves-phase $CaAl_2$ appears to conflict with the usual interpretation of the Laves phases as size-compounds: an ideal Laves-phase packing is achieved for a radius ratio of $R_A/R_B = 1.225$, whereas the ratio of the Goldschmidt radii is $R_{Ca}/R_{Al} = 1.39$ [2.95], thus $CaAl_2$ should be expected to be very unstable. The riddle is easily solved in terms of pair potentials calculated for the alloy $CaAl_2$; here the interatomic distances coincide exactly with the minima of the pair potentials. This result is remarkable in two respects: first, it shows that the intermetallic phase is stabilized by packing as many atoms as possible into the minima of the potential energy function. This is the kind of explanation one would intuitively expect for size-compounds. Second, there is no such thing as a rigid atomic radius; the renormalization of the atomic size is a common effect in many alloys, but it depends in sign and magnitude on the differences in electronegativity and/or difference in electron density of the components.

2.4.2 Chemical Ordering

Many binary alloys, especially those with large differences in the component valency (metallurgists would say that the components tend to be located in both sides of the Zintl border – see e.g. [2.95]), are characterized by a tendency to form a large number of bonds between unlike neighbours. In terms of the pair potentials a preferred heterocoordination means that

---

Fig. 2.8. Screening charge density $n_i(R)$ and interatomic pair potentials for Ca and Al atoms in the pure metals (*curve A*) and in alloys of the composition Ca_2Al (*curve B*) and $CaAl_2$ (*curve C*), demonstrating the "chemical compression" of the electropositive atom in the alloy. The vertical arrows mark the interatomic distances Ca-Ca and Al-Al in the cubic Laves phase $CaAl_2$ [2.89]

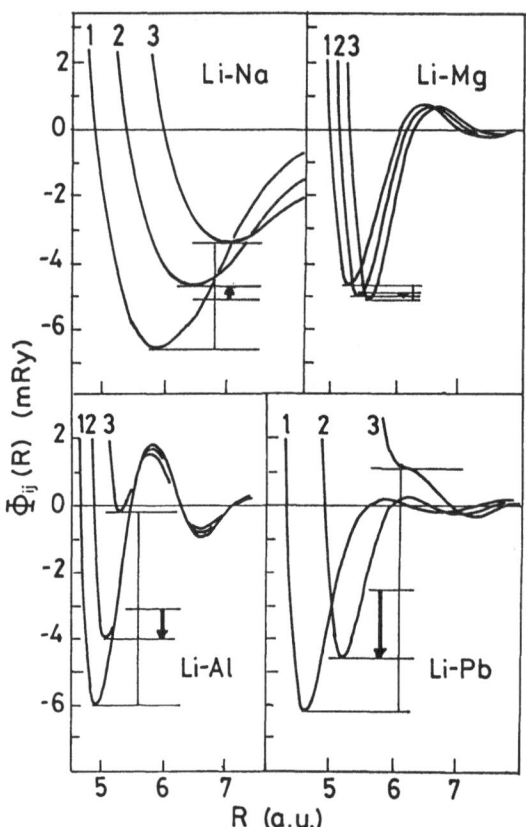

Fig. 2.9. Interatomic pair potentials $\Phi_{ij}(R)$ in equiatomic alloys of Li with Na, Mg, Al, and Pb, showing the increasing trend towards ordering [2.96]. The label 1 refers to the Li-Li interaction; 2, to the Li-X interaction; 3, to the X-X potentials (X =Na, Mg, Al, Pb). The bold vertical arrows emphasize the direction and the strength of the ordering potential [2.96, 101]

the AB interaction must be stronger than the average AA and BB interactions, i.e. $\Phi_{AB}(R) < \frac{1}{2}[\Phi_{AA}(R) + \Phi_{BB}(R)]$ for R around the nearest neighbour distance d_1. For the ordering potential Φ_{cc} this means that the trend towards ordering is expressed by a repulsive $\Phi_{cc}(d_1)$. The inverse relations $\Phi_{AB}(d_1) > \frac{1}{2}[\Phi_{AA}(d_1) + \Phi_{BB}(d_1)]$ and $\Phi_{cc}(d_1) < 0$, lead to clustering (preferred homocoordination) and ultimately to phase separation.

Figure 2.9 shows for the example of equiatomic alloys of Li with Na, Mg, Al and Pb, that the trend is correctly reproduced by the calculations [2.96]. For Li-Na and Li-Mg the pair potentials are almost additive, in the sense $\Phi_{AB}(d_1) \cong \frac{1}{2}[\Phi_{AA}(d_1) + \Phi_{BB}(d_1)]$, with slight deviations in a positive direction for Li-Na, connected with the miscibility gap in liquid Li-Na alloys [2.97, 98], and in a negative direction for Li-Mg leading to a weak short-range order in solid [2.17, 99, 100] and liquid [2.101] Li-Mg alloys. The strong tendency to chemical short-range order in Li-Al and Li-Pb alloys is evident from the pair potentials and well confirmed both by experiment [2.102], and calculations [2.101, 103].

The physical origin of a repulsive ordering potential $\Phi_{cc}(d_1)$ is most easily recognized when we look at the explicit expression for $\Phi_{cc}(R)$. From (2.53) to (2.58) it follows that

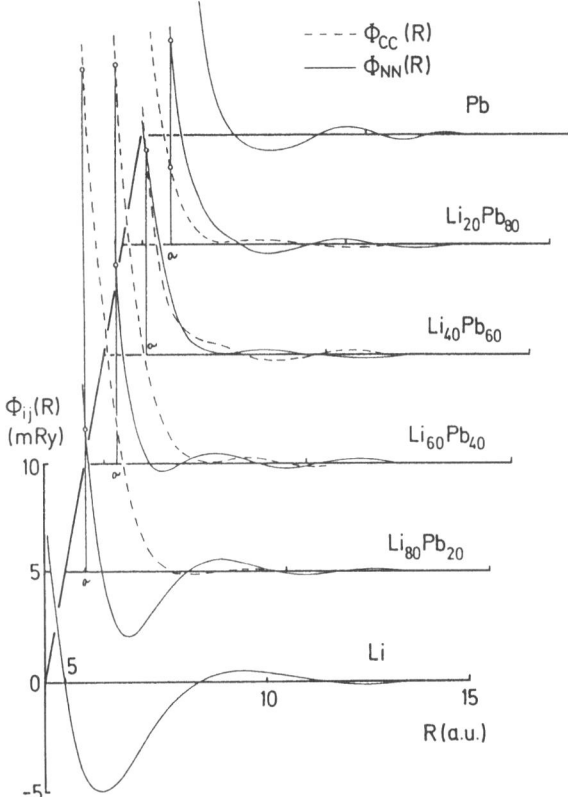

Fig. 2.10. Concentration dependence of the average pair potential $\Phi_{NN}(R)$ and the ordering potential $\Phi_{cc}(R)$ in Li-Pb alloys. The circles mark an average nearest neighbour distance in the liquid alloys [2.101]

$$\Phi_{cc}(R) = c_A c_B \left[\frac{2\Delta Z^2}{R} + \frac{2}{N} \sum_q e^{iqR} F_{cc}(q) \right] . \tag{2.61}$$

Evidently a large ΔZ will yield a strongly repulsive ordering potential – it is rather a question of how these dominant Coulomb effects are reduced to reasonable dimensions by adding the indirect interactions. For this purpose, it is most instructive to look at the concentration dependence of $\Phi_{NN}(R)$ and $\Phi_{cc}(R)$ for the example of Li-Pb alloys (Fig. 2.10). We find, in agreement with the experimentally established trends, that the ordering potential is strongest in the Li-rich regime where, due to the low electron density, the Coulomb repulsion is only weakly screened. As the electron density is increased (by adding Pb), the indirect interactions become more effective and the ordering potential is reduced. Note that both the strength of the ordering potential and the asymmetry of its concentration dependence, increase with ΔZ – again in agreement with the experimentally estabilished trends (for a recent review, see e.g. *van der Lugt* and *Geertsma* [2.104]).

However, upon closer inspection it turns out that the problem is not that simple; a few calculations of pair potentials for binary alloys have been performed on the basis of the effective pseudopotentials of *Dagens, Rasolt* and *Taylor* (DRT) [2.29], assuming that the pseudopotential parameters A_l, R_l in (2.8) *do not change on alloying*. From calculations for Al-based dilute alloys with Be, Mg, Ca, Zn, Li and K [2.105, 106] it appears that the effect of chemical compression is reasonably well reproduced, at least for the unlike-atom interaction (no pair potentials for the minority atoms are given). The problem of the ordering potential appears to be more delicate. *Beauchamp* et al. [2.100] have presented pair-potentials for Li-Mg alloys (Fig. 2.11), using the DRT-effective pseudopotentials. Clearly with this type of potential, the alloy would be expected to exhibit long-range or-

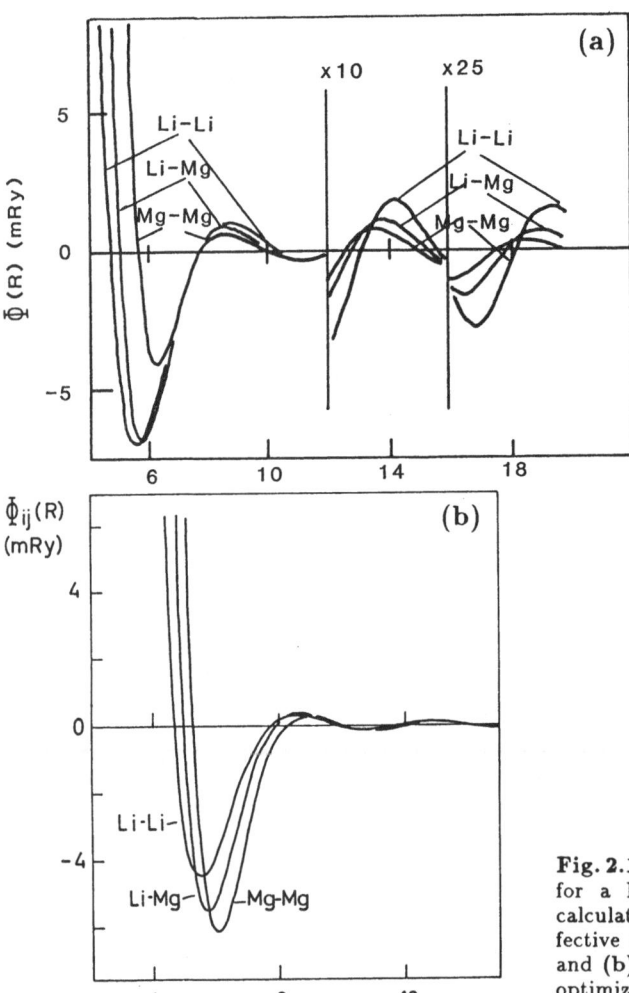

Fig. 2.11a,b. Pair potentials for a $Li_{75}Mg_{25}$ alloy, (a) as calculated using the DRT effective pseudopotentials [2.100] and (b) as calculated using the optimized pseudopotential approach [2.17]

der. However, none is observed experimentally. The Li-Mg potential is found to be much too attractive, in contrast to the pair potentials derived from the optimized pseudopotentials [see Fig. 2.9b]. Calculations using the local-empty-core form of the model potential produce rather unrealistic ordering potentials – both the strength and the shape of the potential are affected. The oscillations in the ordering potential persist down to very short distances, contrary to what has been found in a full nonlocal calculation (see Fig. 2.10) or inferred from experiment [2.103].

Thus we find that both the optimization and the nonlocality of the pseudopotential appear to be even more important for binary alloys than for pure metals. This is certainly not surprising; consider $F_{cc}(q)$ as given by (2.54), for an alloy such as Li-Pb. The s-component of the repulsive part of Δw is relatively weak (since only the difference enters), whereas the p and d components are identical to the full repulsive part of the Pb pseudopotential.

2.5 Interatomic Forces in Non-Simple Metals and Alloys

For the simple metals, the expansion (2.12) in powers of the electron-ion pseudopotential provides a convenient and rapidly convergent expansion of the electronic groundstate energy in terms of one-, two-, three- ... body forces. A corresponding decomposition of the tight-binding d-band energy for transition metals turns out to be poorly convergent. However, one finds that small energy changes with respect to a given reference are often well described by pair forces. *Varma* and *Weber* [2.4] and *Finnis* et al. [2.5] have determined force constants between pairs of atoms in transition metals by an expansion of the tight-binding energy to second order in the atomic displacements and used them in lattice dynamics calculations. *Bieber* and *Gautier* [2.107, 108] have expanded the energy of a binary alloy with respect to a totally disordered configuration, this allows the determination of effective pair interactions (ordering potentials) with respect to a given basic lattice.

Very recently several authors [2.6–8, 109] have presented more general methods for deriving interatomic potentials from tight-binding models for transition metals. All these methods are based essentially on the "force theorem" of *Mackintosh* and *Andersen* [2.110] which shows that the first-order change in the total energy δE may be written, neglecting the Madelung and double-counting contributions, as $\delta E = \sum_\alpha \delta E_\alpha$ where δE_α is the first-order change in the eigenvalues created by a first-order change in the volume or in the interatomic distances while the potential is kept unchanged. (For a more detailed discussion of the force theorem, see [2.111–113] and for its use in calculations of the pressure, see Sect. 3.1.1). The physical basis of this theorem is the cancellation between relaxations in the one-particle potential and changes in the self-energy corrections. Using the moments of the electronic density of states, the cohesive energy projected onto a particular atomic site is obtained as a function of the local environment of this site. It can then be decomposed into a sum over one-, two-, and many-particle contributions. This decomposition can always be achieved in different ways, the most effective being the one that minimizes the many-particle contributions [2.8]. One finds that the effective pair potential describing the change of the total energy for a rearrangement of the atoms at constant volume is given by the sum of three terms:

$$\Phi(R) = \Phi_{\mathrm{sc}}(R) + \Phi_{\mathrm{B}} + \Phi_{\mathrm{R}}(R) \ , \tag{2.62}$$

where $\Phi_{\mathrm{sc}}(R)$ is essentially a screened Coulomb repulsion, $\Phi_{\mathrm{B}}(R)$ is the bonding contribution form the d-band, and $\Phi_{\mathrm{R}}(R)$ is a repulsive contribution arising from the shift of the centre of the d-band [2.7,8,109]. *Wills* and *Harrison* [2.109] combine the s-electron contribution with the screened Coulomb repulsion, so that $\Phi_{\mathrm{sc}}(R)$ becomes identical to an appropriate simple-metal pair potential. Their bonding term is written as

$$\Phi_{\mathrm{B}}(R) = -7.99 Z_d(1 - Z_d/10)(12/N_c)^{1/2} \left(\frac{r_d^3}{R^5} \right) \ , \tag{2.63}$$

where Z_d is the number of d-electrons, r_d is a d-orbital radius (related to the band width) and N_c is the coordination number [so that $\Phi_{\mathrm{B}}(R)$ is not entirely independent of the atomic structure]. The repulsive contribution is approximated by

$$\Phi_{\mathrm{R}}(R) = 22.8 Z_d \left(\frac{r_d^6}{R^8} \right) \tag{2.64}$$

MacDonald and *Taylor* [2.7] propose

$$\Phi_{\mathrm{sc}}(R) = 1.47 \exp\left[-8(R/a) \right] \frac{a}{R} \quad \text{and} \tag{2.65}$$

$$\Phi_{\mathrm{B}}(R) \simeq -0.0073 f_d \frac{(a/R)^5}{(1 + f_d R^{10})} \quad \text{with} \tag{2.66}$$

$$f_d = Z_d(10 - Z_d)/(1 + Z_d/50) \ , \tag{2.67}$$

where a is the lattice constant of an fcc lattice. The repulsive contribution from the shift of the d-band is neglected. The constants in (2.62–66) have been fitted to the calculated band energies.

Expressions of this type have already been used with reasonable success for the calculation of cohesive and structural properties [2.7,109,114], although at the moment these studies remain at a mostly exploratory level.

2.6 Beyond Perturbation Theory

The discussion up to this point has centred on the structural expansion for the ground-state energy introduced at the beginning of this chapter (2.12–14). In practice one assumes that at least after suitable optimization of the pseudopotential the series will be very rapidly convergent so that it will be sufficient to consider only the linear response of the conduction electron distribution to the electron-ion interaction.

An alternative approach to that just described, but one which is difficult to use except in systems with a high degree of symmetry (perfect crystals), is the local density-functional (LDF) method of *Hohenberg, Kohn* and *Sham* [2.115, 116]. The LDF method is devised to calculate the electronic ground-state energy for a given ionic configuration $E_e(\{\boldsymbol{R}_j\})$. It is based on a famous theorem by Hohenberg, Kohn and Sham which shows that the energy of the electron-ion system is a functional of the conduction electron density $\varrho(\boldsymbol{r})$ as follows (we use atomic units)

$$E[\varrho] = \int \frac{\varrho(\boldsymbol{r})\varrho(\boldsymbol{r}')}{|\boldsymbol{r} - \boldsymbol{r}'|} d^3r\, d^3r' + \int \varrho(\boldsymbol{r})W(\boldsymbol{r})d^3r + G[\varrho(\boldsymbol{r})] \ . \tag{2.68}$$

The first term describes the Hartree energy of the electrons, the second their interaction with the ions, and G is a universal functional of $\varrho(\boldsymbol{r})$ describing the kinetic, exchange and correlation energy of a nonuniform electron gas. The ground-state energy E_e is found by minimizing the right hand side of (2.68) with respect to $\varrho(\boldsymbol{r})$, this yields the density-functional equations determining $\varrho(\boldsymbol{r})$. This is in principle an exact relation. However, only for the kinetic energy part is the exact form of the functional $G[\varrho]$ known; the exchange and correlation parts must be approximated. In the linear response approximation for example, the functional $G[\varrho]$ is given by

$$G[\varrho] = E_{eg} + \int K(\boldsymbol{r} - \boldsymbol{r}'; \bar{\varrho})\varrho_{sc}(\boldsymbol{r})\varrho_{sc}(\boldsymbol{r}')d^3r\, d^3r' \tag{2.69}$$

where ϱ_{sc} is the screening charge density and the kernel K is the Fourier transform of

$$K(q) = \frac{4\pi}{q^2} \frac{1}{\varepsilon(q) - 1} \tag{2.70}$$

[see (2.17–19) for comparison]. The *local* density functional method (LDF) proposed by Kohn and Sham consists in replacing the functional G of the inhomogeneous electron gas by that of a homogeneous electron gas of the same local density. In practice it turns out that gradient corrections are important, so that the correct form of the functional is given by [2.117]:

$$G[\varrho] = c_k\varrho(\boldsymbol{r})^{5/3} + \tfrac{1}{4}[\nabla\varrho(\boldsymbol{r})]^2/\varrho(\boldsymbol{r}) - c_x\varrho(\boldsymbol{r})^{4/3}$$

$$- \{0.031 \ln [3/4\pi\varrho(\boldsymbol{r})] - 0.115\}\varrho(\boldsymbol{r}) \ , \tag{2.71}$$

where $c_k = \frac{3}{5}(3\pi^2)^{2/3}$, $c_x = \frac{3}{4}(3/\pi)^{1/2}$. The first two terms give the kinetic energy (the first term is the usual Thomas-Fermi term, the second supplies the gradient corrections), the third is the exchange energy and the final term describes the correlation energy in the Nozières-Pines approximation (1.24).

Recently *Chelikowsky* [2.118, 119] has successfully combined LDF and pseudopotential theory in a study of the cohesive properties of simple metals. Applications to disordered systems appear to be difficult – *Alonso* and coworkers [2.120, 121] have proposed a series of studies of homovalent alloys, but they use a modified virtual crystal approximation, thus eliminating all effects of disorder.

Recent Results

The variation of the effective interatomic interactions with electron density and the compressibility problem (cf. Sect. 2.2) have been reconsidered within the framework of local-density theory [2.122].

Evidence for the influence of the finite electronic mean free path upon the interatomic potentials has been found in investigations of amorphous alloys with extremely high electrical resistivities [2.123].

Non-perturbative techniques for the calculation of effective interatomic potentials continue to be of great interest. *Manninen* [2.124] has proposed an effective medium approach (which is essentially a generalization of the embedded atom method [2.125]) and *Carlsson* [2.126] has presented a new method based on supercell-atomic-structure calculations.

3. Phase Stability of Crystalline Metals

The theory of the cohesion and structure of the crystalline simple metals was reviewed a decade ago by *Heine* and *Weaire* ([3.1] to be referred to as HW). It is not our aim to repeat this very clear overview, so we will present only a very short outline of the main factors governing simple-metal cohesion. Crystal structures and structural changes will be discussed in more detail.

3.1 Simple-Metal Cohesion

If the electron gas is perturbed to first order by the ionic lattice, then the total energy per particle may be written as [see (2.17–19)]

$$E_g = Z(E_{eg} + E_{ion}) \, , \tag{3.1}$$

where Z is the valence. The electron-gas energy

$$E_{eg} = 2.21/R_s^2 - 0.916/R_s - (0.115 - 0.031 \ln R_s) \qquad \text{and} \tag{3.2}$$

$$E_{ion} = -\frac{3Z}{R_a}\left[1 - \left(\frac{R_c}{R_a}\right)\right]^2 + \frac{1.2Z}{R_a} \, , \tag{3.3}$$

(see e.g. HW, p. 261, *Harrison* [3.2], *Girifalco* [3.3], *Pettifor* [3.4]). The first term in (3.3) represents the electron-ion and the second term the electron-electron potential energy – the potential energy has been evaluated in the Wigner-Seitz approximation, neglecting the electrostatic interaction between different cells as they are electrically neutral[1] and using the empty-core model (2.10) for the electron-ion pseudopotential. The equilibrium atomic radius R_a is determined by the condition $p = -(\partial E_g / \partial V) = 0$. This yields an equation relating the equilibrium atomic radius and the core radius R_c,

---

[1] This is equivalent to setting the Madelung constant α_M defining the total electrostatic energy E_{es} [$E_{es} = (Z^2/R_a)\alpha_M$, cf (2.17)] equal to $\alpha_M = -1.8$. The actual values for close-packed metallic lattices are $\alpha_M = -1.79186$ (bcc), $\alpha_M = -1.79175$ (fcc) and $\alpha_M = -1.79168$ (hcp), see Appendix C.

$$\left(\frac{R_c}{R_a}\right)^2 = 0.2 + \frac{0.102}{Z^{2/3}} + \frac{0.0035R_a}{Z} - \frac{0.491}{Z^{1/3}R_a} \quad . \tag{3.4}$$

Both the electron-gas and the ionic terms are composed of attractive and repulsive terms: the kinetic energy and the exchange potential balance each other at an electron density of $R_s = 4.2$ a.u. The electron-ion interaction is attractive, the electron-electron potential of course repulsive.

The core radii calculated from (3.4) are compiled in Table 3.1 and as expected we see that the core radius increases with the atomic number within each group of the periodic table. From (3.3) we expect a decrease in the binding energy with increasing R_c – this again agrees with the general trend observed in the binding energy, but for the polyvalent metals (Z larger than two) the simple first-order expression yields a serious underestimate, because the second order contribution is non-negligible. The bulk modulus, $B = V(d^2E_g/dV^2)$, is easily calculated from (3.3). In units of the bulk

Table 3.1. Equilibrium binding energy and bulk modulus of the simple metals

Metal	R_s [a.u.]	R_c [a.u.] [a]	E_g [Ry/Ion] [b] theor.	exp.	B/B_{eg} [b] theor.	exp.
Li	3.26	1.31	−0.555	−0.518	0.63	0.50
Na	3.93	1.71	−0.476	−0.461	0.81	0.80
K	4.86	2.27	−0.397	−0.388	1.06	1.10
Rb	5.20	2.47	−0.374	−0.372	1.16	1.55
Cs	5.62	2.72	−0.350	−0.347	1.27	1.43
Cu	2.67	0.91	−0.661	−0.825	0.45	2.16
Ag	3.02	1.37	−0.534	−0.778	0.71	2.94
Au	3.01	1.35	−0.537	−0.957	0.69	4.96
Be	1.87	0.76	−2.270	−2.270	0.45	0.27
Mg	2.65	1.31	−1.835	−1.780	0.73	0.54
Ca	3.27	1.73	−1.538	−1.457	0.95	0.66
Sr	3.37	1.80	−1.497	−1.356	0.98	0.78
Ba	3.70	2.02	−1.382	−1.355	1.10	0.84
Zn	2.30	1.06	−2.070	−2.110	0.60	0.45
Cd	2.59	1.26	−1.878	−1.990	0.70	0.63
Hg	2.66	1.31	−1.835	−2.199	0.73	0.59
Al	2.07	1.11	−4.233	−4.161	0.69	0.32
Ga	2.19	1.20	−4.039	−4.423	0.74	0.33
In	2.41	1.37	−3.716	−4.047	0.83	0.39
Tl	2.48	1.42	−3.630	−4.273	0.85	0.39
Si	2.00	1.21	−6.871	−7.924	0.80	0.31
Ge	2.09	1.28	−6.625	−7.910	0.84	0.27
·Sn	2.21	1.38	−6.300	−7.000	0.90	0.30
Pb	2.30	1.45	−6.093	−7.239	0.94	0.32

[a] Calculated using (3.4) from the electron densities given in the first column.
[b] Experimental values from [3.5].

modulus B_{eg} of the noninteracting electron gas $(B_{eg} = 0.586/R_s^5)$, this yields [neglecting the very small correlation contribution and using (3.4)]

$$B/B_{eg} = 0.2 + 0.815 R_c^2/R_s \ . \tag{3.5}$$

It follows from (3.5) that the presence of the ionic core is crucial for determining the bulk modulus, but from the values given in Table 3.1 we find that only for sodium and potassium is the simple first-order expression sufficiently accurate. For all other metals second-order and nonlocal contributions appear to the essential and in addition, s-d hybridization is important for the noble-metal cohesion, as was first pointed out by *Solt* and *Kollar* [3.6].

Before proceeding to the more elaborate expressions, it is worthwhile to consider briefly the real-space formulation of the binding energy (2.20–25). Using again the empty-core model, the explicit expression for the volume energy becomes

$$E_v = \overline{E}_{eg} + \overline{E}_{ion}$$

$$\overline{E}_{eg} = Z \left(\frac{0.982}{R_s^2} - \frac{0.712}{R_s} + 0.031 \ln R_s - 0.110 \right)$$

$$\overline{E}_{ion} = \frac{1}{2} \Phi_{ind}(R = 0) = 16 Z^2 \int_0^\infty \frac{\chi(q)}{\varepsilon(q)} \frac{\cos^2(qR_c)}{q^2} dq \ , \tag{3.6}$$

the first term giving the electron-gas and the second the intra-atomic contribution to the volume energy with the additional pair-interaction contribution E_p to the binding energy being given by the third term of (2.23). The volume energies and pair-interaction energies (for simplicity we consider only nearest-neighbour interactions in an fcc structure) are compiled in Table 3.2. Core radii deduced from the first-order expression and the LDF form of the local field have been used. We find that (i) the pair-interaction energy contributes at most 2–3 % of the total binding energy (even in terms of the cohesive energy this makes at most 10–15 %) – for the polyvalent metals, E_p can even be repulsive. However, it is E_p *alone* that determines the stable crystal structure. (ii) The free-electron contribution E_{eg} to the volume energy is rather small compared to the intra-atomic binding energy $\frac{1}{2}\Phi_{ind}(R = 0)$. As a rule $E_{eg} \sim 0.2$ Ry/el, quite independent of the valence and electron density. Thus we find that it is the variation of the intra-atomic core-valence interaction that determines the trend in the binding energy through the periodic table. *Finnis* [3.7] has pointed out that E_{ion} is related in a very simple way to the properties of the core. Expressing $\Phi_{ind}(R = 0)$ in terms of the normalized energy-wave-number characteristic $F_N(q) = -(q^2 V_a/4\pi Z^2) F(q)$

Table 3.2. Electron-gas contribution $\overline{E}_{eg}$ and intraatomic contribution $\overline{E}_{ion}$ to the volume energy E_v (3.6) and pair-interaction energy E_p, calculated with the empty-core model and LDF-screening

Metal	R_c [a.u.]	$\overline{E}_{eg}$ [a] [Ry/ion]	$\overline{E}_{ion}$ [b] [Ry/ion]	E_v [Ry/ion]	E_p [Ry/ion]	E_g [Ry/ion] theor.	exp.
Li	1.31	−0.200	−0.356 (−0.380)	−0.556	−0.0097	−0.566	−0.518
Na	1.71	−0.186	−0.293 (−0.29)	−0.479	−0.0156	−0.485	−0.461
K	2.27	−0.166	−0.239 (−0.22)	−0.405	−0.0196	−0.425	−0.388
Rb	2.47	−0.159	−0.225 (−0.20)	−0.384	−0.0205	−0.405	−0.372
Cs	2.72	−0.152	−0.211 (−0.18)	−0.363	−0.0215	−0.385	−0.347
Be	0.76	−0.380	−2.170 (−2.63)	−2.550	+0.0539	−2.496	−2.270
Mg	1.31	−0.418	−1.490 (−1.52)	−1.908	−0.0163	−1.924	−1.780
Ca	1.73	−0.397	−1.228 (−1.16)	−1.625	−0.0419	−1.667	−1.457
Sr	1.80	−0.394	−1.196 (−1.11)	−1.590	−0.0453	−1.635	−1.356
Ba	2.02	−0.379	−1.106 (−0.99)	−1.485	−0.0531	−1.538	−1.355
Zn	1.06	−0.416	−1.726 (−1.89)	−2.142	+0.0077	−2.134	−2.110
Cd	1.26	−0.418	−1.529 (−1.59)	−1.947	−0.0124	−1.959	−1.990
Hg	1.31	−0.417	−1.489 (−1.52)	−1.906	−0.0164	−1.922	−2.199
Al	1.11	−0.607	−3.912 (−4.05)	−4.519	+0.0282	−4.491	−4.161
Ga	1.20	−0.618	−3.711 (−3.75)	−4.329	+0.0110	−4.318	−4.423
In	1.37	−0.623	−3.402 (−3.28)	−4.025	−0.0150	−4.040	−4.047
Tl	1.42	−0.642	−3.320 (−3.16)	−3.962	−0.0214	−3.983	−4.273
Si	1.21	−0.796	−6.832 (−6.61)	−7.628	+0.071	−7.557	−7.924
Ge	1.28	−0.812	−6.585 (−6.25)	−7.397	+0.056	−7.341	−7.910
Sn	1.38	−0.826	−6.282 (−5.80)	−7.108	+0.038	−7.070	−7.000
Pb	1.45	−0.832	−6.090 (−5.52)	−6.922	+0.028	−6.894	−7.239

[a]   Calculated using (3.6).
[b]   Calculated according to (3.6); the values in parentheses give the simple estimate according to (3.7b)

$$\Phi_{\mathrm{ind}}(R=0) = -\frac{V_a}{8\pi^3}\frac{4\pi Z^2}{V_a}4\pi\int\limits_0^\infty F_{\mathrm{N}}(q)dq \ , \tag{3.7a}$$

and writing the area under $F_{\mathrm{N}}(q)$ as $\frac{1}{2}q_0$, where q_0 is the first zero in the pseudopotential form factor and in the characteristic, we find

$$\frac{1}{2}\Phi_{\mathrm{ind}}(R=0) \simeq -\frac{Z^2}{\pi}q_0 = -\frac{1}{2}Z^2/R_{\mathrm{c}} \tag{3.7b}$$

and Table 3.2 shows that this rather crude estimate reproduces the trends very well. In summary we find that the simple real-space expression (3.6,7), gives an even better description of the cohesive energies than the first-order expression (3.1–3).

Real-space expressions for the pressure and for the bulk modulus may be derived by differentiating the total energy with respect to the volume (see e.g. [3.7] for details). One finds that the volume energy yields a negative contribution to the pressure (favouring compression), whereas the pair-interaction term gives a positive contribution (favouring expansion); both terms yield positive contributions to the bulk modulus of about equal order of magnitude (Fig. 3.1). Thus the crystal is not – not even approximately – in equilibrium under the action of the pair forces alone. The fact that most of the binding energy comes from the intra-atomic interaction is supported by the recent work of *Chelikowsky* [3.8,9] who related the cohesive energy to the change in kinetic energy which accompanies the transformation of

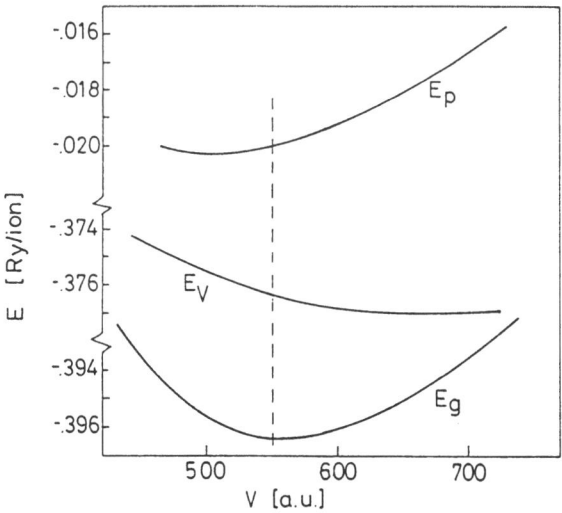

Fig. 3.1. Volume dependence of the volume energy E_{V}, the pair interaction energy E_{p}, and the total energy E_{g} for body-centred cubic potassium, calculated using optimized pseudopotentials. Note that the slopes of the $E_{\mathrm{V}}(V)$ and $E_{\mathrm{p}}(V)$ curves have opposite sign, but similar curvatures

the exponentially damped free-atom wave functions of the valence electrons to plane-wave-like states in the metal.

Of course a fully quantitative analysis of the bulk properties (binding energy, equilibrium lattice constant, bulk modulus, equation of state etc.) necessitates a nonlocal calculation – a glance at Table 3.2 shows that nonlocal corrections will be most important for the first-row elements Li and Be and for the heaviest elements. The task has been undertaken by *Williams, Cohen* and others using a total-energy calculation via the band-structure scheme [3.10–17] and by *Moriarty, Hafner* and others using perturbation theory [3.18–24]. A few representative results are compiled in Table 3.3. Clearly the degree of agreement with experiment increases with the sophistication of the theory. The total-energy calculations based on the band-structure scheme are now capable of predicting the equilibrium interatomic distance, the binding energy and the bulk modulus to within a few percent of the experimental values, with the perturbation calculations lagging not too far behind. The most difficult point in the perturbative scheme is the correct calculation of the equilibrium interatomic distance, since this depends critically on the intra-atomic energy E_{ion} (3.6), which is not very well described in most schemes. *Hafner* [3.19, 20, 25] has pointed out that this is due to the valence-core exchange potential: the LDF functional is a nonlinear functional that applies to the total electron density. Separating the exchange-correlation potential into core, core-valence, and valence contributions necessarily involves some arbitrariness (see Appendix A for a more detailed discussion). Clearly nonlocality is a very important point. Another important point concerns the hybridization of the sp-electrons with d-electrons. *Solt* and *Kollar* [3.6] have shown that the hybridization term accounts for about 50 % of the cohesive energy of the noble metals. *Moriarty* [3.18, 21] and *Jan* and *Skriver* [3.26] have pointed out that the hybridization of the sp-states with the unoccupied d-states plays a very important role in the cohesion of the heavy alkaline-earth metals (Ca, Sr, Ba). Under pressure, s-d hybridization becomes increasingly important in the heavy alkali metals [3.17, 27] and under extremely high pressure even for the lighter metals [3.22].

3.1.1 An Excursion into Transition-Metal Cohesion

Although not directly belonging to the subject of this book, we feel that a brief excursion into the cohesive properties of transition metals is worthwhile if only to emphasize the essential difference in the physics of simple-metal and transition-metal cohesion.

In contrast to the nearly-free-electron-like simple metals, transition metals are characterized by a partially filled d-band which is well described within the tight-binding (TB) approximation by a linear combination of atomic d-orbitals [3.28]. In this LCAO approximation the structure of the

Table 3.3. Predicted and measured equilibrium properties of metals: Atomic radius (R_a), binding energy (E_g), bulk modulus (B_T) and pressure derivative of bulk modulus (B_T')

Metal	Method	R_a [a.u.]	E_g [Ry/ion]	B_T [Mbar]	B_T'
Li	PT	3.19	−0.551	0.126	3.7
	GPT		−0.517		
	APW	3.15	−0.516	0.148	
	exp.	3.26	−0.518	0.131	3.3
Na	PT	4.02	−0.475	0.066	3.8
	GPT		−0.474		
	APW	3.79	−0.460	0.090	
	exp.	3.93	−0.461	0.076	3.7
K	PT	5.19	−0.410	0.028	3.9
	GPT		−0.394		
	APW	4.71	−0.389	0.044	
	exp.	4.86	−0.388	0.037	4.0
Be	PT	2.59	−2.280	0.76	3.8
	GPT		−2.219		
	PBS	2.33	−2.294	1.31	
	APW	2.32	−2.320	1.35	
	exp.	2.36	−2.270	1.16	3.6
Mg	PT	3.38	−1.772	0.331	3.9
	GPT		−1.796		
	APW	3.29	−1.789		
	exp.	3.34	−1.780	0.369	4.0
Ca	PT	4.26	−1.440	0.174	3.9
	GPT		−1.478		
	APW	3.91	−1.487	0.167	
	exp.	4.12	−1.457	0.152	2.6
Al	PT	3.09	−4.152	0.98	3.9
	GPT		−4.177		
	PBS	2.97	−4.189	0.715	
	APW	2.97	−4.209	0.88	
	exp.	2.99	−4.161	0.79	4.3

PT calculated using second order perturbation theory and OPW-based pseudopotentials [3.19, 20, 24]

GPT calculated using second order perturbation theory and generalized pseudopotentials [3.21–22]. The calculations are for the observed lattice constant.

APW from self-consistent bandstructure calculations [3.13] using the augmented-plane-wave method.

PBS from self-consistent bandstructure calculations [3.15, 16] using norm-conserving pseudopotentials

electronic energy bands is obtained by writing the conduction-electron wave function as a linear combination of Bloch sums corresponding to the atomic orbitals forming the band (i.e. to the five d-orbitals in the case of a TB d-band). This results in a secular determinant from which the band structure may be computed. *Andersen* [3.29] has shown that within a spherical approximation to the Wigner-Seitz cell (the so-called atomic-sphere approximation) canonical d-bands may be derived which depend only on the crystal structure and not on the particular transition metal nor lattice constant. The width of the d-band may be derived by applying bonding and antibonding boundary conditions over the Wigner-Seitz sphere of radius R_a. It is found that the band width W and the d-d hopping matrix elements for single, double and triple bonds are related through

$$\left. \begin{array}{rcl} dd\sigma(R) &=& -6 \\ dd\pi(R) &=& 4 \\ dd\delta(R) &=& -1 \end{array} \right\} \times \frac{2}{5} W \left(\frac{R}{R_a} \right)^5 . \tag{3.8}$$

Hence the band width scales uniformly with the hopping integrals.

The tightly bound d-band overlaps and hybridizes with a broad NFE-like sp-band. Therefore the band structure of a transition metal is represented accurately by a hybridized NFE-TB secular equation [3.30]; for the simplified treatment of metallic cohesion that concerns us here, the hybridization effects may be safely neglected.

More important is the variation of the sp- and d-bands with volume. The variation of the position Γ_1 of the bottom of the sp-band with volume is well described by

$$\Gamma_1 = -\frac{3Z}{R_a} \left[1 - \left(\frac{R_c}{R_a} \right) \right]^2 + \frac{2.4Z}{R_a} + V_x , \tag{3.9}$$

[cf. (3.3)], where the three terms represent the electron-ion, Hartree- and exchange potentials. For the d-band, Heine has shown that the Wigner-Seitz boundary conditions imply that the width of the d-band should vary approximately as $W \sim (1/R_a)^5$ [cf. (3.8)]; the same can be expected for the shift of the centre of gravity, E_d, of the d-band with respect to the free-atom d-electron energy E_d^{atom}.

Given the band structure, the binding energy may be calculated by adding up the one-electron energies and subtracting the electron-electron interaction (i.e. the Hartree and exchange-correlation energies) which has been counted twice in the sum over the one-electron states. The presence of the double-counting corrections makes it difficult to interpret the binding energies in terms of individual sp- and d-contributions. *Pettifor* [3.31] has shown that this problem can be circumvented by working not with the total energy, but with the first-order change in energy on a change in volume of the metal. Pettifor has shown that this first-order change δE_g may be written in terms of the change of the one-electron energies E_n alone (see also Sect. 2.5), i.e.

$$\delta E_g = \sum_n \delta E_n \; . \tag{3.10}$$

It follows that the pressure may be decomposed into partial sp- and d-pressures according to

$$p = -\frac{\delta E_g}{\delta V} = -\left(\frac{\delta E_{sp}}{\delta V} + \frac{\delta E_d}{\delta V}\right) = p_{sp} + p_d \; . \tag{3.11}$$

Working again in the atomic-sphere approximation, *Pettifor* [3.31] has shown that these partial pressures may be expressed in terms of the parameters characterizing the bands:

$$3p_{sp}V = 3Z(\Gamma_1 - E_x) + 2E_{sp}^{\text{bond}} \tag{3.12a}$$

$$3p_dV = 3N_d(E_d - E_x)/m_d + 5E_d^{\text{bond}} \; , \qquad \text{where} \tag{3.12b}$$

$$E_{sp}^{\text{bond}} = \int_{-\infty}^{E_F} (E - \Gamma_1)n_{sp}(R)dE \tag{3.13}$$

$$E_d^{\text{bond}} = \int_{-\infty}^{E_F} (E - E_d)n_d(E)dE \; . \tag{3.14}$$

N_d is the number of d-electrons and m_d is the effective mass of the d-band which is related to the bandwidth through $m_d = 25/(W R_a^2)$. The partial sp-pressure consists of a term which gives the variation of the bottom of the band plus the variation of the kinetic energy. [In the approximation of a parabolic sp-band, E_{sp}^{bond} is identical to the kinetic energy, cf. (3.12) and (3.1–3), remembering that $3p_{sp}V = -R_s(dE_g^{sp}/dR_s)$.]

In partial d-pressures may be decomposed in the same fashion. Approximating the d-band density of states $n_d(E)$, by a rectangular one, the d-band energy may be written as

$$E_d^{\text{bond}} = -\frac{1}{20}W N_d(10 - N_d) \; . \tag{3.15}$$

With the approximate volume dependence of W and E_d introduced above, (3.15) may be integrated to yield the d-band contribution to the cohesive energy

$$E_d = N_d(E_d - E_d^{\text{atom}})/4m_d + N_d(\tfrac{4}{3}E_d^{\text{atom}} - E_x)/2m_d$$
$$+ E_d^{\text{bond}} \ . \tag{3.16}$$

The large effective mass of the d-band reduces the importance of the first two terms by about one order of magnitude, so that as proposed by *Friedel* [3.28] the cohesive properties of the transition metal are dominated by the d-band energy. From (3.15) we see that E_d varies parabolically with band filling, reaching its lowest value for $N_d = 5$ when all the bonding but none of the antibonding levels are occupied.

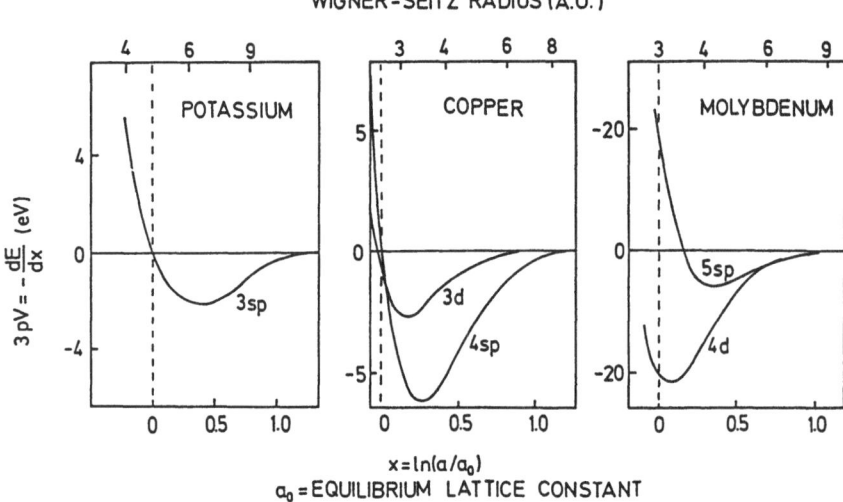

Fig. 3.2. Decomposition of the pressure vs lattice constant curve into sp- and d-electron contributions for the simple metal potassium, the noble metal copper and the transition metal molybdenum. The quantity actually shown is $3pV_i = -\frac{1}{R_a}(dE_i/dR_a)$ $[i = sp, d,$ see (3.11)], equal to the logarithmic derivative of the total energy with respect to the Wigner-Seitz radius of the lattice constant. Equilibrium corresponds to that value of the lattice constant for which the partial pressures cancel to zero; the contribution to the cohesive energy is the area under the curve in the interval $X = (0, \infty)$, the contribution to the bulk modulus is $(9V)^{-1}\frac{d}{dx}(\frac{dE}{dx})$, i.e. proportional to the negative slope of the curve at $x = 0$. After [3.32]

Pettifor's ideas have been taken up and confirmed by full band-structure calculations [3.32]. Figure 3.2 shows the decomposition of the pressure-volume relation for the simple metal potassium, the noble metal copper, and the transition metal molybdenum. In the case of the simple metal, the sp-contribution to the pressure consists of the kinetic energy term (which is always repulsive) and an attractive term form Γ_1; the latter diminishes under compression (because the bottom of the band moves upwards). In the transition metal the d-partial pressure is strongly attractive. The Γ_1 term in the sp-partial pressure is now repulsive since in this case the ion cores occupy a much larger fraction of the atomic volume. Together with the kinetic energy this term provides the necessary repulsion to counteract the strong d-d attraction (note that the d-partial pressure in turn consists of a repulsive center-of-gravity contribution [the first term in (3.12)] and an attractive term associated with the increasing band width. The center-of-gravity term, together with the repulsive sp-partial pressure may be cast approximately into the form of a Born-Mayer potential, see *Pettifor* [3.4] for details).

At the noble-metal end of the transition series, the d-partial pressure is still attractive, and the d-electrons contribute about 50 % to the cohesive energy, as first suggested by *Solt* and *Kollar* [3.6]. Note that the sp-partial pressure also varies much more strongly with volume than for potassium. This, together with the d-electron contribution, explains why the bulk modulus of Cu is about 50 times larger than that of K. For a detailed discussion of noble-metal cohesion within the generalized pseudopotential theory, see e.g. *Moriarty* [3.33] and *Upadhayaya* and *Dagens* [3.34].

3.2 Structural Stability

To study the crystal structure of the simple metals, the homogeneous electron gas must be perturbed at least to second order in the pseudopotential. The pseudopotential theory of the crystal structure was discussed extensively by HW a decade ago. Since then there has been much progress in such calculations, both using the total-energy method [3.10–12, 14, 22, 35] and using perturbation theory [3.19–21, 36–48] (and further references cited therein). In Table 3.4 we compare the structural energy differences calculated for some typical cases with available thermochemical estimates (see e.g. [3.49]). We find that both the total-energy method and the perturbation methods produce quantitatively meaningful structural energy differences. Higher-order contributions to the structural energy have been discussed by *Williams* [3.41] for a series of metals. Although the third-order contribution to ΔE was found to be substantial, it was never found to change the sign of ΔE. In addition Williams emphasized that the local approximation necessary to perform the third-order calculation probably yields an overestimate of ΔE – this agrees with similar conclusions of *Bertoni* et al. [3.50]. Therefore second-order calculations with carefully optimized nonlocal pseudopotentials should be given preference. However, even if the energy difference between two given lattice types is hardly affected, we will find that the third-order contribution (and hence a three-body potential) is vital for the dynamical stability of open structures in group IV. These covalency effects have been discussed in some detail by *Maknovetskii* and *Krasko* [3.46]. Although nonlocality was found to be of extreme importance for getting quantitatively reliable results, the recent history of structural

Table 3.4. Structural energy differences in some simple metals

Metal	Transition	ΔE [cal/g-atom]	Method	Ref.
Be	fcc-hcp	470	exp.	[3.49]
		1290	calc. (PT)	[3.24]
	bcc-hcp	1100	exp.	[3.49]
		1690	calc. (PT)	[3.24]
Mg	fcc-hcp	470	exp.	[3.49]
		165	calc. (PT)	[3.24]
	bcc-hcp	1100	exp.	[3.49]
		610	calc. (PT)	[3.24]
Al	bcc-fcc	2410	exp.	[3.49]
		2980	calc. (PT)	[3.19]
	hcp-fcc	1300	exp.	[3.49]
		950	calc. (PT)	[3.19]
Ga	fcc-Ga I	2000	exp.	[3.52]
		6300	calc. (PT)	[3.40]
Si	bcc-diamond	10600	exp.	[3.53]
		12200	calc. (TE)	[3.11]
	hcp-diamond	12200	exp.	[3.53]
		12600	calc. (TE)	[3.11]
Ge	White tin-dia	6680	exp.	[3.54]
		5750	calc. (TE)	[3.11]
	fcc-diamond	8520	exp.	[3.54]
		10590	calc. (TE)	[3.11]
Sn	White tin-dia	500	exp.	[3.55]
		800	calc. (PT)	[3.36]

exp.   thermochemical estimates
calc.   Pseudopotential calculations
PT   perturbation calculation
TE   total energy calculation

energy calculations has confirmed the validity of the local approximation as a zeroth-order model for the structural stability [3.43–45, 51].

In general, the new results show that the discussion by HW of the crystal structures remains valid. For example the occurrence of distorted structures for Ga and Hg is qualitatively best explained in q-space in terms of the position of the first reciprocal lattice vector relative to the wave vector q_0 at which the pseudopotential passes through zero (HW, pp. 263, 275). If the first reciprocal lattice vector falls close to q_0, its contribution to the second-order energy E_2 is very small [cf. (2.22)] – hence it will be energetically favourable to distort the structure in such a way that no vector of the reciprocal lattice falls close to q_0. Alternatively, the structural stability might be discussed in real space. The discussion of Fig. 20 of HW (we re-

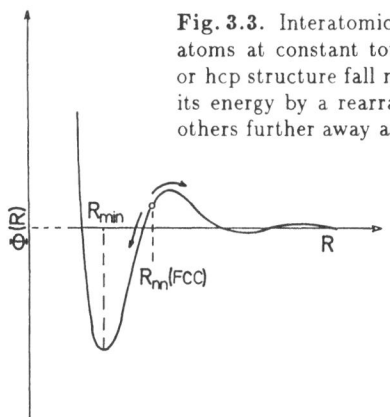

Fig. 3.3. Interatomic potential $\Phi(R)$ for the rearrangement of the atoms at constant total volume. If the nearest neighbours in the fcc or hcp structure fall near the peak where shown, the crystal can lower its energy by a rearrangement which brings some of them closer and others further away as shown by the arrows

produce it here as our Fig. 3.3) is archetypal for this class of arguments. For Ga or Hg, the nearest neighbours of a hypothetical close-packed structure would fall close to a maximum of $\Phi(R)$ – hence a structural rearrangement (at constant volume) which pushes some of the twelve nearest neighbours closer to the central atom and removes the others to a more distant position would lower the energy.

The equivalence of the q-space and R-space argument presupposes that the electrostatic contribution to the structural energy is negligible compared to the band-structure contribution. This is undoubtedly justified if only close-packed structures are considered (see for example the values of the Madelung constants given in Appendix C). On the other hand, if we consider a transition from the diamond ($\alpha_M = -1.670856$) to the white-tin ($\alpha_M = -1.773118$) or to the hcp structure ($\alpha_M = -1.79168$), the differences in the electrostatic energy are no longer negligible. Thus we find the real-space argument to be more complete – $\Phi(R)$ of course incorporates all the electrostatic contributions – and we will use it below to discuss trends in the crystal structures through the periodic table, following the recent work of *Hafner* and *Heine* [3.56].

3.3 Trends in Crystal Structures

The specific trends which we shall discuss are the following: (i) the change from metallic structures in groups I to III to covalent structures in groups IV to VI with decreasing coordination number; (ii) in group IIb the increasing distortion as the atomic number A increases, culminating in a unique structure in Hg; (iii) a similar phenomenon in group III with Al and a unique structure for Ga, but then a return to a less distorted one for In and a close-packed metallic structure for Tl; (iv) the tendency in groups IV to VI towards a more metallic behaviour with increasing atomic number, most pronounced in group IV with Sn and Pb.

Our aim is to explain these trends in terms of the systematic variations of the interatomic potential discussed in Sect. 2.3. There we have found that with a one-parameter pseudopotential, the trends in the pair potentials are discussed most conveniently in terms of the parameters $R_s = R_a/Z^{1/3}$, R_c, and $y = R_a Z^{1/3}$, the last one being irrelevant for the form of $\Phi(R)$ and hence for the structural arguments. Thus the first task is to locate the individual elements in the two-dimensional parameter space R_s vs R_c. *Hafner* and *Heine* [3.56] take the electron density R_s from experiment. Strictly speaking the pseudopotential determines the electron density, and we have seen that in proper calculations good agreement is obtained. But because of the nonlocal nature of the pseudopotential operator, somewhat different combinations of its parts enter the calculation of the equilibrium density and the calculation of the structural energy. Thus taking R_s as given is essentially equivalent to characterizing the nonlocality of the pseudopotential by a second parameter. But even taking R_s as given, one should remember the most important trends in the electron density through the periodic table [3.57]: (i) The size of the atom increases with the principal quantum number. This is because the cancellation theorem pushes each new shell outwards. This trend dominates in groups Ia and IIa. (ii) The existence of a d-electron core makes the pseudopotential strongly attractive. This is responsible for the nonmonotonic variation of R_s within group IIb. (iii) The pseudopotential becomes stronger with increasing valence, thus R_s decreases quite generally with increasing Z.

The parameter R_c is to a first approximation the actual radius of the ionic core (HW, p. 259, see also *Austin* and *Heine* [3.57]), which is then shifted slightly by nonlocality, e.g. in the heavy elements relativistic effects result in an extra binding energy of the s-electrons relative to p-states, which can be easily shown to decrease R_c (see e.g. [3.58]). Values of R_c fitted to the electronic band structure (see Table XVIII of *Cohen* and *Heine* [3.59]) are compiled in Table 3.5. With this in mind, we are now well prepared to discuss the crystal structures.

Figure 3.4 shows the interatomic potentials for the second row elements Na to P, calculated with the R_c and R_s values taken from Table 3.5. We see immediately that for Na, Mg and Al the interatomic distance D_{cp} for close packing (cubic or hexagonal) coincides with the minimum in $\Phi(R)$, and hence the structure is stable. For Si and P, D_{cp} corresponds to a region where both the tangential force constant $T(R) = R^{-1}\Phi'(R)$ and the radial force constant $R(R) = \Phi''(R)$ are negative and a close-packed structure is clearly unstable. We find that this is due to the decrease of R_s at approximately constant R_c/R_s; at high electron density the repulsive core begins to cover the attractive oscillation. Even a nearest-neighbour argument is sufficient to give the correct sign for the structural energy differences fcc vs diamond

Table 3.5. Parameters of the elements. (After [3.56]). Values of R_c/R_s fitted to empirical pseudopotentials; values in parentheses refer to the Heine-Abarenkov pseudopotential

Element	Li		Be	B		
R_s [a.u.]	3.26		1.87	1.58		
R_c/R_s	0.50		(0.58)	(0.53)		

Element	Na		Mg	Al	Si	P
R_s [a.u.]	3.93		2.65	2.07	2.00	1.92
R_c/R_s	(0.47)		0.52	0.54	0.51	0.49

Element	K	Ca	Zn	Ga	Ge	As
R_s [a.u.]	4.86	3.27	2.30	2.19	2.09	1.91
$R_c R_s$	(0.47)	(0.58)	(0.47)	0.48	0.47	0.49

Element	Rb	Sr	Cd	In	Sn	Sb
R_s [a.u.]	5.20	3.57	2.59	2.41	2.21	2.14
$R_c R_s$	(0.41)	(0.58)	0.48	0.45	0.45	0.47

Element	Cs	Ba	Hg	Tl	Pb	Bi
R_s [a.u.]	5.60	3.70	2.66	2.48	2.30	2.25
R_c/R_s	(0.41)	(0.58)	0.44	(0.45)	0.45	0.41

and fcc vs simple cubic. For an explanation of more subtle effects such as the structural energy differences between fcc, bcc and hcp in Na, Mg and Al, interactions with more distant neighbours must be considered, but this too presents no problem (see also [3.4, 22, 60]). The most subtle problem is perhaps the stability of the bcc structure in the heavy alkalis. This will be discussed in some detail in Sect. 3.6.

For the B-group elements Hafner and Heine found that to a first approximation it is sufficient to set $R_s \cong$ const. within each group and to vary only R_c/R_s (Fig. 3.5). At large core radii and low electron densities we find again interatomic potentials with a minimum near the nearest-neighbour distance D_{cp} for close packing, resulting in stable close-packed structures for Mg and Al. As the electron density increases and/or the core radius decreases, the minimum is shifted relative to D_{cp} and flattened – this yields distorted structures for Zn, Cd, Hg, Ga, Si, Ge, Sn and the group V elements. The maximum distortion is clearly limited by the onset of the repulsive core. From the shortest possible nearest-neighbour distance, the coordination number N_c may be estimated (see [3.56] for details). This yields $N_c = 5.2$ for Si (where we have $N_c = 4$ in the diamond structure and $N_c = 6$ in the high-pressure modification with white-tin structure) and $N_c = 3.1$ for P (where $N_c = 3$ is the observed coordination).

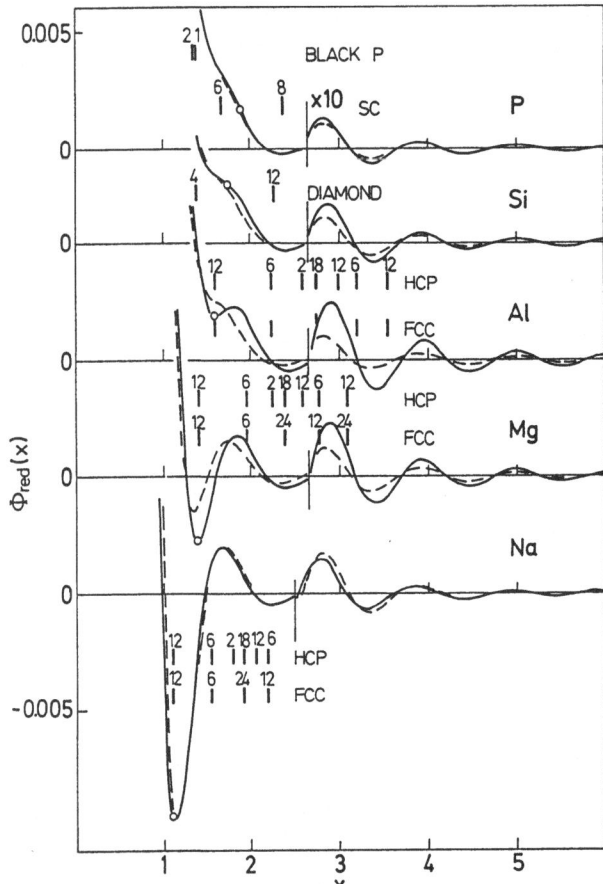

Fig. 3.4. Trends in the reduced interatomic interaction $\Phi_{red}(x)$ for the elements Na to P. (*Solid line*) calculated with R_c from Table 4.5; (*dashed line*) assuming a constant $R_c/R_s = 0.48$. The circles indicate where the n.n. distance for close packing falls on the interatomic potential. The vertical bars indicate near-neighbour distances and coordination numbers for the fcc, hcp, sc and diamond structures and for the structure of black phosphorous (after [3.56])

At even smaller core radii R_c, the amplitude effect becomes important (cf. Sect. 2.3); as q_0 approaches $2k_F$, the leading term of the Friedel oscillations vanishes. Thus for In, Tl and Pb, the interatomic potential is reduced to a screened repulsive potential and of course a close-packed arrangement will be energetically most favourable. Admittedly, the trends in the (R_c/R_s) ratio have to be stretched a bit, but this is well understood in terms of nonlocality.

3.3.1 A Brief Remark on Transition Metal Structures

The crystal structure of the transition metals may be explained by comparing the d-band energy (3.14), calculated for the various structures, since it is

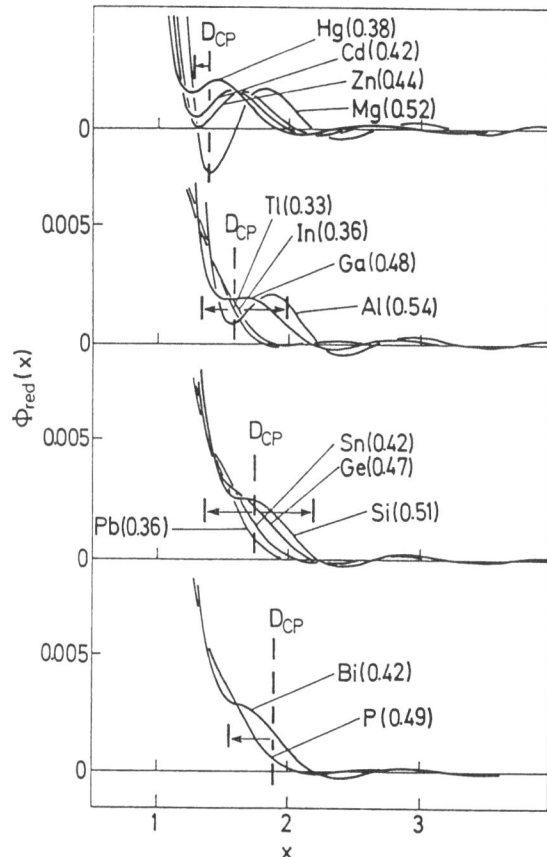

Fig. 3.5. Trends in the reduced interatomic interaction for the B-group elements. The electron density has been set constant for a given Z ($R_s = 2.65$ for $Z = 2$, $R_s = 2.20$ for $Z = 3$, $R_s = 2.05$ for $Z = 4$ and $Z = 5$), the ratio R_c/R_s is given in parentheses in each case. The dashed line marks the n.n. distance D_{cp} for close packing and the arrows show the extent of the rearrangement within the nearest-neighbour shell for the structure with the lowest symmetry in each group (after [3.56])

found that this dominates the cohesive energy. The d-band density of states displays considerable structure characteristic of the crystal lattice, as shown in the pioneering work of *Pettifor* [3.61,62]. As the unhybridized tight-binding d-band is progressively filled up with electrons, Pettifor finds the sequence hcp-bcc-hcp-fcc-bcc of stable structures, in complete agreement with experiment, except for the prediction of bcc stability at the noble-metal end of the series. The stability of the bcc lattice for the half-filled band (V, Nb, Cr, Mo, Ta, W) is related to the larger bonding-antibonding splitting in the bcc relative to the close-packed phases.

The number of d-electrons also influences the structural stability of the heavier alkali and alkaline earth metals; N_d increases with the atomic number and with pressure. The case of the group IIa metals has been studied in detail by *Skriver* [3.63] who showed that the trend hcp-fcc-bcc with increasing atomic number (Be,Mg: hcp; Ca,Sr: fcc; Ba,Ra: bcc) and under pressure (Ca and Sr show pressure-induced transformations from fcc to bcc, see next Section) correlates with an increasing number of d-electrons.

3.3.2 The Crystal Structures of the Lanthanides

The trivalent rare-earth metals and the heavy actinides display an interesting sequence of close-packed crystal structures [3.64]. As a function of increasing pressure for a given metal or as a function of decreasing atomic number the sequence is hexagonal close-packed (hcp) $\rightarrow$ Samarium (Sm) $\rightarrow$ double hexagonal close-packed (dhcp) $\rightarrow$ face centred cubic (fcc). These close-packed structures are usually classified according to the stacking sequence of hexagonal layers which are the closest packing in two dimensions. The individual layers can be labelled A, B, C according relative positions of their atoms. In this notation hcp is AB, Sm ABACACBCB, dhcp ABAC and fcc ABC. Very recently it has been suggested [3.65], on the basis of total-energy calculations in *Andersen*'s [3.29] atomic sphere approximation and of still incomplete experimental evidence [3.66], that there is possibly a fifth rare-earth structure between dhcp and fcc, defined by the stacking sequence ABCBAC and which would consequently be called triple hexagonal close-packed (thcp). *Duthie* and *Pettifor* [3.67] have quantitatively correlated the observed crystal-structure sequence with the d-band occupancy via the d-band energy contribution to the total energy. The effect of pressure on the rare earth is to cause the $5d$ band, which overlaps the Fermi level, to move downwards relative to the $6s$ band. Consequently electrons are transferred from s-like to d-like states. The $s \rightarrow d$ promotion also occurs under zero pressure with decreasing atomic number. The f electrons on the other hand are believed to exhibit a localized character, they play no role in the determination of the stable crystal structure.

The $s \rightarrow d$ transition results in a decrease of the fraction of volume available to the valence electrons outside the core. Even prior to the analysis of Duthie and Pettifor, *Johansson* and *Rosengren* [3.68] had shown that the trend in the rare-earth crystal structures may be modelled using a very simple local pseudopotential. They showed that the observed sequence is correlated to an increase of the ratio (R_c/R_s), i.e. of the radius of the ionic core R_c relative to R_s. Of course one would be hesitant to apply pseudopotential theory to a lanthanide metal and in fact the true microscopic origin of this correlation between (R_c/R_s) and the crystal structure was elucidated only by the work of *Duthie* and *Pettifor* [3.67]. However, the success of Johansson and Rosengren's calculation is just another example of the astonishing ability of pseudopotential theory to grasp the essential effect behind a physical trend. We also refer to the interpretation of the simple-metal structures given earlier in this chapter.

3.3.3 Charge-Density Analysis of Bonding

The indirect part of the effective interatomic interaction consists of the interaction of the screening electron density induced by one ion with the core charge centred on another ion. Hence the interatomic forces and valence electron distributions are closely related. The interatomic potential is not directly accessible to experiment, but the electron density can be determined by diffraction methods [2] [3.69–71]. Therefore the investigation of the electron density can give valuable information on the chemical bond. Note that according to density-functional theory (see Sect. 2.6), all properties of the electronic ground states can be derived from the exact electron density.

Within pseudopotential theory the electron density is described by the superposition of neutral pseudo-atoms. The electron density of each pseudo-atom is composed of a core contribution ϱ_{core}, a contribution of the orthogonalization (or depletion) hole ϱ_{oh} and a screening charge ϱ_{sc},

$$\varrho(r) = \varrho_{core}(r) + \varrho_{oh}(r) + \varrho_{sc}(r) \; . \tag{3.17a}$$

Within linear response theory, the Fourier transform of the screening charge density is related to the screening potential $w_{sc}(q)$ [see (2.37)] via a generalized Poisson equation (HW, p.321, [3.72])

$$\varrho_{sc}(q) = \frac{q^2}{8\pi} \frac{w_{sc}(q)}{1 - G(q)} \; . \tag{3.17b}$$

It turns out that the form factor $\varrho_{sc}(q)$ is a direct test for a pseudopotential matrix element.

Equation (3.17) has been the basis of a number of electron density calculations for simple metals and semiconductors [3.72–76]. Nonlocality of the pseudopotential is found to have a rather strong influence and higher-order corrections are found to be minimized for optimized nonlocal potentials [3.74–76]. On a whole, the pseudo-valence electron density deduced from the linear screening ansatz turns out to be fairly close to the corresponding atomic density at intermediate distances. At large distances the exponential decay of the free-atom density is replaced by the familiar Friedel oscillations.

With a carefully optimized pseudopotential, good agreement is obtained with the measured x-ray scattering form factors of Mg and Al (see Fig. 3.6). The calculated $\varrho(r)$ may be superposed to determine the electron

---

[2] The by now classical methods are the X-N and the X-X difference methods. In the X-N method the structure amplitudes are determined from a neutron scattering experiment, the difference between the neutron and the x-ray scattering intensity is used to calculate the x-ray scattering form factor which is just the Fourier transform of the electron density. In the X-X difference method the structure data are determined from the high-order reflections (i.e. those with large q) where the sharp density maxima of the ionic core dominate and the valence electron contribution can be neglected. The low-order reflections give the necessary information on the scattering form factors.

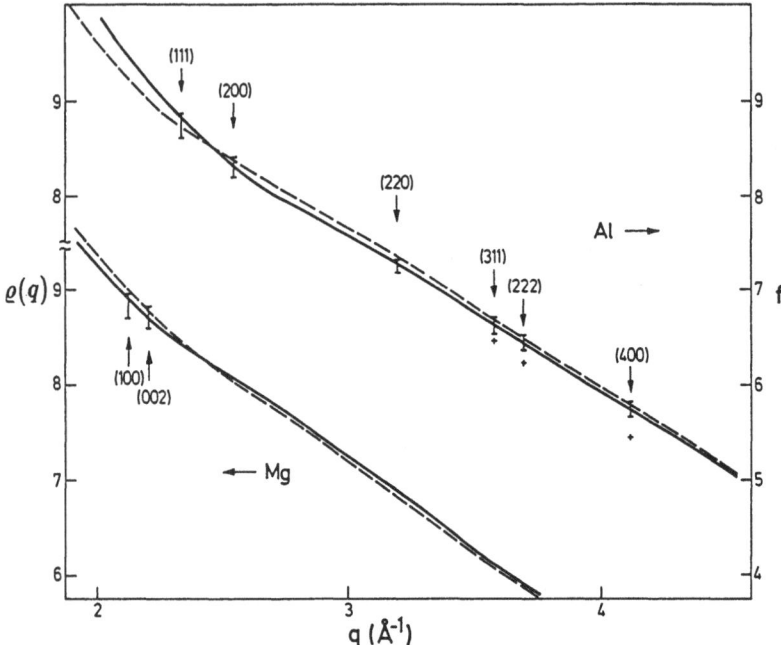

Fig. 3.6. X-ray scattering form factors $\varrho(q) = \varrho_{core}(q) + \varrho_{oh}(q) + \varrho_{sc}(q)$ for Al and Mg. (*Solid line*) calculated for the pure metals, (*dashed line*) calculated for an equiatomic Al-Mg alloy. The vertical bars span the range covered by a number of different experimental determinations of the scattering form factor. After [3.76]

density in the crystal. For the purpose of discussing chemical bonding effects, it is only the pseudo-valence electron density, given by the superposition of the screening charges centred on the individual ions, which is of interest (the orthogonalization hole charge is restricted to the core region and has no direct influence on the electron distribution in the bonding region; the core charge is assumed to be "frozen", i.e. identical in the metal and in the free ion). In Fig. 3.7 we consider the pseudo-valence electron density along a nearest-neighbour bond in Na, Mg, Al and Si. The variation of the electron density distribution very clearly reflects the charateristic trends found in the interatomic interactions. In Na, the charge distribution is nearly constant outside the ionic core. The difference between the interatomic density maximum of 1.14 electrons (per atomic volume) located halfway between nearest neighbours and the minimum density of 1.09 (in the octahedral voids of the bcc structure) is only 5 % of the mean density. In hcp Mg one finds a shallow maximum in the centre of the face shared by two Mg tetrahedra and a shallow minimum in the octhahedral voids of the structure, but the difference is still only of the order of 25 %. In fcc Al the overlap of the valence orbitals is just not large enough to create covalent bonding effects. In Si even a linear screening calculation shows a distinctly expressed bonding charge resulting from the overlap of the spherical pseudo-atoms [3.72], but in that case the

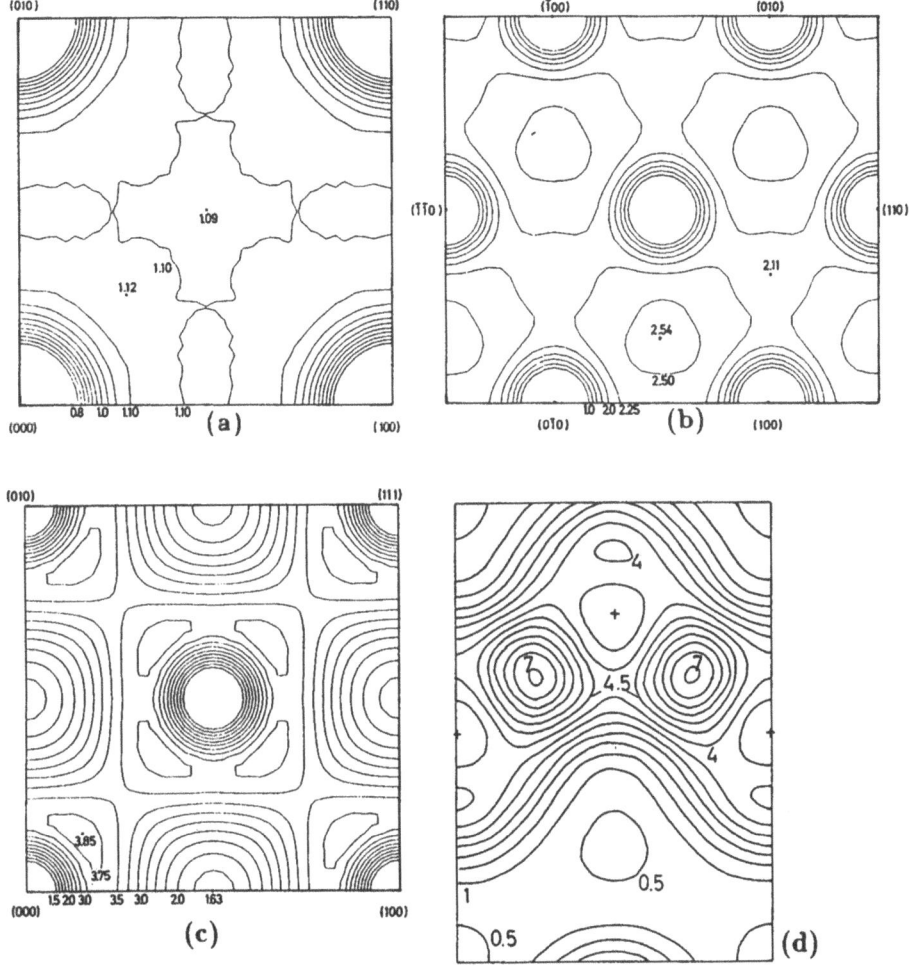

Fig. 3.7a–d. Contour plots of the pseudo-valence electron density (in units of electrons per atomic volume) in Na, Mg, Al and Si: **(a)** body-centred cubic Na, (110) plane, **(b)** hexagonal close-packed Mg, (1000) plane, **(c)** face-centred cubic Al, (100) plane, **(d)** Si, diamond structure, (110) plane; (a–c) after [3.76], (d) after [3.78]

higher-order contributions are non-negligible [3.77, 78] – as they are for the stabilization of the diamond structure against shear distortions. The chemical effects are most distinctly expressed in the difference electron density $\Delta\varrho(r) = \varrho(r) - \varrho_{atom}(r)$, where $\varrho_{atom}(r)$ is the electron density calculated for a superposition of free-atom densities. Figure 3.8 shows such a difference density for Si – the bond charge along the nearest-neighbour bond is clearly visible.

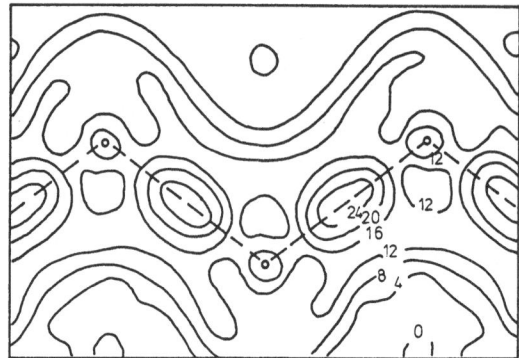

Fig. 3.8. Experimental difference electron density for Si, displaying the bond charges along the nearest-neighbour bonds in the diamond structure. After [3.77]. Values shown are electron density in el./a.u.[3]

3.4 Pressure-Induced Phase Changes

Many simple metals undergo phase changes under elevated pressure: the heavy alkali metals K, Rb and Cs transform from a bcc to a fcc structure (see *Olijnyk* and *Holzapfel* [3.79] for the most recent state of the art) and the alkaline-earth metals Ca and Sr transform from fcc to bcc (at low temperatures) [3.80]. It has been shown that these phase changes are closely related to the d-band occupancy; under pressure the s-levels are shifted upwards, the d-levels are pushed downwards [3.26, 27, 63, 81]. Thus there is an $s \rightarrow d$ transfer and the heavy alkali and alkaline earth metals behave as transition metals (note that the stability of bcc for K, Rb and Cs at $p = 0$ is itself related to vibrational effects as will be discussed below). Very recently it has been suggested that under extreme compressions d-electron driven phase changes will occur even in Na, Mg and Al [3.22]. The origin of a possible bcc $\rightarrow$ fcc transition (at room temperature) in Li [3.82, 83] is still unclear.

The pressure-induced phase changes of the heavy $Z = 3$ elements (Ga, In, Tl) and of the $Z = 4$ and $Z = 5$ elements parallel the change from open to close-packed structures with increasing quantum number (see e.g. [3.84]). For each group a generalized phase diagram may be constructed containing all the information on the pressure-induced phase changes of the group. The phase changes in the group IV materials have been investigated very intensively [3.10–12, 14, 36, 43] and the pressure-induced transitions in Ga have been discussed by *Inglesfield* [3.85] and *Hafner* and *Hittmair* [3.40] in terms of a calculation of the structural energy differences as a function of volume. Figure 3.9 shows the common-tangent construction for the pressure-induced $\alpha \rightarrow \beta$ (diamond $\rightarrow$ white tin) transition in Ge and Sn, and the β-Sn $\rightarrow$ Sn II (white tin $\rightarrow$ body centred tetragonal) transition (for Sn a third transition to a hcp structure is predicted under extreme compression). Note that the calculations yield quantitative results for the transition pressures

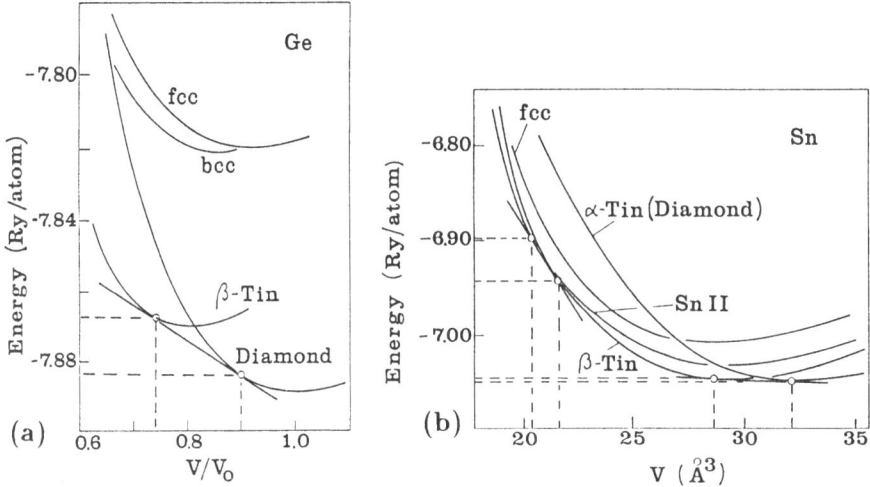

Fig. 3.9a,b. The binding energy of the diamond, β-tin, body-centred tetragonal (Sn II) and fcc structures as a function of the atomic volume for **(a)** Ge (after [3.10–12] using a band structure calculation) and **(b)** Sn (after [3.36] using perturbation theory). The common-tangent constructions for the pressure-induced phase transitions are indicated

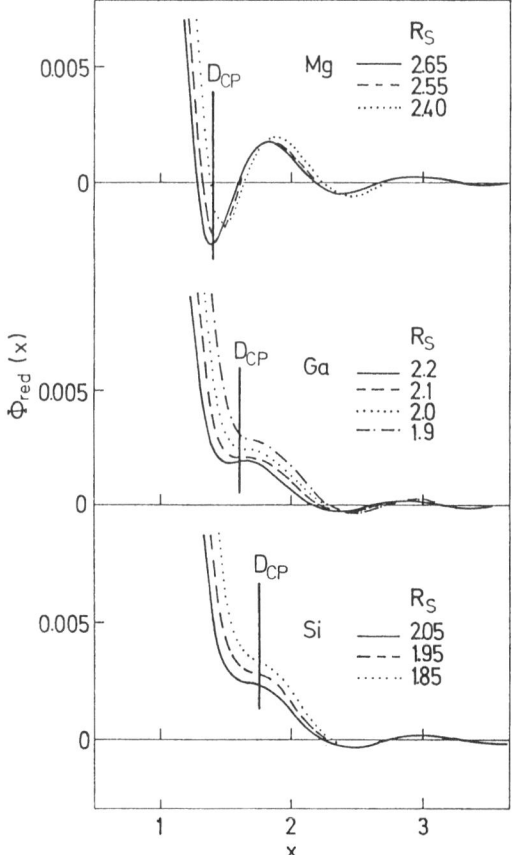

Fig. 3.10. Variation of the reduced interatomic interaction $\Phi_{red}(x)$ under compression for Mg, Ga and Si (after [3.56])

and volume changes at the transitions, and that a perturbation calculation is nearly as accurate as a full band-structure calculation.

The driving mechanism for the phase changes is seen most clearly in terms of the pair potentials (Fig. 3.10); the repulsive core moves over the attractive oscillations and into the nearest-neighbour distance for close packing [3.56]. Hence, whilst compression and chemical substitution by an element with higher atomic number result in the same structural trend, the underlying physics is quite different. Under compression, it is the rapid decrease of the nearest-neighbour distance against a slowly receding core which dominates the trend to a more repulsive potential, while chemical substitution by a heavier element of the same group results in a damping of the oscillations which drives the structural trend.

3.5 Thermodynamics of Crystalline Metals

The dynamics of atoms in the solid is governed by the effective Hamiltonian (2.20). At low temperatures, as long as the thermally induced displacements from the equilibrium positions are small compared to the nearest-neighbour distance, a harmonic calculation (i.e. keeping only terms up to second order in the displacements) is usually sufficient. Anharmonic corrections must be considered for higher temperatures. They arise from the deviation of the interatomic potential from a parabolic form around the nearest-neighbour distance.

3.5.1 Harmonic Lattice Dynamics

The harmonic equations of motion are obtained by expanding the effective ionic Hamiltonian (2.20) to second order in the displacement u_l of the ions from their equilibrium positions R_l, i.e.

$$H = \frac{1}{2} M \sum_{l\alpha} \dot{u}_{l\alpha} + \frac{1}{2} \sum_{\substack{ll' \\ \alpha\beta}} \Phi_{\alpha\beta}(ll') u_{l\alpha} u_{l\beta} \dots \tag{3.18}$$

with the force constants $\Phi_{\alpha\beta}(ll')$ given in terms of the second derivatives of the pair potential [3.86, 87]

$$\Phi_{\alpha\beta}(ll') = -\left[\frac{\partial^2 \Phi(R)}{\partial R_\alpha \partial R_\beta}\right]_{R=R_l-R_{l'}} \quad , \quad l \neq l'$$

$$\Phi_{\alpha\beta}(ll) = \sum_{l' \neq l} \left[\frac{\partial^2 \Phi(R)}{\partial R_\alpha \partial R_\beta}\right]_{R=R_l-R_{l'}} \tag{3.19}$$

(the first derivatives are zero by virtue of the equilibrium condition). From (3.18) we obtain immediately the equation of motion

$$M\ddot{u}_{l\alpha} = -\sum_{l'\beta} \Phi_{\alpha\beta}(ll')u_{l'\beta} \tag{3.20}$$

and a plane-wave ansatz for $u_{l\alpha}$ yields the secular equation for normal modes with frequency $\omega(\boldsymbol{q})$ and polarization $\boldsymbol{e}(\boldsymbol{q})$, see e.g. [3.86, 87]

$$M\omega^2(\boldsymbol{q})e_\alpha(\boldsymbol{q}) = \sum_\beta \overline{D}_{\alpha\beta}(\boldsymbol{q})e_\beta(\boldsymbol{q}) \qquad \text{with} \tag{3.21a}$$

$$\overline{D}_{\alpha\beta}(\boldsymbol{q}) = D_{\alpha\beta}(\boldsymbol{q}) - D_{\alpha\beta}(0) \quad , \tag{3.21b}$$

and the dynamical matrix

$$D_{\alpha\beta}(\boldsymbol{q}) = \sum_{l'} e^{i\boldsymbol{q}\boldsymbol{R}_{l'}}\Phi_{\alpha\beta}(ll') \quad , \tag{3.21c}$$

where the $\Phi_{\alpha\beta}(ll')$ are the force constants related to the pair potential $\Phi(R)$ through (3.19). It follows directly that

$$D_{\alpha\beta}(\boldsymbol{q}) = \frac{1}{2}\sum_{\boldsymbol{Q}}(\boldsymbol{q}+\boldsymbol{Q})_\alpha(\boldsymbol{q}+\boldsymbol{Q})_\beta\,\hat{\Phi}(|\boldsymbol{q}+\boldsymbol{Q}|) \quad , \tag{3.21d}$$

where the sum is over all reciprocal lattice vectors $\boldsymbol{Q}$ and $\hat{\Phi}(q)$ is the Fourier transform of the pair potential

$$\hat{\Phi}(q) = \frac{8\pi Z^2}{V_a q^2} + 2F(q) \quad . \tag{3.22}$$

In practice, the electrostatic contributions are separated from the band-structure term and treated using Madelung techniques (*Brüesch* [3.87], Appendix K). The calculation of phonon dispersion curves for simple metals using effective potentials derived from pseudopotentials and linear response theory has enjoyed a widespread popularity over the last 15 years; about 200 papers have been published in the field and a recent review has been given by *Grimvall* [3.88]. As our primary interest is not in the phonon physics itself, but in the manifestations of the lattice vibrations in the thermodynamics, we shall pass very rapidly over the subject. As an illustrative example of the accuracy that has been achieved in the prediction of phonon frequencies of close-packed metals using first-principles pseudopotentials and second-order perturbation theory, we show in Fig. 3.11 the phonon dispersion calculations for Mg, Li and Be – for a more extensive bibliography we refer to Grimvall's review. *Brovman* et al. [3.89, 90] and *Bertoni* et al. [3.50, 91–93] have pointed out the importance of three-body forces in anisotropic (Be, Zn, Cd) and open (Si, Ge, Sn) structures. For the divalent metals, the corrections

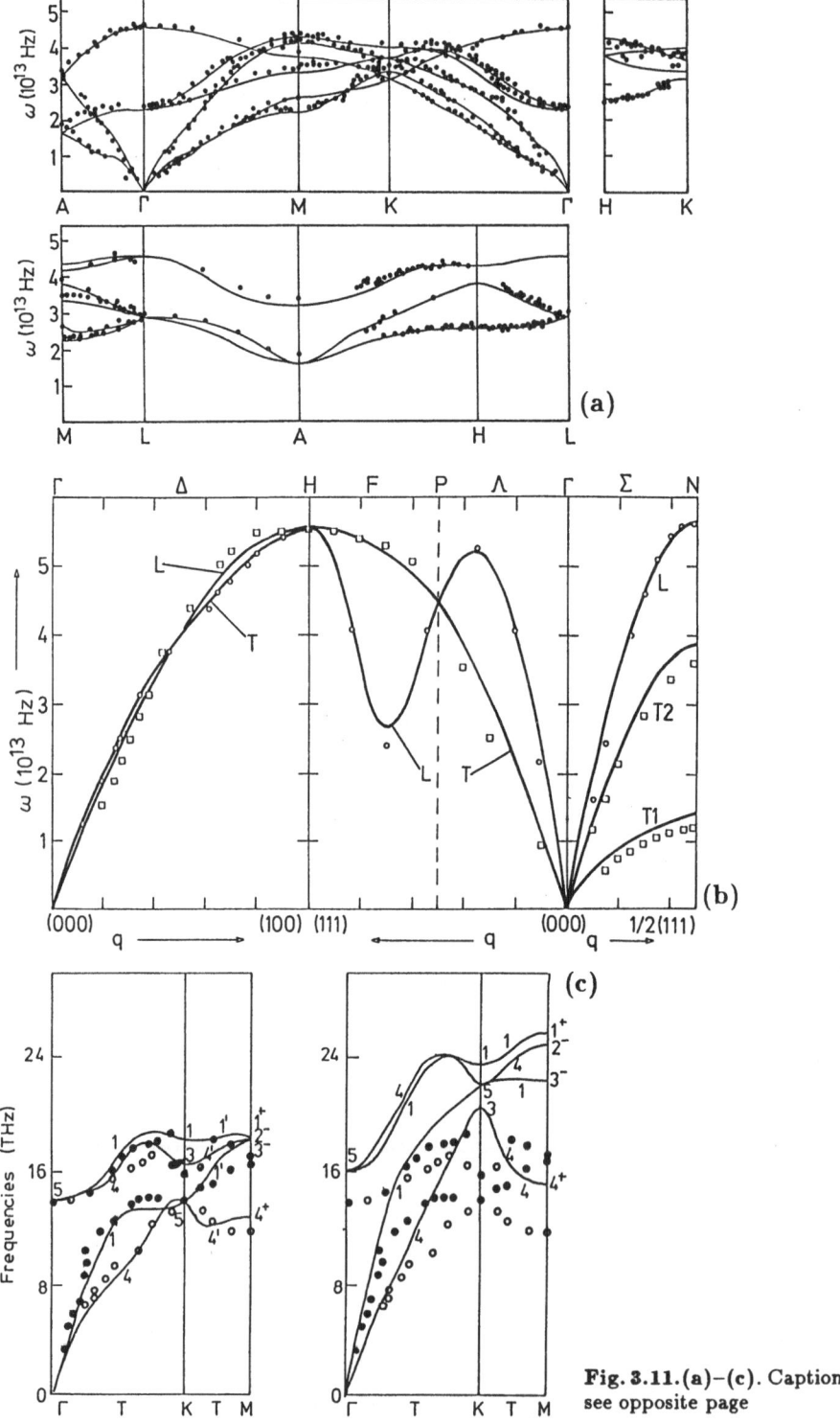

Fig. 3.11.(a)–(c). Caption see opposite page

to the phonon frequencies from three-body forces are globally quite small ($\sim 10\%$), but they may be important for the correct ordering of some of the modes at certain symmetry points (see for example Fig. 3.11c). For the group IV elements free-electron screening is no longer sufficient (for discussing *dynamical* effects) and the effect of a realistic band structure on the screening must be considered [3.93–96]. Effects of *s-d* hybridization on the lattice dynamics of the noble metals [3.34,97], of the heavy IIb metals [3.18,37,98], and of the heavy IIa metals [3.42] have been considered. They have been found to be substantial for the noble metals, less important for Ca, Sr and Ba and almost negligible for Zn.

The quantum-mechanical treatment of lattice vibrations teaches us that the normal modes of the crystal, the phonons, can be treated as quasiparticles of zero spin. Hence the vibrational contribution to the thermodynamic functions may be derived from the statistical mechanics of an ideal Bose gas. The canonical partition function is given by (remember that we use atomic units, $\hbar = 1$) [3.86–88]

$$Z_{\mathrm{ph}} = \prod_{qj} \frac{\exp\left[-\frac{1}{2}\beta\omega(qj)\right]}{1 - \exp\left[-\beta\omega(qj)\right]} \quad , \tag{3.23}$$

where the product is over all normal modes (qj). The thermodynamic functions follow directly from the partition function, e.g. the vibrational free energy is given by

$$\begin{aligned} F_{\mathrm{ph}} &= -k_{\mathrm{B}}T \ln Z_{\mathrm{ph}} = k_{\mathrm{B}}T \sum_{qj} \ln\left\{ 2\sinh\left[\frac{\beta}{2}\omega(qj)\right]\right\} \\ &= 3Nk_{\mathrm{B}}T \int \ln\left\{ 2\sinh\left[\frac{\beta}{2}\omega(qj)\right]\right\} g(\omega)d\omega \quad . \end{aligned} \tag{3.24}$$

The last identity shows that the thermodynamic functions may be expressed in terms of the vibrational density of states $g(\omega)$ so that a detailed knowledge of the individual normal modes is not necessary.

The displacement u_l of the ions from their equilibrium positions R_l also influences the electrostatic and band-structure contributions to the ground-state energy. The structure-factor part of (2.12–19) must now be written in the form

$$|S(q)|^2 = N^{-2} \sum_{lm} \exp\left[iq(R_l - R_m)\right]\langle\exp\left(iqu_l - iqu_m\right)\rangle_{\mathrm{T}}. \tag{3.25}$$

Fig. 3.11. (a) Phonon dispersion relations for hcp Mg, after [3.24]. (*Dots*) experiment, (*solid line*) theory. (b) Phonon dispersion relations for bcc Li, after [3.20]. (c) Phonon dispersion relations in hcp Be, after [3.91]. (*Left*) calculated to second order in the pseudopotential. (*Right*) third-order calculation

The brackets $\langle \ldots \rangle_\mathrm{T}$ express the thermodynamic expectation value. It can be shown [3.99] that for cubic crystals

$$\langle \exp\left(\mathrm{i}\boldsymbol{q}\boldsymbol{u}_l - \mathrm{i}\boldsymbol{q}\boldsymbol{u}_m\right)\rangle_\mathrm{T} = \exp\left[-q^2\varrho(|\boldsymbol{R}_l - \boldsymbol{R}_m|)\right] \tag{3.26a}$$

with

$$\varrho(R) = \frac{1}{2NM} \sum_{\boldsymbol{q}j} \frac{1}{\omega(\boldsymbol{q}j)} \coth\left[\beta\omega(\boldsymbol{q}j)\right][1 - \cos\left(\boldsymbol{q}\boldsymbol{R}\right)] \quad . \tag{3.26b}$$

If the vibrations of different atoms were totally uncorrelated ($\langle \boldsymbol{u}_l\boldsymbol{u}_m \rangle \sim \delta_{lm}$), the quantity given by (3.26a) would reduce to the Debye-Waller factor $\exp(-2W)$. The exponential term $\exp(-q^2\varrho_{lm})$ can be combined with the energy-wave-number characteristic $F(q)$, so that the structural energy of the crystal may be formally written in terms of a static-lattice sum over a temperature-dependent pair potential $\Phi_\mathrm{T}(R)$ defined by

$$\begin{aligned}
\Phi_\mathrm{T}(R) &= \frac{2}{N} \sum_{\boldsymbol{q}} \exp\left[-\mathrm{i}\boldsymbol{q}\boldsymbol{R} - q^2\varrho(R)\right]\left[\frac{4\pi Z^2}{V_a q^2} + F(q)\right] \\
&= \frac{2Z^2}{R}\,\mathrm{erf}\left(\frac{R}{2\varrho(R)}\right) + \frac{1}{2\pi^2 R}\int_0^\infty F(q)\sin\left(qR\right)q\,\mathrm{e}^{-q^2\varrho(R)}\,dq \quad .
\end{aligned} \tag{3.27}$$

Thus in the harmonic approximation the free energy of a crystal is given by

$$F = E_\mathrm{V} + \frac{1}{2N} \sum_{lm} \Phi_\mathrm{T}(|\boldsymbol{R}_l - \boldsymbol{R}_m|) + F_\mathrm{ph} \quad . \tag{3.28}$$

The harmonic approximation is appropriate at not too high temperatures for the calculation of thermodynamic properties of simple metals, see e.g. *Wallace* [3.100] and *Vaks* et al. [3.101], at higher temperatures the increased amplitude of the atomic vibrations forces us to consider terms of higher than second order in the expansion of the potential energy with respect to the atomic displacement − i.e. anharmonicity.

3.5.2 Anharmonicity

The theory needed to describe the effects of anharmonicity is inherently more complicated than the harmonic theory (see *Barron* and *Klein* [3.102] for a general review). The lowest-order approximation − the quasi-harmonic approximation − gave quite good results for the phonon Grüneisen parameters,

$$\gamma(\boldsymbol{q}j) = \frac{\partial \ln \omega(\boldsymbol{q}j)}{\partial \ln V} \quad ,$$

of the alkali metals [3.103, 104] and references cited therein, but is insufficient for describing thermal effects. The most successful (and tractable) theory is anharmonic perturbation theory based on the self-consistent-phonon approximation [3.87, 105, 106]. Like the self-consistent-field method for electrons, the self-consistent-phonon method consists in a variational ansatz of first order for the free energy. One starts by comparing the anharmonic Hamiltonian H (the volume energy E_V is irrelevant for rearrangements at constant volume and can be dropped)

$$H = T + \frac{1}{2N} \sum_{lm} \Phi(|\boldsymbol{R}_l - \boldsymbol{R}_m + \boldsymbol{u}_l - \boldsymbol{u}_m|) \tag{3.29}$$

with a harmonic Hamiltonian H_0

$$H_0 = T + \frac{1}{4N} \sum_{lm} u_{l\alpha} \Phi^0_{\alpha\beta}(lm) u_{m\beta} \quad , \tag{3.30}$$

(summation over repeated greek indices is understood), whose force constants $\Phi^0_{\alpha\beta}(lm)$ are now considered as free parameters. The difference in free energy is then given by

$$\Delta F = \frac{1}{2N} \sum_{lm} \Bigg\{ \langle \Phi(|\boldsymbol{R}_l - \boldsymbol{R}_m + \boldsymbol{u}_l - \boldsymbol{u}_m|) \rangle_T$$

$$- \frac{1}{2} \Lambda_{\alpha\beta}(lm) \Phi^0_{\alpha\beta}(lm) \Bigg\} \quad , \quad \text{where} \tag{3.31a}$$

$$\Lambda_{\alpha\beta}(lm) = \langle u_l u_m \rangle_T \tag{3.31b}$$

is the displacement correlation function. The angular brackets again stand for the thermodynamic expectation value (calculated in the harmonic reference system). Using the identity $f(\boldsymbol{R} + \boldsymbol{u}) = \exp(\boldsymbol{u} \cdot \nabla) \cdot f(\boldsymbol{R})$ and also the fact that for harmonic oscillations the thermal average is given by [3.86, 87], cf. also (3.26)

$$\langle e^{\boldsymbol{u} \cdot \boldsymbol{v}} \rangle_T = \exp\left[\tfrac{1}{2} \langle (\boldsymbol{u} \cdot \boldsymbol{y})^2 \rangle_T \right] \tag{3.32a}$$

we can write the first term in (3.31a) in the form

$$\langle \Phi(|\boldsymbol{R}_l - \boldsymbol{R}_m + \boldsymbol{u}_l - \boldsymbol{u}_m|) \rangle_T$$

$$= \exp\left[\frac{1}{2} \sum_{\alpha\beta} \Lambda_{\alpha\beta}(lm) \frac{\partial^2}{\partial R_\alpha \partial R_\beta} \right] \Phi(R) \Bigg|_{R=\boldsymbol{R}_l - \boldsymbol{R}_m} \quad . \tag{3.32b}$$

The parameters $\Lambda_{\alpha\beta}$ and $\Phi_{\alpha\beta}$ are determined by the condition that ΔF is minimized, thus

$$\frac{\delta \Delta F}{\delta \Lambda_{\alpha\beta}(lm)} = \left\langle \frac{\partial^2 \Phi(R)}{\partial R_\alpha \partial R_\beta} \Bigg|_{R=\boldsymbol{R}_l - \boldsymbol{R}_m + \boldsymbol{u}_l - \boldsymbol{u}_m} \right\rangle_T - \Phi^0_{\alpha\beta}(lm) = 0 \tag{3.33a}$$

$$\frac{\delta \Delta F}{\delta \Phi^0_{\alpha\beta}(lm)} = \frac{\delta F}{\delta \Phi^0_{\alpha\beta}(lm)} - \frac{1}{2}\Lambda_{\alpha\beta}(lm) = 0 \quad . \tag{3.33b}$$

Equation (3.33a) shows that the optimal choice for the harmonic force constants $\Phi^0_{\alpha\beta}$ is the thermal average over the second derivatives of the effective pair potential, and (3.33b) establishes the relationship between $\Phi^0_{\alpha\beta}$ and $\Lambda_{\alpha\beta}$. Equation (3.33) has to be solved self-consistently. An alternative derivation of the self-consistent-phonon equations using diagrammatic techniques shows that the self-consistent approximation is equivalent to a summation of all anharmonicities of even order. In the Hartree method one assumes that each electron moves in the average Coulomb field of its neighbours; here we suppose that each atom moves in the time-averaged force field of its neighbours. The self-consistent-phonon approximation greatly facilitates the calculation of the anharmonic thermodynamic functions. Within this scheme, the entropy is given by the standard harmonic formula

$$S = -k_B \sum_{qj} \left\{ \ln 2 \sinh \left[\beta \overline{\omega}(qj) \right] - \beta \frac{\overline{\omega}(qj)}{2} \coth \left[\beta \frac{\overline{\omega}(qj)}{2} \right] \right\} \tag{3.34}$$

with the harmonic frequencies replaced by the self-consistent frequencies $\overline{\omega}(qj)$, which will now be temperature dependent. Cubic anharmonicities can be incorporated by perturbation theory.

Alternative methods are standard low-order perturbation theory [3.102, 107] and computer simulation [3.108–110].

The anharmonic frequency shifts $\Delta\omega$ and the phonon linewidth Γ have been calculated for a few simple metals: the alkalis Li, Na, K and Rb [3.107, 111–114] and Al [3.115]. Generally the agreement with experiment and computer simulation is good for $\Delta\omega$, but less good for Γ, at least in the vicinity of the melting point.

Anharmonic effects in the elastic constants have been discussed by *Glyde* and *Taylor* [3.112] and *Vaks* et al. [3.116] and anharmonicity in the thermodynamic functions has been investigated by *Shukla* and coworkers [3.117, 118] and *Vaks* et al. [3.119]. Their work shows that anharmonicity can be well described. In fact a full calculation of the anharmonic frequency shifts is more than is really needed to describe the effect of anharmonicity on the thermodynamic functions – simpler and much less time-consuming thermodynamic variational methods can do the same job.

3.5.3 Variational Method for Calculating Thermodynamic Properties (Gibbs-Bogoljubov Method)

The variational condition (3.33) has been introduced in the spirit of the Gibbs-Bogoljubov inequality [3.120, 121] which defines a function $\overline{F}$ that is an upper bound to the exact free energy F of a system described by a

Hamiltonian H;

$$F \leq \overline{F} = F_0 + \langle H - H_0 \rangle_0 \quad , \tag{3.35}$$

where H_0 is a trial Hamiltonian which is a function of a set of adjustable parameters γ_i. F_0 is defined as the free energy of this trial Hamiltonian and the thermal average $\langle \ldots \rangle_0$ is taken with respect to H_0. This inequality is a simple consequence of the convexity of the free energy F,

$$F = -k_B T \ln \mathrm{Tr} \{ \exp(-\beta H) \} \quad , \tag{3.36}$$

see e.g. [3.121]. The minimization of F with respect to the γ_i leads to a set of equations to be solved self-consistently

$$\frac{\partial F}{\partial \gamma_i} \bigg|_{T,V=\mathrm{const}} = 0 \quad . \tag{3.37}$$

If the reference Hamiltonian H_0 is the harmonic Hamiltonian given by (3.30), then clearly the variational parameters are the force constants $\Phi^0_{\alpha\beta}(lm)$ [or equivalently, since

$$H_0 = \sum_{qj} \overline{\omega}(qj)[n(qj) + \tfrac{1}{2}] \quad ,$$

the eigenfrequencies $\overline{\omega}(qj)$] and the variational method is equivalent to the self-consistent-phonon method.

Of course it is tempting to choose a much simpler parametrization – the thermodynamic functions do not depend on the details of the phonon dispersion relation, but only on the vibrational density of states. The simplest possible parametrizations are the well-known Debye and Einstein models. Within the Debye model the approximate free energy is given by [3.99]

$$F = E_{\mathrm{kin}} + E_V + \frac{1}{2N} \sum_{l,m} \Phi_T(|R_l - R_m|) - TS_{\mathrm{ph}} \tag{3.38}$$

with the kinetic energy

$$E_{\mathrm{kin}} = \frac{9}{16} k_B \theta + \frac{3}{2} k_B T I(\theta/T) \quad , \tag{3.39}$$

the phonon entropy

$$S_{\mathrm{ph}} = -k_B[3 \ln(1 - e^{-\theta/T}) - I(\theta/T)] \tag{3.40}$$

and the atomic displacement factor $\varrho(R)$ entering the temperature-dependent pair potential [see (3.25–28)]

$$\varrho(R) = \frac{3}{2} \frac{M}{k_B} \left(\frac{T}{\theta} \right)^2 J \left(\frac{\theta}{T}; R \right) \tag{3.41}$$

expressed in terms of the Debye functions

$$I(x) = \frac{3}{x^3} \int_0^x \frac{t^3 dt}{e^t - 1} \qquad \text{and} \qquad (3.42a)$$

$$J(x; R) = \frac{1}{x} \int_0^x t \coth \frac{1}{2} t \left(1 - \frac{\sin t\mu}{t\mu}\right) dt \qquad \text{with} \qquad (3.42b)$$

$$\mu = \frac{q_D R}{x} \quad .$$

Here θ is the Debye temperature and q_D the Debye wave number ($=$ radius of a sphere with the volume of the Brillouin zone). The variational condition now assumes the simple form

$$\frac{\partial \overline{F}}{\partial \theta}\bigg|_{T,V=\text{const}} = 0 \quad . \qquad (3.43)$$

The solution of (3.43) yields a temperature- and volume-dependent Debye temperature $\theta = \theta(T,V)$ which has to be used in the calculation of the thermodynamic functions.

The corresponding expressions for an Einstein model are even simpler [3.122]. It has been shown that this simple approach yields relatively accurate predictions of entropies, specific heats and thermal expansion coefficients [3.99, 122–124].

3.6 Temperature-Induced Phase Changes

Polymorphism at ordinary pressure is a very common phenomenon among the elements − more than one-third of them exist in more than one allotropic form. In many cases the transformation is from a close-packed (fcc or hcp) low-temperature form to a bcc structure at higher temperatures. It was argued very early by *Zener* [3.125] that this transition is due to the fact that the more open bcc structure has a transverse phonon mode with a very low energy which causes a more rapid increase of the vibrational entropy and hence a more rapid decrease of the Gibbs free energy with temperature for the bcc lattice. Zener's argument seemed plausible enough, but of course microscopic calculations of the phonon frequencies and of the thermodynamic functions are necessary to corroborate the idea. The problem was first taken up by *Animalu* [3.126] who compared the fcc and bcc lattices for the heavy alkaline-earth metals, Ca, Sr, and Ba. He found that the Debye temperature of the bcc modification was generally a few percent lower than that of the close-packed structure and that the difference is sufficient to explain the fcc → bcc transition in Ca and Sr at higher temperature. The same result was also obtained by *Moriarty* [3.97] and *Hafner* [3.127]. The martensitic phase transitions of the light alkali metals Li and Na from an hcp[3] to a bcc structure have been investigated by *Schneider* and *Stoll* [3.132], *Pynn* and *Ebbsjö* [3.133], *Straub* and *Wallace* [3.134], *Hafner* [3.20] and *Kelly* [3.135]. A general discussion of dynamical effects at polymorphic transitions was given by *Grimvall* [3.136]. A Landau theory for martensitic transitions from a bcc phase at high temperatures to close-packed structures with a variety of stacking sequences at low temperatures was proposed by *Kelly* and *Stobbs* [3.137].

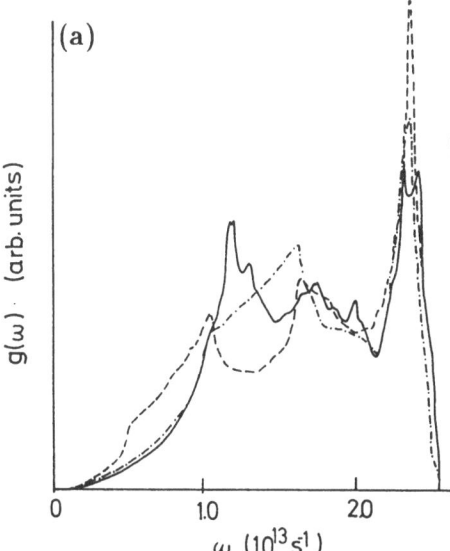

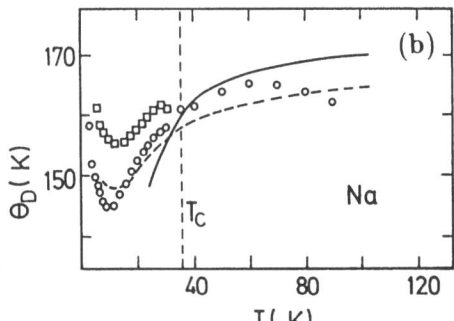

Fig. 3.12. (a) Phonon density of states in hcp (*solid line*), bcc (*dashed line*) and fcc (*dot-dashed line*) sodium from calculations of *Pynn* and *Ebbsjö* [3.133]. (b) Debye temperatures of bcc (*solid line*: calculated, *dots*: experiment) and fcc (*dashed line*: calculated, *squares*: experiment) sodium as a function of temperature, after [3.20]

The phonon density of states calculated by Pynn and Ebbsjö for the three polymorphic modifications of Na is shown in Fig. 3.12. A low-frequency van Hove singularity appears in the bcc frequency spectrum, but is clearly absent in the close-packed structures. The strong weight in the frequency spectrum at low ω stems from the low-lying $[110]$-T_1 mode in the bcc structure (see also Fig. 3.6b); for this mode, the atoms move into the large octahedral holes of the bcc lattice. Anharmonic effects have also been discussed [3.133, 134, 136], and it was concluded that they are not a major cause of polymorphism, although they might have some influence on the quantita-

---

[3] The structure of the low temperature polymorphs of Li and Na was originally described as hcp [3.128, 129]. More recent work [3.130, 131] has shown that the hcp phase is rather poorly developed and never complete. Parts of the low temperature phase might also be fcc with a large number of stacking faults in both structures.

tive prediction of an actual transition temperature. Note that calculations of this type are better based on full phonon calculations – structural effects are rather poorly described by variational calculations with simple reference systems.

Whether or not a transition actually occurs depends on the magnitude of the structural free-energy differences at $T = 0\,\mathrm{K}$. For Li and Na all pseudopotential calculations correctly predict a low-temperature stability of the hcp structure with the fcc phase only slightly higher in energy. The calculated values for Na are – including differences in the zero-point vibrational energies – ΔF (bcc-hcp) $= 23\,\mathrm{cal/g}$ atom [3.20], ΔF (bcc-hcp) $= 19\,\mathrm{cal/g}$ atom [3.133], to be compared with a value of ΔF (bcc-hcp) $= 10\,\mathrm{cal/g}$ atom determined from *Martin's* [3.134] specific heat measurements. This yields quite reasonable values for the transition temperatures: $T_\mathrm{c} = 99\,\mathrm{K}$ [3.133], $T_\mathrm{c} = 125\,\mathrm{K}$ [3.20] and $T_\mathrm{c} = 260\,\mathrm{K}$ [3.134], compared to an experimental value $T_\mathrm{c} \cong 36\,\mathrm{K}$ [3.129], clearly corroborating Zener's picture. The problem is rather that most pseudopotential calculations also predict a stability of an hcp structure for the heavy alkali metals K, Rb, and Cs [3.18, 20, 33, 37, 38] in contrast to the observed bcc structure. The possible reasons for this disagreement have been discussed repeatedly: *s-d* hybridization effects have been suggested as a possible explanation and indeed bcc-stability was predicted by some model-potential calculations [3.48, 139, 140]. However, in view of the much more fundamental calculations of *Moriarty* [3.18], this agreement must be considered as somewhat questionable. It has also been proposed that third- and higher-order contributions might be important [3.140], but the results presented so far are not convincing.

The variation of the transition temperature with pressure has been studied by *Animalu* [3.126], *Moriarty* [3.97], *Pélissier* [3.141] and *Hafner* (unpublished). Figure 3.13 compares these results for the $T_\mathrm{c} = T_\mathrm{c}(p)$ curve with experiment [3.142] for the example of Sr. This shows that at least the solid-state part of a pT phase diagram may be successfully calculated. Mo-

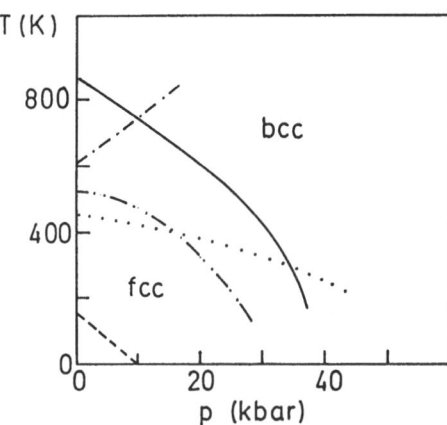

Fig. 3.13. Solid state part of the pT phase diagram of Sr, showing the fcc-bcc transition temperature as a function of pressure. (*Solid line*) experimental, and the results of calculations by Animalu (*dashed line*), Moriarty (*dot-dashed line*), Pelissier (*dot-dot-dashed line*), and Hafner (*dotted line*). See text

riarty also points out the importance of *sd*-hybridization in stabilizing the fcc structure – evidently the appropriate balance between pseudopotential and hybridization terms is very difficult to achieve within the generalized pseudopotential scheme. For a very recent discussion of phase changes in the alkaline-earth metals with further references, see [3.80].

It is clear that the influence of the lattice vibrations on the structural stability should be largest for the lightest elements. Using the self-consistent-phonon method *Straus* and *Ashcroft* [3.142] have shown that a proper inclusion of ion dynamics in the self-consistent-phonon approximation radically alters the structural dependence of the energy of metallic hydrogen, causing lattices with high symmetry to be favoured. In addition they demonstrated that the dynamics reduces the structural sensitivity of the higher-order perturbation terms very markedly.

Recent Results

Pettifor [3.144] has proposed a more accurate parametrization of the pseudoatom self-energy $\frac{1}{2}\Phi_{ind}(R = 0)$, which might be very useful for analyzing trends in binding energies of metals and alloys.

The subject of total energy calculations of the structural properties has been pursued very actively by the Berkeley group [3.145,6] and the Stuttgart group [3.147,8], among many others. We would like in particular, to draw the reader's attention to the detailed work on pressure-induced polymorphic transitions in Si and in Ge [3.145,6] and to the work on the electronic pressures in the noble metals [3.148].

Recent experimental work on the phonon dispersion relations of the heavy alkaline earth metals has stimulated investigations of the influence of the *d*-electrons on their structural and vibrational properties [3.149] and on the vibrational properties close to the martensitic phase transition [3.150]. *Wang* and *Overhauser* [3.151] have attempted to include the influence of the polarization of the outer core shells in a calculation of the lattice vibrations of the very heavy metals.

"Frozen phonon" techniques and the concept of calculation of stresses and forces via the Hellmann-Feynman theorem become more and more important [3.152].

4. Structure and Thermodynamics of Liquid Metals

For a crystal, the atomic structure is completely desribed by a diffraction experiment. The positions of the Bragg reflections specify the reciprocal lattice (and hence the symmetry of the unit cell) and the positions of the atoms within the unit cell may be calculated from the intensities of the Bragg peaks. For a disordered material such as a liquid or a glass, the diffraction pattern consists of diffuse rings with intensity $I(q)$ given by

$$I(q) = f(q)^2 \left\langle \sum_{l,m} \exp\left[i\boldsymbol{q}(\boldsymbol{R}_l - \boldsymbol{R}_m)\right] \right\rangle,$$

where q is the momentum transfer and the average is over all pairs (l, m) of atoms; $f(q)$ is the scattering form factor of an individual atom. Normalizing the intensity $I(q)$ by the number of atoms and by the square of the scattering form factor defines the static structure factor $S(q)$; $S(q)$ is the Fourier transform of the pair correlation function $g(R)$. The latter represents the probability of finding a second atom at a distance R from a given atom:

$$S(q) = N^{-1} \left\langle \sum_{l \neq m} \exp\left[i\boldsymbol{q}(\boldsymbol{R}_l - \boldsymbol{R}_m)\right] \right\rangle - N\delta_{q,0}$$

$$= 1 + 4\pi n \int [g(R) - 1] \exp(i\boldsymbol{q}\boldsymbol{R}) d^3\boldsymbol{R} \quad,$$

where $n = V/N$ is the number density of atoms. Thus for a liquid or a glass, a diffraction experiment yields only a one-dimensional projection of the three-dimensional structure, but this is usually all the structural information that can be obtained.

The central problem of liquid state physics is thus the determination of the pair correlation function $g(R)$ [or of its Fourier transform, the static structure factor $S(q)$] using the methods of classical statistical mechanics. At least for a system described solely in terms of volume and pair forces, the knowledge of $g(R)$ is all we need to know for a calculation of the thermodynamic functions (see e.g. [4.1] and Appendix D.1). Two main approaches are usually employed to achieve this goal: one involves computer simulation, such as the Monte Carlo (MC) and molecular dynamics

(MD) methods; the other approach is semi-analytic and may be divided into two classes: One is based on integro-differential equations derived from the Born-Bogoljubov-Green-Yvon-Kirkwood (BBGYK) hierarchy, relating the pair potential $\Phi(R)$ and the pair correlation function $g(R)$. The most frequently used examples are the Percus-Yevick (PY) and the hypernetted-chain (HNC) equations and the mean-spherical approximation (MSA). The other approach is based on thermodynamic perturbation theory relative to some reference system (essentially hard spheres or a one-component plasma) for which accurate solutions – which must also be sufficiently easy to handle (if possible they should be analytic) – are known.

Note that there is no *exact* theory of the liquid state – all existing theories necessarily involve some approximation. Therefore computer simulations are important in two respects: (i) for testing the reliability of interatomic potentials and (ii) as a test of the liquid state theories for simple model systems (e.g. Lennard-Jones or square-well fluids).

4.1 Computer Simulations

In recent years a number of authors have used computer simulation techniques to study the properties of liquid metals. (There is no room to discuss the technical aspects of computer-simulations – we refer the interested reader to the specialized literature [4.2–5].) The pioneering work of *Pasquin* and *Rahman* [4.6] has been followed by a large number of calculations for all the alkali metals [4.7–16] and for aluminum [4.17–19]. As an example we show in Fig. 4.1 a comparison between the simulated $S(q)$ for Li and Al and the experimental $S(q)$ from diffraction experiments, both calculations are based on pair potentials of the Dagens-Rasolt-Taylor [4.20] type. The

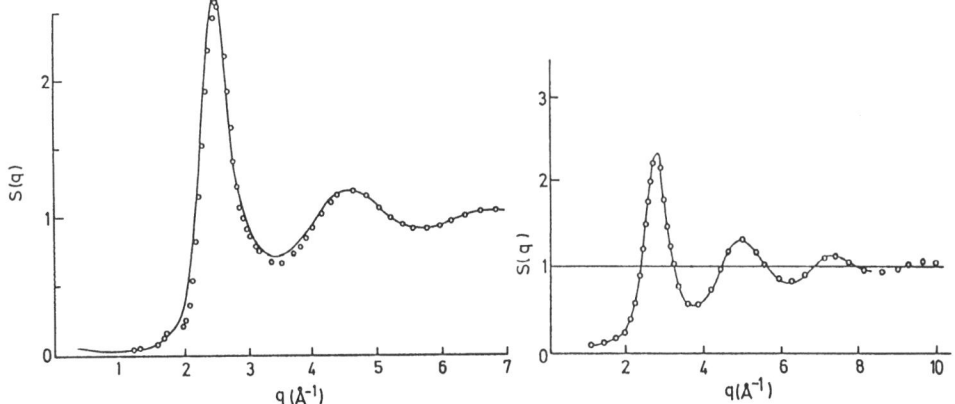

Fig. 4.1a,b. The static structure factor (a) for liquid Li and (b) for liquid Al determined by molecular dynamics (*circles*). The full curves show the experimental results. [4.10, 19]

agreement with experiment can certainly be considered as excellent – except perhaps for very low q. This shows us that a pair-potential description based on pseudopotentials and linear screening is undoubtedly accurate for the liquid simple metals. The limited accuracy of the MD $S(q)$ for low q stems from the need to use a finite "sample" (typically at most a thousand atoms) in the computer simulation. This introduces boundary effects and termination corrections into the calculated static structure factor.

Another important application of coumputer simulation is in testing the various integral equations and perturbation methods proposed in liquid state theory.

4.2 Integral Equation Approach

Integro-differential equations relating the interatomic potential $\Phi(R)$ and the pair correlation function $g(R)$ may be derived by two different techniques: (a) by differentiating the equation for $g(R)$ within the canonical ensemble one arrives at [4.1], see also eqs. (1.71–74)

$$
\begin{aligned}
\nabla_1 g(R_{12}) = & - \beta g(R_{12}) \nabla_1 \Phi(R_{12}) \\
& - \beta n \int g^{(3)}(R_1, R_2, R_3) \nabla_1 \Phi(R_{13}) d^3 R_3
\end{aligned}
\tag{4.1}
$$

and since the last term contains the three-particle correlation function this is only the first term of a hierarchy of equations (the BBGYK hierarchy) relating n-particle correlation functions to $(n+1)$ particle correlations functions. The closure of such equations, that is, the imposition of additional physical approximations relating higher correlation functions to lower, results in integro-differential equations for a single distribution function. The most elementary closure, the superposition relation

$$
g^{(3)}(R_1, R_2, R_3) = g(R_{12})g(R_{23})g(R_{31})
\tag{4.2}
$$

yields the Born-Green equation for the pair-correlation function – many other different closures have been discussed in the literature [4.1].

(b) The alternative approach starts from the Ornstein-Zernike equation

$$
h(R) = c(R) + n \int c(R')h(|R - R'|)d^3 R'
\tag{4.3a}
$$

or in Fourier space

$$
\hat{h}(q) = \frac{\hat{c}(q)}{1 - n\hat{c}(q)} \quad ,
\tag{4.3b}
$$

which decomposes the total correlation function $h(R) = g(R) - 1$ into the direct correlation function $c(R)$ and the totality of the indirect effects mediated by a third particle which is itself correlated to the original two. The point is that structure in $h(R)$ is usually much more long-ranged than that of $\Phi(R)$, hence the desire to introduce a correlation function $c(R)$, which has a range representative of $\Phi(R)$. Again the Ornstein-Zernike approximation must be supplemented by a closure relating $c(R)$ and $h(R)$. Usually one proceeds by writing $h(R)$ as an infinite series of integrals over the Mayer functions [4.1] $f(R_{ij}) = \exp\left[-\beta\Phi(R_{ij})\right] - 1$. A diagrammatic language has been developed to facilitate the burdensome combinatorial analysis. The Ornstein-Zernike equation is used to derive a diagrammatic expression for $c(R)$ (see Appendix D.2). Approximate expressions for $c(R)$ correspond to the neglect of certain topological classes of diagrams. Eliminating the very complicated "bridge" diagrams gives the HNC-closure [4.21].

$$c(R) = h(R) - \beta\Phi(R) - \ln g(R) \quad , \tag{4.4}$$

while neglecting a larger class of diagrams produces the PY approximation (*Percus* and *Yevick* [4.22])

$$c(R) = \{1 - \exp[\beta\Phi(R)]\}g(R) \quad . \tag{4.5}$$

From comparison with computer simulations one learns that the PY approximation is far superior to the HNC approach for hard-sphere fluids or other systems with strongly repulsive potentials, while the HNC approximation is considerably better for charged systems at low to moderate density[1] [4.23]. For other model systems such as the square-well fluid [4.24,25] or the Lennard-Jones fluid [4.26] it has been shown that neither the PY nor the HNC approach provide acceptable theories (for both systems thermodynamic perturbation theories are shown to be superior to the PY and HNC theories). *Taylor* and *Watts* [4.27] draw the same conclusion for liquid metals, based on comparisons with the computer simulations of *Jacucci* et al. [4.10,19].

Nevertheless, the PY and HNC equations have been applied by a number of workers [4.28–32], with moderate to reasonable agreement with experiment for the PY equation which turns out to be more appropriate to liquid metals then the HNC equation [4.31], but the author is unable to see the superiority claimed by *Bratkovsky* et al. [4.30] for the PY equation over the perturbation-theoretic approaches.

A new approach proposed by *Rosenfeld* and *Ashcroft* [4.33] seems to be very promising. They recognized that within the accuracy of present-

---

[1] The thermodynamic functions may be derived from $g(R)$ via one of the equations of state (pressure, energy or compressibility equation, see Appendix D). If the exact $g(R)$ were known the result should be the same either way. But as one calculates only an approximate $g(R)$, the internal consistency of the thermodynamic functions determined in different ways is an important test for any theory.

day computer simulations, the bridge functions (i.e. the sum of the bridge diagrams, assumed zero in the HNC approach) are universal characteristics, irrespective of the assumed pair potential. Thus they can be derived from the known parametrized results for hard spheres and this yields a modified HNC equation including bridge-function corrections. Rosenfeld and Ashcroft have shown that this new method yields good results for a wide variety of model systems, but a systematic application to liquid metals is still lacking.

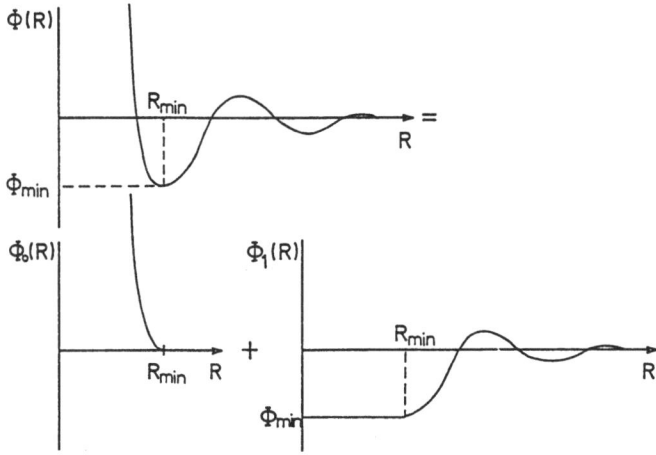

Fig. 4.2. Separation of the full interatomic potential $\Phi(R)$ into a repulsive part $\Phi_0(R)$ and a long range part $\Phi_1(R)$ according to (4.6)

Given the state of the art of the integral-equation approach, the philosophy behind the thermodynamic perturbation approach, namely to divide the full interatomic potential $\Phi(R)$ into a steeply repulsive short-ranged term $\Phi_0(R)$ and longer ranged part $\Phi_1(R)$ (see also Fig. 4.2) according to

$$\Phi(R) = \Phi_0(R) + \Phi_1(R) \quad \text{and} \tag{4.6a}$$

$$\Phi_0(R) = \begin{cases} \Phi(R) - \Phi(R_{\min}) & R < R_{\min} \\ 0 & R > R_{\min} \end{cases} \tag{4.6b}$$

$$\Phi_1(R) = \begin{cases} \Phi(R_{\min}) & R < R_{\min} \\ \Phi(R) & R > R_{\min} \end{cases} \tag{4.6c}$$

and to treat each part separately with the most appropriate method seems to be more promising. An integral equation closely related to thermodynamic perturbation theory which has been applied successfully to a variety of systems is the mean-spherical approximation (MSA) [4.34, 35]. For a system with a continuous reference potential $\Phi_0(R)$, the MSA closure reads [4.35]

$$c(R) = \{1 - \exp[\beta\Phi_0(R)]\}g(R) - \beta\Phi_1(R) \quad . \tag{4.7}$$

The MSA is mostly used in conjunction with a hard-sphere reference system. In that case the closure relation reduces to (σ is the hard-sphere diameter)

$$c(R) = \begin{cases} -\beta \Phi_1(R) & R > \sigma \\ -\beta \Phi_1^*(R) & R < \sigma \end{cases} , \qquad (4.8)$$

where $\Phi_1^*(R)$ is a renormalized potential chosen to satisfy the condition $g(R) = 0$ for $R < \sigma$. The MSA will be discussed in more detail in conjunction with thermodynamic perturbation theory. Very recently mixed closures (switching from an MSA or PY form at short to a HNC form at large distances) have been proposed. They are discussed in Appendix D.

4.3 Thermodynamic Perturbation Theories

4.3.1 Thermodynamic Variational Method for Liquids (Gibbs-Bogoljubov-Method)

The simplest possible thermodynamic perturbation theory is a variational method based on the Gibbs-Bogoljubov inequality (cf. Sect. 3.5.3). The free energy is expressed in terms of the free energy F_0 of a reference system plus a perturbation term $\langle V \rangle_0$ which represents the difference between the Hamiltonians of the real and reference systems evaluated with respect to the distribution functions of the latter; the sum $F_0 + \langle V \rangle_0$ is then an upper bound of F,

$$F \leq F_0 + \langle V \rangle_0$$
$$= F_0 + E_V + 2\pi n \int \Phi(R) g_0(R) R^2 dR$$
$$= F_0 + E_0 + E_1 + \frac{1}{2\pi^2 n} \int \hat{\Phi}(q) S_0(q) q^2 dq \qquad (4.9)$$

in either a real- or reciprocal-space formulation [$\hat{\Phi}$ is the total interatomic potential in q-space, $\hat{\Phi}(q) = (8\pi Z^2/V_a q^2) + 2F(q)$, see Chap. 2]. This permits the development of a variational principle analogous to (3.27) where the parameters of the reference system are the variational parameters.

The next step must be the choice of an appropriate reference system. It was first pointed out by *Ashcroft* and *Lekner* [4.36] and *Mansoori* and *Canfield* [4.37] that a hard-sphere system might be a good reference system for classical liquids: analytical expressions for the static structure factor are available from the solution of th PY-equation [4.36] and the free energy is available analytically as a fit to an exact machine calculation [4.38], see Appendix D for details. The variational parameter is the hard-sphere diameter σ or the packing fraction $\eta = \pi \sigma^3 n/6$. The hard-sphere variational method was first applied to liquid metals by *Jones* [4.39] and *Stroud* and *Ashcroft* [4.40]; for an application to a wider series of metals and further

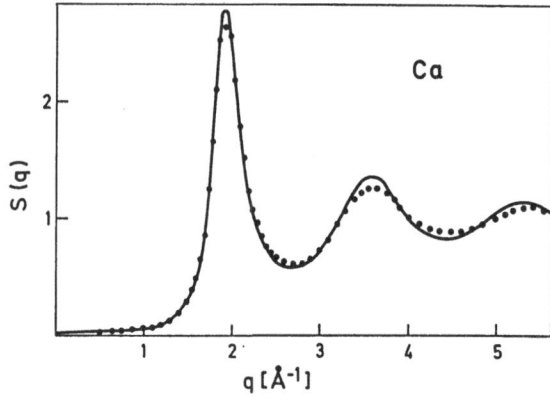

Fig. 4.3. Static structure factor for liquid Ca. (*Full line*) hard-sphere variational calculation, (*dots*) experiment (after [4.44])

Table 4.1. Internal energy E, excess entropy S_E, isothermal bulk modulus B_T, and thermal expansion coefficient α for some liquid simple metals close to the melting point, caculated with the HS variational method (after [4.42]

Metal	T [K]		E [kcal/g-atom]	S_E/k_B	B_T [Mbar]	α [10^{-4} K^{-1}]
Li	453	calc.	171.36	-3.75	0.081	3.9
		exp.	164.18	-3.61	0.115	
Na	373	calc.	149.97	-3.47	0.037	5.2
		exp.	145.43	-3.45	0.053	2.8
Mg	938	calc.	548.82	-4.10	0.216	1.71
		exp.	563.55	-3.45	0.204	1.66
Al	938	calc.	1292.06	-4.18	0.374	1.19
		exp.	1309.77	-3.60	0.413	1.22

references see [4.41–46]. As an example we show in Fig. 4.3 the variationally determined HS structure for liquid Ca, compared with experiment. It has been shown [4.45] that the variationally determined hard-core diameter has a very clear physical significance: it obeys the simple relation $\Phi(\sigma) = \Phi_{min} + \langle E_{kin} \rangle$ (Φ_{min} is the value of the pair potential at the first minimum and $\langle E_{kin} \rangle = 1.5\,k_B T$ is the average kinetic energy of an ion), hence it represents a mean collision distance. In Table 4.1 a few representative thermodynamic data for simple metals are compiled. We find that the agreement is excellent for Mg and Al, good for Li, but only fair for Na. Clearly this is connected with the difference in the interatomic potentials – the polyvalent metals Mg and Al have rather steeply repulsive pair potentials (but they are not as "hard" as Lennard-Jones potentials) whereas those for the alkali metals are much softer and a HS-reference system is less appropriate (see Fig. 4.4).

Ross et al. [4.47] have proposed to use a one-component plasma (OCP) as the reference system. The OCP is a liquid of positive point ions interact-

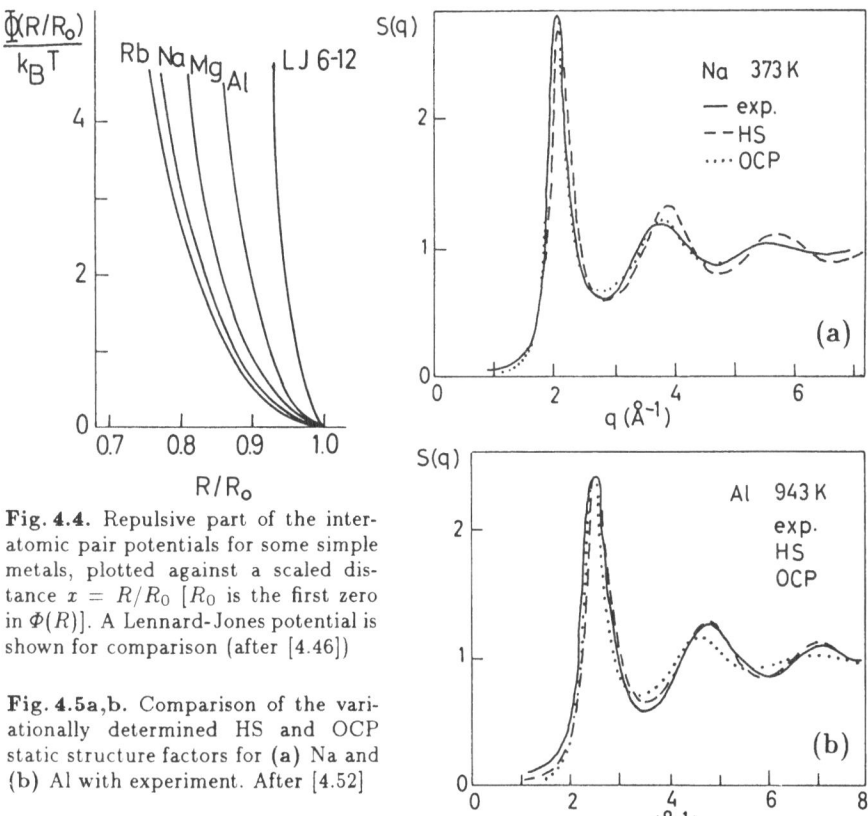

Fig. 4.4. Repulsive part of the interatomic pair potentials for some simple metals, plotted against a scaled distance $x = R/R_0$ [R_0 is the first zero in $\Phi(R)$]. A Lennard-Jones potential is shown for comparison (after [4.46])

Fig. 4.5a,b. Comparison of the variationally determined HS and OCP static structure factors for (a) Na and (b) Al with experiment. After [4.52]

ing via a pure Coulomb potential in a negative compensating background. The OCP has been studied extensively by Monte Carlo simulation and accurate fits to the MC data for the thermodynamic functions are available [4.48,49]; tables of the static structure factor $S(q)$ are available from accurate solutions of the HNC [4.50] and HNC with bridge function corrections [4.51]. The variational parameter in this model is the plasma parameter $\Gamma = (2Z^2)/R_a k_B T$. A comparative study of the HS versus OCP reference systems for Na and Al has been presented by *Mon* et al. [4.52]. They showed that for liquid Na the OCP yields a lower variational upper bound to the free energy and a better description of the static structure than the HS system, but that for Al the situation is just the reverse (see Fig. 4.5). This conclusion is also corroborated by the recent work of *Moriarty* et al. [4.53], so in general the applicability of the OCP seems to be restricted to the alkali metals.

For metals with a somewhat harder pair potential *Ross* [4.54] has proposed the so-called soft-sphere (SS) reference system. He keeps the HS form of the static structure factor, but parametrizes the free energy according to the MC simulation of the R^{-12} potential. For Al, the SS reference yields a lower variational upper bound to F than both the HS and OCP calculations [4.53], but it seems, however, to lack internal consistency.

The ever increasing number of models for which analytic solutions are available (charged hard spheres with bare or screened Coulomb interactions [4.55] inter alia) offers the possibility of constructing variants of the variational methods. However, at present this approach is still at the level of multiparameter fits [4.56, 57].

4.3.2 Repulsive Forces: Weeks-Chandler-Andersen Theory

Weeks, Chandler and *Andersen* (WCA) [4.58–60] have proposed a functional Taylor expansion of the free energy of a system with a soft repulsive interaction $\Phi_0(R)$ in terms of the "blip function" $B_\sigma(R)$

$$B_\sigma(R) = y_\sigma(R)\{\exp[-\beta\Phi_0(R)] - \exp(-\beta\Phi_\sigma)\} \quad \text{with} \qquad (4.10)$$

$$y_\sigma(R) = g_\sigma(R)\exp(\beta\Phi_\sigma)$$

measuring the departure of the actual potential Φ_0 from a HS reference potential Φ_σ with diameter σ [$g_\sigma(R)$ is the paircorrelation function of this reference system] (Fig. 4.6). Thus

$$F = F_0 - \frac{n}{2}k_\mathrm{B}T \int B_\sigma(R)d^3R + \dots \quad . \qquad (4.11)$$

The HS diameter is determined by the WCA condition

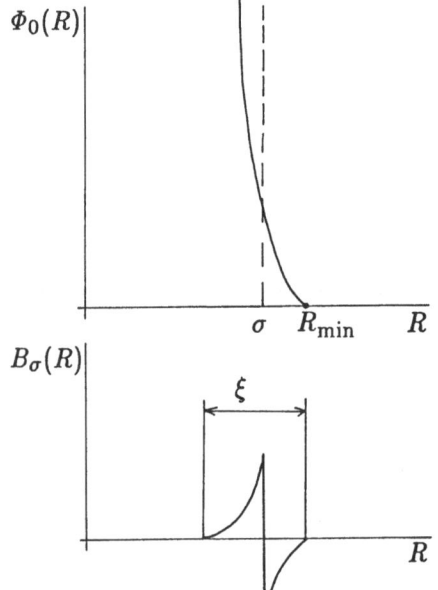

Fig. 4.6. The "blip function" $B_\sigma(R)$ measuring the departure of a soft repulsive potential $\Phi_0(R)$ from an effective hard-sphere potential Φ_σ. The spatial extension ξ of the blip function measures the softness of the potential

$$\int B_\sigma(R) d^3R = 0 \qquad (4.12)$$

in order to optimize the convergence of the functional Taylor series. Thus the WCA theory yields

$$F = F_0 \quad \text{and} \qquad (4.13a)$$

$$F_0 = F_\sigma + E_V + 2\pi n \int_0^\infty \Phi_1(R) g_\sigma(R) R^2 dR$$

$$= F_\sigma + E_0 + E_1 + \frac{1}{2\pi^2 n} \int_0^\infty \hat{\Phi}_1(q) S_\sigma(q) q^2 dq \qquad (4.13b)$$

in either a real- or reciprocal-space formulation $[\hat{\Phi}_1 = (8Z^2/V_a q^2) + 2F(q)]$.

WCA show that with condition (4.12) the expansion (4.11) is correct up to terms of fourth order in the "softness parameter" ξ which measures the range over which $B_\sigma(R) \neq 0$ (see Fig. 4.6). The pair correlation function and the static structure factor are given by

$$g_0(R) = y_\sigma(R) \exp\left[-\beta\Phi_0(R)\right] \qquad (4.14)$$

$$S_0(q) = S_\sigma(q) + n\hat{B}_\sigma(q) \quad . \qquad (4.15)$$

Jacobs and *Andersen* [4.61] and *Telo da Gama* and *Evans* [4.62] have shown that for low q, (4.15) is advantageously replaced by

$$S_0(q) = S_\sigma(q)[1 - nS_\sigma(q)\hat{B}_\sigma(q)] \quad . \qquad (4.16)$$

This corresponds to a partial summation of higher-order terms. The WCA theory has been thoroughly tested for a number of model systems [4.26] and it certainly constitutes an important improvement over a simple HS variational calculation. The WCA approach was applied to liquid metals by *Wehling* et al. [4.63], *Hasegawa* and *Watabe* [4.64], *Hasegawa* [4.65] and *Kumaravadivel* and *Evans* [4.46], among others.

Figure 4.7 shows that the WCA corrections not only give a more realistic shape of the first peak in the pair-correlation function, but also provide the necessary damping of the large-q oscillations in $S(q)$. The results compiled in Table 4.2 show that the accuracy of the thermodynamic functions is also improved (see Table 4.1 for the corresponding variational results). However, there is still a slight difference between the computer-simulated $S(q)$ and the WCA approach in form of a small phase shift in the oscillations. As we shall see shortly, this is the signature of the attractive part of the interatomic potential.

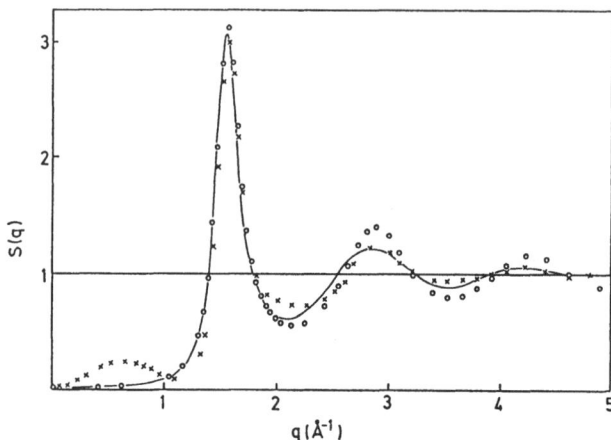

Fig. 4.7. Static structure factor for liquid Rb, calculated using the HS variational method
($\cdots$) and the WCA technique ($\times$ $\times$ $\times$). The full curve is the molecular dynamics result
using the same potential. After [4.46]

Table 4.2. Excess entropies of simple liquid metals, calculated in th WCA approximation
(after [4.46])

Metal	$-S_E/k_B$	
	theor.	exp.
Na	3.61	3.45
K	3.84	3.45
Rb	3.73	3.63
Cs	3.78	3.56
Mg	3.68	3.45
Zn	3.80	3.78
Al	3.42	3.60

The work of *Hasegawa* [4.65] is worth mentioning, because it constitutes the only serious attempt to investigate the influence of third-order
terms in the electron-ion pseudopotential on the liquid structure and thermodynamics. However, since – at least at the level of the HS-variational and
WCA theories – one can handle only central pair interactions, he considers only the third-order contribution to the pair potential [see (2.28)] and
neglects the genuine three-body contributions. It is therefore not surprising
that he finds that the inclusion of third-order terms does not improve the
accuracy of the calculations. This is just another illustration of the fact that
the separation of higher-order terms into pairwise and non-pairwise potentials is not effective: the two-body and three-body contributions to the third
order energy are of the same order of magnitude and largely cancel. Again
we find that it is more advantageous to "fold" higher-order contributions
down to lower order by optimizing the pseudopotential.

4.3.3 Attractive Forces: Random Phase Approximation, Optimized Random Phase Approximation and Mean Spherical Approximation

Up to now, the attractive forces have entered into the theory only as a perturbative correction to the free energy; their role in determining the liquid structure has not been considered. A more complete theory is based on a decomposition of the direct correlation function $c(R)$ and the total correlation function $h(R)$ analogous to (4.6), i.e.

$$c(R) = c_0(R) + c_1(R) \tag{4.17}$$

$$h(R) = h_0(R) + h_1(R) \quad , \tag{4.18}$$

where $c_0(R)$ and $h_0(R)$ are those parts of the correlation functions which are determined by the repulsive forces alone. For a soft repulsive potential, $h_0(R)$ can be replaced by an effective hard-core correlation function $h_\sigma(R)$ by means of a WCA expansion.

This leaves us with the problem of finding an appropriate solution for $h_1(R)$ and $c_1(R)$. Introducing (4.14) into the Ornstein-Zernike equation (4.3) and subtracting the purely repulsive couplings – which are solutions of the OZ equation

$$h_\sigma(R) = c_\sigma(R) + n \int c_\sigma(R') h_\sigma(|\boldsymbol{R} - \boldsymbol{R'}|) d^3 \boldsymbol{R'} \tag{4.19}$$

– yields a residual OZ equation for $c_1(R)$ and $h_1(R)$ of the form [4.66]

$$
\begin{aligned}
h_1(R) = c_1(R) \ &+ n \int c_\sigma(R') h_1(|\boldsymbol{R} - \boldsymbol{R'}|) d^3 \boldsymbol{R'} \\
&+ n \int c_1(R') h_\sigma(|\boldsymbol{R} - \boldsymbol{R'}|) d^3 \boldsymbol{R'} \\
&+ n \int c_1(R') h_1(|\boldsymbol{R} - \boldsymbol{R'}|) d^3 \boldsymbol{R'}
\end{aligned}
\tag{4.20}
$$

or in Fourier space

$$\hat{h}_1(q) = \frac{S_\sigma(q) \hat{c}_1(q) S_\sigma(q)}{1 - n\hat{c}_1(q) S_\sigma(q)} \quad . \tag{4.21}$$

The next step must be the determination of a closure condition for this residual OZ equation. The simplest possible choice is the random phase approximation (RPA) which postulates the assymptotic low-density limit of the direct correlation function,

$$c_1(R) = -\beta \Phi_1(R) \tag{4.22}$$

as the closure condition. From (4.18) the total correlation function may be calculated directly. The essential physical assumption of the RPA is that the interference of all correlations with third particles is destructive at *all* distances. It is therefore not surprising that the RPA yields a pair-correlation function which is unphysical at short distances − it is non-zero inside the hard core. *Andersen* et al. [4.67–69] proposed to replace the RPA by the optimized random phase approximation (ORPA) defined by

$$c_1(R) = -\beta\Phi_1(R) \quad R>\sigma \tag{4.23a}$$

$$h_1(R) = 0 \quad\quad\quad R<\sigma \tag{4.23b}$$

or equivalently

$$c_1(R) = -\beta\Phi_1^*(R) = \begin{cases} -\beta\Phi_1(R) & R>\sigma \\ -\beta\psi(R) & R<\sigma \end{cases} \tag{4.24}$$

with the optimized potential $\Phi_1^*(R)$ [$\psi(R)$ is some arbitrary function determined by the condition that $g_1(R)$ vanishes in the hard-core region]. If $c_\sigma(R)$ is taken to be the analytical solution of the HS-PY equation, the ORPA is entirely equivalent to the mean spherical approximation (MSA) of *Waisman* and *Lebowitz* [4.34, 55]; see (4.7,8). If more accurate semiempirical descriptions of the HS reference system are used [Andersen et al. use the parametrized $g_\sigma(R)$ of *Verlet* and *Weis* [4.70] and the technique of *Grundke* and *Henderson* [4.71] to construct the corresponding $c_\sigma(R)$], then the MSA and the ORPA are no longer equivalent. Introducing the closure relation into the residual OZ equation, we find that $h_1(R)$ and $\Phi_1^*(R)$ are related through the integral equation

$$h_1(R) = -\frac{\beta}{(2\pi)^3} \int e^{i\boldsymbol{q}\cdot\boldsymbol{R}} \frac{S_\sigma^2(q)\hat{\Phi}_1^*(q)}{1 - nS_\sigma(q)\hat{\Phi}_1^{*2}(q)} d^3q \quad . \tag{4.25}$$

The optimization condition [$h_1(R) = 0$ for $R<\sigma$] represents an integral equation for $\Phi_1^*(R)$. It is conveniently replaced by an equivalent, but simpler variational condition for the RPA contribution to the free energy [4.69] (see Appendix D for a general discussion of the liquid state integral equations and their equivalent variational principles)

$$F_{\text{ORPA}} = \frac{1}{(2\pi)^3} \int \{n\beta S_\sigma(q)\hat{\Phi}_1^*(q) - \ln\left[1 + n\beta S_\sigma(q)\hat{\Phi}_1^*(q)\right]\}d^3q \tag{4.26}$$

with respect to Φ_1^*. The functional derivative of F_{ORPA} is just

$$\frac{\delta F_{\text{ORPA}}[\hat{\Phi}_1^*]}{\delta\hat{\Phi}_1^*(q)} = -\frac{n\beta}{(2\pi)^3}\hat{h}_1(q) \tag{4.27a}$$

and after Fourier transforming one obtains

$$\frac{\delta F_{\mathrm{ORPA}}[\Phi_1^*]}{\delta \Phi_1^*(R)} = -nh_1(R) \quad . \tag{4.27b}$$

Combined with (4.23) and (4.24) we find that the correct behaviour of the optimized potential is to make the free energy F stationary with respect to changes in $\Phi_1^*(R)$ for $R < \sigma$.

Using the solution of (4.27) and taking (4.12–15) to describe the effect of the soft repulsive forces we arrive at the following final result for the free energy and the static structure factor

$$F = F_\sigma + E_{\mathrm{V}} + 2\pi n^2 \int \Phi_1(R) g_\sigma(R) R^2 dR$$

$$- \frac{1}{4\pi^2 \beta n} \int \{S_\sigma(q) n\beta \hat{\Phi}_1^*(q) - \ln [1 + n\beta S_\sigma(q)\hat{\Phi}_1^*(q)]\} d^3 R \tag{4.28}$$

and

$$S(q) = S_\sigma(q) + n\hat{B}_\sigma(q) + n\hat{h}_1(q) \tag{4.29}$$

[with $\hat{h}_1$ given by (4.25)].

The application of this technique to systems with continuous potentials poses a problem of self-consistency. The solution of (4.27) defines an optimized attractive potential $\Phi_1^*(R)$ given by

$$\Phi_1^*(R) = \begin{cases} \Phi(R) & R > R_{\mathrm{min}} \\ \Phi(R_{\mathrm{min}}) & \sigma < R < R_{\mathrm{min}} \\ \Phi_1^*(R) & R < \sigma \end{cases} \quad . \tag{4.30}$$

As the total interaction is still described by $\Phi(R)$, we are forced to define a new repulsive potential $\Phi_0^*(R)$ via the relation

$$\Phi(R) = \Phi_0^*(R) + \Phi_1^*(R) \tag{4.31}$$

and the WCA + ORPA procedure has to be iterated until a self-consistent separation of the potential has been achieved. This aspect of the problem had been noticed by *Andersen* and *Chandler* [4.67], but for the rather harshly repulsive Lennard-Jones potential the change of the potentials upon iteration turned out to be numerically insignificant. This is entirely different for the rather soft metallic repulsions where the iteration to self-consistency turns out to be essential.

Summing up higher-order diagrams yields to the so-called optimized cluster theory (OCT). The OCT expression for the pair-correlation function reads $g(R) = g_0(R) \exp [h_1(R)]$ and is sometimes referred to as the exponential approximation. For details we refer the reader to the review of *Andersen* et al. [4.26], and also to [4.25]. The main advantage of the OCT over the ORPA is to improve the thermodynamic consistency.

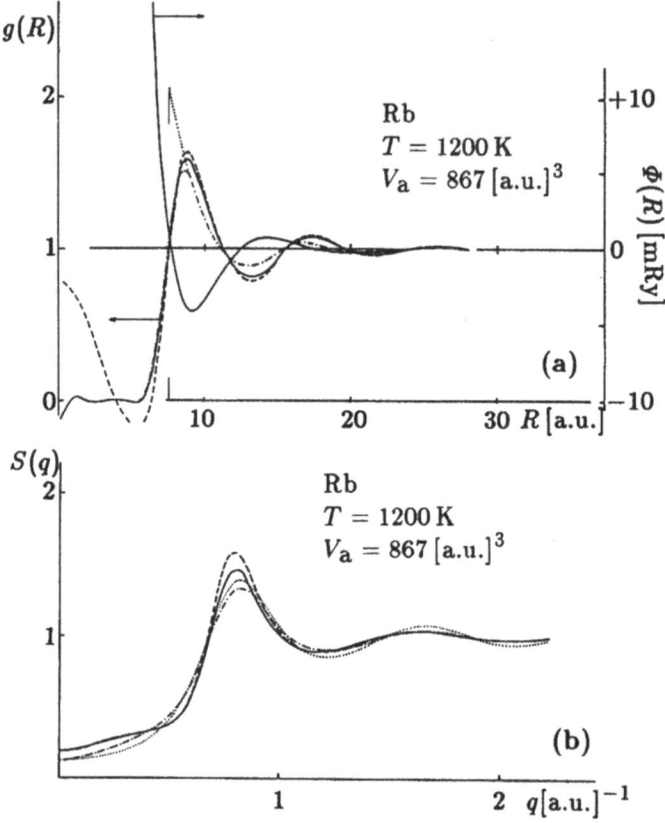

Fig. 4.8. (a) Interatomic potential $\Phi(R)$ and pair correlation function $g(R)$ for liquid Rb at $T = 900\,\mathrm{K}$. (*Dotted line*) hard-sphere reference $g(R)$, (*dot-dashed line*) WCA, (*dashed line*) RPA, (*full line*) ORPA. (b) Static structure factors, same symbols (after [4.77])

The ORPA has been applied to a variety of models, such as the Lennard-Jones potential [4.26] and the square-well fluid [4.24, 25] and it was found to be an important improvement over the PY and HNC approaches. Applications to metallic systems are still rather rare [4.72–79]. The effect of the attractive interactions on the liquid structure is seen very clearly if we consider the successive HS, WCA, RPA and ORPA approximations to the structure of a liquid metal (Fig. 4.8): the HS model is a reasonable first approximation, the WCA approximation brings the rounding of the first-peak of $g(R)$ (which is otherwise left unchanged) – in the static structure factor this is reflected by a damping of the large-q oscillations (see Fig. 4.8). The attractive forces bring an important modification of the correlation function: compared to the WCA approximation, $g(R)$ is enhanced at distances where $\Phi(R)$ is attractive and reduced where $\Phi(R)$ is repulsive. In $S(q)$ this appears in the form of a sharper first peak and a slight phase shift of the high-q peaks. Both effects improve the agreement with experiment. Thus we find that the effect of the oscillatory potential is to order the liquid to a greater extent than in the absence of the Friedel oscillations by forcing the atoms into the attractive wells of the potential. The same conclusion had been reached by *Cummings* et al. [4.80,81] on the basis of analytic solutions of the MSA for a model liquid metal potential with an oscillatory tail. These effects are correctly described by the ORPA, but over-emphasized by the RPA[2], because the latter neglects the interference between the ordering tendency enforced by the oscillatory potential and the geometrical packing requirements.

This last point is particularly important for the correct description of polyvalent metals. Many of them display features in the static structure factors – asymmetry of the first peak, high angle shoulders to this peak, intermediary maxima – which cannot be accommodated by a HS model (see *Waseda* [4.83] for a review of experimental results). As an example we show in Fig. 4.9 the pair correlation function and the static structure factor of liquid Ge. Solid Ge is a semiconductor with an open covalent structure, the diamond structure with coordination number $N_c = 4$ at normal pressure and the white-tin structure with $N_c = 6$ at elevated pressure. Liquid Ge is metallic, but the coordination number determined by integrating over the symmetric part of the first peak of $g(R)$,

$$N_c = 2n \int\limits_0^{R_{\max}} g(R)4\pi R^2 dR \qquad (4.32)$$

is about $N_c = 6.5$ (at $T = 1250\,\mathrm{K}$ [4.84]), in contrast to $N_c \sim 10$–11 for the "normal" liquid metals like Na, Mg or Al. This fact has often been interpreted in terms of microcrystallite models, assuming a certain fraction of the Ge atoms to form tetrahedrally coordinated "molecular units" or "associates" (this choice is not obvious – as the volume contracts upon melting, the high-pressure polymorph would seem to be a more appropriate reference). The results shown in Fig. 4.9 give a much more natural explanation of the structure of liquid Ge: the repulsive slope sets a hard-sphere diameter which yields an approximately correct position of the first peak of $g(R)$, but otherwise the HS model is a rather poor approximation to the actual

---

[2] The RPA is least reliable for strongly attractive potentials with deep minima and at temperatures close to the melting point. In this case the RPA produces very strong peaks in $S(q)$, which are appropriately damped in the ORPA. In some cases the RPA produces even a divergence of $S(q)$ near Q_p – a phenomenon which has been discussed under the name of the "RPA catastrophe" [4.82]. Note that the RPA catastrophe does not represent a real physical phenomenon, but simply reflects the fact that the RPA is no longer a good starting approximation for the solution of the ORPA or MSA. Because the relation $c(R) = -\beta\Phi(R)$ has an asymptotic validity (see e.g. [4.1]), it has often been claimed that the RPA $S(q)$ is exact in the limit $q \to 0$, which is demonstrably wrong (see later in this section).

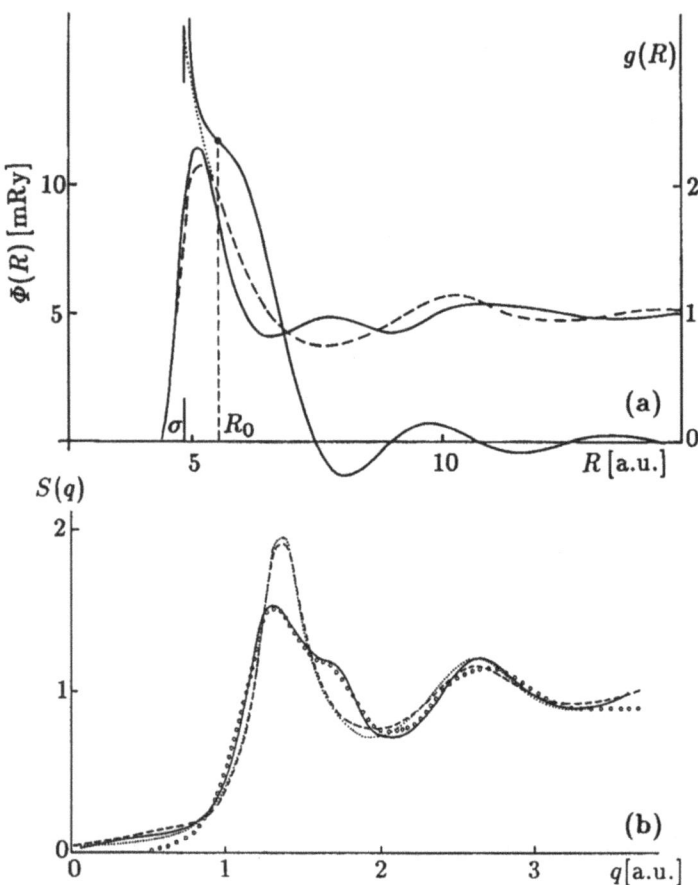

Fig. 4.9a,b. The structure of liquid germanium at a temperature close to its melting point ($T = 1250\,\mathrm{K}$). (a) Pair potential $\Phi(R)$ (*left hand scale*) and pair correlation function $g(R)$ (*right hand sale*). The circle on the curve representing $\Phi(R)$ marks the inflection point R_0 used to separate $\Phi(R)$ into a short-range repulsive part $\Phi_0(R)$ and a long-range oscillating part $\Phi_1(R)$. (b) Static structure factor. (*Dotted line*) HS approximation, (*dashed line*) WCA approximation, (*solid line*) ORPA and potential, (*circles*) experiment (after [4.78, 79])

structure. The WCA description brings no essential improvement. The essential corrections come from the oscillatory potential – again we find that $g(R)$ is enhanced at distances where $\Phi(R)$ is attractive and reduced where $\Phi(R)$ is repulsive. The intermediary maximum in $g(R)$ near $R = 7.6$ a.u., which is so typical for the molten tetravalent elements is a direct consequence of this effect. Thus we find that in real space the structure of molten Ge is determined by two characteristic distances: the effective hard-core diameter σ expressing the geometrical requirements of sphere packing, and the Friedel wavelength $\lambda_\mathrm{F} = 2\pi/2k_\mathrm{F}$ setting the periodicity preferred by the intermediate- and long-range potential. The complex structure of many polyvalent metals results from the interplay of these two length scales – in the static structure factor they appear as the peak near $q \sim 2\pi/\sigma$ and the high-q shoulder at $q \sim 2k_F$. The fact that these shoulders appear preferentially near $q \sim 2k_F$ has already been pointed out by *Oberle* and *Beck* [4.74].

The structure of the liquid noble metals has been discussed by *Regnaut* et al. [4.75] on the basis of the generalized model potential and ORPA approaches. They show that both the *s-d* hybridization and the overlap interactions reduce the short-range attractions (see Fig. 4.10a) resulting in a reasonable agreement of the calculated structure factor with experiment (Fig. 4.10b).

Before we turn to a more general discussion of liquid structures, we would like to comment on an interesting attempt to combine the RPA and ORPA with a reference system having a continuous potential, namely the OCP. *Tosi* and coworkers [4.85–88] start from the explicit expression of the interionic potential in q-space [see (2.21, 22)]

$$\hat{\Phi}(q) = \frac{8\pi Z^2}{V_a q^2} + 2F(q) \quad , \tag{4.33}$$

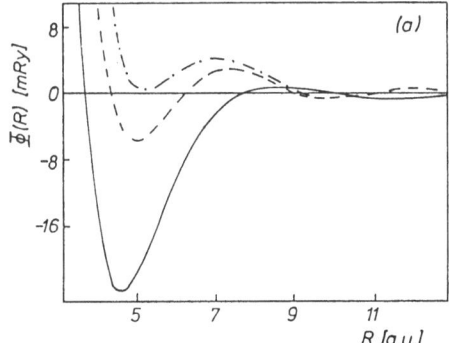

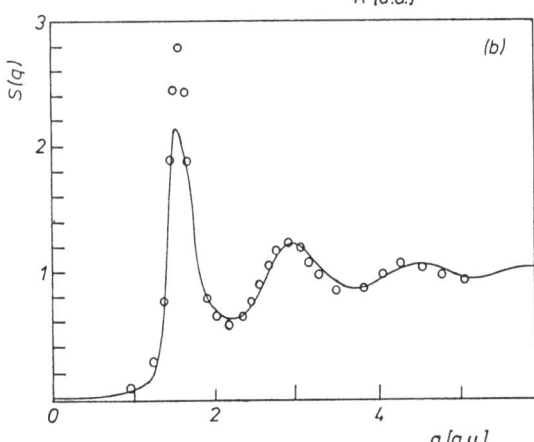

Fig. 4.10. (a) Interatomic potential for Cu from the generalized pseudopotential approach. (*Full line*) simple metal limit, (*dashed line*) including *s-d* hybridization, (*dot-dashed line*) including *s-d* hybridization and overlap. (b) Static structure factor $S(q)$ of liquid Cu; (*solid line*) calculated using the full potential and the ORPA (after [4.75])

where the first terms is evidently identical to the interionic interaction in the OCP. Therefore they write the direct correlation function of a liquid metal in the form

$$\hat{c}(q) = \hat{c}_{OCP}(q) - 2\beta F(q) \quad , \tag{4.34}$$

where the first term on the right-hand side stands for the direct correlation function of the OCP (which they describe by an intermediate charged-hard-sphere description, see [4.85, 86] for details). The second term represents the RPA for the indirect ion-electron-ion interaction. Equation (4.31) yields to the following expression for the static structure factor $S(q)$,

$$S(q) = S_{OCP}(q)/[1 + 2n\beta F(q)S_{OCP}(q)] \quad . \tag{4.35}$$

Two points should be noted: (i) In order to obtain good agreement with the experimentally measured $S(q)$, the plasma parameter $\Gamma = 2Z^2\beta/R_a$ has to be lowered somewhat below the value corresponding to the density and temperature of the liquid metals. (ii) With the full energy-wavenumber characteristic $F(q)$, (4.35) yields a static structure factor which diverges at the position of the first peak. Tosi et al. found it necessary to introduce an artificial cutoff in $F(q)$ at wave numbers close to the first mode of the pseudopotential form factor. In a subsequent publication [4.88] *Senatore* and *Tosi* argue that the truncation is "in the spirit of the ORPA". To support their argument they show that the truncation of $F(q)$ modifies the indirect interaction $\Phi_{ind}(R)$ mainly inside the diameter of their charged hard spheres and that the resulting $g(R)$ has the correct behaviour (see Fig. 4.11). But of course this is not sufficient to show that $\Phi_{ind}(R)$, as derived from the truncation of $F(q)$, is really an optimized potential is the sense that it is a solution of the ORPA integral equation. Evidently the intermediate charged-hard-sphere description of the OCP used by Tosi et al. is an important step in obtaining this result. In principle the application of the MSA to an OCP reference would require the solution of the general MSA equation as given by (4.7) and this has not been undertaken as yet, although it may prove to give interesting results.

The inclusion of the oscillatory forces for liquid metals is certainly a very important ingredient of the theory. Only a proper consideration of the attractive forces allows the trends in liquid structures to be followed throughout the periodic table and from the melting point into the near-critical region.

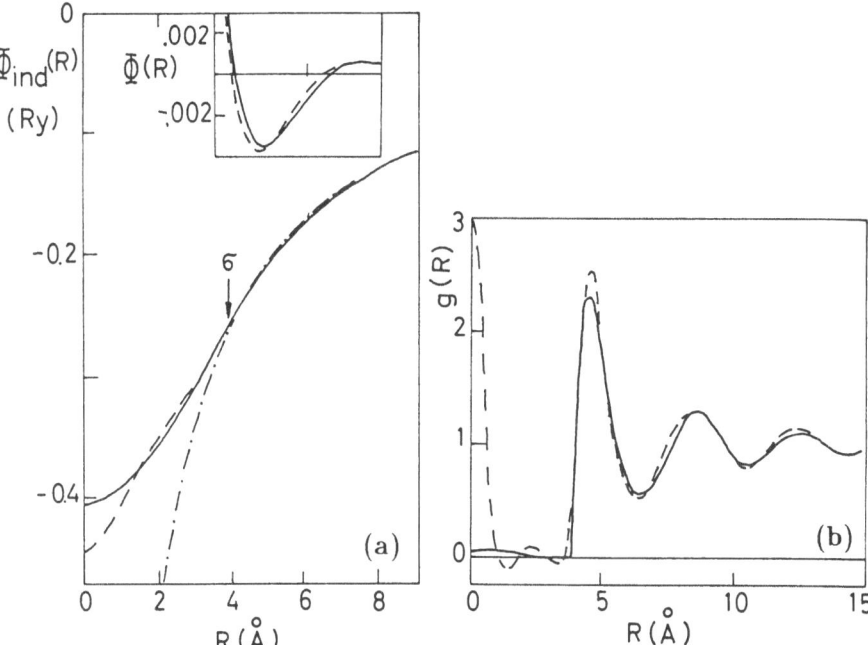

Fig. 4.11. (a) Illustration of the "ORPA property" of the truncated indirect ion-electron-ion potential $\Phi_{ind}(R)$; (*solid line*) $\Phi_{ind}(R)$ as calculated with the truncated energy-wavenumber characteristic $F(q)$, (*dashed line*) as calculated with the full $F(q)$, (*dot-dashed line*) Coulomb-potential $2Z/R$. The full interatomic potential $2Z/R + \Phi_{ind}(R)$ is shown in the inset on a reduced scale. (b) The pair-correlation function $g(R)$ for liquid potassium. (*Solid line*) as calculated using (4.35), (*dashed line*) experimental (after [4.47])

4.4 Trends in Liquid Structures

The systematic trends in the crystal structures of the metallic elements and their explanation in terms of the variations of the interatomic potentials with electron density and pseudopotential have been discussed in Sect. 3.3. It was suggested very early [4.89, 90] that most of these trends persist in the liquid state, where they appear as distortions of the first few peaks of the static structure factor or of the pair correlation function. These effects where formerly rather controversial owing to the unreliability of many of the early measurements, but now they seem to be soundly established (for a recent compilation of liquid structure data see *Waseda* [4.83] and *Steeb* [4.91]). Following Waseda it is possible to distinguish essentially three different types of liquid structures:

a) Hard-sphere-like structures. They show the following characteristic features: (i) The first peaks of both $g(R)$ and $S(q)$ are symmetric. (ii) The ratio of the positions of the second and first peaks of these functions is about $Q_2/Q_1 = 1.86$ and $R_2/R_1 = 1.91$. (iii) The number of atoms in the first coordination shell [calculated using (4.32)] lies between 9.0 and 11.0. The structures of the liquid group Ia and IIa elements and of Be, Mg, Al, and Pb belong to this class; Tl is a border line case.

b) Asymmetric structures, characterized by (i) an asymmetric peak in $S(q)$ and/or $g(R)$, (ii) slightly non-ideal ratios R_2/R_1 and Q_2/Q_1, and (iii) essentially metallic coordination numbers, i.e. $N_c \sim 9\text{--}11$. Examples of these asymmetric structures are Zn, Cd, Hg, In, and to some degree Tl. In the group IIb elements the distortion increases from Zn to Hg.

c) Loosely packed structures with even stronger anomalies which show up as (i) shoulders on the high-angle (i.e. high-q) side of the static structure factor and subsidiary maxima in $g(R)$ located between the main peaks, (ii) strongly non-ideal values for the peak positions ($Q_2/Q_1 \cong 1.96\text{--}2.1$, $R_2/R_1 \cong 2.0\text{--}2.3$), and (ii) exceptionally low coordination numbers ($N_c \simeq 6.5\text{--}7$ for Si, Ge). These loosely packed structures are found in liquid Ga, Si, Ge and Sn.

Thus, as in the crystalline state we find in group IIb a tendency towards an increasing distortion of the symmetric, hard-sphere-like structure, in group III a transition from a hard-sphere-like structure in Al to an anomalous one in Ga, but then a return to a less distorted structure in In and Tl, and finally in group IV we find again that the "anomalous" features become weaker going from Si to Sn and have completely disappeared in Pb (Fig. 4.12). Finally we illustrate in Fig. 4.13 the trend from Na to Si – i.e.

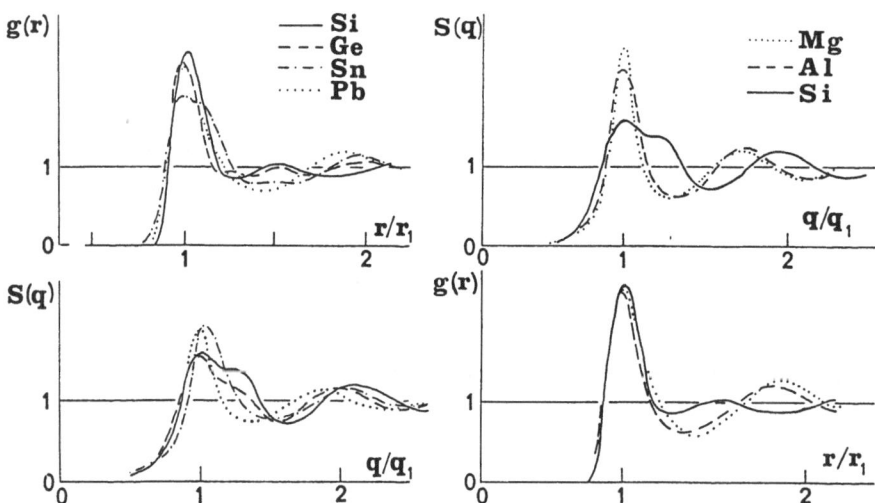

Fig. 4.12. Trend from "covalent" to more metallic structures in the group IV elements with increasing atomic number A (from Si to Pb). Experimental data as compiled by *Waseda* [4.83]

Fig. 4.13. A comparison of the hard-sphere-like structures for Mg, and Al with the more complicated structure of liquid Si. Experimental data as compiled by *Waseda* [4.83]

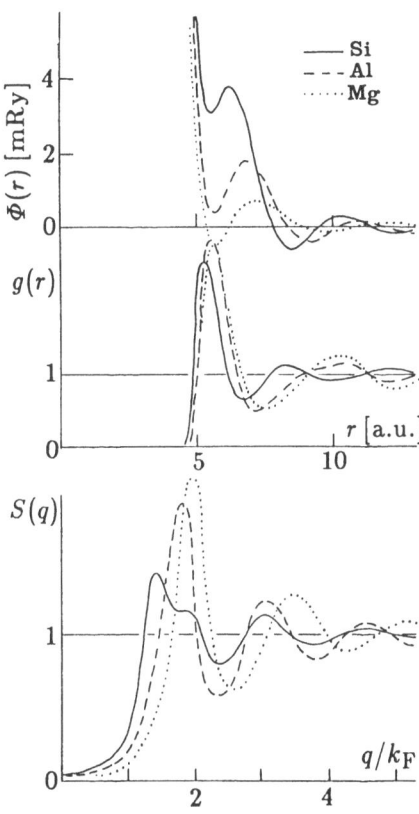

Fig. 4.14. Variation of the interatomic potential $\Phi(R)$ and of the liquid structure [as calculated from $\Phi(R)$ using the ORPA] with the valency Z at essentially constant $R_c/R_s \sim 0.48$, reflecting the structural trend from liquid Mg to liquid Si (after [4.79])

from a typically metallic hard-sphere-like to an apparently at least partially covalent more open structure and from high to low coordination numbers ($N_c = 8.9$, 10.2, 9.9 and 6.4, for Na, Mg, Al, and Si).

For the explanation of these trends we follow again the strategy that was so successful for the crystalline structures. In Sect. 3.3 we found that the main effect in the series from Na to Si comes from the different (R_a/R_s) due to the different valencies, at an approximately constant (R_c/R_s) : at low R_s the repulsive core begins to cover the attractive oscillations, and since $D_{cp} \propto R_a$ (the role of the close-packing distance D_{cp} is now taken over by the effective hard core diameter σ) the close-packing distance moves on a repulsive hump of the interatomic potential. Figure 4.14 demonstrates that this argument also serves to explain the trends in the liquid structures: for Na, Mg, and Al the first peak of a HS-like $g(R)$ falls into the main minimum of the pair potential, and its higher minima fit quite well into the oscillations of the pair potential. For $Z = 4$, however, the first peak of $g(R)$ falls on the repulsive slope. Therefore, it is energetically favourable to reduce the height of this peak (i.e. to lower the coordination number) and to place these atoms into the next minimum of $g(R)$ (see also Fig. 4.9). The covering up of the first minimum dominates the trend to smaller coordination numbers. Using (4.32) we calculate $N_c = 8.8$, 9.5, 9.6, and 6.6 for Na, Mg, Al and Si. The higher oscillations mainly follow the pair potential but due to interference with the geometrical requirements of sphere packing they are strongly damped. In the static structure factor the "constructive interference" between the HS structure and the potential is illustrated by the fact that the first peak of $S(q)$ falls very close to $2k_F$, i.e. $Q_1 \simeq 2k_F$. For Si we have $Q_1 > 2k_F$ and the shoulder near $2k_F$ expresses a periodicity in the potential and in the mean atomic positions with the Friedel wavelength, $\lambda_F = 2\pi/2k_F$, which interferes with the geometrical requirements of sphere packing (cf. the preceding section).

The trends within one group of the periodic table are dominated by the variation of R_c/R_s at an essentially constant electron density R_s. Figure 4.15 shows the variation of the liquid structure with R_c/R_s at constant $Z = 4$ and constant atomic volume. Again we find the "amplitude effect":

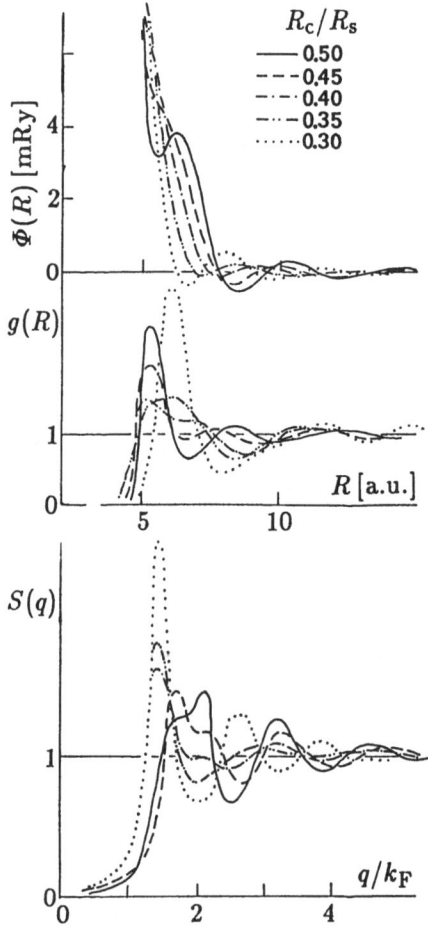

Fig. 4.15. Variation of the interatomic potential $\Phi(R)$ and of the liquid structure [as calculated from $\Phi(R)$ using the ORPA] with the ratio R_c/R_s and constant $Z = 4$ and constant atomic volume – reflecting the return from open to hard-sphere like structures in the liquid tetravalent elements with increasing atomic number (after [4.79])

for high atomic numbers q_0 approaches $2k_F$, hence the oscillations in $\Phi(R)$ are strongly damped and therefore not effective.

Put most simply, one finds that the basic question is the compatibility of the geometric and the electronic factors influencing the liquid structures: at low electron densities and large core radii both are compatible with HS-like structures. As the electron density increases and/or the core radius decreases, this "constructive" interference disappears and the oscillations in the potential are phase shifted relative to the repulsive core. The diameter of the repulsive core and the Friedel-wavelength which gives the periodicity of the intermediate- and long-range oscillations, now set two different characteristic distances: the complex structures of Zn, Cd, Hg, Ga, In, Si, Ge, Sn result from the interplay of these two different length scales. But finally, the last trace of an oscillation around the nearest neighbour distance disappears and only the influence of geometric factors remains – this explains the return to HS-like structures in Pb.

4.5 Expanded Fluid Metals

We have seen that at conditions near to their triple point, liquid metals exhibit a rather well-defined short-range order, several coordination spheres are well expressed in the pair correlation function. If the temperature is

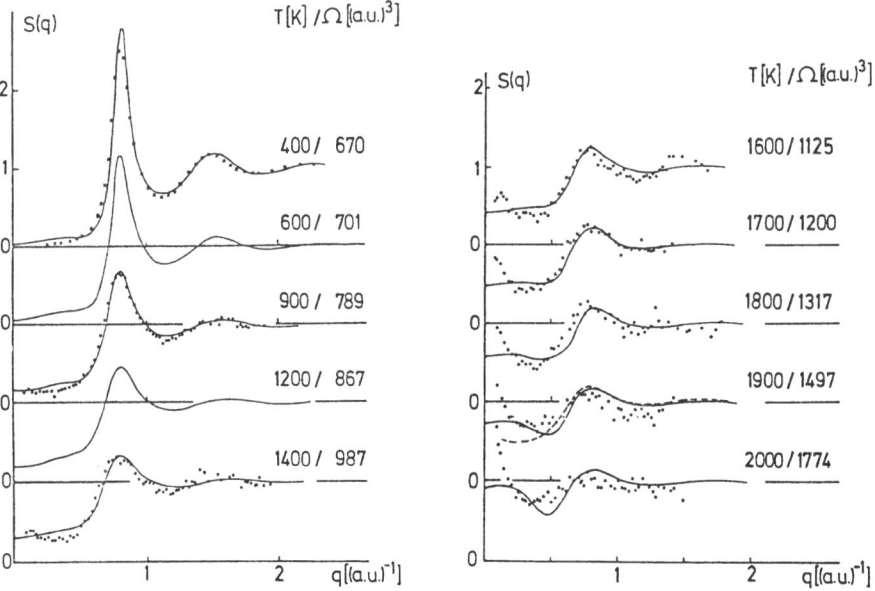

Fig. 4.16. Static structure factor of expanded fluid rubidium for a series of thermodynamic states along the saturated vapour-pressure curve. (*Solid line*) ORPA calculation, (*circles*) neutron scattering results of *Franz* et al. [4.106], (*crosses*) neutron scattering results of *Waseda* [4.83] (after [4.77])

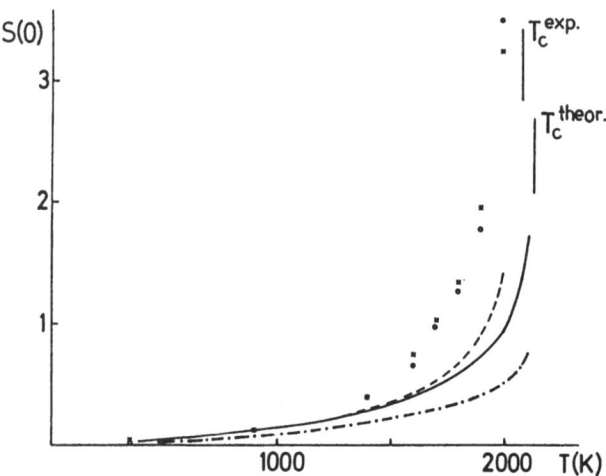

Fig. 4.17. Temperature dependence of $S(O)$ for liquid Rb along the vapour-pressure curve, (*solid line*) ORPA, (*dashed line*) RPA, (*dot-dashed line*) WCA (repulsive core only). Experiment; (*circles*) from compressibility data, (*crosses*) from diffraction data (after [4.77])

increased, this short-range order is decreased by thermal collisions; at a density halfway between the triple point and the critical point the higher-order oscillations have virtually disappeared and the height of the first peak has been strongly reduced (Fig. 4.16). At the same time a pronounced small-angle scattering begins to show up. Figure 4.17 shows the temperature dependence of $S(O)$ close to the critical point. Via the compressibility relation $S(O) = nk_BT\chi_T$, the small-angle scattering can be related to the critical divergence of the isothermal compressibility. The microscopic description of the critical region of the liquid-vapour transition is still an open problem, even within the theory of classical fluids (see [4.92–95] for recent attempts to find a solution). In metallic systems there are additional complications [4.96,97]: at densities corresponding to about twice the critical density a strong maximum in the magnetic susceptibility is observed – this gives an indication that at low density, the correlation of the electron gas is a dominant factor in the expanded fluid metals. At even lower densities, the electronic transport coefficients change from a clearly metallic conductivity, ($\sigma \sim 1000\,\Omega^{-1}\,cm^{-1}$, $d\sigma/dT<0$) to a non-metallic characteristic ($\sigma \sim 250\,\Omega^{-1}\,cm^{-1}$, $d\sigma/dT>0$ at the critical point [4.96]). Recent work of *Jüngst* et al. [4.98] has shown that there are characteristic differences in the critical behaviour of classical fluids and molten metals: in classical systems the "diameter" of the liquid-vapour coexistence curve, $n_d = n_l - n_v$ is a linear function of temperature, in expanded metals it shows a $|\tau|^{1-\alpha}$, $\tau = (T_c - T)/T_c$ singularity where $\alpha = 0.11$ is the critical exponent of the specific heat. *Goldstein* and *Ashcroft* [4.99] have suggested that the origin of the critical diameter singularity is the strong state-dependence of the effective interatomic potential in metals.

The important change in the interatomic interactions and in the physical properties with varying density should be reflected in the structural data. Evidently any attempt to describe the low-angle structure factors is bound to fail without a proper description of the attractive forces. Simple theories such as the RPA can at best offer a qualitative insight; see [4.100, 101] and further references therein. Another approximation which is apparently quite successful for classical fluids [4.102] is the so-called mean-density approximation (MDA) of *Henderson* and *Ashcroft* [4.103]. Within the MDA, the static structure factor is given by

$$S_{MDA}^{-1}(q) = S_{HS}^{-1}(q) + \beta n \hat{\Phi}_1(q) + n \left(\frac{\partial I}{\partial n} + \frac{1}{2} n \frac{\partial^2 I}{\partial n^2} \right) \tag{4.36}$$

with

$$I(q) = \int \hat{\Phi}_1(k) [S_{HS}(|q - k|) - 1] \frac{d^3 k}{(2\pi)^3} \quad . \tag{4.37}$$

The MDA has been derived by means of a functional differentiation of the free energy of an inhomogeneous classical fluid with respect to local fluctuations in the density. Therefore (i) the applicability of the MDA is restricted to the low-q region and (ii) the MDA is immediately applicable only to classical fluids – the density dependence of metallic pair potentials requires a modification of the MDA. *McLaughlin* and *Young* [4.104] have fitted the temperature dependence of $S(O)$ for liquid Rb using (4.36) and a temperature-dependent pseudopotential, but in view of the above remarks it remains to be seen whether the MDA is really a good description of the expanded fluid metals (intuitively one would expect the density dependence of the potential to become even more important on approaching the critical point).

The ORPA has been applied to expanded fluid Rb by *Kahl* and *Hafner* [4.77] and their results for a series of thermodynamic states along the saturated vapour-pressure curve are shown in Fig. 4.16 together with the experimental results. Good agreement is found for temperatures up to $T = 1400$ K. From the divergence of $S(O)$ the critical temperature is estimated to be $T_c = (2120 \pm 10)$ K, compared to an experimental value of $T_c = (2090 \pm 20)$ K. The calculation reproduces the strong increase of the small-angle scattering with temperature (see Fig. 4.17), but it is found to be confined to a narrower temperature interval below T_c than observed experimentally. It is significant that the theory begins to deviate from experiment precisely at the temperature where the electronic mean free path drops below the average interatomic distance. Thus the nearly-free-electron approximation breaks down,[3] and the response functions have to be corrected for finite mean free

---

[3] This conclusion is supported by a recent calculation of *Pelissier* [4.105] for the expanded alkali metals from Na to Cs.

path effects, strong electronic correlation etc. Evidently refinements of the present state of the art theory are required.

The results obtained with the ORPA are corroborated by calculations using the OCP + RPA scheme by *Chaturvedi* et al. [4.86] and using the PY integral equation [4.30, 31]. Again the good agreement with experiment is restricted to temperatures $T \lesssim 1400 \, \mathrm{K}$ (although the ORPA turns out to be somewhat more accurate than the other methods). In view of the fact that the PY and ORPA yield totally different results for the critical behaviour (the PY equation yields classical critical exponent for all systems [4.92], whereas the ORPA gives the critical exponents of a spherical model [4.95] – again this seems to be a universal characteristic), we conclude that the failure of the existing theories to describe the near critical region is not to be blamed on the liquid-state theory, but on a breakdown of the linear response theory used for the construction of the interatomic potential.

4.6 Structure and Thermodynamics of Liquid Transition and Rare-Earth Metals

It is surprising that the structure of liquid transition and rare-earth metals differs hardly at all from that of a prototype simple metal such as Al (for a compilation of structural data see again *Waseda* [4.83]). In terms of the classification introduced in Sect. 4.4, all liquid transition and rare-earth metals belong to group (a) – hard-sphere-like structures. Consequently, a parametrization of the structure and thermodynamics of liquid transition metals in terms of a hard-sphere model has been shown to be quite adequate [4.106, 107]. More recently, *Khanna* and *Cyrot-Lackmann* [4.108] have shown that the OCP has some advantages as a reference system, since it accounts better for the softness of the pair potential. *Itami* and *Shimoji* [4.109] came to the same conclusion on the basis of an analysis of the thermodynamic properties. In spite of our restricted knowledge of the interatomic potentials some progress has been achieved just recently (see "Recent results", p. 132).

4.7 Atomic Motion in Liquid Metals

Up to now we have considered only the static structure of the liquid, as expressed by the pair correlation function $g(R)$ and the static structure factor $S(q)$. The thermodynamic functions were expressed in terms of these static correlation functions. This is in contrast to the crystalline case, where the thermodynamic properties are given in terms of the vibrational spectrum, i.e. the dynamical properties of the atoms. In order to discuss the interrelations, we turn now very briefly to a discussion of the atomic motion in a liquid metal.

The time-dependent pair correlation function $g(R,t)$ gives the probability of finding a particle at R at a time t, given that the same *or* another particle had been at the origin at $t = 0$; $g_s(R,t)$ is the probability of finding the *same* particle at R that was originally at the origin. This means that $g(R,t)$ describes the collective motion of particles, while $g_s(R,t)$ describes the individual motion of a particle. The double Fourier-transforms (with respect to distance and time) of $g(R,t)$ and $g_s(R,t)$ are just the coherent- and the incoherent-inelastic neutron-scattering functions [4.110, 111, 112] $S(q,\omega)$ and $S_s(q,\omega)$. The relationship to the static structure factor is established by the sum rules

$$\int S(q,\omega)d\omega = S(q) \tag{4.38a}$$

$$\int \omega^2 S(q,\omega)d\omega = q^2 \frac{k_B T}{M} \quad , \quad \text{and} \tag{4.38b}$$

$$\int S_s(q,\omega)d\omega = 1 \quad . \tag{4.38c}$$

If one looks at the motion of an individual particle in a liquid, one is usually interested in the diffusion coefficient D of that particle. In the hydrodynamic regime (i.e. when the wavelength $\lambda = 2\pi/q$ is very large compared to an average interatomic distance) the particle number conservation law yields the following relation between $S_s(q,\omega)$ and D

$$S_s(q,\omega) = \frac{1}{\pi} Dq^2 [\omega^2 + (Dq^2)^2]^{-1} \quad , \tag{4.39}$$

i.e. $g_s(R,t)$ is simply the solution of the classical diffusion equation. In the high-q regime the scattering is predominantly incoherent, i.e. $S_s(q,\omega) = S(q,\omega)$, and it approaches that of an ideal gas.

The most interesting question concerning the collective motion of particles in a liquid is for which range of momentum transfers do non-overdamped collective excitations exist. In $S(q,\omega)$ such collective excitations manifest themselves as peaks at finite frequency – remember that for an ideally harmonic solid a normal mode of vibration corresponds to a δ-function peak in $S(q,\omega)$. In the hydrodynamic regime momentum conservation ensures that the sound damping constant is proportional to q^2 and this yields long-lived compressional (longitudinal modes) with an approximately linear dispersion law $\omega(q) = v_L q$ where v_L is the longitudinal sound velocity. Transverse excitations, however, cannot exist in the hydrodynamic regime. Entropy fluctuations produce a peak at $\omega = 0$ ("quasi-elastic" or Rayleigh peak) which is again of the form (4.39) with D replaced by the thermal diffusivity D_T. At larger momentum transfer the inelastic peaks in $S(q,\omega)$ will broaden and finally become overdamped. In analogy to the crystalline case, one usually defines a "phonon dispersion relation" in the liquid as the locus of the maxima in $S(q,\omega)$.

The first experimental evidence for phonon-like collective excitations has been found by *Copley* and *Rowe* [4.113] for liquid Rb, very recently similar collective excitations have been detected in liquid K [4.114]. These data could be reproduced very nicely by computer simulations based on pseudopotential-derived interatomic potentials and the molecular dynamics technique [4.115,116]. Computer simulations [4.18,117] have also predicted similar collective excitations in polyvalent liquid metals such as Al. The density fluctuations in liquid metals have also been the subject of many analytical investigations [4.118,119].

Figure 4.18 shows a typical $S(q,\omega)$ and the measured and the calculated phonon dispersion curves for longitudinal excitations in liquid Rb. One finds a pronounced maximum in the dispersion curve at a wave vector $q \sim Q_p/2$ where Q_p is the wave vector at which the static structure factor has its first maximum. This shows that to some extent the region $0 \leq q \leq Q_p/2$ plays a role similar to the first Brillouin zone in crystals. The computer simulations also enable one to study transverse collective excitations, which are not observable in a neutron scattering experiment. It appears that the dispersion of shear waves may be written in the form $\omega_T(q) = v_T(q - q_T)$, where q_T is

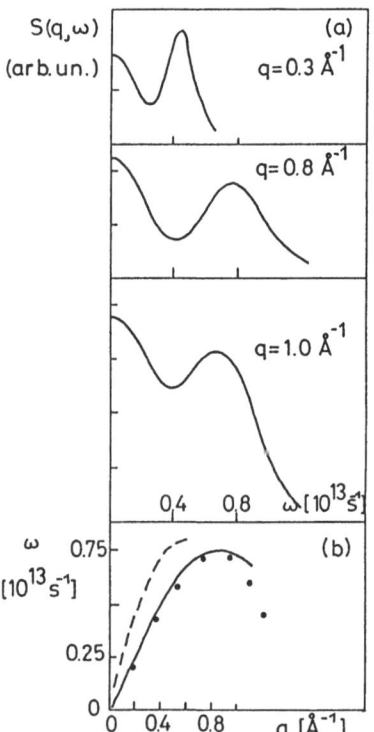

Fig. 4.18. (a) Dynamical structure factor $S(q,\omega)$ for liquid Rb. (*Solid line*) molecular dynamics calculation of *Rahman* [4.115], (*full points*) inelastic coherent neutron scattering [4.113]. (a) Dispersion law for collective excitations in liquid Rb. The dashed line shows for comparison the dispersion relation for longitudinal phonons along the [100] direction in crystalline Rb

a critical wave number which marks the boundary of the shear wave regime. v_T is the speed of propagation of the shear wave. In each case it is found that $v_T < v_L$, and the value of the ratio v_T/v_L is consistent with the known velocities of longitudinal and transverse sound waves in the corresponding crystals [4.116]. q_T and the upper limit q_L at which longitudinal collective excitations can be detected, may be used to define two characteristic wavelengths: $\lambda_T = 2\pi/q_T$ and $\lambda_L = 2\pi/q_L$, an upper limiting wavelength at which shear waves disappear and a lower limiting wavelength at which sound wave propagation ceases. *Jacucci* and *McDonald* [4.116] found that in metals (Na, K, Rb) $\lambda_T/d \simeq 14$ and $\lambda_L/d \simeq 1.4$ (d is an average interatomic distance), whereas in liquid rare gases (Ar) $\lambda_T/d \simeq 7$ and $\lambda_L/d \simeq 6$. This means that in liquid metals we have a fairly wide regime of wavelength (roughly from one to fourteen interatomic spacings) in which the liquid can support propagating collective excitations of both longitudinal and transverse character but for the liquid rare gases this range is extremely small. Clearly this suggests that the atomic motions in liquid metals are of a much more pronounced collective character then in other monatomic liquids.[4]

These results can be regarded as justification for the "phonon-theory" of the thermodynamic properties of liquid metals proposed very early by *Eisenschitz* and *Wilford* [4.121] and further developed by *Bratby* et al. [4.122]. In their approach the entropy of the liquid is expressed as a sum over $3N$ normal modes of vibration (neglecting effects of finite lifetime), in complete analogy to the crystalline case [see (3.24)], plus a small term arising from phonon-phonon interactions (i.e. anharmonic effects). This of course raises the question of whether this approach, which is based on the notion of collective motions of the atoms, is consistent with the usual approach to the liquid entropy (see Sects. 4.2, 3 and Appendix D) which is based on the concept of packing. This question has been addressed by *Young* and coworkers [4.123, 124]. They showed that if lifetime effects are neglected, i.e. $S(q, \omega) \simeq S(q)\delta[\omega - \omega(q)]$, the moment relation (4.38b) may be used to define an approximate (but reasonable) dispersion relation for degenerate phonons of the form (see also [4.125])

$$\omega(q) = q\{k_B T / [M S(q)]\}^{1/2} \tag{4.40}$$

and using (4.40) in (3.24), the entropy may be evaluated. The harmonic term is found to contribute 90 % of the entropy and to be in good agreement with the entropy derived from the integral equations. This is not unexpected: equation (4.40) just expresses the fact that the frequencies of the collective excitations are again determined mainly by packing effects.

---

[4] Very recently, *Egelstaff* and *Gläser* [4.120] have proposed a new method for analyzing the coherent dynamic structure factor $S(q, \omega)$ of liquids. Their technique is very suitable for investigating the characteristic differences between the collective atomic motions in liquid metals and in classical liquids such as noble-gas fluids.

Computer simulations for liquid Si and Ge have been performed using classical molecular dynamics and empirical two- and three-body forces [4.126, 127], and pseudopotential-derived volume- and pair-forces [4.128]. *Car* et al. [4.129] have presented a novel density-functional molecular dynamics technique which includes both atomic and electronic degrees of freedom in the simulation. The application of this new technique to the simultaneous calculation of the atomic *and* the electronic structure of liquid and amorphous Si [4.130] is certainly a pioneering piece of work.

The new mixed-closure integral equations with imposed thermodynamic consistency (see Appendix D) were first applied to liquid metals by *Pastore* and *Kahl* [4.131].

Charged hard spheres with Yukawa interactions have been used as a reference system for thermodynamic variational calculations for liquid simple metals [4.132, 133] and liquid transition metals [4.133]. *Russier* et al. [4.134] have calculated the structure of some liquid transition metals in the optimized random-phase approximation [4.134].

5. The pT Phase Diagram of Pure Metals

In the preceding chapters we have discussed theoretical foundations for the calculation of the interatomic forces in metals and we have learned how to calculate the structure and the thermodynamic properties of a metal in both the crystalline and the liquid states. This theory is sufficiently advanced to attack a calculation of the complete pressure-temperature phase diagram of a simple metal.

5.1 Solid-Liquid Transitions: The Total Energy Approach

When a solid coexists with the melt, we can equate the Gibbs free energies at a given temperature and pressure and write

$$F_s + pV_s = F_l + pV_l \quad , \tag{5.1}$$

where the subscripts s and l refer to the solid and liquid states. We have introduced the free energy F because this is the quantity on which the Gibbs-Bogoljubov variational technique focuses. In fact, the simplest possible calculation of a melting curve is based on a variational calculation for both the solid and the liquid states: *Stroud* and *Ashcroft* [5.1][1] used a Debye reference system (3.17, 28–31) for the solid and a hard-sphere reference system (see Appendix D.3.1) for the liquid. At a given temperature and density, the free energy is then determined from the variational conditions

$$\left(\frac{\partial F_s}{\partial \theta}\right)_{T,V} = 0 \quad \text{and} \quad \left(\frac{\partial F_l}{\partial \sigma}\right)_{T,V} = 0 \quad . \tag{5.2}$$

The pV term is obtained separately as the volume derivative

$$p_{s,l} = -\left(\frac{\partial F_{s,l}}{\partial V}\right)_T \quad . \tag{5.3}$$

---

[1] See also the early calculation of the latent heat of fusion by *Hartman* [5.2] and of the zero pressure melting properties by *Jones* [5.3].

In addition to the equality of the Gibbs free energies (5.1) we also have to ensure the equality of the pressures:

$$p_s(V_s, T_m) = p_l(V_l, T_m) = p_m(T_m) \quad . \tag{5.4}$$

At a given temperature T and fluid volume V_l, the equations (5.1) and (5.4) have to be solved for p_m and the corresponding solid volume V_s.

Stroud and Ashcroft achieve a quite remarkable agreement of the calculated melting curve with experiment (Fig. 5.1). The consistency of the calculation may be checked via the Clausius-Clapeyron relation which determines the slope of the melting curve. Their calculations also predict a Lindemann law, i.e. the ratio of the mean-square displacements and the nearest-neighbour distance is approximately constant along the melting curve.

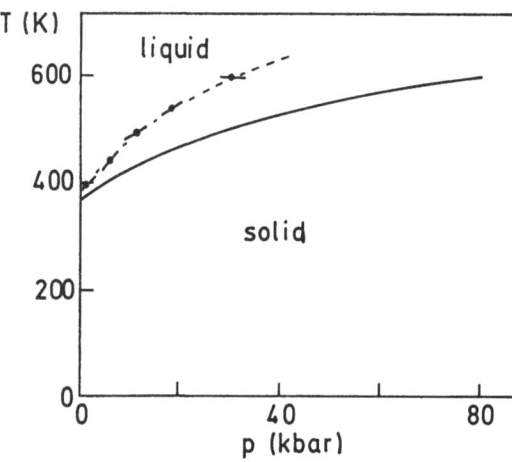

Fig. 5.1. Melting curve of sodium. (*Solid line*) experiment, (*dots*) calculated points, (*short bars*) slope calculated via the Clausius-Clapeyron relation; the dashed line is merely a guide to the eye (after [5.1])

Similar phase-diagram calculations have been performed by *Young* and *Ross* [5.4] for the alkali metals Li, Na, and K, by *Bratkovsky* et al. [5.5] for Na, K, Rb, and Cs, and by *Moriarty* et al. [5.6] for Al. The calculation of Young and Ross is based on an OCP, HS, and soft-sphere (SS) variational theory for the liquid metal and quasi-harmonic lattice dynamics for the solid. The dynamical matrix (3.20) was diagonalized for a sufficient number of k-points in the irreducible Brillouin zone and the free energy calculated using (3.24). Their results (which also include the low-temperature polymorphic transitions) are compiled in Fig. 5.2. Admittedly the success of any such calculation owes a lot to the cancellation of errors: the methods chosen to describe the solid and the liquid state must be of comparable accuracy. This point has also been emphasized by *Pelissier* and *Angelie* [5.7, 8]. They demonstrated that on the whole the variationally determined free energy for Na is respectively about 1.5 mRyd (solid-Einstein reference system), 2 mRyd (liquid-HS) and 0.75 mRyd (liquid-OCP) too high com-

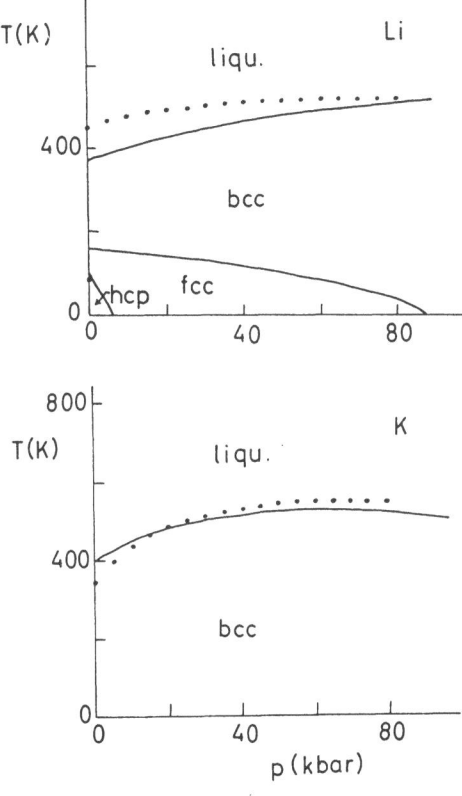

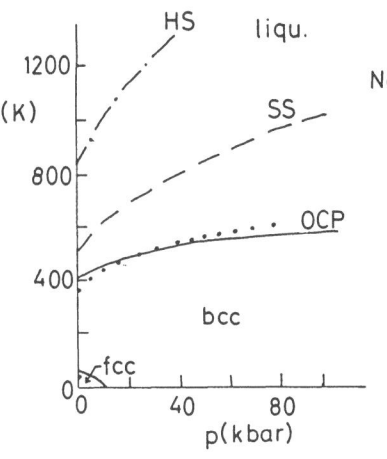

Fig. 5.2a–c. Comparison of experimental and theoretical phase diagrams for the alkali metals. (a) Lithium: a bcc-hcp transition is observed at $T = 77$ K and $p = 0$, and a bcc-fcc (?) transition at $T = 300$ K and $p = 69$ kbar (see also Sect. 3.4). (b) Sodium: a bcc-hcp transition is observed at $T \simeq 36$ K and $p = 0$. The theoretical melting curves based on three different models are shown. (c) Potassium (after [5.4])

pared to experiment. The relevant energy scale is the latent heat of fusion (about 1 mRyd typically). This explains the good accuracy of Stroud and Ashcroft's original calculation and shows that to improve the theory only on one side (e.g. by replacing the HS by an OCP reference system) would only worsen the agreement with experiment. *Pelissier* [5.8] has presented a calculation of the melting curve of sodium which is based on the Heine-Abarenkov model potential, Vashishta-Singwi screening (see Sect. 2) and a full quasi-harmonic phonon spectrum calculation (plus approximate corrections for anharmonicity) on the solid side, and the optimized cluster theory (OCT), (see Sect. 4.3.3) on the liquid side. This lowers the errors in the free energy to about 0.1 mRyd for the solid and about 0.04 mRyd for the liquid state (at 373 K and the melting density) and yields a melting curve in perfect agreement with experiment up to pressures of about 60 kbar. Later work extended these calculations to the heavy alkali metals K, Rb, and Cs [5.9].

Very recently Pelissier has also presented a full calculation of the phase diagrams (including the solid-solid (fcc → bcc) and solid-liquid transitions)

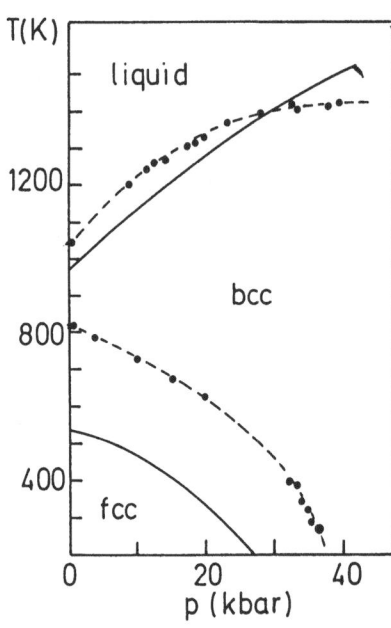

T(K)

liquid

1200

bcc

800

fcc

400

0   20   40
        p (kbar)

Fig. 5.3. The pT phase diagram of strontium. (*Dots*) experiment, linked by a dashed line; (*solid line*) theory (after [5.11])

of the divalent metals Sr and Yb [5.10] and a calculation of the high-pressure melting curves of Al [5.11] and Pb [5.12]. For the melting curve an accuracy similar to the alkali-metal calculations was achieved (Fig. 5.3), and in addition, the solid-solid transition is correctly described. Pelissier points out that in the high-temperature limit the transition temperature T_c for the polymorphic transition is given by

$$T_c \simeq \Delta E(0)/\Delta S_{ph} \quad , \tag{5.5}$$

where $\Delta E(0)$ and ΔS_{ph} are the differences in the zero-temperature electronic ground-state energies and in the phonon entropies of the two phases. In the same approximation, the Clausius-Clapeyron relation reads

$$\frac{dT_c}{dp} = \frac{\Delta V}{\Delta S_{ph}} = \frac{V}{\Delta S_{ph}} \frac{\Delta P(0) + 3(T_c/V)\Delta\gamma(V)}{B(V,T_c)} \quad , \tag{5.6}$$

[$P(0)$ stands for the zero-temperature pressure and γ for the Grüneisen parameter.] Equations (5.5) and (5.6) show that not only does the actual value of the transition temperature depend rather critically on a balance between electronic and vibrational energies but also the very sign of the slope of the transition curve depends on the factor $[\Delta P(0)+3(T_c/V)\Delta\gamma(V)]$, i.e. on a subtle equilibrium between differential zero- and high-temperature contributions to the pressure.

The calculation of the melting curve of Al by *Moriarty* et al. [5.6] is based on quasi-harmonic lattice dynamics, and HS, SS and OCP variational theory (the lowest free energy is obtained for the soft-sphere calculation); again good agreement with experiment is achieved over the entire experimentally accessible range. The theoretical calculation has been performed for pressures up to 2 Mbar (corresponding to melting temperatures of about 6000 K) and good agreement with shock compression data was claimed.

The results of the semi-analytical calculations are also supported by recent molecular dynamics calculations of the melting of sodium and lithium [5.13]. Here again the main difficulty is that the statistical uncertainty in the calculated free energies of the competing phases is nearly of the order of the heat of transformation. Attempts to observe the onset of the spontaneous solid/liquid transformation are handicapped by large hysteresis effects arising from the small number of atoms in the molecular dynamics cell.

From the foregoing discussion it is clear that the calculation of pT phase diagrams for simple metals is definitely within the reach of present-day theories.

5.2 Microscopic Theories of Melting and Freezing

In the preceding section we have shown how thermodynamic perturbation theories can be used to derive the conditions under which the crystalline-solid $\rightarrow$ classical-fluid transition will take place, but evidently this does not provide us with a microscopic structural theory for the transformation. Experiments [5.14] and computer experiments [5.15–17] have together brought out some salient features of this transition, but in general it is still insufficiently understood.

The first important contribution to the melting problem was that of *Lindemann* [5.18] who pointed out that melting occurs if the ratio of the mean-square displacement $\langle u^2 \rangle$ to the square of the interatomic separation d_1 has a value of $\langle u^2 \rangle / d_1^2 \sim 0.01$. In the liquid phase, the analogue of Lindemann's melting criterion is perhaps "Verlet's rule". *Verlet* [5.17, 19] observed that near freezing the height of the first peak of the static structure factor is nearly $S(Q_p) \approx 2.85$ for all classical fluids at melting and that the $S(q/Q_p)$ of the most classical fluids are nearly identical, at least for intermediate values of the momentum transfer q. Both Lindemann's criterion and Verlet's rule point to the geometrical nature of this transition.

The open challenge is to provide a microscopic explanation of the apparent universality of the melting/freezing phenomenon. Note that although the thermodynamic perturbation calculations have been shown to produce results consistent with Lindemann's criterion, they do not immediately offer the microscopic explanation we are looking for. It is clear that any attempt to formulate such a microscopic theory should follow a strategy quite different from that used in the previous sections: the explicit use of individual interatomic potentials $\Phi(R)$ would make an explanation of the universal character of the transition quite difficult, if not impossible.

Following *Born's* [5.20] original suggestion, Lindemann's criterion has often been interpreted as an absolute crystal instability with respect to shear deformations. Indeed, self-consistent phonon theory predicts such an instability [5.21], but it was shown [5.1] that this absolute instability occurs only at a temperature which is much higher than the actual melting temperature, a conclusion supported by recent computer simulations [5.22]. In the dislocation model [5.23,24] the crystal is assumed to melt when it becomes unstable with respect to the spontaneous generation of dislocations. This has been backed up by a detailed computer analysis [5.25]. Very recently *Novikov* [5.26] published an interesting paper in which he interprets melting as a transition to a curved space. The idea behind this approach is briefly the following: Fluids (or glasses) are considered as crystals in a properly curved three-dimensional space [5.27]. The crystallographic groups in this curved space allow a local order – for example rings containing five or seven bonds – which is forbidden in the flat physical three-dimensional space, where only rotations by $2\pi/n$, $n = 2, 3, 4, 6$ are compatible with translational symmetry. One can show that these rings containing an odd number of bonds surround continuous lines which close as loops or terminate at the surface [5.28]. These are disclination lines. The mapping from the curved to the flat space causes disordering because of the appearance of a large number of disclinations. In Novikov's theory the lattice free energy of a harmonically vibrating crystal with arbitrary pair interactions is calculated as a function of temperature where the free energy difference $\Delta F = F(\chi) - F(0)$ (χ is the curvature) vanishes. The role of the curvature here is analogous to that of the order parameter in the Landau theory of phase transitions. However, the curvature is zero in the ordered and non-zero in the disordered phase, therefore it is rather a disorder parameter. Novikov's theory seems to summarize very nicely what we know about the melting transition: it relates the melting temperature T_m to the pair potential, explains Lindemann's rule and relates curvature to the density of disclinations. (See also the curved-space description of the structure of amorphous solids to be discussed in Sect. 11.1.1.)

Starting from the liquid side, *Ramakrishnan* and *Yussouff* [5.29] have developed an order-parameter theory of freezing. The first basic idea of their approach is that the correlation functions of the solid phase (with some assumed structure)[2] are approximated by correlation functions of the bulk liquid in a thermodynamic perturbation theory. The central assumption is that the system is either entirely liquid or entirely solid – no phase coexistence is permitted. As a result, fluctuations are not treated in a fully correct manner. In the sense of this "homogeneity assumption" the Ramakrishnan-Yussouff theory is a mean field theory. The second important point to note

---

[2] Note that none of the existing molecular theories of freezing and melting addresses the more fundamental problem of spontaneous symmetry breaking.

is the use of the density-functional theory of classical systems pioneered by *Mermin* [5.30], *Saam* and *Ebner* [5.31], and *Evans* [5.32] (which is closely related to the Hohenberg-Kohn-Sham [5.33,34] density-functional theory of quantum systems): in the grand canonical ensemble there exists a functional $F(n(\boldsymbol{R}))$, independent of the external potential $V(\boldsymbol{R})$, such that

$$\Omega \equiv \int V(\boldsymbol{R}) n(\boldsymbol{R}) d^3 \boldsymbol{R} + F(n(\boldsymbol{R})) \tag{5.7}$$

is minimum and equal to the grand thermodynamic potential associated with $V(\boldsymbol{R})$ when $n(\boldsymbol{R})$ is the equilibrium density in the presence of $V(\boldsymbol{R})$. The functional derivatives of $\Phi = F_{\mathrm{id}} - F$ [where F_{id} is the functional defined in (5.7) for an ideal, noninteracting system] are related to the direct correlation functions [5.35, 36]

$$c(\boldsymbol{R}) = \frac{\delta \Phi}{\delta n(\boldsymbol{R})} \tag{5.8}$$

$$c^{(2)}(\boldsymbol{R}_1, \boldsymbol{R}_2) = \frac{\delta^2 \Phi}{\delta n(\boldsymbol{R}_1) \delta n(\boldsymbol{R}_2)} \tag{5.9}$$

and so on. The single-particle direct correlation function $c(\boldsymbol{R})$ is related to the single-particle density through

$$n(\boldsymbol{R}) = \Lambda^{-1} e^{\beta \mu} e^{c(\boldsymbol{R})} \quad , \tag{5.10}$$

where $\Lambda = (\beta/\pi M)$ with M the particle mass (in atomic units). Equations (5.8–10) provide the starting point for the thermodynamic perturbation theory. Taking the ratio of (5.10) for the solid and the liquid phase, one obtains with $\mu_{\mathrm{L}} = \mu_{\mathrm{S}} = \mu$ and $T_{\mathrm{L}} = T_{\mathrm{S}} = T$

$$\ln \left[n_{\mathrm{S}}(\boldsymbol{R}) / n_{\mathrm{L}} \right] = c_{\mathrm{S}}(\boldsymbol{R}; \mu, T) - c_{\mathrm{L}}(\mu, T) \quad . \tag{5.11}$$

Expanding the right hand side of the exact relation (5.11) in powers of the density difference $\Delta n(\boldsymbol{R}) = n_{\mathrm{S}}(R) - n_{\mathrm{L}}$ yields (to first order in the density difference)

$$\ln \left[n_{\mathrm{S}}(\boldsymbol{R}_1) / n_{\mathrm{L}} \right] = \int d^3 R_2 c_{\mathrm{L}}^{(2)}(\boldsymbol{R}_1, \boldsymbol{R}_2; \mu, T) \Delta n(\boldsymbol{R}_2) \quad , \tag{5.12}$$

and from an analogous expansion of the grand thermodynamical potential of the solid and the liquid phases we find for their difference

$$\begin{aligned}
\beta \Delta \Omega &= \beta \Omega_{\mathrm{S}} - \beta \Omega_{\mathrm{L}} = V(p_{\mathrm{S}} - p_{\mathrm{L}}) \\
&= - \int d^3 R_1 \Delta n(\boldsymbol{R}_1) \\
&\quad + \frac{1}{2} \int d^3 R_1 \int d^3 R_2 c_{\mathrm{L}}^{(2)}(\boldsymbol{R}_1, \boldsymbol{R}_2) \Delta n(\boldsymbol{R}_1) \Delta n(\boldsymbol{R}_2) \quad . \tag{5.13}
\end{aligned}$$

Equation (5.12) is an implicit equation for the single-particle density of the solid phase which can be solved in at least two different ways, both of which must assume a certain crystal structure. One can either Fourier-expand $n_S(R)$ according to

$$n_S(R) = n_L[1 + \eta + \sum_n \mu_n \exp(iK_n R)] \tag{5.14}$$

where $\eta = (n_S - n_L)/n_L$ is the fractional density change on freezing. The μ_n's are the order parameters that measure the degree of translational periodicity in terms of the amplitude of the diffraction intensity at a reciprocal lattice vector K_n [5.29, 36]. By exponentiating (5.12), inserting (5.14) on both sides, and Fourier transforming, we get a set of nonlinear equations for the order parameters

$$\mu_j + \delta_{j0} = \frac{1}{V} \int d^3 R \exp(-iK_j R) \exp\left[\sum_{n=0} c_n \mu_n e^{iK_n R}\right], \quad j = 0, 1 \ldots \tag{5.15}$$

with $\mu_0 \equiv \eta$, $|K_0| \equiv 0$ and $c_n = \hat{c}(|K_n|) = [1 - S(|K_n|)]^{-1}$ is the Fourier transform of the direct correlation function of the liquid measured at the sites of the reciprocal lattice vectors of the crystalline lattice. Thus the only input required is the static structure factor $S(k)$ of the liquid which is obtainable from integral-equation theories, perturbation theories, or computer simulation. The eqs. (5.15) has to be solved for the μ_n's. Freezing occurs if a nontrivial solution exists and when, at constant temperature and constant chemical potential, the solid and the liquid phases have the same grand potential, i.e. $\Delta\Omega = 0$. The order parameter driving the transition is essentially (but see below) that corresponding to the first reciprocal lattice vector[3]. The transition appears as a purely structural one which occurs when the structural correlations in the liquid [as measured by the lattice periodic components $\hat{c}(|K_n|)$ of the direct correlation function and hence by the amplitudes of the peaks in $S(q)$] are strong enough – see Verlets rule.

The alternative approach would be a real-space expansion of the solid density in the form of a sum over Gaussians centred at the lattice sites R_i [5.37–39]

$$n_S(R) = \sum_i \left(\frac{\alpha}{\pi}\right)^{3/2} \exp[-\alpha(R - R_i)^2] \quad . \tag{5.16}$$

This has the advantage of simplicity, but the disadvantage that it *assumes* a quasi-harmonic atomic dynamics with a Debye-Waller factor independent of the wavevector.

---

[3] Note the self-consistency requirement inherent in these equations: the fractional density change η calculated from (5.15) must be consistent with the magnitude of the reciprocal lattice vector K_1 entering the equation.

The numerical approximations that go into the theory are (i) the truncation of the functional expansions (5.12) and (5.13) at the level of the two-particle direct correlation function and (ii) the truncation of the order parameter expansion (5.14) or (5.16) after a finite number of terms. The second approximation is less serious since the relevant form of the Fourier expansion is (5.15) which converges rapidly. But nonetheless it turns out that for a hard-sphere or a Lennard-Jones system the sum must be extended over about forty stars of reciprocal lattice vectors in the case of freezing into an fcc structure [5.36, 40] and even more for a less symmetric structure such as hcp to obtain a result which is quantitatively converged – qualitative results are sometimes obtained with only one or two order parameters (see [5.41] for a detailed investigation of this point). The first approximation is perhaps more important: *Haymet* [5.36, 42] has considered partial contributions from higher-order terms (without a significant improvement of the results); *Igloi* and *Hafner* [5.40] have shown that the introduction of a reference liquid with a density $n_R \neq n_L$ (n_R is determined by a variational criterion) corresponds to a resummation of terms up to infinite order. This approach seems to be most promising; it avoids the spurious "remelting" phenomenon of the high-density hard-sphere solid found by *Haymet* [5.42], it is self-consistent (see Footnote 3), and it is the first approach to produce a quantitatively meaningful equation of state for the high-temperature solid phase.

Application of the density-functional theory of freezing has not been restricted to simple model systems (hard spheres [5.36–42], Lennard-Jones fluids [5.43], the one-component plasma [5.44, 45]), it has also been applied to realistic models of molten salts [5.46, 47] and liquid metals [5.29, 48–50]. The results obtained so far for model systems indicate that the theory is a quantitatively useful tool for predicting the freezing transition; changes in the density, the entropy etc. on freezing and the properties of the hot equilibrium solid are calculated with good accuracy. A remarkable result is that the Debye-Waller factor [which is just equal to $\bar{\mu}_n = \mu_n/(1+\eta)$] is found to be fairly close to the quasi-harmonic prediction $\bar{\mu}_n = \exp\left(-u_n^2 K_n^2/6\right)$ with u_n independent of n – at least for fcc structures. The Lindemann ratio is another quantity predicted by the theory and it is found to be consistently lower than the expected 10–15 % at melting.

The calculations for realistic models of metals (Na, Mg, Al) are still at a rather rudimentary stage of one- and two-order-parameter theories [5.29, 48, 50], with the noteable exception of [5.49]. Here it is shown that the density functional theory predicts a crystallization into the correct structure (fcc for Al, hcp for Mg) and yields quantitatively accurate results for the structural energy differences and the melting curve. As an example we show in Fig. 5.4 the freezing curves for fcc Al, as calculated using the density functional theory [5.49] and the total energy approach [5.6, 11]. Freezing into a hypothetical hcp structure is predicted to occur at a tem-

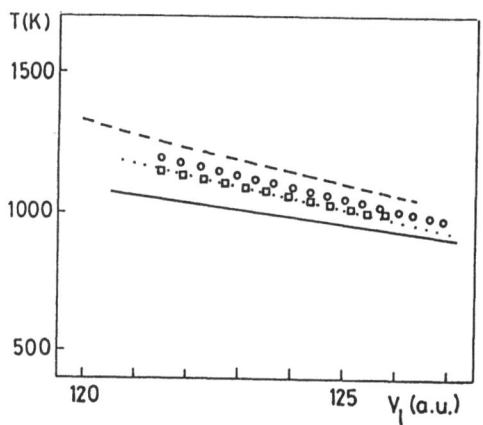

Fig. 5.4. Freezing curve for fcc Al: (*solid line*) density functional calculation [5.49], (*dashed line*) total energy calculation [5.6], (*dotted line*) total energy calculation [5.11]; (*circles and squares*) experiment

perature that is about 360 K lower than the freezing temperature of fcc Al. This difference in the melting temperatures is in reasonable agreement with the structural energy difference estimated by thermochemical methods (cf. Sect. 3.2). These calculations also establish a clear correlation between the precise form of the liquid structure factor and the reciprocal lattice vectors of the stable crystal structure: a liquid will freeze into that crystal structure for which the lattice-periodic components $c(|\mathbf{K}_n|)$ of the direct correlation function are largest [5.49]. Thus the theory provides a quantitatively accurate link between the trends in the liquid and in the crystalline structures (see Sects. 3.3 and 4.4) and is at present the most satisfactory theory of the liquid/solid transition.

5.3 The Liquid-Vapour Transition

Little attention has been paid in general to the liquid-vapour transition. The many difficult new problems which arise in the vicinity of the critical point have already been mentioned, but at lower pressure a calculation of the liquid-vapour coexistence curve should now be possible. So far, however, it seems that the work of *Minchin* et al. [5.51] remains the only attempt in that direction. Starting from

$$F_l + pV_l = F_g + pV_g \quad , \tag{5.17}$$

the next step must involve appropriate approximations to the free energies of the liquid and of the vapour. For the liquid, Minchin et al. choose again a HS-variational method, i.e.

$$F_l = \tfrac{3}{2}k_{\mathrm{B}}T - TS_{\mathrm{gas}}(V_l) + F_{\mathrm{ex}}(V_l) \tag{5.18}$$

where

$$F_{\text{ex}}(V_l) = E_{\text{v}} + 2\pi n \int_0^\infty \Phi(R) g_\sigma(R) R^2 dR - T S_{\text{ex}}(V_l) \qquad (5.19)$$

is the excess free energy relative to an ideal gas at the same volume and $g_\sigma(R)$ is the pair-correlation function of the variationally determined HS reference system. S_{gas} is the ideal-gas entropy and S_{ex} the excess entropy of a hard-sphere fluid. The metallic vapour is considered as an ideal gas of neutral atoms, i.e.

$$F_{\text{g}} = \tfrac{3}{2} k_{\text{B}} T - T S_{\text{gas}}(V_{\text{g}}) + E_{\text{ion}} \quad , \qquad (5.20)$$

where $(-E_{\text{ion}})$ is the energy required to remove all the valence electrons from the free atom (it is necessary to add E_{ion} to ensure that the same zero of energy is taken in the two phases). The pressures are again derived by differentiating the free energy, i.e.

$$\frac{pV_l}{k_{\text{B}}T} = 1 - \frac{V_l}{k_{\text{B}}T} F'_{\text{ex}}(V_l) \qquad (5.21a)$$

(the prime indicates differentiation with respect to volume) and

$$\frac{pV_{\text{g}}}{k_{\text{B}}T} = 1 \quad . \qquad (5.21b)$$

Equating the pressures and the free energies of both phases then gives

$$\begin{aligned}
&1 - V_l F'_{\text{ex}}(V_l)/k_{\text{B}}T \\
&= \exp\left\{ [F_{\text{ex}}(V_l) - V_l F'_{\text{ex}}(V_l) - E_{\text{ion}}]/k_{\text{B}}T \right\} \quad .
\end{aligned} \qquad (5.22)$$

Equation (5.20) has to be solved numerically for V_l and this in turn gives the pressure p along the saturated vapour pressure curve.

With a very simple model potential (but allowing for a slight adjustment of the volume energy), Minchin et al. calculate a one-atmosphere boiling point of $T_{\text{b}} = 1125\,\text{K}$ (exp. $T_{\text{b}} = 1156\,\text{K}$), an associated entropy change on boiling of $\Delta S = 19.2\,\text{cal}\,(\text{kg g-atom})^{-1}$, [exp. $\Delta S = 20.0\,\text{cal}\,(\text{kg-g-atom})^{-1}$], and a vapour pressure curve in good agreement with experiment, at least at lower pressures.

Of course there are limitations to such an exceedingly simple theory – the limitations of a HS-variational theory have already been pointed out. More serious, however, is that the interparticle potential is in reality *qualitatively* different in the liquid and in the gaseous state: in the liquid we have ions whose interactions are determined by Coulomb forces and by the polarization of the electron gas whereas in the vapour we have neutral atoms interacting by van der Waals forces (in the approach sketched above these have been neglected). Any attempt to improve upon the present theory must start by tackling this important point. A more accurate description of the gaseous state should also be able to account for dimerization, which is in fact observed even at elevated temperatures.

6. Alloy Formation and Stability

The stability of a binary alloy depends on the excess free energy ΔG given by

$$\Delta G = \Delta H - T\Delta S + p\Delta V \quad ,$$

where ΔH, ΔS and ΔV are the heat, entropy and volume of formation. Under equilibrium conditions ($p = 0$) the volume of formation makes no direct contribution, but we shall see that ΔV has an important – though often neglected – indirect influence through the volume dependence of the heat and also of the entropy of formation. It is immediately clear that the prediction of the stability of an $A_x B_{1-x}$ alloy formed by the two compounds A and B is an extremely complex problem: the alloy can be either a random solid solution (with complete substitutional disorder and with a crystal structure corresponding either to that of one of the components or an entirely new structure), or a stoichiometric intermetallic compound (which would then be completely ordered), but of course any intermediate situation is possible as well (partial substitutional disorder in an intermetallic compound, long or short range order in a solid solution, etc.). Moreover, the differences in the atomic sizes and in the interatomic potentials will induce static lattice distortions (and thus a certain degree of topological disorder) for any alloy which shows substitutional disorder.

It is therefore not surprising that up to now ab initio calculations have not provided us with a global theory of alloy formation, although their success in predicting heats of formation for alloys of a given composition and a given structure is certainly ever increasing. At present, semiempirical metallurgical concepts continue to play a very important role: alloy formation is discussed in terms of "alloy chemical parameters" such as size, valence, electronegativity etc. – although nature of course needs only one parameter, the nuclear charge [6.1]. The ab-initio methods allow us to see how the alloy chemical parameters are interrelated and give insight into their physical significance.

Therefore we will begin with a very brief discussion of the most important general concepts. The subsequent chapters will supplement this with detailed discussions of the electronic theory of solid solutions, ordering phenomena, intermetallic compounds and liquid alloys. In Chap. 3 we found that even the lowest-order pseudopotential approach is surprinsingly successful in explaining the trends in metallic cohesion – therefore our first step will be to explore how far this method will take us in the case of alloys.

6.1 Nearly-Free-Electron Approach to Alloy Formation

In Sect. 4.1 we have calculated the total energy of a metal using the approximation that the electron gas is perturbed only to first order by the presence of the ionic lattice [see (3.1–3)]. We found that this very simple approach is sufficient to reproduce the main trends in metallic cohesion. It is therefore tempting to explore whether this method can also serve as a guide to alloy formation. We begin by considering the simplest problem – solubility in homovalent alloys.

6.1.1 Criteria for Solubility in Homovalent Systems

To extend the simple first-order theory to a disordered A-B alloy, we further assume that all atomic cells have the same volume (for a solid solution this is equivalent to postulate that the ions occupy the sites of an undistorted lattice and for a liquid solution it means that local fluctuations in the number density are neglected). For a pure metal the electron-ion potential energy has been calculated in the Wigner-Seitz approximation from the boundary condition at the surface of a single unit cell (3.1–3). In our alloy with a homogeneous electron distribution, this analysis applies independently to each of the unit cells, so that the electron-ion potential energy E_0 [the first term on the r.h.s. of (3.3)] is now given by [6.2]

$$E_0^{AB} = -c_A \frac{3Z}{R_a} \left[1 - \left(\frac{R_{c_A}}{R_a} \right)^2 \right] - c_B \frac{3Z}{R_a} \left[1 - \left(\frac{R_{c_B}}{R_a} \right) \right]^2 \quad . \tag{6.1}$$

The kinetic, Coulomb, exchange and correlation energies are still determined only by the electron density parameter R_s, so that the binding energy of the alloy is simply given by

$$E_g^{AB} = c_A E_g^A(R_s) + c_B E_g^B(R_s) \quad , \tag{6.2}$$

where E_g^A and E_g^B are given by expressions identical to (3.1–3) applied to the A and B metals, respectively. The equilibrium density of the alloy is again determined by applying the equilibrium condition

$$P = -(\partial E_g^{AB}/\partial V) = 0 \quad ,$$

resulting in an equation for the equilibrium atomic radius R_a which is a direct generalization of (3.4), i.e.

$$c_A \left(\frac{R_{c_A}}{R_a} \right)^2 + c_B \left(\frac{R_{c_B}}{R_a} \right)^2 = 0.2 + \frac{0.102}{Z^{2/3}} + \frac{0.0035 R_a}{Z} - \frac{0.491}{Z^{1/3} R_a} \quad .$$

(6.3)

As the core radii are determined by the equilibrium condition for the pure A and B metals [see (3.4)], we can eliminate R_{c_A} and R_{c_B} from (6.3). If the small contribution from the correlation energy in (6.3) (the third term on the r.h.s.) is neglected, this results in the following simple relation determining the density of a homovalent alloy [6.3]

$$\overline{R}_s^2 = c_A^2 \overline{R}_{s_A}^2 + c_B^2 \overline{R}_{s_B}^2 \quad \text{with}$$

(6.4a)

$$\overline{R}_s = -1 + 1.229 R_s \quad \text{and}$$

(6.4b)

$$\overline{R}_{s_i} = -1 + 1.229 R_{s_i}$$

(6.4c)

where R_{s_i} $(i = A, B)$ are the equilibrium R_s values of the pure constitutents. Note that (6.4) predicts neither a linear variation of the atomic radius (as postulated by *Vegard*'s law [6.4]), nor a linear variation of the atomic volume (as postulated by *Zen*'s law [6.5]). For a discussion of these two "laws" which are perhaps the only laws in physics for which it is much harder to find a system to agree with, than to disagree with, see e.g. *Pearson* [Ref. 6.6, p. 174 ff]. There are only a few simple-metal systems for which lattice-parameter data over the entire composition range are available: the miscible alkali-metal alloys K-Rb, K-Cs, and Rb-Cs and the alkaline-earth alloy Ca-Sr [6.7, 8]. For these alloys (6.4) predicts only minimal ($\Delta R_s < 0.1\%$) deviations form Vegard's law – the experimentally detected deviations span a range between $+0.2\%$ and -0.2%. The data base for liquid solutions is not much more extensive: for alloy systems with solid solubility as well, the observed deviations tend again to be very small (in accordance with our intuition, ΔR_s is as a rule smaller for liquid than for solid solutions). For systems that are miscible only in the liquid state, somewhat larger deviations are observed (e.g. Na-K, Na-Cs [6.9, 10]). As an example, we show in Fig. 6.1 the deviations from Vegard's law for some solid and liquid alkali alloys – as predicted by the simple relation (6.4) (and by more complete calculations) and as observed experimentally. It is clearly evident that even the simplest theory describes the variation of the interatomic distance in homovalent solutions better than Vegard's law. Note that the asymmetry in ΔR_s is well reproduced.

With the density of the alloy given by (6.4), the heat of formation ΔH can be written as

$$\Delta H = c_A [E_g^A(R_s) - E_g^A(R_{s_A})] + c_B [E_g^B(R_s) - E_g^B(R_{s_B})] \quad ,$$

(6.5)

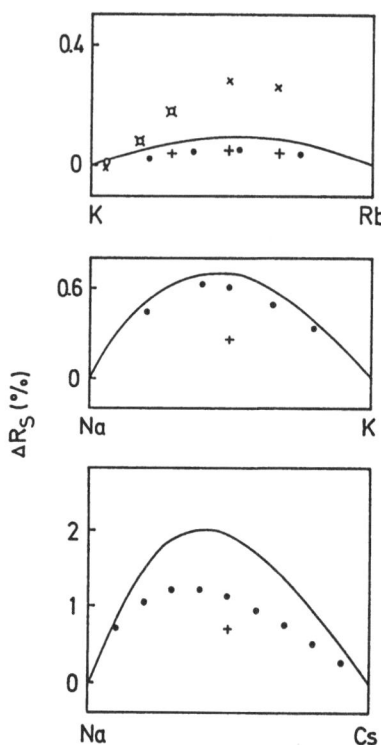

Fig. 6.1a–c. Deviations from Vegard's law in **(a)** K-Rb, **(b)** Na-K and **(c)** Na-Cs solutions; *(solid line)* according to (6.4), (✕): calculated for bcc solid solutions (by minimizing the ground-state energy calculated to second order in the pseudopotential and considering the effect of static lattice distortions, see [6.11], (○): experimental for bcc solid solutions [6.8], (+): calculated for liquid solutions close to the melting point using a thermodynamic variational method [6.12], (●): experimental for liquid solutions [6.9]

thus we find that within this simple approximation the heat of formation of a homovalent alloy consists merely of the change in energy which results from the compression (expansion) of the Wigner-Seitz cells of the two components from their sizes in the pure metals to their actual size in the alloy – clearly ΔH can only be positive. Equation (6.5) has been used by *Girifalco* [6.2] to investigate the solid solubility of 205 homovalent alloys formed by the alkali metals, the alkaline-earth metals, the rare-earth metals, the trivalent metals and the tetravalent metals (Pb, Sn). With ΔH given by (6.5), ΔS was approximated by the ideal entropy of random mixing; a system was called soluble if ΔF at $T = 300\,\mathrm{K}$ was zero or negative for the 50-50 alloy. The results are in good agreement as shown in Figs. 6.2 and 6.3 – all systems that were predicted to be soluble do indeed show solid solubility. In a plot of $R_{c_A}^2$ against $R_{c_B}^2$ the soluble and the insoluble systems are separated by straight lines defining a solubility criterion of the form

$$R_{c_A}^2 \leq R_{c_B}^2 \leq R_{c_A}^2 \left(1.40 - \frac{0.40}{R_{c_A}^2} \right) \quad \text{for} \quad Z = 1 \quad \text{or} \tag{6.6a}$$

$$R_{c_A}^2 \leq R_{c_B}^2 \leq R_{c_A}^2 \left(1.26 - \frac{0.40}{R_{c_A}^2} \right) \quad \text{for} \quad Z = 2 \ . \tag{6.6b}$$

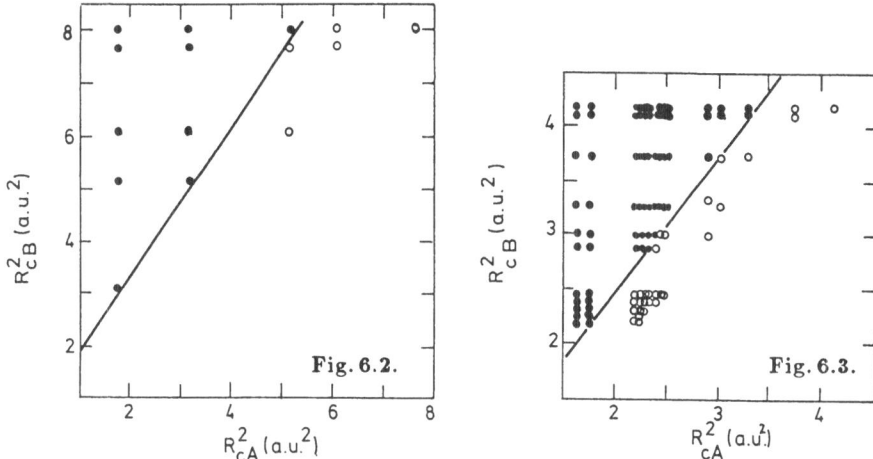

Fig. 6.2. Empty-core radii R_c of random alloys of the alkali metals. Circles indicate that the constituents are predicted to form solid solutions; dots indicate the opposite (after [6.2])

Fig. 6.3. Same as Fig. 6.2, but for the group II and rare-earth metals

Since the second term in the parentheses is small compared to the first, this means that solid solubility will not occur if the difference in the empty core radii exceeds $\sim 18\,\%$ for the alkalis and $\sim 12\,\%$ for the divalent metals. Thus the criterion (6.6) is essentially equivalent to Hume-Rothery's rule that the atomic sizes must differ by less than $15\,\%$ for solid solutions to form. That the solubility, which one would think to be determined by the ratio of the *atomic sizes*, is found to be dependent on the radii of the *ion cores* might seem surprising at first. However, remember that in our analysis of the inter-atomic potentials we found that the position of the first minimum (which is a good measure of the atomic size) is determined by R_c alone (see Sect. 2.3 and [6.13]). Thus we find that at least for homovalent alloys, the simplest approach already represents a valuable guide to the stability of homovalent solid solutions. The situation is more complex in the heterovalent case.

6.1.2 Variations of the Atomic Volume in Heterovalent Alloy Systems

Formally, the theory is easily generalized to heterovalent alloys: we stick again to the approximation of spherical cells, but we contract one type of cell (say those of the A ions, if the valence Z_A is lower than Z_B) and expand the B cells until the electron density in both is the same. The cell radii are then given by

$$\frac{R_{a_A}}{\sqrt[3]{Z_A}} = \frac{R_{a_B}}{\sqrt[3]{Z_B}} = R_s \tag{6.7}$$

149

with R_s referring to the mean electron density in the alloy. Calculating in the usual way the mean value of the ionic pseudopotential plus the potential energy of the homogeneous electron gas, and applying the double-counting correction for the electron-electron interactions, we arrive at

$$E_{ion}^i = -\frac{3Z_i}{R_{a_i}}\left[1 - \left(\frac{R_{c_i}}{R_{a_i}}\right)^2\right] + \frac{1.2Z_i^{2/3}}{R_s} \quad , \quad i = A, B \quad , \tag{6.8}$$

which is a straightforward generalization of (3.3) and (6.1). The A cells contain Z_A electrons, and the B cells Z_B, so that the ground-state energy is finally given by

$$\begin{aligned}
E_g &= c_A Z_A[E_{eg}(R_s) + E_{ion}^A(R_s)] + c_B Z_B[E_{eg}(R_s) + E_{ion}^B(R_s)] \\
&= Z\left[\frac{2.21}{R_s^2} - \frac{0.916}{R_s} + E_{corr}(R_s)\right] \\
&\quad - c_A\frac{3Z_A}{R_s}\left[Z_A^{2/3} - \left(\frac{R_{c_A}}{R_s}\right)^2\right] - c_B\frac{3Z_B}{R_s}\left[Z_B^{2/3} - \left(\frac{R_{c_B}}{R_s}\right)^2\right] \\
&\quad + 1.2(c_A Z_A^{5/3} + c_B Z_B^{5/3})/R_s \quad ,
\end{aligned} \tag{6.9}$$

where $Z = c_A Z_A + c_B Z_B$ is the average valence. Note that the A and B cells now have different volumes – for packing reasons the alloy will have to form some type of ordered lattice. At the equilibrium density of the alloy, we have again $(\partial E_g^{AB}/\partial R_s) = 0$ and this establishes a relation between the core radii and R_s. Eliminating the core radii using the equilibrium conditions for the pure metals yields, in analogy to (6.4) a simple relation for the equilibrium density of the alloy[1]

$$(\overline{R}_s^2 - 1) = c_A d_A(\overline{R}_{s_A}^2 - 1) + c_B d_B(\overline{R}_{s_B}^2 - 1) \quad , \tag{6.10a}$$

with

$$\overline{R}_s = -1 + 0.452R_s\left(0.916 + 1.8\frac{c_A Z_A^{5/3} + c_B Z_B^{5/3}}{Z}\right) \quad , \tag{6.10b}$$

---

[1] A slightly different expression has been proposed by *Ashcroft* [6.3]. He writes the ground state energy in the form

$$E_g = ZE_{eg} + Z\left(c_A\frac{3R_{c_A}^2}{R_s^3} + c_B\frac{3R_{c_B}^2}{R_s^3}\right) - 1.8(c_A Z_A^{5/3} + c_B Z_B^{5/3})/R_s \quad .$$

Evidently the electrostatic contributions are treated in the same way, but the second term is evaluated under the assumption that the conduction electrons see an average electron-ion potential (or equivalently that the electrons are found with equal probability in the A and in the B cells). We think that this is not entirely consistent with the way the electrostatic contributions are calculated. Ashcroft obtains further an expression entirely analogous to (6.10), except that the factor Z_i/Z is missing in the expression for d_i.

$$\overline{R}_{s_i} = -1 + 0.452 R_{s_i}(0.916 + 1.8 Z_i^{2/3}) \qquad (6.10c)$$

and

$$d_i = \frac{Z_i}{Z} \frac{\left(0.916 + 1.8 \dfrac{c_A Z_A^{5/3} + c_B Z_B^{5/3}}{Z}\right)}{(0.916 + 1.8 Z_i^{2/3})} . \qquad (6.10d)$$

A similar analysis of the variation of the density in various heterovalent alloys, but this time based on a slightly more complicated two-parameter pseudopotential has been presented by *van der Broek* [6.14].

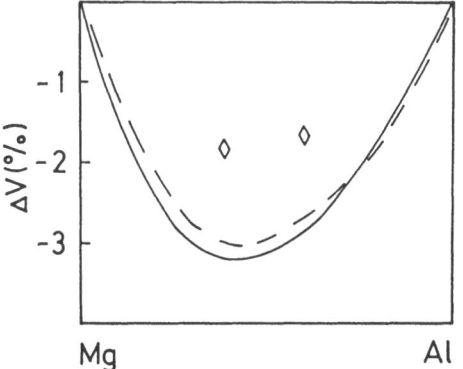

Fig. 6.4. Volume of formation of Al-Mg alloys: (——) calculated using (6.10), (- - -) observed ΔV for liquid alloys [6.15], ($\diamond$) observed ΔV in solid alloys [6.7]

The simple relation (6.10) is found to be surprisingly successful — it reproduces the observed volume contractions not only for alloys with small volume of formation (as an example, ΔV for Al-Mg is shown in Fig. 6.4), but is even able to reproduce the trend towards a large negative excess volume in alloys with strong chemical interactions. For this category we show in Fig. 6.5 the volume of formation of a series of Li-based alloys with polyvalent metals (Mg, In, Pb, and Bi). In view of the extreme simplicity of the approach the agreement with experiment is certainly quite remarkable. [A calculation using Ashcroft's ansatz (Footnote 1) yields somewhat larger volume contractions.] The model also works when the volume increases upon mixing. For a BaLi$_4$ alloy, for example, we calculate a volume expansion of $\Delta V = 3.45\,\%$, whereas experimentally one finds $\Delta V \sim +3.2\,\%$ [6.7].

It is interesting that the variation of the atomic volume with composition is almost linear over a range of about 30–50 at. %, starting from the electropositive metal, and that the intercept of this straight line with the $c_{Li} = 1$ line is nearly independent of the valence of the electronegative element. From Fig. 6.5 we find that the intercept occurs at about $\Delta V \sim -30\,\%$, corresponding to an apparent atomic volume (which is called the "increment" in metallurgy) of $V_{Li} \simeq 100$ a.u. This is the rule of "additive constant increments" [6.20], $V_{AB} = c_x V_x + c_{Li} V_{Li}$ — the "increment" of the poly-

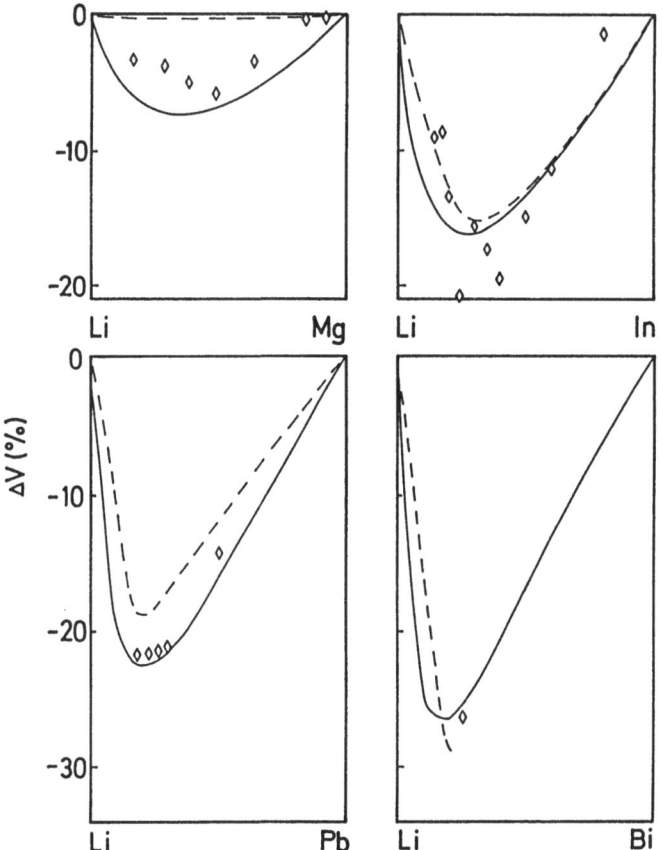

Fig. 6.5. Volume of formation ΔV of Li-based alloys: (——) as calculated using (6.10), (- - -) observed ΔV for liquid alloys, ($\diamond$) ΔV in solid alloys (the experimental data have been compiled in the following references: Li-Mg [6.16], Li-In [6.17], Li-Pb [6.18] and Li-Bi [6.19])

valent metal is practically identical with the atomic volume. This means that the Li atoms are dissolved in the polyvalent metal at a reduced volume which varies only a little with the valence of its partner. As can be seen from Fig. 6.5, the calculation is in good agreement with this empirical rule (the "increment" of the Li atom is found to vary slightly between a value of $V_{\mathrm{Li}} \simeq 87$ u.a. in Li-Bi alloys and $V_{\mathrm{Li}} \simeq 110$ a.u. in Li-Cd in good agreement with the empirical data), but it does not immediately give a simple explanation for the mathematical form of this rule nor for the value of the increment.

The physical mechanism behind these volume changes is rather simple: in our approximation – as expressed by (6.9) – the energy change upon alloying is given merely by the elastic energy necessary to expand or compress the individual atomic cells. Now the energy required to expand the cells occupied by the ions with the higher valence varies much more rapidly

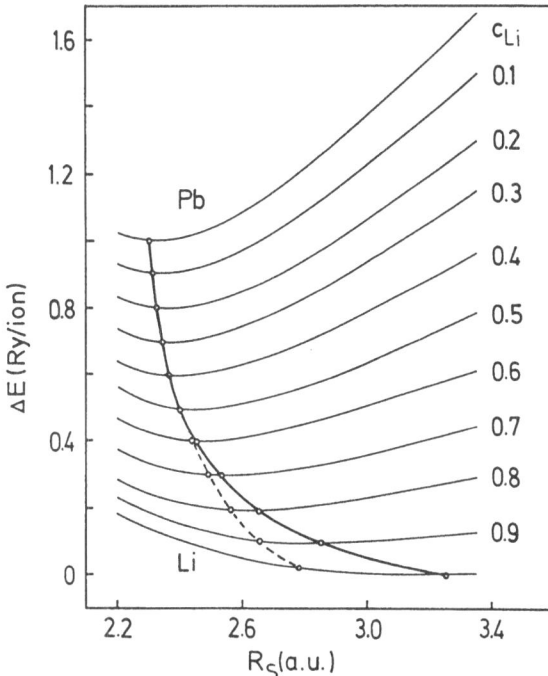

Fig. 6.6. Variation ΔE of the ground-state energy of Li-Pb alloys with the electron density relative to the equilibrium density (the zeros are displaced by steps of 0.1 Ry/ion). The heavy line connects the equilibrium R_s values, the dashed line represents the rule of "additive constant increments", see text

with R_s than the energy required to compress the cells occupied by the ions with the lower valence (Fig. 6.6). Therefore the mean electron density is only little changed upon addition of a small concentration of atoms with a low valence and this shows up as a large volume contraction.

Thus we find that the simple first order perturbation approach is as successful in predicting the volume effects in alloy formation as in predicting the trends in simple metal cohesion. However, it does not yield good results for the heat of formation; it is clear that (6.9) yields only positive values for ΔH. The reason for this failure is simple: the homogeneous deformation of the individual atomic cells until they have reached the same electron density, is only the first step in alloy formation. In the second step the electronic charge distribution is permitted to relax to reach its ground-state distribution by transferring electrons between the cells. The driving force of the charge transfer is the difference in the electrochemical potentials of the prepared cells. Realistic results for the heat of formation can only be obtained by including these charge transfer effects.

6.1.3 The Chemical Potential Model for the Heat of Formation

The electrochemical model for the heat of formation of simple-metal alloys was first considered by *Hodges* and *Stott* [6.21] and *Alonso* and *Girifalco* [6.22, 23]. Their fundamental result may be stated as follows: if the electron density $\varrho_p(r)$ of the prepared alloy (prepared in the sense described at the beginning of the last section) is close to the ground-state density $\varrho_g(r)$, then the difference in energy between the two states, and hence the electrochemical contribution ΔH_{ec} to the heat of formation is

$$\Delta H_{ec} = E[\varrho_g] - E[\varrho_p] \quad . \tag{6.11}$$

One can make a cell separation of the energy according to

$$E[\varrho] = \sum_A E_A[\varrho] + \sum_B E_B[\varrho] + \sum_{i,j} \frac{Q_i Q_j}{R_{ij}} \quad , \tag{6.12}$$

where the single sums involve only the in-cell energies and the third term gives the leading contribution to the intercell energy, the Madelung energy. For $\varrho = \varrho_p$, all the cells are electrically neutral and the Madelung energy is zero. Thus we find for ΔH_{ec}

$$\Delta H_{ec} = \frac{1}{2} \sum_A \int_A \mu_p(r)[\varrho_p(r) - \varrho_g(r)]d^3r$$

$$+ \frac{1}{2} \sum_B \int_B \mu_p(r)[\varrho_p(r) - \varrho_g(r)]d^3r + \sum_{i,j} \frac{Q_i Q_j}{R_{ij}} \quad , \tag{6.13}$$

where $\mu_p(r)$ is the electrochemical potential of the state with electron density $\varrho_p(r)$. If one assumes $\mu_p(r)$ to be a constant in the A cells of the prepared alloy and another constant μ_B in the B cells, then this equation reduces (for an equiatomic alloy) to

$$\Delta H_{ec} = -Q(\mu_A - \mu_B)/4 - Q^2 \sum_{i,j} \frac{1}{R_{ij}} \tag{6.14}$$

per atom, where Q is the charge transferred per cell. If one assumes, as one would expect, that $Q \propto (\mu_A - \mu_B)$, then from (6.14) $\Delta H_{ec} \propto - (\mu_A - \mu_B)^2$ and thus we find an attractive contribution to the heat of formation which increases with the difference in the electrochemical potentials of the constitutents. Up to this point, this is a rather formal result: the important point is the actual calculation of the electrochemical potentials. In a naive approximation one would set μ_A and μ_B equal to their values in the pure metals. In a Na-Al alloy for example, this would result in an electron transfer from near Al to near Na, because E_F is much higher for Al then for Na. This is in direct contradiction to the direction of the charge transfer expected from the electronegativity difference and hence in contradiction to chemical common sense. A method to calculate the chemical potentials of the prepared cells as a function of the electron density has been proposed by *Alonso* et al. [6.24]. It is based on a density-functional version of pseudopotential theory as proposed by *Young* et al. [6.25] (for a recent discussion of the density-functional pseudopotential method for alloys see also [6.26, 27]). The functional for the ground-state energy is written (in atomic units) as

$$E[\varrho] = \int f[\varrho]d^3r - \frac{1}{2}\int \varrho V_{\mathrm{H}}d^3r + V_{\mathrm{ii}} \quad , \tag{6.15}$$

where V_{H} is the Hartree potential of the electron density distribution ϱ, V_{ii} is the Coulombic ion-ion interaction and the functional f is given by

$$f[\varrho] = 5.74\varrho^{5/3} + \frac{1}{4}\frac{(\nabla\varrho)^2}{\varrho} - 1.48\varrho^{4/3}$$
$$- [0.031 \ln (3/4\pi\varrho) - 0.115] - \varrho W \quad . \tag{6.16}$$

The first term is the usual Thomas-Fermi term, the second the Weizäcker inhomogeneity contribution, the third and fourth the exchange and correlation functionals and the last, the electron-ion-interaction – most simply in the form of an empty-core pseudopotential W (2.10, 11). In the spirit of the density-functional formalism, $E[\varrho]$ has to be minimal with respect to variations $\delta\varrho(\mathbf{r})$ of the electron density distribution, subject to the normalization condition $\int \varrho(\mathbf{r})d^3r = NZ$. The Euler equation of this variational problem reads $(\delta f/\delta\varrho) - (V_{\mathrm{H}} - \mu) = 0$ (where μ is a Lagrange multiplier arising from the normalization requirement) or explicit

$$9.57\varrho^{2/3} - (\nabla^2\varrho^{1/2}/\varrho^{1/2}) - 1.97\varrho^{1/3}$$
$$+0.031 \ln (3/4\pi\varrho) - 0.115 - v = 0 \quad . \tag{6.17}$$

Here $v = (W + V_{\mathrm{H}} - \mu)$ and is coupled to ϱ through Poisson's equation $\nabla^2 v = 8\pi\varrho$. These two coupled equations have to be solved subject to the conditions that the only discontinuities are those introduced by the pseudopotentials and that both $\nabla\varrho$ and ∇v vanish at the cell boundaries and at the ionic centers. The LDF pseudopotential equations have been solved by *Alonso* et al. [6.24] in the approximation of spherical atomic cells. Their results for the chemical potential μ (measured relative to the electrostatic potential at the cell boundary) as a function of the electron density at the cell boundary ϱ_{b} (or equivalently the atomic volume) for a series of simple metals are shown in Fig. 6.7. The results show very clearly that the chemical potential depends rather strongly on the atomic volume. The effective $\Delta\mu = \mu_A - \mu_B$ has to be evaluated at the electron density of the alloy, as sketched for the example of Li-Mg and Li-Pb. We see that although in the pure metals we have $\mu_{\mathrm{Pb}} > \mu_{\mathrm{Li}}$ and $\mu_{\mathrm{Mg}} > \mu_{\mathrm{Li}}$, in the alloy the

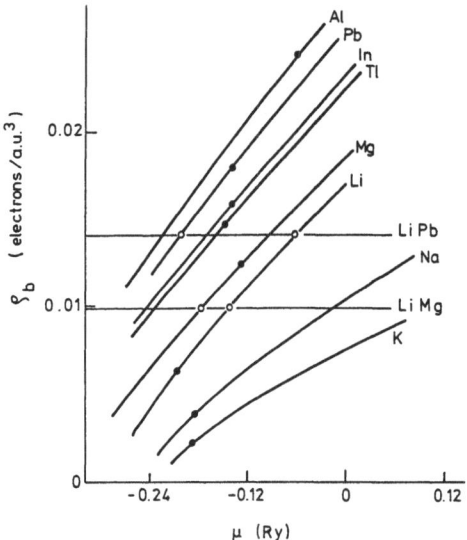

Fig. 6.7. The electrochemical potential μ versus the electron density ϱ_b at the cell boundary, as calculated by *Alonso* et al. [6.22] using the density-functional pseudopotential method. The dot on each curve corresponds to the experimental volume of the pure metal. The circles correspond to the electron densities in Li-Mg and Li-Pb alloys, respectively

sign is reversed and $\mu_{Pb} < \mu_{Li}$ and $\mu_{Mg} < \mu_{Li}$. Thus we find that the direction of the charge transfer is given by the ordering of the μ vs ϱ_b curves (i.e. the order of the μ values at a common value of ϱ_b) and in agreement with chemical common sense. This ordering is dominated by the valence Z and for elements with the same valence by the empty core radius – altogether one finds that the stronger the mean ion-electron potential, the lower is the chemical potential. Moreover we find that except for the heavy alkali metals, the curves are almost parallel – thus the effective $\Delta\mu$ will change very little with volume and the charge redistribution will have only a small influence on the equilibrium volume. This fits very nicely with the results of the preceding section where reasonable values for the volume of formation have been calculated from the elastic deformation energies of the atomic cells alone.

A more difficult point concerns the calculation of the amount of charge actually transferred. In spite of several attempts [6.28–30] we still do not have a simple (but at the same time reliable) theory for calculating the charge transfer for a given difference in the electrochemical potentials. Most schemes rely upon the results of band structure calculations and thus they cannot provide us with the simple picture we are looking for.

If we try to estimate the charge transfer from the difference between the experimental heat of formation ΔH_{exp}, and the "elastic" contribution ΔH_{el} (calculated according to (6.9) for the correct equilibrium density) with $\Delta\mu$ determined from Fig. 6.7 we obtain (assuming a CsCl $B2$ structure) $Q \sim 0.2$ for Li-Mg and $Q \sim 0.43$ for Li-Pb. These are reasonable values and suggest that the physical picture behind the chemical potential model could be correct (we will come back to this point in the next section), but a priori calculations of the heat of formation seem to be very difficult.

6.1.4 Band Picture for Simple-Metal Alloys

An alternative approach to understanding the heat of formation of simple metal alloys has been proposed by *Schlüter* and *Varma* [6.31] and is based on the following general expression for ΔH

$$\Delta H = \frac{1}{2} \sum_i \left[\tilde{V}_i \int_{-\infty}^{E_F} E\tilde{n}_i(E)dE - V_i \int_{\infty}^{E_F^i} En_i(E)dE \right.$$
$$\left. - \frac{1}{2}(\tilde{W}_i - W_i)(\tilde{N}_i + N_i) \right] \quad , \tag{6.18}$$

where $n_i(E)$ is the local density of states (per unit volume) on a site of type i with cell volume V_i, N_i is the number of valence electrons in this cell $[N_i = V_i \int_{-\infty}^{E_F} n_i(E)dE]$, and W_i is the self-consistent mean potential within that cell. All quantities without a tilde refer to the pure metals A and B, those with a tilde to equiatomic A-B alloy. Thus the first two terms in (6.18) are the sum of the one-electron energies in the alloy and in the pure metals and the last term stems from the double-counting corrections for the overcounted electrostatic, exchange and correlation energies. Equation (6.18) is accurate in density-functional theory up to and including terms of order ΔN_i^2, the only assumption being that within a given cell the electrostatic and exchange-correlation potentials are approximately constant. The relation between this band picture and the chemical-potential picture of *Hodges* and *Stott* [6.21] has been discussed by *Pettifor* and *Varma* [6.32]. For the pure metals the mean potential W_i is calculated from a simple pseudopotential and the local density of states $n_i(E)$ is approximated by a free-electron parabola. The key to the calculation of ΔH is then the determination of the local density of states $\tilde{n}_i(E)$ and of the self-consistent mean potential in the alloy. Generally one would expect the following changes in the alloy relative to the pure metals: The lowest state will lie at the self-consistent mean potential $\tilde{W} = (\tilde{V}_A\tilde{W}_A + \tilde{V}_B\tilde{W}_B)/(\tilde{V}_A + \tilde{V}_B)$ of the alloy (i.e. the possiblitiy of having bound states is neglected). Then, if B is the constituent with the more attractive pseudopotential $(\tilde{W}_B < \tilde{W}_A)$, states near the bottom of the conduction band will have large amplitudes in the B cells and small amplitudes in the A cells, i.e. $\tilde{n}_B(E) > \tilde{n}_A(E)$ for $E \sim \tilde{W}$, but for states with a large kinetic energy $\tilde{n}_B(E)$ and $\tilde{n}_A(E)$ will converge to their free electron values[2]. The situation is illustrated schematically in Fig. 6.8. Schlüter and Varma write the $\tilde{n}_i(E)$ in the form

$$\tilde{n}_i(E) = \begin{cases} f_i(E, \alpha, \varepsilon_0)\sqrt{E} & E > \tilde{W} \\ 0 & E < \tilde{W} \end{cases} \quad , \tag{6.19}$$

---

[2] This simple picture is well confirmed by both band-structure calculations (see e.g. *Zunger* [6.33] and *Robertson* [6.34] for Li-Al, *Taut* and *Radwan* [6.35] and *Nagel* et al. [6.37] for Ca-Al) and spectroscopic measurements (see *Tagle* et al. [6.36] for Li-Mg and Li-Al alloys, for Ca-Al alloys see [6.35, 37]).

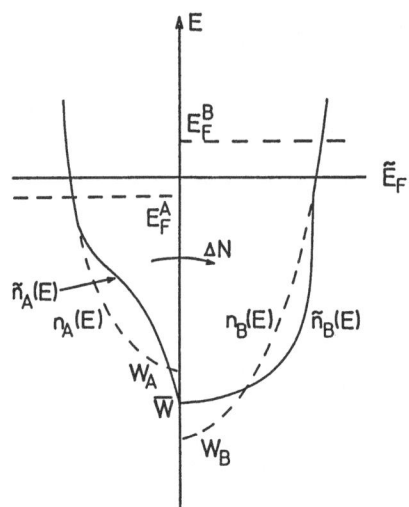

Fig. 6.8. Schematic illustration of the changes in the local densities of states of two metals upon alloying. Before alloying the mean potentials of the two metals are W_A, W_B; their free-electron-like densities of states $n_A(E)$, $n_B(E)$ with Fermi levels E_F^A, E_F^B. After alloying, the bottom of the conduction band is at the mean potential $\overline{W}$, with a common Fermi level $\tilde{E}_F$, the local densities of state are $\tilde{n}_A(E)$ and $\tilde{n}_B(E)$, respectively. In the case shown here, B is the more electronegative element (the one with the more attractive pseudopotential) and electrons are transferred from A to B upon alloying (from [6.31])

where the f_i are simple two-parameter functions which promote the required modulation of the local density of states in the alloy, i.e. $f_B(E) > 1$, $f_A(E) < 1$ for $E \sim \tilde{W}$ and $f_A(E)$, $f_B(E) \to 1$ for $E \gg \tilde{W}$. The two parameters α and ε_0 are determined by the following criteria: (i) a self-consistency criterion requiring that within a B-type cell embedded in the overall mean potential $\tilde{W}$, the lowest state must be an eigenstate with zero energy for a spherical attractive potential of depth $\Delta \tilde{W}_B = \tilde{W}_B - \tilde{W}$, and (ii) a variational criterion in the spirit of density-functional theory requiring the total energy to be a minimum with respect to the electron density $\varrho(\mathbf{r})$ and therefore N_A/N_B.

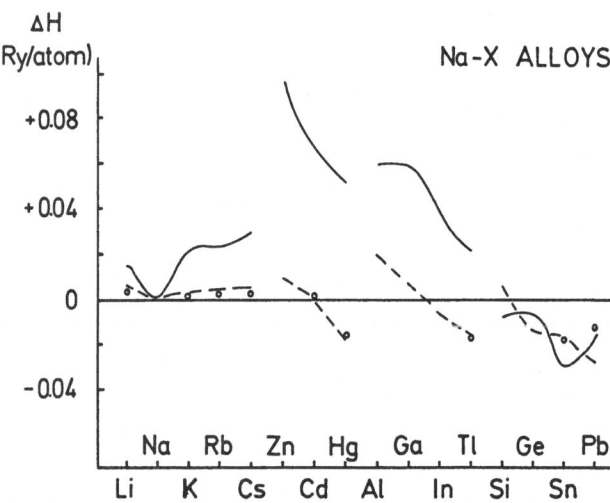

Fig. 6.9. Calculated heat of formation (*solid line*) of equiatomic Na-based alloys with other simple metals, compared with Miedema's semiempirical parametrization (*dashed line*). For most alloys, experimental data are available only for the liquid alloys, they are represented by the open circles. (From [6.31])

As an example of Schlüter and Varma's results we show in Fig. 6.9 the calculated heat of formation of Na-based alloys, compared to the ΔH from a semiempirical model[3]. Two characteristic trends are easily recognizable: for the homovalent alloys the heat of formation is always positive and increases with the difference in the atomic volume. This parallels the trend discussed in 6.1.1. For heterovalent alloys the heat of formation decreases as the atomic number and the valence of the polyvalent component increase. For most alloys of simple metals, the difference in the Fermi energies of the two constituents is much smaller than the difference in the mean potentials. Thus it is the latter which predominantly determines the distortion of the local densities of states and hence the charge transfer. This agrees well with the result discussed in the previous paragraph. Schlüter and Varma find a charge transfer from Na to Al, Ga, In, and Tl of, respectively, 0.14, 0.20, 0.16 and 0.22 electrons (i.e. of the same order of mangitude as estimated from the chemical potential model). Given a certain amount of charge transfer, the heat of formation is composed of a decrease in the kinetic energy ΔE_{kin} (which directly reflects the change in the density of states and the requirement of a common Fermi level, see Fig. 6.8) and a change in the potential energy ΔE_{pot}. ΔE_{pot} has an attractive contribution determined by the difference in the one-electron potentials $\Delta \tilde{W}_i$ and a repulsive contribution roughly proportional to $[(\Delta N/V_A)^{1/3} - (\Delta N/V_B)^{1/3}]$. The latter dominates except for very large $\Delta \tilde{W}_i$ ($\sim$ large difference in valence). The calculated heats of formation reproduce the trends in the experimental or semiempirical data quite nicely, but the absolute values are distinctly too large. Schlüter and Varma attribute this to the fact that the parametrization of the local densities of states is an oversimplification. We feel that the weak point of their approach is their use of Zen's law to calculate the atomic volume of the alloy. Since it is known that most alloys show significant volume contraction this must inevitably lead to an overestimate of the dominantly repulsive ΔE_{pot}.

6.1.5 Real-Space Picture for Alloy Formation

In Chap. 2 we showed that the binding energy of a metal consists of a volume energy E_{V} and a pair-interaction energy E_{p}. The volume energy in turn consists of an electron-gas contribution $\overline{E}_{\mathrm{eg}}$ and of the intra-atomic binding energy $\overline{E}_{\mathrm{ion}}$ (the electrostatic energy of the screening charge in the field of the ion). In Chap. 3 we saw that the trends in simple-metal cohesion are dominated by the volume term – the pair interaction energy contributes at most 2–3 % of the binding energy. It is therefore tempting to explore whether the variation of the volume energy on alloying can also serve as a guide to alloy formation.

---

[3] The semiempirical model for ΔH proposed by Miedema will be discussed in the following section.

The electron-gas term is simple to handle [see (3.6)]; the difficulties lie in the calculation of the intra-atomic term. However, we also found (see Table 3.2) that the very simple expression proposed by *Finnis* [6.38], $\overline{E}_{ion} = \frac{1}{2}\Phi_{ind}(R = 0) \simeq - Z^2/2R_c$,correctly reproduces all the important trends. Note that in this approximation $\overline{E}_{ion}$ is independent of the atomic volume. Furthermore we know that the pseudoatoms remain electrically neutral (this is not a contradiction to the concept of charge transfer), thus at this level of approximation, the intra-atomic term varies linearly with concentration in the alloy, $\overline{E}_{ion} = -\frac{1}{2}(c_A Z_A/R_{c_A}^2 + c_B Z_B/R_{c_A}^2)$ and does not contribute to the heat of formation (in a more realistic calculation, the self-energy of the pseudo-atoms would be allowed to change with the electron density). In this case ΔH would be given simply by an electron-gas term and a pair-interaction term. The idea that the heat of formation of simple-metal alloys is dominated by the variation of $\overline{E}_{eg}$ was first put forward by *Pettifor* [6.39]. This electron-gas contribution ΔH_{eg} is easily calculated from (3.6),

$$\Delta H_{eg} = \overline{E}_{eg}(R_s) - c_A \overline{E}_{eg}(R_{s_A}) - c_B \overline{E}_{eg}(R_{s_B}) \tag{6.20}$$

with

$$\overline{E}_{eg} = Z\left(\frac{0.982}{R_s^2} - \frac{0.712}{R_s} + 0.031 \ln R_s - 0.110\right) \quad,$$

where R_s is the electron-density parameter of the alloy. The physical content of (6.20) becomes more evident if we use an approximate result stated by Pettifor,

$$\Delta H_{eg} = - Z(1.228 - 0.356 R_s - 0.031 R_s^2)[\Delta(1/R_s)]^2 \quad, \tag{6.21}$$

which is valid to order $[\Delta(1/R_s)]^2$ for equiatomic alloys that follow Zen's law. Note that $1/R_s \sim \varrho^{1/3}$. Thus the electron-gas contribution to the heat of formation goes as the square of the difference in the cube roots of the electron density $(\Delta \varrho^{1/3})^2$, with a prefactor which depends on the average electron density. The three terms in (6.21) are the kinetic, exchange and correlation contributions. This shows that ΔH_{eg} consists of an attractive contribution from the relaxation of the electron-gas kinetic energy and a repulsive contribution from the loss of exchange and correlation energy on alloying – at higher electron densities the attractive contribution dominates.

In Figure 6.10 the heat of formation per electron, normalized by Z. $[\Delta(1/R_s)]^2$, is plotted against $(1/R_s)$: in part a) we show the result of local-density-functional band-structure calculations by Pettifor and Gelatt (as cited in [6.39]) for CsCl-type binary alloys of Na, Mg, Al, Si and P, and in part b) we show the experimental ΔH for a large series of liquid simple-metal alloys (these values are mostly taken from the compilation of *Hultgren* et al. [6.40]). The results for the solid alloys fall approximately on a single curve and follow broadly the trend in the electron-gas contri-

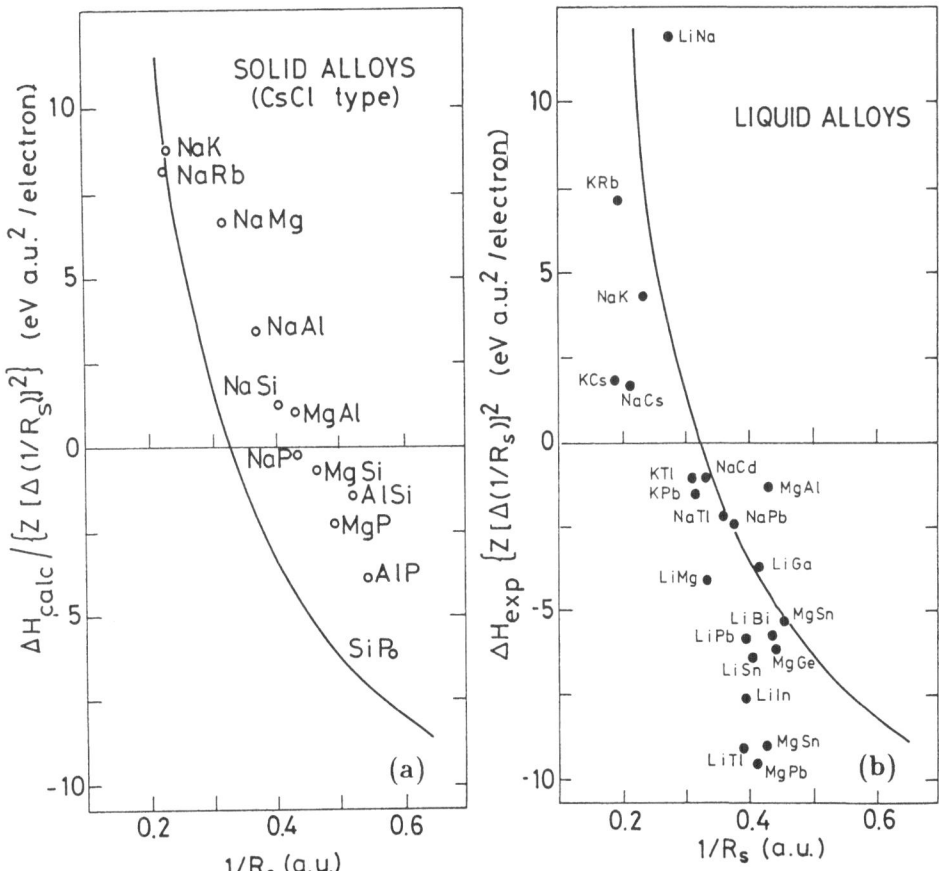

Fig. 6.10a,b. The heat of formation per electron normalized by $Z[\Delta(1/R_s)]^2$ as a function of $1/R_s$: **(a)** LDF calculations for solid alloys fo the Na series, assuming a Cs-Cl structure (after [6.39]). **(b)** Experimental heat of formation of liquid alloys (most of the results are taken from the compilation of *Hultgren* et al. [6.40], see also [6.41]). The solid line represents the electron gas contribution as given by (6.21)

bution. The pair-interaction contributions are of course responsible for the structural energy. None of the systems really forms a CsCl-type alloy, so the fact that the LDF results are higher than ΔH_{eg} is not unreasonable. The heats of formation of the liquid alloys show somewhat more scatter, but again they follow essentially the trend set by the electron-gas contribution ΔH_{eg}. Structural effects and the change of the self-energies of the pseudo-atoms on alloying, are responsible for the variation of ΔH with respect to the master curve given by (6.21). The increasingly negative ΔH in a series such as Li-Ga, Li-In, Li-Tl is associated with an increasing chemical short-range order (see e.g. [6.41] for a detailed discussion of the thermodynamic and structural manifestations of chemical short-range order in liquid alloys). This is just one example which illustrates that structural effects can be very important for the heat of formation of simple-metal alloys – much more so than for transition-metal alloys. Structural effects are dominant especially in those cases where small electron-density differences mean that the electron-gas contribution is small. A full second-order calculation for Ca-Mg alloys [6.42] yields $\Delta H > 0$ for solid solutions and $\Delta H < 0$ for liquid alloys and for the intermetallic compund $CaMg_2$; similar results have been obtained for homovalent alkali alloys [6.43]. In all cases the structural differences in ΔH can be explained in terms of the arrangement of the nearest neighbour atoms with respect to the minimum in the pair potentials. For the heterovalent alloys the effect of the electron gas is more important and sets the general trend.

For the alloys as for the pure metals, the equilibrium density can be determined only by considering both the volume and the pair-interaction forces.

6.1.6 A Brief Summary

The foregoing sections have been devoted to the discussion of three simple models for alloy formation, so it might be useful to stop for a while and to discuss their interrelations. Model (i) (Sect. 6.1.3) is based on a two-step alloying process and decomposes the heat of formation, ΔH, into a repulsive contribution representing the elastic energy necessary to deform both types of atomic cells until they have reached the same electron density and attractive contribution from the subsequent relaxation of the electron density distribution to equilibrate the chemical potentials. Model (ii) (Sect. 6.1.4) starts from a band picture: in the alloy the difference in the strength of the pseudopotentials creates a distortion of the local density of states which results in a lowering of the kinetic energy and in general in an enhancement of the potential energy (this includes electrostatic, exchange and correlation energies). Model (iii) (Sect. 6.1.5) considers only the electron-gas contribution and again one finds an attractive contribution from the relaxation of the kinetic energy and a repulsive contribution from the loss in potential energy, but this time of a homogeneous electron gas.

Compared to model (ii), the much simpler model (iii) neglects both the difference in the mean potentials W_i and the deformation of the local density of states. We find that the electron-gas contribution enables us to explain the trend of ΔH of e.g. Na alloys with the elements of groups II and IV (the values are $\Delta H_{eg} = 0$, -0.12, and $-0.36\,eV$ for alloys of Na with Cd, Tl and Pb, to be compared with Fig. 6.9), which arises from the lowering of the electron gas kinetic energy on alloy formation. The model also reproduces the variation of ΔH in the alloys of Na with different elements of its own group. The latter is associated with a change in the relative size of the A and B cells.

Comparing model (i) and model (iii) we have to remember that when we transform from the reciprocal-space to the real-space representation the large first-order and intra-cell electrostatic terms, E_{ion}, [see (3.3)] cancel against the volume average of the pair interaction, except for the term $-\frac{1}{2}V_a B_{\text{eg}}$ which appears as a consequence of the compressibility sum rule (2.22–25). It is precisely E_{ion} which dominates the repulsive "elastic" contribution ΔH_{el}. For alloy formation to occur, this has to be more than compensated for by the electrochemical term ΔH_{ec} (given by the second- and higher-order contributions). Model (iii) takes advantage of this compensation which largely explains its success.

To understand the heat of formation, perhaps the most attractive approach is via the simple electron gas model. This correctly sets the important trends and allows us to express the deviations from ΔH_{eg} in terms of changes in the pair interactions on alloying. Volume effects on the other hand are best discussed in reciprocal space [model (i)] – both the volume- and the pair-interaction terms are strongly volume-dependent and in this case the cancellation of the large effects is seen better in q-space.

Pettifor [6.44] has remarked that the electrochemical model is essentially *classical*, being based on the charge flowing from one atom to the other to equilibrate the local chemical potentials. In its original form it is also based implicitly on a *rigid-band* approximation. The second-order pseudopotential theory on the other hand is *quantum mechanical*. The ions are screened by the response functions of the *common band* of free electrons.

6.2 Miedema's Semiempirical Theory of Alloy Formation

Over the last few years, *Miedema* and coworkers [6.45] have developed an exceedingly simple, but nonetheless very successful semiempirical scheme for calculating the heats of formation of binary alloys. Each element is characterized by two chemical coordinates, ϕ^* and $\varrho^{1/3}$ and the heat of formation is then given by

$$\Delta H = (Q\Delta\varrho^{1/3} - P\Delta\phi^* - R)f(c) \quad , \tag{6.22}$$

where P and Q are positive constants (in fact they are slightly different for transition-metal and simple-metal alloys and for liquid and solid alloys) and R is a parameter which accounts for of p-d hybridization effects in alloys of simple metals with transition metals. The function $f(c)$ represents the usual concentration dependence appropriate e.g. to a regular solution [in that case $f(c) = c(1-c)$]. The attractive contribution to ΔH depends on the difference $\Delta\phi^*$ in the work functions ϕ^* of the elements and is essentially equivalent to *Pauling*'s [6.46] electronegativity term. The repulsive term depends on the difference in the cube root of the electron density at the boundary of the Wigner-Seitz cells of the individual elements and was argued to arise from the charge-density mismatch at the interface of the A and B cells. Thus the physical basis of this expression is closely analogous to the two-step model discussed in Sect. 6.1.3: the first term describes the "elastic" energy necessary to deform the individual atomic cells until the electron-density mismatch has been eliminated, and the second term is the energy associated with the charge transfer occurring when two metals with different workfunctions ($\sim$ chemical potentials) are brought into contact. Allowing for some adjustment of the individual chemical coordinates, Miedema's expression has been extremely useful in predicting quantitative values for ΔH.

Miedema was especially successful in treating both solid and liquid alloys of transition metals. A single pair of values for Q and P were found to be sufficient, suggesting that structural effects are rather unimportant. For transition metals structural effects enter only in the function $f(c)$, (see [Ref. 6.45, p. 14] for details). For non-transition metals the situation is more complicated. Firstly, the constants Q and P are about 30 % smaller than for transition metals but with this adjustment good results for ΔH in liquid alloys of non-transition metals are achieved (typically within a factor of 2–3 if $\Delta H > 1$ kcal/g-atom). For solid alloys of non-transition metals, however, even the sign of ΔH is incorrectly predicted in a relatively large number of cases. Upon closer inspection one finds that these exceptions concern systems either of a rather unusual stoichiometry and structure (e.g. KZn_{13}), or those containing at least one semiconducting or semimetallic element (Si, Ge, Sb, Bi). Examples in the second category are Na-Si, Zn-Ge, Al-Si, Al-Sb etc. Miedema concludes – and we tend to agree – that structure-dependent contributions to the heat of formation cannot be neglected for non-transition metal alloys.

The interesting problem of course is the microscopic interpretation of Miedema's expression for the heat of formation.

6.2.1 Microscopic Interpretation of Miedema's Alloying Rules: Simple Metals

For the simple metals, the electrochemical model discussed in the previous section, seems to provide a natural explanation of Miedema's rule. *Alonso* and *Girifalco* [6.22] have shown that the "elastic" energies needed for the expansion/compression of the atomic cells agree in approximate magnitude with Miedema's repulsive terms, and using (6.9), their result is easily confirmed for simple-metal alloys. The more difficult question concerns the microscopic interpretation of Miedema's electronegativity scale ϕ^*. Here again the density-functional calculation of the chemical potentials [6.24] is helpful: it is shown that the chemical potentials evaluated at a *common* electron density for both constituents correlate very well with Miedema's work func-

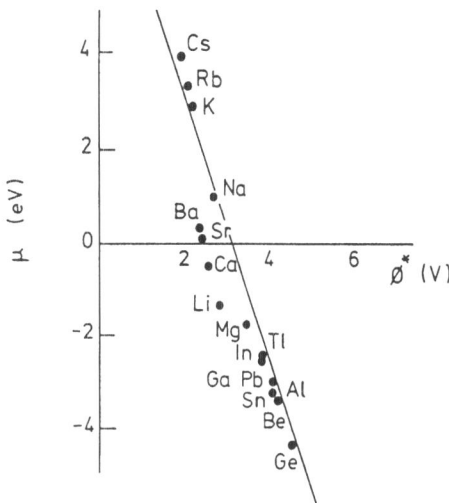

Fig. 6.11. Calculated chemical potential μ of the simple metals (for states with a cell-boundary electron density of $\varrho_b = 0.013\,e/a.u.^3$) versus Miedema's electronegativity ϕ^* (after [6.24]). The straight line is a guide to the eye

tion (see Fig. 6.11). The correlation is quite convincing and suggests that $\phi^* \alpha \mu$ and hence the proportionality of the attractive terms in (6.11) and (6.22), i.e. $P\Delta\phi^{*2} \propto \Delta\mu^2$. The problem is that the chemical-potential model provides no means of calculating the charge transfer of the alloy. Only when this is known can one assign the constant of proportionality and thus test this interpretation not only qualitatively, but also quantitatively. On the other hand, the quantitatively successful real-space model shows no direct correlation to the ideas underlying Miedema's model.

6.2.2 Microscopic Interpretation of Miedema's Alloying Rules: Transition Metals

For transition-metal alloys the chemical-potential picture has recently been criticized by *Williams* et al. [6.47] on the basis of first-principles band-structure calculations. Both the electrochemical contribution and the "elastic" contribution can be tested against the band-structure results and the conclusion is that the chemical-potential picture is inappropriate for transition metals although expected to hold for simple-metal alloys.

For transition-metal alloys, *Pettifor* [6.48] has shown that by analogy to Friedel's [6.49, 50] tight-binding treatment of the transition-metal cohesion, the d-band of a transition-metal alloy may be modelled by a rectangular density of states whose width W_{AB} is given by [6.51]

$$W_{AB} = (W^2 + 3\Delta E_d^2)^{1/2} \quad . \tag{6.23}$$

The first term is the contribution to the alloy band width which arises from nearest-neighbour bonding and the second term gives the additional bonding due to the chemical differences of the constituent metals measured by the difference $\Delta E_d = E_d^B - E_d^A$ of their d-electron energies.[4] Filling up the alloy band with the average number $\overline{N}_d$ of d-electrons yields the d-band energy of the alloy and subtracting the average d-electron cohesion energy of the pure metals gives an attractive contribution to the heat of formation which may be written to second order in $\Delta N_d = N_d^B - N_d^A$ as [6.48]

$$\frac{\Delta H_0}{W} = -\frac{1}{80}(\overline{N}_d)^2 - \frac{1}{4}\Delta N_d\left(\frac{\Delta E_d}{W}\right)$$
$$- \frac{3}{40}\overline{N}_d(10 - \overline{N}_d)\left(\frac{\Delta E_d}{W}\right)^2 \quad . \tag{6.24}$$

In addition there is another contribution ΔH_1 to the heat of formation which arises from the fact that the d-band widths of the elemental metals are generally not equal, due to the different atomic volumes V_A and V_B. Assuming that the d-d transfer integral (and consequently the band width) follows an inverse fifth power law with interatomic distance [6.52] and that the alloy volume follows Zen's law, then

$$\Delta W = W_B - W_A = -\tfrac{3}{5}W(\Delta V/\overline{V}) \tag{6.25}$$

(to first order in ΔV). The energy change due to the change of the band width of the elemental metals from W_A and W_B to W, gives the following contribution to ΔH

$$\frac{\Delta H_1}{W} = -\tfrac{1}{24}(5 - \overline{N}_d)\Delta N_d \Delta V/V \quad . \tag{6.26}$$

Pettifor's simple model achieves very good agreement with Miedema's prediction (see Fig. 6.12). There is no one-to-one correspondence between ΔH_0, ΔH_1 and the attractive and repulsive terms in Miedema's formula. Essentially ΔH_1 stands for Miedema's repulsive contribution but is determined by the mismatch in the d-band width rather than in the boundary electron-density. A refined version of the d-band model has been presented by *Watson* and *Bennett* [6.53,54] who also tabulate heats of formation of the 3d, 4d, and 5d series. The d-band model is fully supported by first-principles band-structure calculations of *Williams* et al. [6.47]. These calculations also

---

[4] In a dimer the energy levels E_A, E_B are shifted to [6.51]

$$E_{AB}^{\pm} = \overline{E} \pm \tfrac{1}{2}(4h^2 + \Delta E^2)^{1/2},$$

$[\overline{E} = (E_A + E_B)/2]$ where h is the transfer integral (hopping matrix element). Thus the bonding-antibonding splitting is

$$W_{AB} = (4h^2 + \Delta E^2)^{1/2} = (W^2 + \Delta E^2)^{1/2} \quad ,$$

the first term describing the covalent, the second the ionic contribution to the bond. Equation (6.23) is the direct generalization of this result to a bulk alloy [6.48].

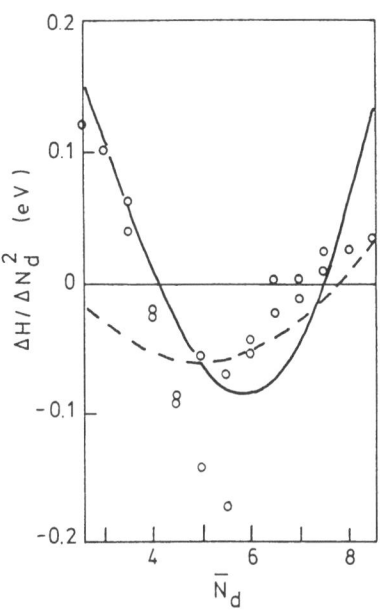

Fig. 6.12. $\Delta H/(\Delta N_d)^2$ as a function of the average band filling across the $4d$ transition metal series (*solid line*); the dashed line is the contribution of ΔH_0 (6.24) alone. The circles represent Miedema's values (from [6.48])

demonstrate that the complete neglect of s- and p-electron contributions to the heat of formation is justified – the reason lies in a cancellation of inter-atomic and intra-atomic Coulomb electrostatic effects. In view of these results it might seem puzzling that Miedema's scheme works so well. Williams et al. have shown that this is due to the fact that any combination of parameters will do the same, as long as they satisfy certain boundary conditions; evidently this is the case for Miedema's ϕ^* and $\varrho^{1/3}$ as for Pettifor's $\Delta N_d, \overline{N}_d$. With regard to the utility, both are equivalent, but only Pettifor's argument provides the correct physical picture.

Thus, somewhat ironically, one finds that the ideas on which Miedema's model was originally based, are not really applicable to those systems for which the model works best. On the other hand the treatment is not unreasonable for those systems (the simple-metal alloys) where its quantitative success is limited due to the importance of structural effects. But here again we tend to side with *Pettifor*'s [6.44] quantum mechanical critique of the Miedema rules in preferring the common-band picture based on second-order pseudopotential theory. One of the reasons is that this allows a unifying view of alloy formation in simple- and in transition-metal systems: The heat of formation arises from the creation of a common band within the alloy.

Therefore Miedema's alloy-chemical coordinates should be considered as extremely useful empirical parameters allowing predictions with an easiness that will not be reached by the quantum-mechanical calculations for some time to come.

A discussion of structural effects in alloy formation requires a full calculation of the ground-state energy to at least second order in the pseudopotential and a detailed consideration of problems such as transferability, optimization and nonlocality of the pseudopotentials used. This is the subject of the following chapters.

Recent Results

We would like to draw the readers attention to *Pettifor*'s [6.44] fundamental quantum-mechanical critique of the Miedema rules for alloy formation. This article also contains an interesting extension of the real space picture for alloy stability. Pettifor shows that the pseudoatom-self energies give an important (generally repulsive) contribution to the heat of formation.

7. Solid Substitutional Alloys

For a disordered substitutional alloy, the structure-dependent contributions to the total energy are most conveniently written in the form [see (2.52–54)]

$$E_2 = \sum_{q \neq 0} \{ |S(\boldsymbol{q})|^2 F_{NN}(q) + [S(\boldsymbol{q})D^*(\boldsymbol{q}) + S^*(\boldsymbol{q})D(\boldsymbol{q})] F_{Nc}(q)$$
$$+ |D(\boldsymbol{q})|^2 F_{cc}(q) \}$$

and analogously for the electrostatic energy. For primitive lattices and for non-primitive lattices in which all sites are equivalent the product of the average structure factor $S(\boldsymbol{q})$ and the difference structure factor $D(\boldsymbol{q})$ vanishes, i.e.

$$D^*(\boldsymbol{q})S(\boldsymbol{q}) = S^*(\boldsymbol{q})D(\boldsymbol{q}) = 0 \tag{7.1}$$

and $D(0) = 0$ (the proof is given by *Inglesfield* [7.1], see also [7.2]). For a *completely* random alloy we find easily that

$$D^*(\boldsymbol{q})D(\boldsymbol{q}) = c(1 - c) \tag{7.2}$$

(for partially ordered alloys, $D(\boldsymbol{q})$ may be expressed in terms of the order parameter, see below). In the disordered case the second-order energy may be written as

$$E_2 = \overline{E}_{bs} + \Delta E_{bs} \quad \text{where} \tag{7.3}$$

$$\overline{E}_{bs} = \sum_{q \neq 0} |S(\boldsymbol{q})|^2 F_{NN}(q) \tag{7.4}$$

is the band-structure energy of the *virtual crystal* described by the average pseudopotential $W = c_A W_A + c_B W_B$ and the difference term ΔE_{bs} given by

$$\Delta E_{bs} = c(1 - c) \frac{V_a}{(2\pi)^3} \int F_{cc}(q) d^3 \boldsymbol{q} \tag{7.5}$$

is structure insensitive. The electrostatic energy is simply that of the virtual crystal (it is easily shown that the difference term vanishes [7.3]),

$$E_{es} = \alpha_M \overline{Z}^2 / R_a \quad . \tag{7.6}$$

In the real-space representation, the only structure-dependent term is again the virtual-crystal part of the pair-interaction energy

$$E_P = \frac{1}{2N} \sum_{lm} \phi_{NN}(R_{lm}, V) \tag{7.7}$$

with the volume energy E_V given by (2.56). Thus, for completely random alloys, the structure-dependent part of the total energy is well described in the virtual-crystal approximation[1] and a discussion of the stable alloy structures is completely analogous to the discussion of the crystal structures of the pure metals. Equations (7.1–7) provide the formal background for a second-order perturbation calculation of the ground-state energy of substitutional alloys to which we will turn now.

The basic empirical rules governing the phase stability and structure of many substitutional alloys were first proposed in the early 1930's by *Hume-Rothery* and coworkers [7.4,5]. Three rules concern the formation of primary solid solutions. These are: (1) the size rule, (2) the electrochemical factor, and (3) the relative valence effect. The first rule states that the formation of solid solutions is not to be expected if the atomic sizes of the solvent and the solute differ by more than 15 %, and that solid solutions *may* form if the size difference is less than 15 %, provided the other factors are favourable. Rule (2) states that the electrochemical nature of the two elements must be similar for solid solution formation to occur (if the contrary is true, compound formation is likely). Rules (1) and (2) may be combined in what is called a Darken-Gurry map [7.6]: each element is represented by a point in a plot of the atomic radius (abscissa) versus the electronegativity (ordinate). An ellipse centred at the solvent and with one axis of ± 15 percent of the solvent's size and another axis of ± 0.4 electronegativity units circumscribes most of the solutes with a solubility greater than 5 %. Rule (3) states that a disparity in valence leads to a lower solubility and that this disparity has an especially pronounced effect when the valence of the solute is less than that of the solvent. Evidently rules (2) and (3) are rather qualitative in nature and not very useful in a quantitative analysis. Rule (3) especially has been often criticized – evidently its weakness is that it completely neglects the different nature of sp- and d-electrons. For alloys formed by two sp-bonded elements, however, rules (1–3) seem to apply quite well, provided that both elements have one of the common metallic structures (otherwise the interatomic forces are sufficiently different to inhibit the formation of an extended solid solution).

But the most famous Hume-Rothery rule is the one which relates the structure of a substitutional alloy to the mean valence (or valence electron

---

[1] But remember that $\phi_{NN}(R)$ is calculated for the alloy, and is not equal to the concentration-average of the component pair interactions.

concentration) Z. For example, in the brass-type alloys of the noble metals with B-group metals the disordered α (fcc) phase has a stability limit near $Z = 1.36$ and the disordered β (bcc) phase is the only stable structure at high temperatures in a narrow concentration range around $Z = 1.48$. Attempts to explain such correlations in terms of the electronic structure date back to the pioneering work of *Mott* and *Jones* [7.7]. The following sections are devoted to a discussion of the physics behind Hume-Rothery's rules in the light of modern pseudopotential theory.

7.1 Primary Solid Solutions

7.1.1 The Homovalent Case

The simplest possible alloys are the homovalent alloys formed by the group Ia and IIa metals. Given the rather small differences in the electrochemical potentials and in the electronegativities between the metals of one group, we would expect the solid solubility to be governed by the size factor alone. Experimentally one finds that this is indeed the case: a continuous series of solid solutions is formed in the systems (the number in parentheses gives the ratio of the atomic radii) K-Rb (1.07), K-Cs (1.15)[2], Rb-Cs (1.08), Ca-Sr (1.03), Ca-Ba (1.13), Sr-Ba (1.10), Mg-Cd (1.03)[3], Mg-Hg (1.01)[3], Cd-Hg (1.03)[3] – for all other systems the radius ratio exceeds the critical value of 1.15 and the solid solubility is virtually zero.

A number of pseudopotential calculations have been presented for these systems. The simplest possible approach which goes beyond a first-order calculation is the "pseudo-alloy atom" model proposed by *Tanigawa* and *Doyama* [7.11]; in fact this is a virtual-crystal approximation in which the electron-ion interaction is described by an average pseudopotential – in their case by an empty-core model potential whose core radius R_c is assumed to vary linearly with concentration, i.e. $R_c = c_A R_{c_A} + c_B R_{c_B}$ (a similar approach for a two-parameter model has been proposed by *Soma* et al. [7.12]). Hence all effects of disorder, size difference etc. are eliminated and it is therefore not surprising that the predictions of this model are essentially identical to the electron-gas contribution ΔH_{eg} to the heat of formation [see (6.20–21) and Table 7.1]. The same kind of virtual-crystal approximation has also been used within the framework of a pseudopotential local-density-functional calculation by *Iniguez* and *Alonso* [7.13]. They find

---

[2] K-Cs is a border-line case: at ambient temperature the solid solubility is unlimited, at temperatures $T < 180$ K two intermetallic compounds of a topologically close-packed type are formed (K_2Cs, K_7Cs_6 [7.8]).

[3] These systems show a continuous series of solid solutions at normal temperatures and ordered compounds at lower temperature (see the discussion in *Inglesfield* [7.9] and *Leung* et al. [7.10]).

Table 7.1. Disordered bcc alloys of the alkali metals: heat of formation ΔH

Alloy	ΔH_{eg}, eq. (6.5) [cal/g-atom]	PP[a]	VC[b]	VC-LDF[c]	exp.[d]
NaK	1882	1150	1063	2435	
NaRb	3130	3160	1788	3855	
NaCs	4985	5133	2853	6727	
$K_{0.3}Rb_{0.7}$	147	130	81	154	46
KCs	782	700	100	1258	166
RbCs	219	120	63	478	13

[a] Second-order pseudopotential calculation with a nonlocal first-principles pseudopotential [7.8].

[b] Second-order pseudopotential calculations, using a virtual-crystal approximation for a local model potential [7.11].

[c] Local-density-functional calculation in a virtual-crystal approximation [7.13].

[d] Exp., estimated from ΔH for liquid solution and heat of fusion [7.14].

very large positive values for ΔH which, in view of the preceding remarks, must appear of questionable significance. Full second-order calculations using first-principles pseudopotentials optimized for each concentration of the alloy, have been peformed by *Hafner* [7.8]. The predictions of this calculation are reproduced in Fig. 7.1 and Table 7.1. We find that the heat of formation is only slightly lower than the elastic contribution (6.5) based on

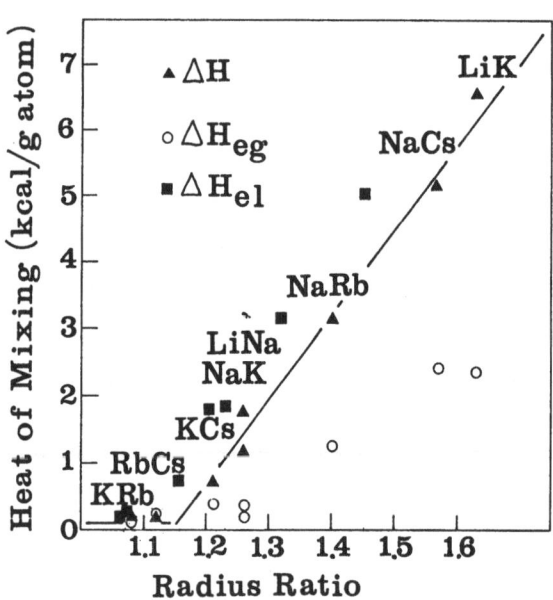

Fig. 7.1. Heat of formation of disordered bcc alkali alloys versus radius ratio. ΔH – result of a second-order calculation using an optimized first-principles pseudopotential; ΔH_{eg} – electron gas contribution (6.21), ΔH_{el} – elastic (first order + electrostatic) contribution (6.4–5). Note that ΔH is plotted versus the radius ratio calculated from the theoretical equilibrium densities of the pure metal – which are slightly lower than the experimental data

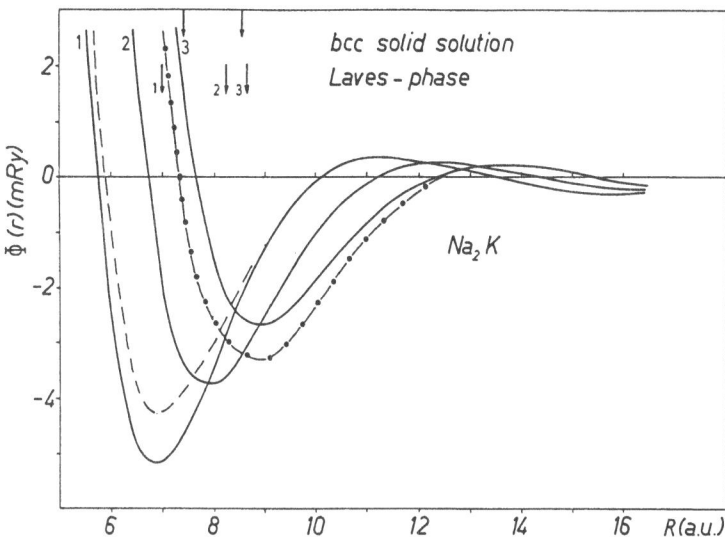

Fig. 7.2. Interatomic pair potentials in a Na_2K alloy [(solid lines) the curves labelled 1, 2, 3 refer to the Na-Na, Na-K, and K-K potentials, respectively] and in pure Na (*dashed line*) and pure K (*dot-dashed line*). The vertical arrows mark the interatomic distances in the bcc metals and solid solution and in the Laves phase Na_2K

a first-order expression; thus a full calculation essentially corroborates Girifalco's simple argument. In a real-space picture we find that the electron-gas term ΔH_{eg} is responsible only for a part of the actual heat of formation – were ΔH given only by ΔH_{eg}, Li-Na and Na-K would show substantial solid solubility, contrary to experiment. Thus the size factor is reflected in the self-energies of the pseudo-atoms and in the pair-potential term. Figure 7.2 shows that indeed the interatomic distances of a bcc solid solution do not fit very well into the minima of the pair potentials of the alloy, whereas they do fit nicely with the pair potentials of the pure metals. The heat of formation is found to vary linearly with the radius ratio with an intercept of $R_A/R_B = 1.15$ on the abscissa – this is again Hume-Rothery's size factor rule. Unfortunately, no experimental values for the heat of formation are available for a quantitative test of these predictions. All we can do is to estimate ΔH for solid solutions from the known ΔH for liquid solutions and the heat of fusion ΔH_f [7.14]. This is only possible of course for those compositions at which the solid solution melts congruently (i.e. at the minimum of the liquidus curve). We find that the theoretical values for the heat of formation lie distinctly above the experimental estimates, but not high enough to suppress the solid solubility at ambient temperatures (assuming an essentially ideal entropy of mixing). *Solt* and *Zhernov* [7.15] and *Hafner* and *Punz* [7.16] have shown that a large part of the discrepancy between theory and experiment is due to the neglect of lattice relaxation – in fact the relaxation energy is found to be of the same order of magnitude as the heat of mixing, a point to which we shall return in Sect. 7.3.

For the divalent metals, the elastic contribution ΔH_{el} is again a valuable guide to solid solubility. For all systems that are known to show unlimited solubility, we find $\Delta H_{el} \lesssim 500\,\text{cal/g-atom}$, except for Ca-Ba. Full second-order perturbation calculations have been performed for Ca-Mg [7.17], Cd-Mg [7.18], Mg-Zn [7.19], Cd-Zn [7.20] and Ba-Cd [7.19], using various types of pseudopotentials; virtual-crystal LDF pseudopotential calculations have been presented for the alkaline-earth alloys [7.21] and for Mg-Zn [7.22]. Again the heats of formation obtained from the LDF calculations appear to be unrealistically large – both for a prediction of the correct trends, and in comparison with the more realistic calculations which include disorder. As an example, Fig. 7.3 shows the heat of formation of disordered hcp Mg-Cd alloys [1.25] compared with the experimental results. In contrast to the case of monovalent alloys, the band-structure contribution is of fundamental importance for obtaining realistic values for the heat of formation – remember that the first-order contribution can only give positive values of ΔH!

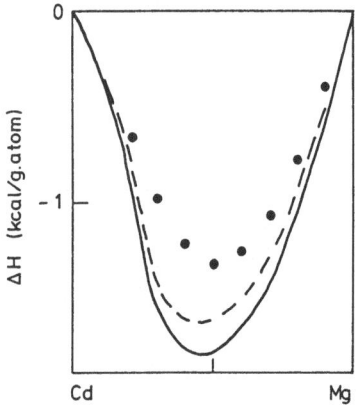

Fig. 7.3. Heat of formation of solid disordered Mg-Cd alloys. The dots indicate the experimental results; the curves correspond to the result of calculations using different choices for the empty core radii (after [7.18])

The formation of solid solutions between the heavier group IIb and the tri- and tetravalent metals are complicated by structural effects (see Sect. 3.3). The interesting structural changes in Cd-Mg and In-Tl alloys have been discussed by *Finnis* and *Heine* [7.23] and *Gunton* and *Saunders* [7.24]. In both cases the change in the crystal structure entails a softening of certain elastic modes. In the case of Cd-Mg alloys the hcp phase with axial ratio close to ideal is stable in the Mg-rich region but as the Cd content increases beyond 30–40 at.%, a shear distortion of the hexagonal phase takes place (Fig. 7.4) which is accompanied by a softening of all the elastic moduli. In the case of In-Tl, the hcp phase is stable up to ∼10 at.% In and the fcc phase (which is also the high-pressure polymorph of Tl) for concentrations beyond 55 at.% In. As the In content exceeds 80 at.%, a tetragonal distortion takes place. The shear modulus $\frac{1}{2}(C_{11} - C_{12})$ which corresponds to such a distortion of the cubic phase tends to zero at this concentration, while the other elastic moduli stay finite.

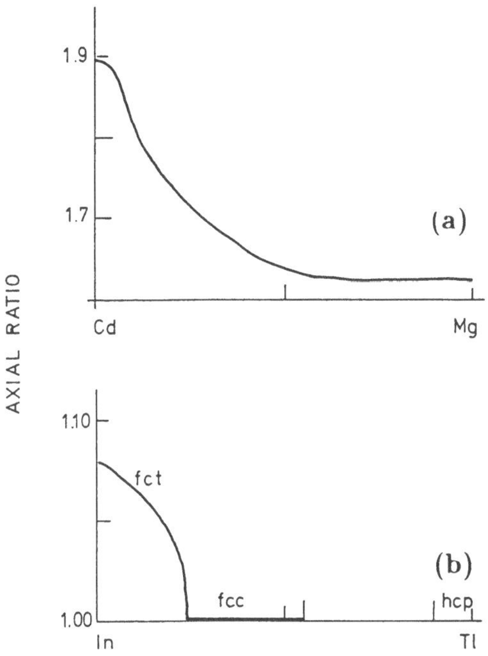

Fig. 7.4a,b. Axial ratio c/a in hexagonal Mg-Cd alloys (a) and in cubic and tetragonal In-Tl alloys (b) (after [7.24])

The instabilities may be discussed within the same framework which we used to treat trends in the crystal structures of the elements – in fact, the analysis of the alloys has preceded and partially guided the study of the elements. A crystal is stable against all homogeneous deformations if the elastic moduli are positive. They can be evaluated by standard methods. In the case of an hcp crystal, the four elastic moduli may be expressed by the nearest-neighbour tangential and radial force constants T and R via

$$C' = \tfrac{1}{2}(C_{11} - C_{12}) = (4V_a)^{-1/3}(19T/6 + 5R/6)$$

$$C = \tfrac{1}{6}(C_{11} + C_{12} + 2C_{33} - 4C_{13}) = (4V_a)^{-1/3}(3T + R)$$

$$C_{44} = (4V_a)^{-1/3}(10T/3 + 2R/3)$$

$$B = (4V_a)^{-1/3}(-8T/3 + 4R/3) + \text{volume terms} \quad . \tag{7.8}$$

(C, C' and C_{44} are shear constants and B is the bulk modulus.) Equations (7.8) relate the elastic moduli to the form of the pair potential $\Phi_{NN}(R)$ at the nearest-neighbour distance. The relevant trends in $\Phi(R)$ have been shown in Figs. 3.4 and 3.5 for the elements; the $\Phi_{NN}(R)$ in the alloys shows a smooth transition between these limiting cases. In the case of Mg we find a minimum at the nearest neighbour distance D_{cp} of the close packed phase. With increasing Cd content the oscillation is shifted with respect to D_{cp}.

Once D_{cp} has moved beyond the inflection point, $\mathcal{T}$ becomes small and $\mathcal{R}$ negative – clearly this will lead to a drastic decrease of all the elastic constants and phonon frequencies and to a possible instability. The first elastic modulus to vanish will be the one with the smallest coefficient in front of $\mathcal{T}$ – i.e. the first instability is $C = 0$, corresponding to a change in the axial ratio.

In the case of an fcc crystal the shear moduli may be expressed as

$$C' = \tfrac{1}{2}(C_{11} - C_{12}) = (4V_a)^{-1/3}(7\mathcal{T}/2 + \mathcal{R}/2)$$

$$C_{44} = (4V_a)^{-1/3}(3\mathcal{T} + \mathcal{R}) \quad . \tag{7.9}$$

As shown in Sect. 3.3, the close-packed structures of Tl and Pb are stabilized by a large positive curvature $\mathcal{R}$ against a negative gradient $\mathcal{T}$. Going from Tl to In one finds a strong decrease of $\mathcal{R}$, while $\mathcal{T}$ stays negative. Due to the different numerical coefficients in (7.9), the softening will be much more pronounced for the shear mode C' connected with the instability towards tetragonal distortion. *Gunton* and *Saunders* [7.24] have calculated the phonon dispersion curves for In and fcc and fct In-Tl alloys using the optimized model potential. The presence of a soft acoustic mode along [110], polarized in [1$\bar{1}$0] has been established. The closer the alloy composition is to the phase boundary, the softer the mode becomes. The stabilization of fcc against hcp at intermediate concentrations is a more subtle effect. As in the case of the pressure-driven hcp $\rightarrow$ fcc transition in pure Tl, it can only be discussed with reference to interactions with more distant neighbours [7.25].

Structural effects also dominate the solubility trends in groups III and IV in general. From Table 3.4 we see that the structural energy differences for e.g. the Ga I $\rightarrow$ fcc transition in Ga ($\Delta E = 2000$ cal/g-atom) or the Sn I $\rightarrow$ fcc transition in Sn ($\Delta E = 1400$ cal/g-atom, [7.26]) are large enough to suppress the formation of Al-Ga or Sn-Pb solid solutions in spite of a favourable size ratio ($R_{Ga}/R_{Al} = 1.06$, $R_{Pb}/R_{Sn} = 1.05$) and no difference in the electronegativity (according to the Gordy-Thomas [7.27] scale). This leaves only the Ge-Si system as a possible candidate for the formation of a continuous series of solid solutions. A pseudopotential calculation to second order, including some leading third-order corrections has been performed by *Bublik* et al. [7.28]. They predict a small but positive heat of formation for a disordered solid solution ($\Delta H \simeq 300$ cal/g-atom). At low temperatures, this should lead to phase separation and Bublik et al. claim to have found experimental evidence for this effect by diffuse x-ray diffraction. However, *Logan* et al. [7.29] conclude from phonon spectra measurements using inelastic neutron scattering that Ge-Si forms an ordered diamond-type lattice with preferred heterocoordination. The heats of solution at infinite dilution have been calculated by *Singh* and *Young* [7.30] for a large number of simple-metal alloys using empty-core pseudopotentials. For homovalent systems, the calculated heats of solution are consistent with those found experimentally.

7.1.2 The Heterovalent Case

In the previous section we found that even for homovalent alloys, a general analysis of solubility trends is considerably complicated by structural effects. In heterovalent alloys additional complications arise from the differences in electron density (valence) and electronegativity. We begin by considering the relatively simple case of Li-Mg alloys.

Detailed investigations of the Li-Mg system have been presented by *Hafner* [7.31] using optimized first-principles pseudopotentials, by *Beauchamp* et al. [7.32] using DRT-pseudopotentials, and by *Belenkij* et al. [7.33] using model potentials. Hafners calculation yields reasonably accurate results for the volume and heat of formation and for the crystal structures (see Fig. 7.5), Beauchamp et al. obtain a ΔH which is about twice as large (assuming Vegard's law) and the ΔH and ΔV given by Belenkij et al. are too large by at least a factor of five. This shows that rather subtle effects such as the concentration-dependent changes in the pseudopotential as described in Sect. 2.4 are indeed important. The results shown in Fig. 7.5 are remarkable in two respects: (i) the structural energy difference ΔE changes abruptly at those concentrations at which the Fermi surface touches a Brillouin-zone plane in one of the structures. (ii) The fact that a structure has the lowest energy at a given composition does not necessarily mean that it is actually stable. For example, fcc is the structure with the lowest energy for concentrations between about 15 at. % and 35 at. % Mg, but the ΔH curve has a positive curvature at precisely these compositions and a common-tangent construction shows that the fcc phase is unstable against a decomposition into a Li-rich hcp phase and a bcc phase with ~ 50 at. % Mg. Therefore, the fcc phase can only be formed as a metastable phase – indeed at concentrations between 13.4 and 19.7 at. % Mg, a martensitic transition to an fcc phase can be induced by cold working [7.34]. Hence the calculation shown in Fig. 7.5 not only correctly reproduces the observed sequence of hcp-bcc-hcp phases (at low temperatures), but also the existence of a metastable fcc phase in the correct composition region.

The stabilization of the bcc phase at intermediate concentrations ($Z \simeq 1.5$) is the first example of Hume-Rothery's valence electron concentration (VEC) rule and of its traditional Fermi surface interpretation – we shall come back to this point in a moment. Point (ii) is evidently related to a certain asymmetry of the $\Delta H(c)$ curve – the heat of solution is larger for Li dissolving in Mg than for Mg dissolving in Li, and this is clearly reminiscent of rule (3). The effect becomes even more pronounced if the disparity in the component valence increases. As a further example Fig. 7.6 shows the heat

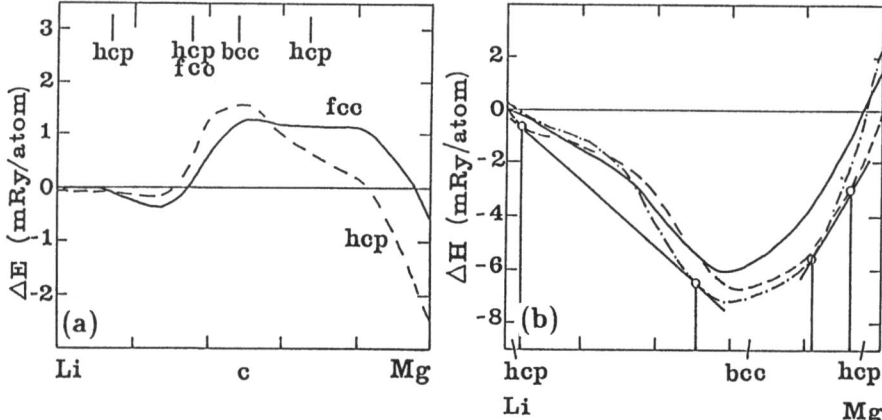

Fig. 7.5a,b. Structural energy differences relative to the bcc phase (a) and heat of formation ΔH (b) of disordered Li-Mg solid solutions (—— fcc, -·-·- bcc, - - - hcp). The common tangent construction gives the stability limit of the different phases. The vertical bars mark those concentrations for which the absolute value of a reciprocal lattice vector equals twice the Fermi momentum, $|Q| = 2K_F$ (after [7.31])

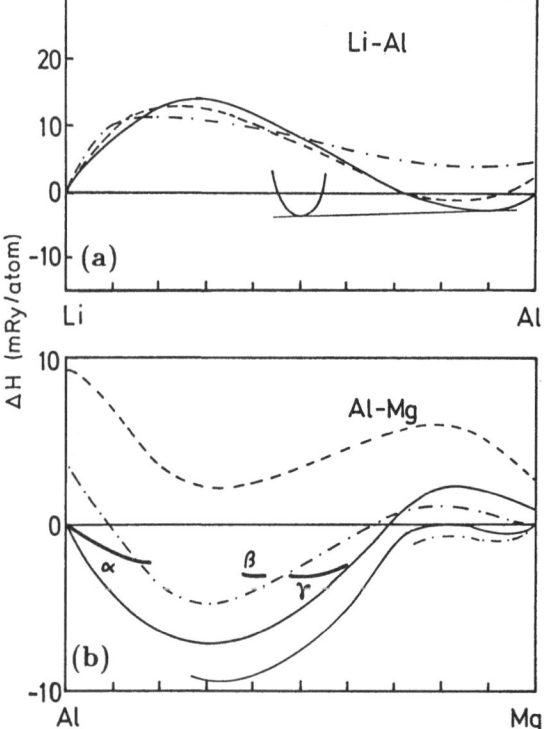

Fig. 7.6. (a) Heat of solution of disordered Li-Al solid solutions (—— fcc, -·-·- bcc, - - - hcp) and of a LiAl compound with Na-Tl structure. **(b)** Heat of solution of disordered Al-Mg solid solutions (——fcc, - - - bcc, -·-·- hcp). The upper solid line gives ΔH as calculated for $T = 0\,\mathrm{K}$, the lower line ΔG for $T = 300\,\mathrm{K}$ (calculated with the assumption of a temperature-independent ΔH and an ideal entropy of mixing ΔS). The curves labelled α, β, γ show the experimental heat of formation of the terminal solid solution and of the intermetallic β and γ phases. The common-tangent construction shows the limits of solid solubility. Part (a): unpublished results of Hafner, Part (b): after [7.35]

of formation of disordered solid Li-Al solutions (calculated using optimized pseudopotentials). First we note that the sequence of the structures with the lowest energy is hcp-fcc-bcc-hcp-fcc – the first half parallels the sequence found in the Li-Mg system. The $\Delta H(c)$ curve is S-shaped with negative values occuring only at the Al-rich end. Thus we predict a considerable solid solubility of Li in Al, but no solubility of Al in Li, in agreement with experiment.

As a final example we consider solid solutions in the Al-Mg system [7.35]. The results are shown in Fig. 7.6 (b) and fit rather well into the general framework: the structural sequence is hcp-fcc, the solubility of Mg in Al is large and that of Al in Mg is small. Experimentally, the Al-Mg system is characterized by two stable intermetallic compounds with complex cubic structures [7.36], but metastable phases have also been investigated. According to *Murray* [7.37], supersaturated fcc solid solutions may form up to a composition of 38 at. % Mg. Thus we find that the calculated diagram reproduces the solubility trends in this system reasonably well.

To summarize, we have found that (i) the energetically most favourable structure is determined largely by the valence electron concentration (VEC) in the following sequence: VEC $\cong$ 1 – 1.15 : hcp, VEC $\cong$ 1.15 – 1.35 : fcc, VEC $\cong$ 1.35 – 1.80 : bcc, VEC $\cong$ 1.8 – 2.25 : hcp, VEC $\cong$ 2.25 – 3 : fcc. Note that a change in the crystal structure sets an absolute limit to the stability of a phase, but it does not immediately imply that the next phase is going to be stable. The stability depends on the non-structural contribution to the heat of formation as well. (ii) The shape of the $\Delta H(c)$ curve is asymmetric. The asymmetry increases with the disparity in the component valence and also with the electron density. Its origin is mainly in the "elastic" term ΔH_{el} (see Sect. 6.1.2); it simply reflects the differences in the bulk moduli of the components (see Sect. 3.1 for a discussion of the general trends): the component with the lower valence is much more compressible and thus it fits relatively easily into the host lattice. The component with the higher valence has a larger bulk modulus and if it is introduced into the host lattice it produces a relatively large strain energy.

Generally we find that the Hume-Rothery rules (1) to (3) are of little use for predicting the stability of heterovalent solid solutions of simple metals. Among the A-group elements, a Darken-Gurry plot predicts the formation of Li-Mg, Na-Ca and K-Ba solid solutions. In fact only Li-Mg forms extended solid solutions, the other systems are immiscible both in the solid and in the liquid states. The essential feature which distinguishes between the Li-Mg system on one hand and the Na-Ca and K-Ba systems on the other, is the strength of the pseudopotentials: in the Li-Mg system the difference in the pseudopotentials is sufficiently strong to produce an appreciable difference in the chemical potential and thus the second-order effects can overcompensate the repulsive elastic contribution. In the other two systems these effects are too small and the elastic contribution dominates. Among the *B*-group elements, the formation of extended solid solutions is strongly influenced by structural effects; extended solid solutions are found only in those systems where both components have one of the common metallic structures, Al-Zn, (In,Tl)-(Sn,Pb) – here again it is characteristic that the range of stable substitutional alloys is much smaller in (In,Tl)-Sn than in (In,Tl)-Pb. The reason is that in the Sn alloys the heat of formation is reduced by the amount of energy necessary to transform Sn from its equilibrium structure to a metallic close-packed one. In this connection the extensive work of *Giessen* [7.38, 39] on metastable *B*-metal alloy phases may be of interest. This should allow a test of theoretical predictions of alloy structure – but evidently this is a vast field where much remains to be done.

7.2 Hume-Rothery Phases

The origin of the Hume-Rothery rules relating the range of stability of the certain so-called "electron-compounds" to characteristic values of the valence electron concentration is one of the most famous problems of physical metallurgy. The rule was originally formulated for alloys of the noble metals with polyvalent metals, [4] but we have already seen for a few examples of simple-metal alloys that here again the sequence of the stable structures is dominated by the valence electron concentration.

The first interpretation of the Hume-Rothery idea was given by *Mott* and *Jones* [7.7]. The idea was that an alloy phase should have a special stability when its Fermi surface just touches a Brillouin-Zone plane, i.e. if $|Q| = 2k_F$ where Q is a vector of the reciprocal lattice. Later, the argument fell into doubt when detailed model calculations showed that even a fairly large gap in the dispersion relation $E(k)$ results only in a fairly small van Hove singularity in the density of states (evidently the strength of the singularity depends on how many Brillouin zone planes touch the Fermi surface), which would then tend to be further reduced and smeared out after integrating up the total energy.

What then can account for the success of the Hume-Rothery rule? *Blandin* [7.41, 42] and *Heine* [7.43] were the first to propose an explanation based on pseudopotential theory. The band-structure energy depends on the pseudopotential form factors and the polarizability $\chi(q)$ of the uniform electron gas, evaluated at the reciprocal lattice vectors Q. The polarizability $\chi(q)$ (2.16) drops sharply as q approaches $2k_F$ and its derivative has a logartihmic singularity at this point. Blandin and Heine speculated that the energy of a particular crystal structure will drop rapidly as $2k_F$ passes through $|Q|$ and could fall below that of other crystal structures (see Fig. 7.7). Subsequently *Stroud* and *Ashcroft* [7.44] and *Evans* et al. [7.45] [5] performed cal-

---

[4] For a recent review on Hume-Rothery phases see *Mizutani* and *Massalski* [7.40].

[5] Later work by *Rahman* [7.46] has extended their work to finite temperatures.

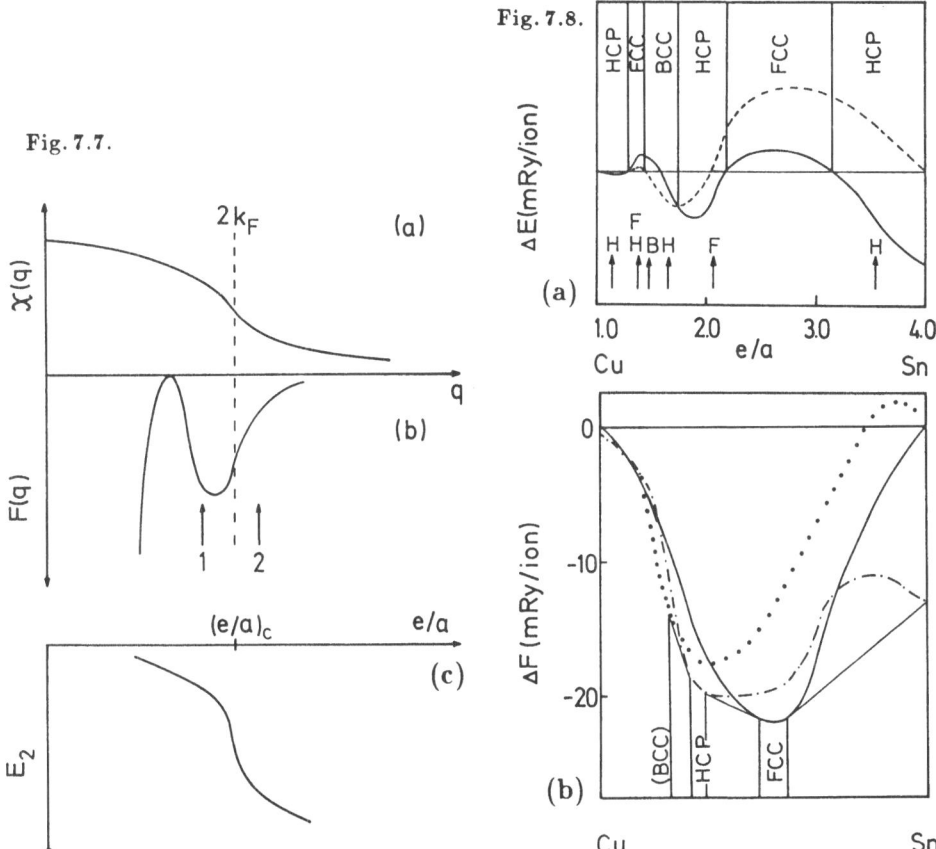

Fig. 7.7.

Fig. 7.8.

Fig. 7.7. Pseudopotential explanation of the origin of the critical valence-electron concentrations: *(a)* Negative polarizability $-\chi(q)$ of the free-electron gas. *(b)* Typical form of the energy-wavenumber characteristic $F(q) = (V_a/2)[\chi(q)/\varepsilon(q)]w(q)^2$. The arrows labelled 1 and 2 show the relative positions of a reciprocal lattice vector $\boldsymbol{Q}$ below *(1)* and above *(2)* the critical value $(e/a)_c$ of the valence electron concentration for which $|\boldsymbol{Q}| = 2k_F$. *(c)* Corresponding behaviour of the second-order band-structure energy E_2 as a function of the valence-electron concentration

Fig. 7.8. (a) Structural energy differences ΔE (bcc-fcc) *(dashed line)* and ΔE (hcp-fcc) *(solid line)* for Cu-Sn alloys. After [7.45]. The vertical arrows mark the concentrations at which $|\boldsymbol{Q}| = 2k_F$ for the structure indicated. (b) Free energy of formation ΔF (at $T = 298$ K) for the fcc (——), bcc ($\cdots$) and hcp (- - -) phases of Cu-Sn alloys. The common tangent constructions for the stability limits are indicated. After [7.46]

culations on brass-type alloys (analogous to the simple-metal calculations discussed in the last section) and concluded that the Hume-Rothery rule arises from the rapid variation of the polarizability in the vicinity of $2k_F$. A typical result is shown in Fig. 7.8. For the system Cu-Sn the predicted sequence of energetically favourable structures is hcp-fcc-bcc-hcp-fcc-hcp (note the agreement with the simple-metal calculations, both as regards the sequence of structures and the critical values of the electron/atom ratio).

The prediction of hcp stability for pure Cu is a spurious result, evidently the pure noble metals are treated correctly only with proper consideration of the d-band, but otherwise the agreement with experiment is satisfactory: the predicted bcc stability corresponds to the β and γ high-temperature phases, hcp stability appears as the ζ phase (hexagonal) and the ε phase (orthorhombic, but may be regarded as a superlattice on a hcp basis), fcc stability would correspond to the η (possibly NiAs type) phase, but here the correlation is not so clear. These calculations are based on empty-core pseudopotentials and completely neglect the hybridization of the s-band with the noble-metal d-band. *Kogachi* [7.47] has used Moriarty's d-band pseudopotentials to investigate Ag-Mg, Au-Mg and Ag-Al alloys. His calculations qualitatively reproduce the experimental phase diagrams, i.e. the existence of the β and ε phases in the Ag-Mg system, of the β phase in Au-Mg and of the β and η phases in Ag-Al. Kogachi also discussed ordering effects (see Sect. 7.4).

Very recently *Borisova* et al. [7.48] have presented a detailed study of the structures of intermetallic CuZn and AuCd phases, including a discussion of the martensitic transformations B2 → L1₀ (tetragonally distorted fcc) and B2 → B19 (orthorhombically distorted hcp) and a variational determination of the reaction path for these transformations.

Evans et al. [7.45] also made an attempt to put the argument on a firm mathematical basis. It is easy to see that the second-order energy may be decomposed into a one-electron contribution and a term arising from the double-counting corrections according to [see (2.19, 38, 39) and Appendix B]

$$
\begin{aligned}
E_2 &= -\frac{V_a}{2} \sum_q {}' \left(\frac{\chi(q)}{\varepsilon(q)} |w(q)|^2 \right) \delta_{q,Q} \\
&= -\frac{V_a}{2} \sum_q {}' (\chi(q)|\tilde{w}(q)|^2 + v_{\text{eff}}(q)|\varrho_{\text{sc}}(q)|^2) \cdot \delta_{q,Q} \quad .
\end{aligned}
\tag{7.10}
$$

where $\tilde{w}(q) = w(q)/\varepsilon(q)$ is the screened pseudopotential and $\varrho_{\text{sc}}(q) = \chi_0(q)w(q)$ the screening charge density. The first term represents the one-electron contribution and the second the electron-electron interaction. One finds that the one-electron term dominates the structural energy differences. Using an integral representation of the polarizability, Evans et al. show that the structural energy difference is given essentially by

$$
\Delta E(\text{bcc-fcc}) \sim \int_0^{2k_F} H'(k)dk
\tag{7.11a}
$$

and analogously for other combinations of structures.

$$
H'(k) \sim (4k_F^2 - k^2)w(k)^2 [G^{\text{BCC}}(k) - G^{\text{FCC}}(k)]
\tag{7.11b}
$$

and the function $G(k)$,

$$G(k) = \left(\sum_q{}' \delta_{q,Q} - \frac{V_a}{(2\pi)^3} \int d^3q \right) (k^2 - q^2)^{-1} \tag{7.12}$$

is a structure constant which has simple poles whenever $k = |k|$ is equal to the modulus of a reciprocal lattice vector[6]. Only the poles with $k \leq 2k_F$ contribute to the structural energy differences – they arise from the analytical properties fo the Lindhard function. In the integrand $H'(k)$, the poles of the structure constants are weighted with the square of the screened pseudopotential; those in the immediate vicinity of $2k_F$ are damped by the prefactor. The form of the $H'(k)$ for different values of the electron/atom ratio is shown schematically in Fig. 7.9. The structural energy difference is given by the *integral* over $H'(k)$. When $2k_F > Q_1^{FCC}$, $H'(k)$ has a negative contribution in the range $Q_1^{FCC} < k < 2k_F$ which grows with Z. A further negative contribution develops as k approaches Q_1^{BCC} showing that the form of the integrand in the vicinity of $2k_F$ is as important as the singularity itself.

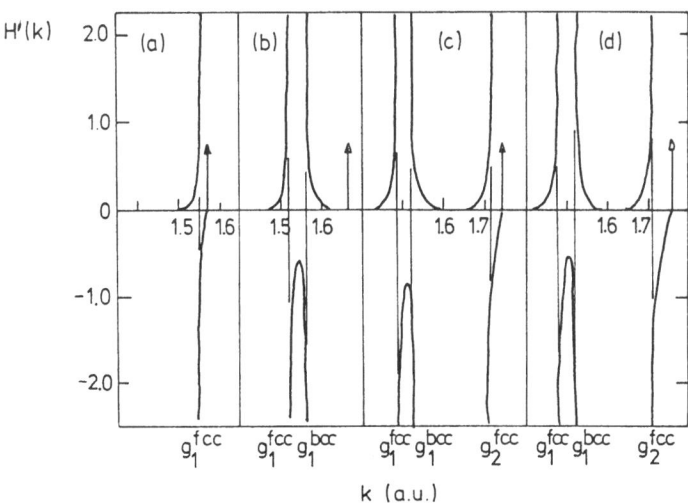

Fig. 7.9. The form of the integrand $H'(k)$ in (7.11b) as a function of k for different values of the e/a ratio equal to the mean valence $\overline{Z}$ (schematically, for CuAl, see text). (a) $\overline{Z} = 1.4$ (fcc), (b) $\overline{Z} = 1.8$ (bcc), (c) $\overline{Z} = 2.2$ (bcc), (d) $\overline{Z} = 2.4$ (fcc). After [7.45]

From the foregoing discussion it is clear that the microscopic origin of the Hume-Rothery phases is now quite well understood. *Rahman's* [7.46] extension of the calculations to finite temperatures (the techniques used in this generalization will be discussed later) shows that the theory is also quite reliable in predicting phase boundaries.

---

[6] The structure constant converges slowly, in practice it is necessary to use an Ewald-type transformation for its evaluation.

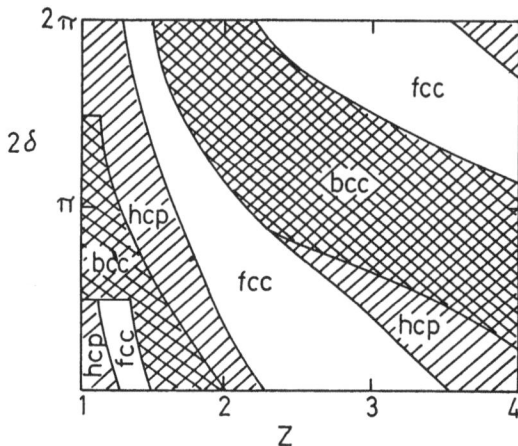

Fig. 7.10. Stable crystal structures among the fcc, bcc, and hcp polytypes as a function of the mean valence $\overline{Z}$ and the phase shift δ, calculated from the exact summation of the Friedel potential (after [7.50])

Up to now the discussion has involved a reciprocal space picture – how can we relate this to our discussion of the crystal structures in terms of the interatomic potentials? The singularity in the response function gives rise to the Friedel oscillations and hence in real space the Hume-Rothery rules can only be explained by summing up the oscillatory part of the pair potentials. *Cousins* [7.49] and *Krause* and *Morris* [7.50] have evaluated lattice sums of the type

$$E = \sum_i \cos\left(2k_F R_i + 2\delta\right)/(2k_F R_i)^3 \qquad (7.13)$$

for various crystal structures. This allows one to map out the stable crystal structures as a function of the mean valence Z and of the phase shift δ. The result given by Krause and Morris is shown in Fig. 7.10. In the limit of a small phase shift the sequence of the stable crystal structures agrees with that found from a full second-order calculation in reciprocal space. The relation $|Q| = 2k_F$ has a clear physical significance: it means that the interatomic distance $D = 2\pi/|Q|$ corresponding to the reciprocal lattice vector Q agrees with the Friedel wavelength $\lambda_F = 2\pi/(2k_F)$. If the nearest neighbours fall into a minimum of the pair potential, there will also be some neighbours in the next minimum and so on. In Chaps. 9 and 11 we shall find that this real-space argument will allow us to generalize the Hume-Rothery arguments to liquid and amorphous phases.

7.3 Static Lattice Distortions

The random distribution of two different kinds of atom over the positions of a regular lattice will, no matter how small the difference in their interatomic potentials, invariably cause local elastic strains. The crystal can react to these strains in two different ways: either the atoms will depart locally from the sites of the idealized lattice, even in the absence of thermal vibrations, or the distribution of the two atomic species will no longer be random, the crystal assumes some degree of short- or long-range order. Both effects lead to a reduction of the internal energy of the crystal, thus they are important for an accurate treatment of the thermodynamic properties of substitutional alloys. The techniques for calculating static lattice distortions in simple metal alloys are presented in this section and ordering effects will be discussed in the next section.

In spite of their obvious importance for any property of the substitutional alloy which depends on the configurational energy, the static lattice distortions have received only little attention over several decades (for a review of earlier work, see [7.51]). Recent theoretical interest has been stimulated by progress in the experimental techniques (diffuse and anomalous x-ray scattering, diffuse neutron scattering, extended x-ray absorption fine structure etc.) for the investigation of the fluctuations in the interatomic spacings.

Very recently, *Froyen* and *Herring* [7.52] have calculated the departures of the atomic positions from the idealized lattice as a linear harmonic response to the interatomic forces generated by the replacement of the mean atoms by particular atoms of one or the other species. The first and second moments of the static lattice displacement are evaluated from the elastic spectrum of the crystal and from $\partial d/\partial c$, the variation of the lattice constant with composition. For the case of random K-Rb alloys, the predictions of this simple theory are well confirmed by a computer simulation based on pseudopotential-derived interatomic forces [7.16] – a typical result is shown in Fig. 7.11. As is to be expected, the nearest-neighbour atoms relax in the direction of the minimum in the interatomic potentials, their distances are distributed over a finite range, the first and second moments of the displacements are found to be in good agreement with the predictions of the linear-harmonic-response theory.

A microscopic theory of static lattice distortions has been developed by *Solt* and coworkers [7.15, 53, 54]. Their method combines *Kanzaki*'s [7.55] lattice statics theory with the pseudopotential formalism for obtaining both the elastic response function of the solvent and the impurity-host interaction. The energy change $\Delta E_{\rm rel}$ which occurs as a consequence of static lattice displacements $u_\alpha(l)$ is expanded in the form

$$\Delta E_{\rm rel} = \sum_{l,\alpha} F_\alpha(l) u_\alpha(l) + \frac{1}{2} \sum_{\substack{l,\alpha \\ l,\beta}} \Phi_{\alpha\beta}(ll') u_\alpha(l) u_\beta(l') \quad . \tag{7.14}$$

$F(l)$ is the force acting upon the lth atom and $\Phi_{\alpha\beta}(ll')$ is the force-constant matrix of the host lattice (the change of the restoring forces due to the presence of the impurity is neglected). Equilibrium occurs at those displacements that minimize $\Delta E_{\rm rel}$, i.e. from (7.14)

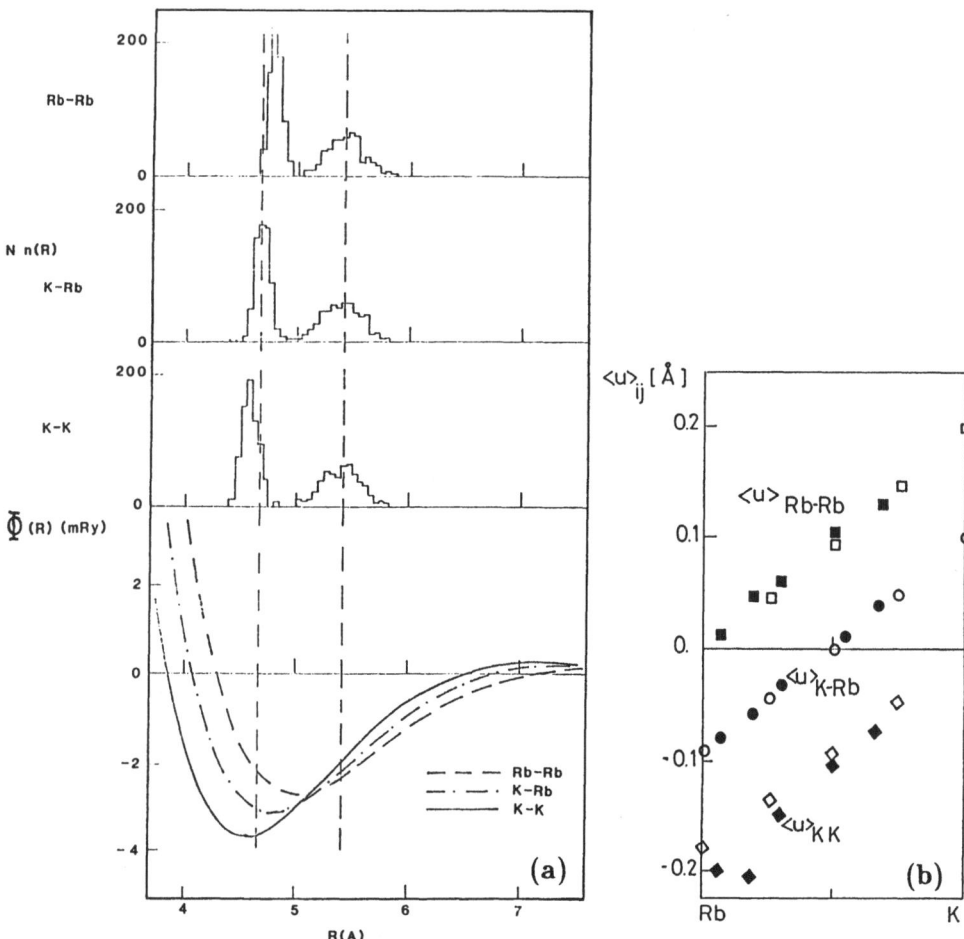

Fig. 7.11. (a) Effective interatomic potentials and partial radial distribution functions for a $K_{0.5}Rb_{0.5}$ alloy, evaluated with a cluster-relaxation technique. (b) Deviation of the interatomic spacing in nearest-neighbour pairs from the ideal distance in a bcc lattice as a function of concentration. (*Full symbols*) results of the cluster-relaxation; (*open symbols*) as calculated from the linear harmonic-response theory of Froyen and Herring (after [7.16])

$$u_\alpha(l) = - \sum_{l',\beta} g_{\alpha\beta}(ll') F_\beta(l') \quad , \tag{7.15}$$

where $g_{\alpha\beta}(ll') = [\Phi_{\alpha\beta}(ll')]^{-1}$ is the Green's function of the host lattice. A perturbation series for the displacing forces $F(l)$ may be constructed by expanding the individual terms of the perturbation series for the total energy (2.19) in terms of the atomic displacements, with the pseudopotential matrix element $W(q)$ now being given by

$$W(q) = w(q)S(q) + N^{-1}\Delta w(q)\,e^{iqR_l} \tag{7.16}$$

instead of $W = w \cdot S$ as previously. Here $W(q)$ is the pseudopotential form factor of the host and $S(q)$ the structure factor of the mean lattice; $\Delta w(q)$ is the difference between the pseudopotential of the impurity located at the site R_l and that of the host. In Fourier space the resulting expansion may be written in the form [see again (2.19)]

$$F_\alpha(l) = \frac{\mathrm{i}}{N} \sum_k k_\alpha \hat{F}(k) \, \mathrm{e}^{\mathrm{i}kR_l} \tag{7.17a}$$

$$\hat{F}(k) = \hat{F}^{(2)}(k) + \hat{F}^{(3)}(k) + \ldots \quad \text{with} \tag{7.17b}$$

$$\hat{F}^{(2)}(k) = -V_a w(k) \Delta w(k) \chi(k)/\varepsilon(k) \quad , \tag{7.18}$$

$$\hat{F}^{(3)}(k) = 6V_a \frac{w(k)}{\varepsilon(k)} \left[\frac{1}{2N} \sum_q \frac{\Lambda^{(3)}(k, q, |-k-q|)}{\varepsilon(q)\varepsilon(k+q)} \Delta w(q) \Delta w(|-k-q|) \right.$$
$$\left. + \sum_{Q \neq 0} \frac{\Lambda^{(3)}(k, q, |-k-Q|)}{\varepsilon(Q)\varepsilon(k+Q)} \Delta w(Q) \Delta w(|-k-Q|) \right]. \tag{7.19}$$

Solt and coworkers point out that in the long-wavelength limit $(k = 0)$ the third-order term $\hat{F}^{(3)}(0)$ is not negligible in comparison to the second-order term. This is a consequence of the density dependence of the interatomic forces in metals and entirely analogous to the compressibility problem (see the discussion in Chap. 2). The effect is especially important for a correct description of the impurity-induced change in the atomic volume.

The static lattice distortion may be probed by diffuse elastic neutron scattering in which the scattering amplitude from the lattice distortions is given by $K(q) = -\mathrm{i}q u(q)$ where $u(q)$ is the Fourier transform of the equilibrium lattice distortions calculated using (7.15–19). Figure 7.12 shows the calculated scattering cross-section for Li impurities in Al, together with the experimental results, both from *Solt* and *Werner* [7.54]. Their calculation is based on simple model potentials and the Toigo-Woodruff [7.56] screening function and it produces a quite impressive degree of agreement with experiment. A similar analysis for Al-3.2 % Mg alloys has been performed by *Werner* et al. [7.57]. Compared with more phenomenological fits with only nearest- or next-nearest-neighbour interactions, these studies show that the pseudopotential model for comparatively long-range solute-solvent interactions is an important improvement for the theory of lattice distortions in substitutional alloys.

The influence of the relaxation energy ΔE_{rel} on the alloy stability has also been investigated. *Solt* and *Zhernov* [7.15] find that ΔE_{rel} gives a substantial contribution to the heat of solution of dilute alkali alloys. *Hafner* and *Punz* [7.16] find that the relaxation effect reduces the heat of formation

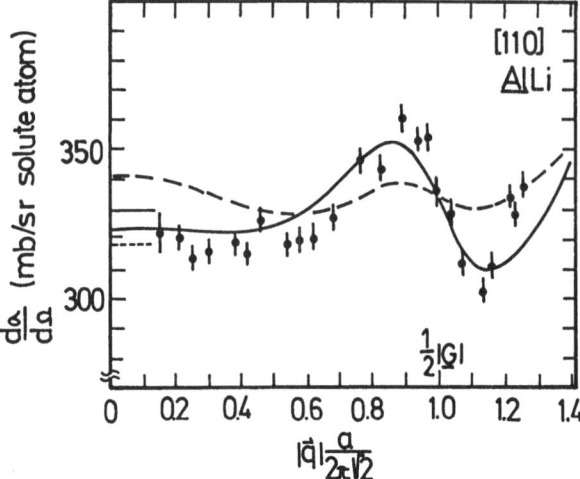

Fig. 7.12. Diffuse elastic neutron scattering cross section of the lattice distortions in Al-Li solid solutions (less than 1 at.% Li). The theoretical result is calculated in the linear-screening approximation $(F = F^{(2)})$. The difference between the solid and the dashed curves is due to different choices of the local-field corrections (Toigo-Woodruff and Geldart-Vosko screening, respectively, see Sect. 2.2.2). If the nonlinear contributions where taken into account $(F = F^{(2)} + F^{(3)})$, the curves should arrive for $q = 0$ at the point marked by the horizontal lines. (*dots*) experimental results (after [7.54])

calculated for an ideal bcc K-Rb solution by about 30 % – this is certainly a non-negligible effect.

7.4 Ordering in Substitutional Alloys

The transition from a completely random to a partially or completely ordered distribution of the two atomic species over the sites of the mean lattice has been repeatedly treated using pseudopotential methods since the pioneering work of *Hayes* and coworkers [7.2] and of *Inglesfield* [7.1,58]. In an ordered alloy, the structural energy depends no longer on the mean pair potential $\Phi_{NN}(R)$ alone, but is a function of the complete set $\Phi_{AA}(R)$, $\Phi_{BB}(R)$ and $\Phi_{AB}(R)$ of pair potentials. That part which couples to the degree of order is the ordering potential $\Phi_{cc}(R)$ [see (2.58)]. The chemical trends in the ordering potential have been discussed in Sect. 2.4.2. A comparatively easy problem is the calculation of the ordering energy, i.e. the energy difference between a completely random solid solution and an ordered compound with the same mean lattice. If the sublattices occupied by the A and B atoms are described by the structure factors $S_A(q)$ and $S_B(q)$, respectively (2.49), the second-order (i.e. bandstructure) contribution to the ordering energy ΔE_{ord} is given by

$$\Delta E_{\mathrm{ord}}^{\mathrm{bs}} = E_{\mathrm{bs}}(\mathrm{ord}) - \Delta E_{\mathrm{bs}}(\mathrm{disord}) \quad \text{with} \tag{7.20}$$

$$E_{\mathrm{bs}}(\mathrm{ord}) = \frac{1}{N} \sum_{q} {}' F_{\mathrm{cc}}(q) |S_A(q) - S_B(q)|^2 \tag{7.21}$$

and $\Delta E_{\mathrm{bs}}(\mathrm{disord})$ given by (7.5). To this, one has to add an electrostatic contribution $\Delta E_{\mathrm{ord}}^{\mathrm{es}}$ corresponding to an ionic lattice with charges $\pm \Delta Z/2$. Thus $\Delta E_{\mathrm{ord}} = \Delta E_{\mathrm{ord}}^{\mathrm{bs}} + \Delta E_{\mathrm{ord}}^{\mathrm{es}}$.

Ordering energies for a variety of simple-metal alloys and noble-metal simple-metal alloys have been calculated by *Hayes* et al. [7.2], *Inglesfield* [7.1,58], *Hafner* [7.31], *Beauchamp* et al. [7.32] and *Kogachi* and collaborators [7.47,59–61]. Invariably, the magnitude of the ordering energy turns out to be very sensitive to details of the pseudopotential and of the screening function – at least if simple model potentials are used. The conclusion is that at present the predictions are of a qualitative rather than a quantitative nature and will remain so until a considerable improvement in the pseudopotential for *binary* alloys is achieved. The order in the solid solution may either be long ranged or short ranged. A short-range order exists if the partial coordination numbers deviate from those corresponding to a completely random site occupancy. The Warren-Cowley short-range order (SRO) parameters are defined by [7.62,63]

$$P_{BA}(\boldsymbol{R}_j) = (1 - c)[1 - \alpha(\boldsymbol{R}_j)] \tag{7.22a}$$

$$P_{AB}(\boldsymbol{R}_j) = c[1 - \alpha(\boldsymbol{R}_j)] \tag{7.22b}$$

where $P_{\alpha\beta}$ is the conditional probability of finding an α atom at $\boldsymbol{R}_j$ if the origin is occupied by a β atom. Since $P_{BA} + P_{AA} = 1$ and $P_{AB} + P_{BB} = 1$, we have in addition

$$P_{AA}(\boldsymbol{R}_j) = (1 - c) + c\alpha(\boldsymbol{R}_j) \tag{7.22c}$$

$$P_{BB}(\boldsymbol{R}_j) = c + (1 - c)\alpha(\boldsymbol{R}_j) \quad . \tag{7.22d}$$

For preferred heterocoordination we have $P_{BA} > (1 - c)$ (c is the concentration of the A atoms) and $P_{AB} > c$ at the nearest neighbour sites, and hence $\alpha_1 < 0$. The condition that $P_{\alpha\beta} \geq 0$ sets the limits to the SRO parameters.

Long-range order (LRO) is related to the existence of a superstructure, in which the atoms of one kind occupy preferentially a sublattice of the mean lattice of the alloy. The Bragg-Williams LRO parameter η [7.64] measures the preferential occupation of a sublattice,

$$\eta = \frac{f_A - x}{1 - x} \quad , \tag{7.23}$$

where f_A is the fraction of the sites of the A sublattice actually occupied by A atoms and x is their concentration. The pseudopotential theory of SRO and LRO will be briefly sketched in the following.

7.4.1 Long-Range Order

Based on the fundamental work by *Landau* and *Lifshitz* [7.65], *Khatchatu-ryan* [7.66] has developed a statistical thermodynamic self-consistent-field theory of order-disorder phenomena in alloys. This theory has been formulated in terms of pseudopotential perturbation theory by *Krasko* and *Maknovetskij* [7.67,68]. The starting point is the following formulation of the Gibbs free energy of the alloy in terms of the ordering potential $\Phi_{cc}(R)$ and the equilibrium probability $n(R_i)$ that an atom A occupies a site R_i : [7]

$$G = E - TS + \mu \sum_i n(R_i) \tag{7.24a}$$

$$E = \overline{E} + \frac{1}{2} \sum_{i,j \neq i} \Phi_{cc}(|R_i - R_j|) \frac{n(R_i)n(R_j)}{c(1-c)} \tag{7.24b}$$

$$S = -\sum_i \{ n(R_i) \ln n(R_i) - [1 - n(R_i)] \ln [1 - n(R_i)] \} \quad , \tag{7.24c}$$

where $\overline{E}$ is the internal energy of the disordered alloy. The chemical potential μ is determined from the normalization condition $\sum_i n(R_i) = 1$ and $n(R_i)$ is determined by the condition for the Gibbs free energy to be stationary with respect to variations of $n(R_i)$: $\delta G[n(R)]/\delta n(R) = 0$. This results in the following self-consistent equation for the probability $n(R_i)$

$$n(R_i) = \left\{ \exp \left[-\beta\mu + \beta \sum_{j \neq i} \Phi_{cc}(|R_j - R_i|) \frac{n(R_j)}{c(1-c)} \right] + 1 \right\}^{-1} \tag{7.25}$$

The right hand side of (7.25) looks like a Fermi function, which is a consequence of a kind of exclusion principle: there cannot be more than one atom at each lattice site. *Khatchaturyan* [7.66] has shown that at and below a temperature T_c the spatially homogeneous solution $n(R_i) \equiv c$ becomes thermodynamically unstable with respect to the formation of a superstructure characterized by a wavevector k_s. The critical temperature T_c is determined by the condition

$$\frac{\delta^2 G[n(R)]}{\delta n(R)^2} \bigg|_{n(R)=c} = 0 \tag{7.26}$$

and is found to be [7.66]

$$T_c = -\tilde{\Phi}_{cc}(k_s) \quad \text{with} \tag{7.27}$$

---

[7] Equation (7.24) describes a static approximation. Any vibrational contribution to the thermodynamic functions is neglected at this stage.

$$\tilde{\Phi}_{cc}(k) = \sum_i{}' \Phi_{cc}(R_i) e^{-ikR_i} = \hat{\Phi}_{cc}(k) - \Phi_{cc}(R = 0) \quad . \qquad (7.28)$$

Here $\hat{\Phi}_{cc}(k) = c(1-c)[8\pi^2 \Delta Z^2/k^2 + 2F_{cc}(q)]$ is the usual Fourier transform of the ordering potential $\Phi_{cc}(R)$. The wave vector k_s characterizes the super structure appearing at T_c. It is clear from (7.25) that the most stable super-structure (the one with the highest critical temperature) is the one which minimizes the function $\tilde{\Phi}_{cc}(k)$. The allowable superlattice vectors k_s are determined by the well-known Landau-Lifshitz criteria: (i) The space group of symmetry elements of the ordered structure must be a subgroup of the space group of the disordered solution. This is equivalent to the condition that the ordering transition is a purely replacive one. Conditions (ii) and (iii) provide necessary (but not sufficient) conditions for the order-disorder transition to be of second order: (ii) k_s must be located at one of the special points in the first Brillouin zone where any function having the symmetry of the reciprocal lattice has extrema by virtue of symmetry requirements alone. Of course certain wave vectors may accidentally minimize $\tilde{\Phi}_{cc}(k)$ for a certain temperature, pressure and average concentration, but as these parameters are altered the wave vector which minimizes the ordering energy is expected to change. Therefore, stable ordered structures (in particular structures stable with respect to long period modulations) correspond to a superposition of static concentration waves whose wave vectors are all special points, i.e.

$$n(R_i) = c + \sum_s \eta_s \varepsilon_s(R_i) \quad , \quad \text{where} \qquad (7.29)$$

$$\varepsilon_s(R_i) = \frac{1}{2} \sum_{j_s} \gamma_s(j_s) \exp\left(ik_{j_s} R_i\right) \quad . \qquad (7.30)$$

The summation in (7.30) is over the vectors of the star of k_s. The η_s are the long-range-order parameters and the coefficients $\gamma_s(j_s)$ are determined by the condition that in a completely ordered alloy all the order parameters are equal to unity when $n(R_i)$ equals either zero or unity. Condition (iii) concerns the necessary vanishing of the third-order coefficient in an expansion of the free energy with respect to the order parameter η. It is found that this coefficient vanishes identically whenever [Ref. 7.51, p. 158 ff]

$$k_{s_1} + k_{s_2} + k_{s_3} \neq Q \quad .$$

i.e. if it is impossible to combine wave vectors of the star in such a way that they add up vectorially to a reciprocal lattice vector.

As a concrete example [7.66] we consider the case of layer-type su-perstructures. In this case, $k_s = Q/2$ and the function $n(R_i)$ takes only the two values $c + \gamma\eta$ and $c - \gamma\eta$ in alternating crystallographic places normal to the vector k_s [examples are CuAu-(I)-type ordering in fcc lattices where $k_s = 2\pi/a \left(\frac{1}{2}\frac{1}{2}\frac{1}{2}\right)$ and CsCl-type ordering in bcc lattices with $k_s = 2\pi/a \,(111)$]. Inserting (7.29) in (7.25) yields an equation for determining the long-range order parameter as a function of temperature and concentration:

$$\ln \frac{(c - \gamma\eta)(1 - c - \gamma\eta)}{(c + \gamma\eta)(1 - c + \gamma\eta)} = \frac{2\beta\gamma\eta\tilde{\Phi}_{cc}(k_s)}{c(1 - c)} \quad . \tag{7.31}$$

The condition that $\eta = 1$ for the completely ordered case fixes the value of the coefficient γ at $\gamma = 1/2$. In the nearest-neighbour approximation, (7.31) reduces to the corresponding expressions of the Bragg-Williams theory.

The simplest example considered by *Krasko* and *Maknovetskij* [7.67] is the ordering transition in bcc Ca-Ba alloys. For a bcc lattice, the special points are (in units of $2\pi/a$): (100), corresponding to the B2 (CsCl-type) lattice; $\frac{1}{2}\frac{1}{2}\frac{1}{2}$ and $\frac{\bar{1}}{2}\frac{\bar{1}}{2}\frac{\bar{1}}{2}$, corresponding to the B32 (NaTl-type) lattice; and $\left(\frac{1}{2}\frac{1}{2}0\right)$ and $\left(\frac{1}{2}0\frac{1}{2}\right)$, $\left(0\frac{1}{2}\frac{1}{2}\right)$, $\left(\frac{1}{2}\frac{\bar{1}}{2}0\right)$, $\left(\frac{1}{2}0\frac{\bar{1}}{2}\right)$, corresponding to a new type of superstructure for which no example is known. The function $\tilde{\Phi}_{cc}(k)$ is shown in Fig. 7.13. Accidental minima do not exist in this case and the deepest "symmetry minimum" corresponds to a CsCl-type ordering. The depth of this minimum gives a transition temperature of $T_c = 800\,\mathrm{K}$. Experimentally one has $T_c \simeq 400\,\mathrm{K}$, with an as yet not completely defined superstructure. *Gurskij* and *Baranitskij* [7.69] have performed a similar calculation for K-Rb alloys. They predict a B32-type ordering with $T_c = 95\,\mathrm{K}$. However, the ordering energy turns out to be smaller than the relaxation energy of the disordered bcc lattice (see Sect. 7.3) and therefore we expect the distorted disordered bcc lattice to remain the stable configuration below this temperature.

For a given superstructure, the task of calculating the order parameter as a function of temperature reduces to the minimization of the free energy (7.24) with respect to the order parameter η, with $n(R_i)$ given by (7.29). The

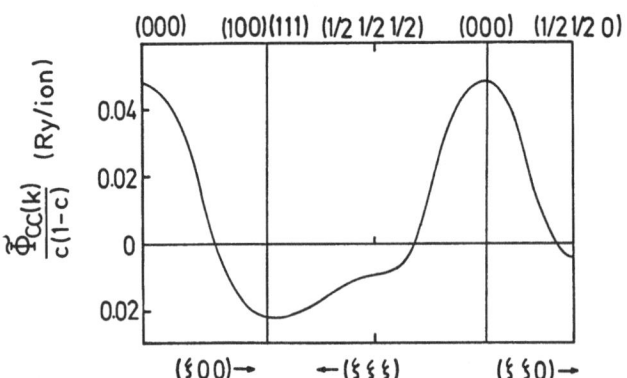

Fig. 7.13. The function $\tilde{\Phi}_{CC}(k)$ along the main symmetry directions of the Brillouin zone of a bcc lattice, as calculated for a $Ba_{0.5}Ca_{0.5}$ alloy (after [7.67])

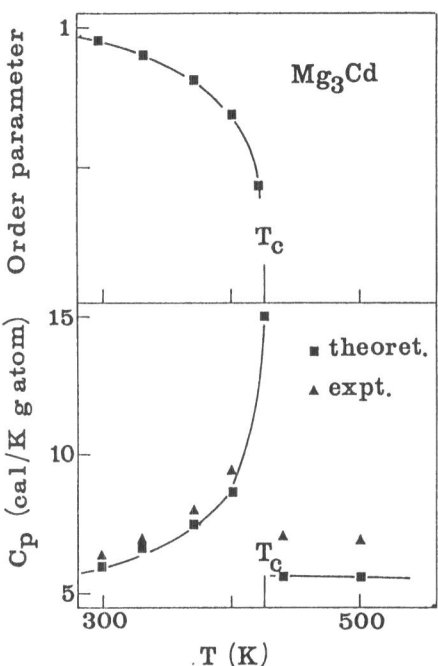

Fig. 7.14. Long-range order parameter η and specific heat C_p of a Mg_3Cd alloy at the disordered hcp (ε) to ordered hcp (DO_{19}) phase transition (after [7.10])

formalism has been applied by *Leung* [7.10] to Mg_3Cd and $MgCd_3$ alloys (DO_{19} superstructure) and by *Rahman* [7.70] to Mg_3Cd, $AuCu_3$ ($L1_2$ superstructure), and $CuZn$ (B2 Cs-Cl-type superstructure). *Kogachi* and *Katada* [7.60] have investigated the stability of long-period stacking order in Mg_3In and $Mg_3In_{1-x}Cd_x$. Their calculations reproduce well the observed variation of the stacking sequence with composition and pressure. *Fuks* and coworkers [7.71] studied the variation of the lattice parameters of alloys during ordering. In most of these calculations, simple model potentials of the empty-core type have been used. After some adjustment of the parameters, it is possible to achieve a good fit to the experimental results, but generally it was found that the calculated transition temperature (and especially its concentration dependence) is rather sensitive to the choice of the pseudopotential[8]. As an example of these results Fig. 7.14 shows the temperature dependence of the order parameter and of the specific heat C_p of Mg_3Cd as calculated by *Leung* [7.10]; good agreement with experiment is found. In this calculation the effect of the lattice vibrations on the order-disorder transition has also been considered – a point to which we will return in the next section.

The physical mechanisms explaining the stability of the B32 (NaTl-type) versus the B2 (CsCl-type) ordering are still a matter of controversy [7.9,67,72,73]. In the B2 structure each atom has 8 unlike nearest neighbours and 6 like next-nearest neighbours; in the B32 structure there are

---

[8] In Sect. 2.4.2 we have already emphasized that simple model potentials do not describe the chemical bond in binary systems with sufficient accuracy. Somewhat better quantitative predictions should be possible with optimized nonlocal potentials.

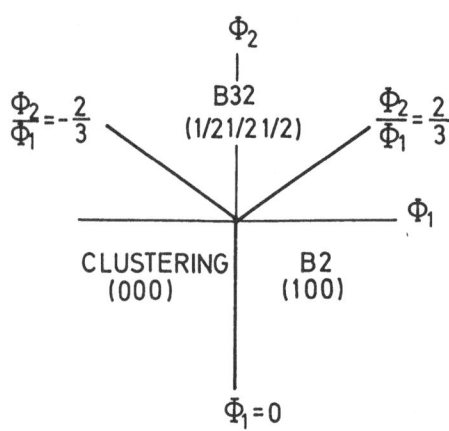

Fig. 7.15. Structural phase diagram for bcc binary alloys in the next-nearest-neighbour approximation (see text)

4 like and 4 unlike nearest neighbours and 6 unlike next-nearest neighbours and both structures have 12 like third-nearest neighbours. In the next-nearest neighbour approximation this results in the structural diagram shown in Fig. 7.15 which displays the stability of the B32, B2 and bcc two-phase mixture as a function of $\Phi_{cc}(d_1) \equiv \Phi_1$ and $\Phi_{cc}(d_2) \equiv \Phi_2$ at zero temperature (note that a disordered solid solution can be stable relative to the ordered phase only at finite temperature). The B32-B2 phase boundary is given by the relation $\Phi_2/\Phi_1 = 2/3$. Thus pairwise interactions can stabilize the B32 ordering only if the strength of the ordering potential for second-nearest neighbours is at least two-thirds of that at the nearest neighbour distances. *Inglesfield* [7.9] has pointed out that this is rather unlikely, because in general $\Phi_{cc}(R)$ is a short-range potential – this is certainly the case for the Li-Pb alloys discussed in Sect. 2 (Fig. 2.11). Indeed *Krasko* and *Maknovetskij* [7.69] have predicted B2 to be stable relative to B32 for LiAl, LiTl and NaTl alloys – only for LiTl is this actually the case. Inglesfield proposed to explain the B32 structure found in many simple-metal alloys by a filling of the Jones zone, in complete analogy to the method which *Heine* and *Jones* [7.72] originally used for treating covalent effects in the diamond-type semiconductors Si, Ge, and α-Sn. In lowest-order perturbation theory the band gap at the zone boundary is $2|W(Q)|$ where $W(Q)$ is the pseudopotential matrix element of the corresponding reciprocal lattice vector. Heine and Jones showed that for open structures like the diamond structure the condition $|W(Q)|/E_F \ll 1$ is no longer satisfied and higher-order terms must be considered in calculating the semiconductor band gap. For an ordered structure the relevant reciprocal lattice vector is the super-structure vector k_s. The calculations of *Inglesfield* [7.9] and of *Krasko* and *Maknovetskij* [7.67] show that the higher-order terms do indeed tend to widen the gap at the Jones zone surface for the B32, but not for the B2 structure. Hence there is a covalent bonding mechanism which stabilizes the B32 structure relative to B2. However, this "covalent" contribution is not expressible in terms of central pair forces alone. Maknovetskij and Krasko have shown that the multi-ion forces dominate.

Another difficult point is that the validity of the band-filling argument depends upon the electron-atom ratio – in the I-III alloys $e/a = 2$ and there are just enough electron to fill the Jones zone. But what about I-II alloys such as Li-Zn and Li-Cd? These will receive much less band-gap stabilization as there are not enough electrons to fill the zone, but they nonetheless form the B32 structure. *McNeil* et al. [7.73] have pointed out that geometrical arguments may be important as well: in the B32 structure both kinds of atoms occupy equivalent-lattice sites and there is an equal number of *A-A*, *B-B* and *A-B* contacts. A consideration of the observed interatomic distances shows that in achieving the contact between the smaller atoms (say *B-B* contacts), the *A-A* and *A-B* contacts are considerably compressed [7.74]. In the B2 structure on the other hand, there are only *A-B* contacts, with an *A-B* distance which is generally close to the sum of the atomic radii. The smaller atomic volume will confer a lower volume energy to the B32 structure, so that it forms in preference to the B2 lattice provided that (i) the radius ratio is close to unity and (ii) the larger atom is the more compressible one. If these conditions are not met, the elastic energy associated with the compression of the *A-B* and *A-A* contacts exceeds the free-energy advantage of B32 over B2 arising from the smaller atomic volume and from the band-gap terms. Examples supporting these arguments are Li-Cd, Li-Zn, Na-Tl, Li-Al, Li-Ga, Li-In which meet the requirements (i) and (ii) and are stable in the B32 structure and Li-Tl, K-Tl, Rb-Tl which violate them and do not form this structure.

A reliable pseudopotential should of course include both effects, but in all these alloys the differences in electron density and electronegativity are quite large and therefore we have to expect relatively large changes in the pseudopotential and alloying. The influence of these changes on the ordering problem has not as yet been explored. Hence we conclude that the pseudopotential approach yields a qualitatively correct picture although it does not yet provide us with a tool to predict quantitatively accurate ordering energies, transition temperatures etc.

7.4.2 Short-Range Order

The existence of a short-range order simply expresses the preference for a certain nearest, next-nearest etc. neighbour coordination, hence it is not subject to symmetry restrictions. With the definition of the short-range order parameters $\alpha(R_i)$ (i refers to the coordination shell) given in (7.22), the ordering energy may be expressed as

$$\Delta E_{\text{SRO}} = \sum_i N_i \alpha(R_i) \Phi_{\text{cc}}(R_i) \quad , \tag{7.32}$$

where R_i is the radius and N_i the number of atoms in the ith shell. An expression for the configurational entropy as a function of the SRO parameters has been given by *Cowley* [7.62]. The equilibrium value of the $\alpha(R_i) \equiv \alpha_i$ may be determined by minimizing the free energy

$$\frac{\partial F}{\partial \alpha_i} = \frac{\partial \Delta E_{SRO}}{\partial \alpha_i} - T \frac{\partial S}{\partial \alpha_i} = 0 \quad . \tag{7.33}$$

Using this method, the SRO was calculated by *Kogachi* [7.60] for In-Mg, Li-Mg and Al-Zn alloys, by *Hayes* et al. [7.2] for Li-Mg, and *Khwaya* et al. [7.75, 76] for Al-Zn alloys. In agreement with experiment, preferred hetero-coordination ($\alpha_1 < 0$) was found in Li-Mg and In-Mg alloys, whereas a tendency to clustering ($\alpha_1 > 0$) is predicted for Al-Zn alloys. That the actual results depend very sensitively on the value of the ordering potential at the nearest neighbour distance, and therefore on the details of the pseudopotential and the local field corrections in the screening function, is of course not surprising.

7.4.3 Ordering in Substitutional Transition-Metal Alloys

For binary transition-metal systems, the Strasbourg group has shown [7.77–79] that the ordering interactions may be derived using a generalized perturbation theory starting from the completely disordered state

$$E(\{n_\alpha(R_l)\}) = \overline{E} + \sum_{\alpha l} n_\alpha(R_l) V_l^\alpha$$

$$+ \frac{1}{2} \sum_{\substack{\alpha\beta \\ ll'}} n_\alpha(R_l) n_\beta(R_{l'}) V_{ll'}^{\alpha\beta} + \dots \quad . \tag{7.34}$$

Here $n_\alpha(R_l) = 1$ if the site R_l is occupied by an α-atom and zero otherwise. $\overline{E}$ is the energy of the completely disordered state representing the periodic reference system. The coefficients of this expansion are called "cluster interactions" and in a tight-binding approximation they may be expressed in terms of the moments of the electronic density of states. Clearly the cluster interactions will be concentration dependent. In the pair approximation only the V_l^α terms are retained in the series (7.30). This gives directly the ordering pair potentials for the nearest, next-nearest and more distant neighbours in terms of the squared difference between the corresponding hopping matrix elements for A and B atoms times the sum of the squared Green's functions all integrated over the d-band. These pair interactions can then be used in any of the variants of the statistical thermodynamic theories to obtain the order parameters. For a recent review and further references see [7.80, 81].

7.5 Thermodynamics of Alloys

The thermodynamic functions of disordered substitutional alloys consist of a configurational and a vibrational contribution. The description of the configurational part of the thermodynamic functions is implicit in the treatment of the order-disorder transition given in the preceding section; the calculation of the vibrational contribution requires the determination of the phonon spectra.

7.5.1 Lattice Dynamics of Substitutional Alloys

As a consequence of the disorder, the phonon spectra of random substitutional alloys can differ considerably in character from those of the pure metals. Localized vibrational modes may be present, and all phonons acquire a broadening and a shift in frequency. Besides their importance for the thermodynamic properties, the lattice vibrations in random alloys provide an ideal testing ground for any theory of elementary excitations in disordered systems because the energy wave vector relationship can be measured directly by coherent inelastic neutron scattering.

The theory of elementary excitations such as phonons, electrons, magnons etc. in substitutional alloys has been considerably developed in recent years, notably through the use of the coherent potential approximation (CPA) and related approaches [7.82–84]. In the CPA, elementary excitations are described by means of a configurationally averaged effective Green's function $\langle G \rangle$ with a complex self-energy which possesses the full translational invariance of the mean lattice. The self-energy is determined self-consistently such that the average scattering from a single real atom in this effective medium vanishes, symbolically $c_A T_A + c_B T_B = 0$, where T_A and T_B are the t-matrices of the two kinds of atoms (cf. Sect. 1.9.1).

In its original form the CPA was formulated for site-diagonal disorder only; i.e. for phonons, only the difference in the masses [7.82] and for electrons only the difference in the atomic energy eigenvalues [7.83, 84], was considered; the differences in the force constants or the hopping matrix elements were neglected. The mass-defect CPA has been applied to a number of alloys which have been studied by inelastic neutron scattering. These include Cu-Au, Ge-Si [7.85], Cu-Al [7.86], and Rb-K [7.87]. Except possibly for the Ge-Si system, the results of the mass-defect CPA were not in quantitative agreement with the experimental data, clearly suggesting the need to include force constant disorder in the CPA.

Several such extensions have been proposed [7.88–92]. In all cases a solution of the CPA with force constant disorders, (or CPA-F) is possible only using some simplifying assumptions. *Kaplan* and *Mostoller* [7.88, 89] assume that the force constants in the alloy superimpose linearly, i.e. that the A-B force constants are the arithmetic average of the A-A and B-B force constants. *Grünewald* et al. [7.90–92] assume that the A-B force constant is the geometrical mean of the A-A and B-B force constants, which allows one to write the force-constant matrix $\Phi_{\alpha\beta}(ll')$ in the separable form:

$$\Phi_{\alpha\beta}(ll') = \lambda(l)\Phi^0_{\alpha\beta}(l-l')\lambda(l') \quad \text{with} \tag{7.35}$$

$$\lambda(l) = \begin{cases} 1 & A \text{ atom at } l \\ \lambda & B \text{ atom at } l \end{cases},$$

where $\Phi^0_{\alpha\beta}(ll')$ is an effective force-constant matrix with the full translational invariance of the lattice and which includes changes in the A-A interactions caused by the alloying. λ is determined by the condition that the force constant for an A-A pair is just $\Phi^0_{\alpha\beta}(ll')$ and that for a B-B pair $\lambda^2\Phi^0_{\alpha\beta}(ll')$.

In the following, we shall sketch very briefly the approach followed by Mostoller and Kaplan. One-phonon Green's functions are defined in the usual way as the Fourier transforms (with respect to time) of retarded displacement-displacement correlation functions. The Green's function of an ideal crystal satisfies the equation

$$\sum_{l''\gamma}[M\omega^2\delta_{\alpha\gamma}\delta_{ll''} - \Phi^0_{\alpha\gamma}(ll'')]G^0_{\gamma\beta}(l''l',\omega) = \delta_{\alpha\beta}\delta_{ll'} \quad, \tag{7.36}$$

where $\Phi^0_{\alpha\gamma}(ll'')$ is the force constant matrix of a perfect reference crystal. With the assumption of linear superposition, the Green's function of the alloy can then be written in the form

$$\sum_{l''\gamma}\{M\omega^2\delta_{\alpha\gamma}\delta_{ll''} - \Phi^0_{\alpha\gamma}(ll'') - \sum_{\kappa}\mathcal{D}^\kappa_{\alpha\gamma}(ll'';\omega)\}$$

$$\times G_{\gamma\beta}(l''l';\omega) = \delta_{\alpha\beta}\delta_{ll'} \quad, \tag{7.37}$$

where $\mathcal{D}^\kappa_{\alpha\gamma}(ll'';\omega)$ is the perturbation (with respect to the reference crystal) at site κ. If the perturbation extends over a range of sites s, s' around κ (the interaction radius), it may be formulated as

$$\mathcal{D}^\kappa_{\alpha\gamma}(ll''\omega) = \sum_{ss'}D_{\alpha\gamma}(ss',\omega)\delta_{l,\kappa+s}\delta_{l'',\kappa+s'} \tag{7.38}$$

with the local perturbation

$$D_{\alpha\gamma}(ss';\omega) = \Delta M\omega^2\delta_{\alpha\gamma}\delta_{s0}\delta_{s'0} + \delta\Phi_{\alpha\gamma}(ss') \quad, \tag{7.39}$$

which includes both mass and force-constant changes. In the CPA one defines a translationally invariant effective medium through the relation (cf. Sect. 1.9.1)

$$\sum_{l''\gamma}\{M\omega^2\delta_{\alpha\gamma}\delta_{ll''} - \Phi^0_{\alpha\gamma}(ll'') - \Sigma_{\alpha\gamma}(ll'';\omega)\}$$

$$\times G^{\text{eff}}_{\gamma\beta}(l''l';\omega) = \delta_{\alpha\beta}\delta_{ll'} \quad. \tag{7.40}$$

198

Here $\Sigma_{\alpha\gamma}(ll'';\omega)$ is a complex self-energy which may be decomposed, in analogy to (7.38) into local self-energy matrices according to

$$\Sigma_{\alpha\gamma}(ll'';\omega) = \sum_{\substack{\kappa \\ ss'}} S(ss';\omega)\delta_{l,\kappa+s}\delta_{l'',\kappa+s'} \quad . \tag{7.41}$$

The configurationally averaged Green's function of the alloy $\langle G_{\alpha\gamma}(ll';\omega)\rangle$ can be expressed as a series involving $G^{\text{eff}}_{\alpha\gamma}(ll';\omega)$ and the t-matrices for scattering from the perturbations. In the spirit of the CPA the self-energy and hence the Green's function of the effective medium are chosen so that on average the single-site scattering, i.e. the compositionally averaged t-matrices, vanish. The resulting equation for the self-energy is [7.89]

$$
\begin{aligned}
S_{\alpha\beta}(ss';\omega) &- cD_{\alpha\beta}(ss';\omega) \\
&+ \sum_{tt'} S_{\alpha\gamma}(st;\omega)G^{\text{eff}}_{\gamma\delta}(tt';\omega)[S_{\delta\beta}(t's';\omega) - D_{\delta\beta}(t's';\omega)] = 0 \quad . \tag{7.42}
\end{aligned}
$$

Equations (7.40–42) represent a closed set of equations for G^{eff}, Σ, and S and must be solved self-consistently. Based on the approximation (7.31) to the force-constant matrix, *Grünewald* [7.90] has derived a different variant of the CPA.

Of all the alloys investigated using inelastic neutron scattering, only for the Rb-K alloys studied by *Kamitakahara* and *Copley* [7.87] may reliable force constants be derived from a microscopic theory. The Rb-K system has been studied using all three variants of the CPA (mass-defect CPA and CPA-F of *Mostoller* and *Kaplan* [7.89] and of *Grünewald* et al. [7.90–92]). In Fig. 7.16(a) the calculated neutron-scattering cross-sections for a fixed different momentum transfer are compared with experiment. Mostoller and Kaplan use a virtual-crystal approximation for $\Phi^0_{\alpha\beta}(ll')$ and describe the force-constant changes in the nearest-neighbour approximation. Thus $D_{\alpha\beta}(s,s';\omega)$ reduces to a 3×3 matrix which depends only on two parameters, the mass defect ΔM and the force-constant defect $\Delta\Phi$; the latter is treated as an adjustable parameter. The calculation of Grünewald et al. avoids the nearest-neighbour approximation, but here λ (7.34–35) is treated as an adjustable parameter.

A typical neutron-scattering cross-section shows two peaks, in the case of Rb-K the lower one corresponds to the band mode (which becomes the Rb mode in the limit of low K concentrations), the higher one to the local mode of the minority atoms. The mass-defect CPA overestimates considerably the splitting of the local and the band modes, but this can be improved by the inclusion of force-constant disorder. Of the two versions of CPA-F, that of Grünewald et al. proves to be somewhat better. The common faults shared by both approaches are that the calculated minimum separating the peaks is too shallow, and the experimental local modes extend to higher frequencies

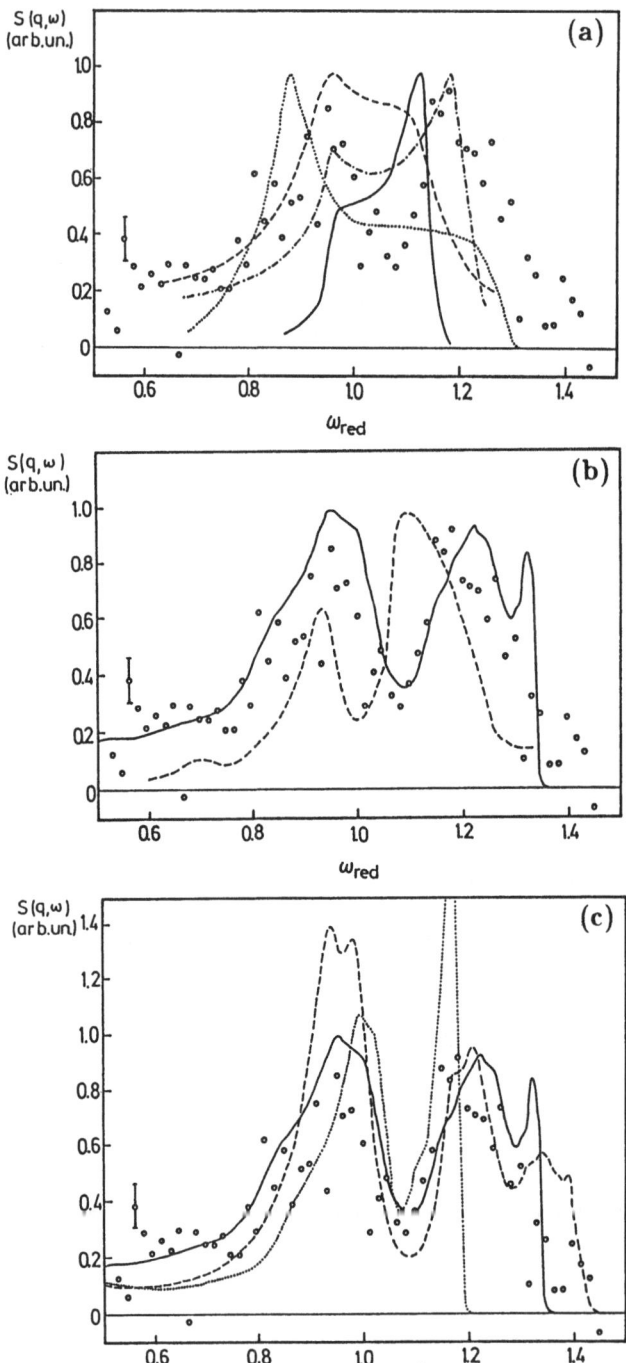

Fig. 7.16(a)–(c). Caption see opposite page

than the calculated ones. Presumably this is due to the mean-field character of the CPA. At higher concentrations the environment of the impurity is highly variable and this has stronger effects on the local mode than on the band mode. To generalize the CPA to include local environment effects is extremely difficult – to date no cluster version of the CPA-F has been presented.

Local environment effects are best studied using computer simulation. *Jacucci* et al. [7.93] have carried out molecular dynamics calculations for a fully relaxed 432-atom cluster representing $K_{29}Rb_{71}$; *Hafner* and *Punz* [7.94] used the recursion method of *Haydock* et al. [7.95] (see also [7.96] for a recent review) to calculate the coherent neutron scattering cross section for various concentrations of the Rb-K system, using fully relaxed clusters with up to 1024 atoms and periodic boundary conditions. The results of both computer simulations are now in reasonable agreement with experiment, as shown in Fig. 7.16(b). They also reproduce the dispersion relations quite well: the band mode has, even at this rather high concentration, a dispersion relation quite close to that of the host metal and the local mode shows only a very weak dispersion. The relaxation of the alloy (see also Sect. 7.3) has introduced a new element: topological disorder, which is superposed on the mass and force-constant disorder. Hafner and Punz have made a detailed study of the influence of all three types of disorder on the lattice vibrations [see Fig. 7.16(c)]: Mass disorder yields a large splitting between the band mode and the local mode, which is drastically reduced on the inclusion of force-constant disorder. Upon relaxation the interatomic spacings vary in such a way that the distance in each nearest-neighbour pair is close to the minimum of the corresponding pair potential (see Fig. 7.11) and as a consequence, the effective force-constant differences are strongly reduced and this results again in a larger splitting of the two peaks. Thus local environment effects, and especially a proper consideration of the static lattice distortion, turn out to be absolutely essential for a correct description of the lattice dynamics of random substitutional alloys. Somewhat paradoxically we find that the CPA works reasonably well, purely because it *cannot* account for the *full* force-constant disorder. To go beyond the present level of the CPA-F would probably destroy the agreement with experiment, unless the theory

---

Fig. 7.16a–c. Neutron scattering cross sections for a $Rb_{71}K_{29}$ alloy for $q = (1, 2, 2)*(2\pi/a)$. The experiments are from [7.87]. (a) Comparison with a mass-defect CPA (*dotted line*) and the CPA-F of [7.89] (*dashed line*) and the CPA-F of [7.91,92] [(*solid line*) calculated using an empty-core potential; (*dot-dashed line*) calculated using the DRT potential]. (b) Comparison with the molecular dynamic calculation of [7.93] [(*dashed line*) based on the DRT potential] and the recursion calculation of [7.94] [(*solid line*) based on OPW pseudopotentials]. Experimental and theoretical curves have been matched with respect to the background level and magnitude so as to provide a good overall match. (c) Comparison with recursion calculations [7.94] considering only mass disorder (*dotted line*), mass and force-constant disorder (*dashed line*), and mass, force-constant, and topological disorder (*solid line*)

is developed sufficiently far to enable the inclusion of topological disorder as well.

Once the phonon spectrum of the alloy is known, the calculation of the vibrational contribution to the thermodynamic functions is straightforward (3.13–14). It is achieved most simply by the use of the virtual-crystal approximation. This approach has been employed by *Straus* et al. [7.97] in what might be called an exploratory study in astrophysical metallurgy: they studied the thermodynamic stability of metallic hydrogen-helium alloys under high temperatures and pressures (temperatures up to 19 000 K and pressures up to 90 Mbar are considered in this study). For hydrogen and helium the electron-ion potentials are purely Coulombic and third-order contributions have to be included. Straus et al. found that these mixtures of Coulombic systems are immiscible at temperatures up to $T = 4000$ K and that a miscibility gap remains at the highest temperatures and pressures considered.

Studies that go beyond the virtual-crystal approximation are rather scarce. *Hafner* and *Punz* [7.94] have studied the influence of the different types of disorder on the thermodynamic excess functions of K-Rb alloys and found it to be rather small.

Martensitic phase transitions from a low-temperature hcp to a high-temperature bcc phase in disordered Li-Mg alloys have been studied by *Punz* and *Hafner* [7.98]. Experimental studies by *Barrett* and *Trautz* [7.99] had found that the martensite temperature T_M rises initially upon alloying Mg to Li, but decreases rapidly once the Mg concentration exceeds a critical value of about 20 %. The calculations based on the recursion technique for phonons show that the increase in T_M is due to the hardening of the transverse [110] mode (polarized along [1$\bar{1}$0]) which drives the martensitic transition (see Sect. 3.6). The reversal of the trend with increasing Mg concentration stems from the valence electron concentration effects on the ground-state energy as discussed in Sect. 7.1.2.

A simpler approach is again that based on thermodynamic perturbation theory.

7.5.2 Thermodynamic Perturbation Theory

The standard Gibbs-Bogoljubov variational technique introduced in Sect. 3.5.3 is easily generalized to binary systems. In the case of pure metals, the Debye and Einstein models offered a choice between two equally convenient reference systems. In the alloy case, the Einstein model (each atom of type A and B vibrates independently with frequency ω_A or ω_B in a harmonic potential well) deserves preference because of its simplicity [9]. An upper

---

[9] For pure metals, a comparison of the two reference systems shows that at least at high temperatures, the Einstein model is as useful as a Debye model. The idea behind the Einstein model – each type of ion gives rise to a separate phonon branch which is then replaced by a constant – seems to reflect the results discussed in the last section rather well.

bound to the free energy of the alloy is given by

$$\overline{F} = E_{\text{kin}} + E_{\text{v}} + \frac{1}{2N} \sum_{i,j=A,B} \sum_{l(i),m(j)}' \Phi_{ij,T}(R_{lm})$$

$$- T(S_{\text{ph}} + S_{\text{conf}}) \quad . \tag{7.43}$$

where E_{v} is the usual volume energy and the kinetic energy and vibrational entropy are expressed in terms of the Einstein temperatures $\theta_i = h\omega_i/k_B$ $(i = A, B)$ by

$$E_{\text{kin}} = \tfrac{3}{4}k_B[c_A\theta_A \coth{(\theta_A/2T)} + c_B\theta_B\coth{(\theta_B/2T)}] \tag{7.44}$$

$$TS_{\text{ph}} = 3k_B \sum_{i=A,B} c_i\theta_i \left(\frac{1}{\exp{(\theta_i/T)} - 1} - T\frac{\ln{[1 - \exp{(-\theta_i/T)]}}}{\theta_i} \right) \quad .$$

$$\tag{7.45}$$

The structure-dependent part of the internal energy is expressed in terms of temperature-dependent partial pair potentials $\Phi_{ij,T}(R)$ [cf. (3.27) and (3.38–41)]

$$\Phi_{ij,T}(R) = \frac{2}{N} \sum_q \exp{[i\mathbf{q}\mathbf{R} - q^2(\varrho_i + \varrho_j)]} \left[\frac{4\pi Z_i Z_j}{V_a q^2} + F_{ij}(q) \right]$$

$$= \frac{2Z_i Z_j}{R} \operatorname{erf}\left(\frac{R}{2(\varrho_i + \varrho_j)} \right)$$

$$+ \frac{1}{2\pi^2 R} \int_0^\infty F_{ij}(q)(\sin{qR})q \exp{[-q^2(\varrho_i + \varrho_j)]}dq \quad , \tag{7.46}$$

where the ϱ_i, ϱ_j stand for the mean-square amplitudes of vibrations. Within the Einstein model they are given by

$$\varrho_i = \frac{\coth{(\theta_i/2T)}}{4M_i k_B \theta_i} \quad , \quad i = A, B \quad . \tag{7.47}$$

Equivalently, the free energy might be expressed in reciprocal space in terms of the second-order energy E_2 and the electrostatic energy E_{es}. The relevant expressions are easily derived from the static case $(2.48\,\text{ff})$ by replacing each phase factor $\exp{(-i\mathbf{q}\mathbf{R}_i)}$, which enters the structure factors by its thermal expectation value evaluated in the Einstein reference system,

$$\langle\exp{(-i\mathbf{q}\mathbf{R}_i)}\rangle_E = \exp{(-i\mathbf{q}\mathbf{R}_{i0})}\exp{(-q^2\varrho_i)} \quad , \tag{7.48}$$

where $\mathbf{R}_{i0}$ is the equilibrium position of atom i ($\varrho_i = \varrho_A$ or ϱ_B depending on the occupation of this site). The configurational entropy depends on the degree of ordering of the system; for a completely random alloy it is simply given by the well-known expression for the ideal entropy of mixing

$$S_{\text{conf}} = -k_B(c_A \ln c_A + c_B \ln c_B). \tag{7.49}$$

Since (7.43) is an upper bound to the exact free energy of the alloy for all possible values of θ_A and θ_B, it follows that the best values of F, θ_A, and θ_B are found by minimizing the right hand side of (7.43) with respect to θ_A and θ_B, i.e.

$$\left(\frac{\partial \overline{F}}{\partial \theta_i}\right)_{T,V} = 0 \quad , \quad i = A, B \quad . \tag{7.50}$$

This technique has been applied by *Leung* et al. [7.18] to disordered hcp Cd-Mg alloys at $T = 543$ K (i.e. at temperatures above the ordering transition). Their result for the heat of formation has already been shown in Fig. 7.3 and found to be in reasonable agreement with experiment. Here we consider the excess entropy of formation, $\Delta S^E = \Delta S - S_{\text{id}}$ (i.e. the entropy of formation minus the ideal mixing entropy). It follows from (7.50) that

$$S = -\left(\frac{\partial \overline{F}}{\partial T}\right)_V = S_{\text{ph}} + S_{\text{conf}}, \tag{7.51}$$

i.e. for a random alloy ΔS^E measures directly the change of the vibrational entropy S_{ph}, on alloying. Leung et al. found that the concentration dependence of the Einstein temperatures and consequently the excess entropy ΔS^E depends critically on the change of volume on alloying (Fig. 7.17).

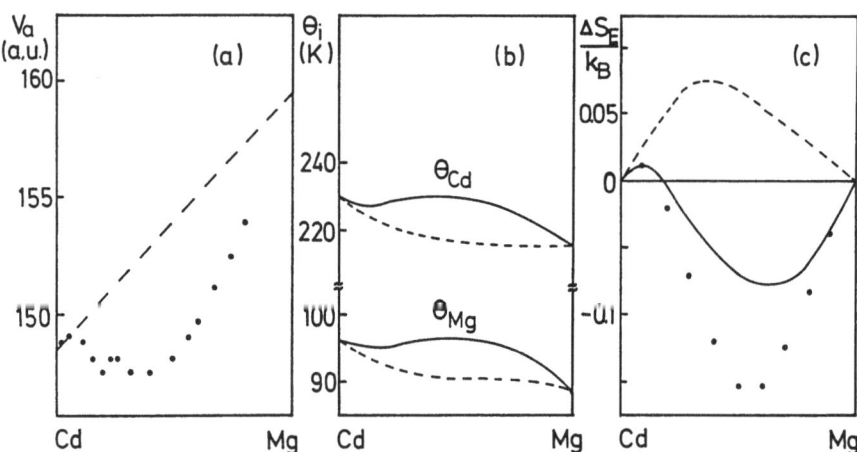

Fig. 7.17. (a) Atomic volume V_a, **(b)** Einstein temperatures θ_{Cd}, θ_{Mg} and **(c)** excess entropies ΔS^E per atom in disordered hcp Cd-Mg alloys at $T = 543$ K. The dots represent the experimental volume and excess entropy, the solid lines refer to θ_i and ΔS^E calculated with the experimental volume and the dashed lines correspond to the hypothetical zero-excess-volume case (after [7.18] and [7.100])

This correlation has been further explored by *Hasegawa* and *Young* [7.100] for Hg-Cd, Al-Mg, and Al-Zn alloys. The calculations show that with a zero excess volume, the Einstein temperatures vary essentially linearly with concentration. If there is any nonlinearity at all, it corresponds to a slightly lower Einstein temperature, reflecting a somewhat softer phonon spectrum and resulting in a positive excess entropy. If the volume is reduced, the Einstein temperatures are raised by an amount related to the Grüneisen parameter of the material. The harder phonon spectrum naturally results in a reduction of the entropy of the alloy. A similar entropy-volume correlation is also well established for liquid alloys (see Chap. 9 and [7.101]).

7.5.3 Vibrational Dynamics and the Ordering Transition

The interrelation between the vibrational properties and the ordering transition has been discussed by *Leung* [7.10]. The reference system now depends on the Bragg-Williams order parameter and on the two Einstein temperatures; the expectation value with respect to the reference system should now be written explicitly as $\langle \ \rangle_{\text{ref}} = \langle\langle \ \rangle_E\rangle_c$ where $\langle \ \rangle_E$ stands for the thermal average over the Einstein model, and $\langle \ \rangle_c$ for the configurational average over all the allowable configurations specified by the Bragg-Williams order parameter. The configurational entropy is now given by

$$
\begin{aligned}
S_{\text{conf}} = \ & - k_B\{c_A[f_\alpha \ln f_\alpha + (1 - f_\alpha) \ln (1 - f_\alpha)] \\
& + c_B[f_\beta \ln f_\beta + (1 - f_\beta) \ln (1 - f_\beta)]\} \quad ,
\end{aligned}
\tag{7.52}
$$

[see (7.23) for a definition of f_α]. The explicit form of the upper bound F to the free energy is given in [7.10]. The optimal values of the parameters of the reference system are given by the conditions

$$
\left(\frac{\partial \overline{F}}{\partial \theta_A}\right)_{T,V} = \left(\frac{\partial \overline{F}}{\partial \theta_B}\right)_{T,V} = \left(\frac{\partial \overline{F}}{\partial \eta}\right)_{T,V} = 0 \quad .
\tag{7.53}
$$

This method has been used to investigate the influence of the lattice vibrations upon the ordering transition. The calculation is based on empty-core potentials and we have already mentioned that with this oversimplified choice of pseudopotential, an empirical adjustment of the ordering potential is necessary to get a transition temperature of the right order of magnitude. In the static limit (the Bragg-Williams theory), the calculated T_c's are T_c (Mg$_3$Cd)$= 421$ K and T_c (MgCd$_3$)$= 438$ K. The inclusion of the lattice vibrations reduces the transition temperatures to 380 K and 363 K respectively, the experimental values are T_c (Mg$_3$Cd)$= 423.8$ K and T_c (MgCd$_3$)$= 356$ K. Note that in the static limit the transition temperatures would be equal if both alloys had the same lattice constant. Thus the predicted lowering of T_c in MgCd$_3$ relative to Mg$_3$Cd – which is related to the lower Einstein-temperature (expressing the lower vibrational frequencies) of the Cd-rich alloy – is supported by experiment and confirms the important role of lattice vibrations in determining T_c. A fully quantitative treatment of this effect, however, will undoubtedly necessitate more refined pseudopotentials.

Recent Results

The coherent potential approximation (CPA, see Sect. 1.9.1) for the calculation of the electronic structure of disordered alloys has been extended to calculate total electronic energies [7.102]. This new approach has been applied to several Cu-based Hume-Rothery phases [7.103]. The results essentially confirm the interpretation for the valence-electron rule given on the basis of the perturbation calculations.

8. Intermetallic Compounds

The theory of the structure and stability of stoichiometric intermetallic compounds is an extremely wide field. A large amount of crystallographic data have been accumulated [8.1–3], but their explanation in terms of the electron theory has hardly been tackled as yet. The complexity of most of the relevant crystal structures renders band structure calculations quite difficult and the strong chemical bonding effects make a perturbation approach based on simple local model potentials virtually impossible.

8.1 Structure Maps

As a result of the difficulties mentioned above, the theoretical approaches to the structure of intermetallic compounds are mostly still at the level of *structure maps*. *Chemical coordinates* such as the valence, the core size, the atomic energy levels, the electronegativity etc. of the constituent elements are used as the basis for the construction of two-dimensional structure maps with the aim of separating the different crystal structures of binary compounds of a given stoichiometry $A_x B_y$ into distinct characteristic domains. For example, *St. John* and *Bloch* [8.4] have used the s- and p-core radii r_p and r_s of their nonlocal empirical pseudopotential to construct a structure map for the simple metals. *Zunger* [8.5] has shown how these orbital radii may be derived from a first-principles pseudopotential calculation (the orbital radii are interpreted as the expectation values of $\langle r^2 \rangle$ for free-atom Hartree-Fock orbitals). A good separation of a wide range of crystal structures is achieved in terms of the two structural coordinates $R_\sigma = (r_p^A - r_s^A) + (r_p^B - r_s^B)$ and $R_\pi = (r_p^A + r_s^A) - (r_p^B + r_s^B)$ (Fig. 8.1). According to *Phillips* [8.6] $(r_p^\alpha + r_s^\alpha)$ corresponds to the formation of s-p hybrids on atom α, and $(r_p^\alpha - r_s^\alpha)$ to p^2-sp hybrids, i.e. their ratio measures the strength of σ bonding relative to that of π bonding. Note, however, that even if a quite remarkable separation is achieved, this still doesn't provide us with a physical argument explaining structural stability. There remains for example an open question as to what causes an "island" of B32 stability in a parameter field that otherwise shows B2 stability.

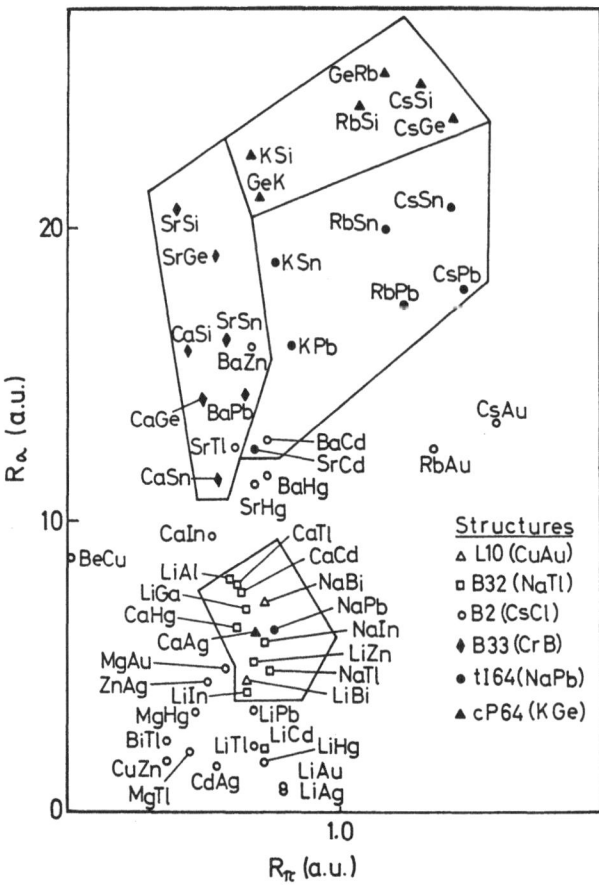

Fig. 8.1. R_σ versus R_π plot for binary intermetallic compounds of simple and noble metals with equiatomic composition. The separation lines have been drawn to emphasize the structural separation (after [8.5])

Zunger and [8.7] and *Burdett* et al. [8.8] have extended this approach to transition metals. Other structure maps for transition metals have been based on valence ($\overline{Z}$ vs. ΔZ, [8.9–11], atomic energy levels and valence (ΔE vs. Z, [8.12]) metallic radii (R_A/R_B vs. $\overline{R}$ [8.13]) or electronegativity [8.14].

Each of these coordinates has its specific limitations. For example, valence does not differentiate between the elements of a group, core size overestimates the differences and does not differentiate very well between elements of the same row. Nature, of course, needs only one coordinate per element to specify the crystal structure of a binary compound. *Pettifor* [8.9] has undertaken the ambitious job of setting up a single chemical scale characterizing each element in the periodic table. The new scale is set up by requiring that the variation of the new chemical coordinate χ within a group does

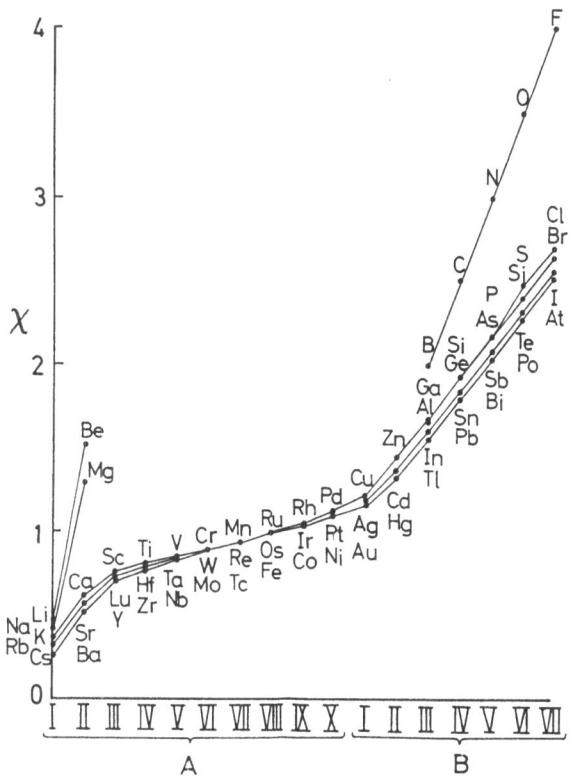

Fig. 8.2. The chemical scale χ for all the elements of the periodic table (after [8.9, 15])

not overlap with that of a neighbouring group (except for the second row elements from B to F which behave in a way different from the other elements of the group). Furthermore, χ is constrained to vary linearly across the transition-metal series and also through a row of p elements. This is motivated by the linear variation of the atomic energy levels, of the inverse core size and of the number of valence electrons. The magnitude of χ is fixed by requiring it to take the Pauling electronegativity values for the second row elements from Be to F. Figure 8.2 shows the chemical scale for all the elements of the periodic table. Pettifor shows that a structure map based on a two-dimensional plot of χ_A vs χ_B achieves an excellent structural separation of the 574 known A-B compounds. In Fig. 8.3 we show only that part of the map which is most relevant to our present subject, namely the 69 sp-sp non-octet compounds. The new coordinate turns out to be distinctly more sensitive then Zunger's R_σ and R_π radii (cf. Fig. 8.1); there is now for example a good separation between the regions of B2 and B32 stability, whereas these structures are rather intermingled in the Zunger plot. The Pettifor plot also achieves a perfect separation of the sp-sp octet compounds into four regions of Ba (NaCl), B2 (CsCl), B3 (ZnS), and B4 (ZnO) stability and is quite successful for d-d and sp-d compounds as well. However, as

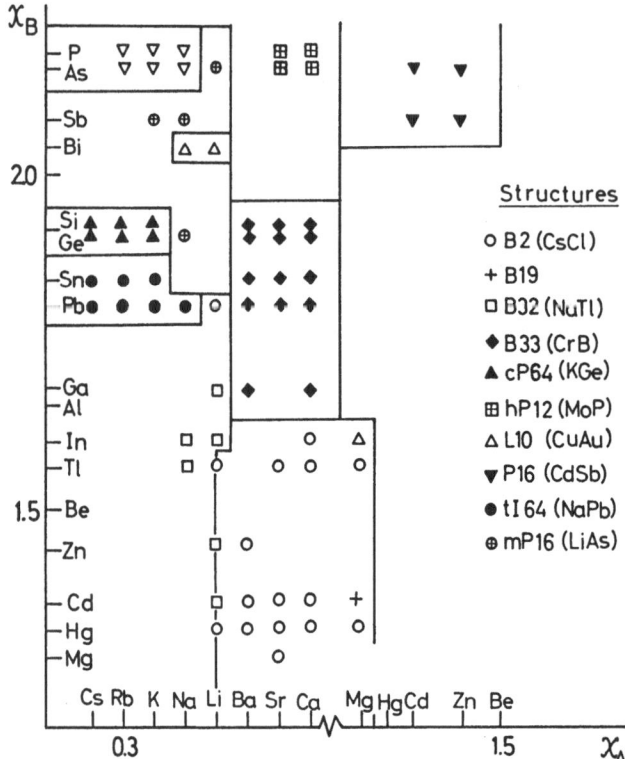

Fig. 8.3. The (χ_A, χ_B) structure map for the binary *sp-sp* non-octet compounds (after [8.9])

for any other of the proposed chemical coordinates, the physical origin of the different structural domains in such a map remains to be detailed. A first attempt in this direction has been undertaken very recently by *Pettifor* and *Podloucky* [8.16]. They have attempted to construct structure maps for *p-d* intermetallic compounds from total energy calculations based on a simplified tight-binding Hamiltonian and the recursion method.

Another problem with such structure maps is that although they may be drawn for compounds of any stoichiometry, they do not give a clue as to the relative stability of $A_x B_y$ compounds of different composition. Therefore, the structure maps should be complemented by more fundamental calculations. As, at least for some time to come, first principles calculations will not be able to provide an appropriate coverage of so vast a subject, semiempirical schemes such as the pair-potential analysis to be discussed in the next section will still be valuable.

8.2 Empirical Pair-Potential Analysis of Intermetallic Phases

In a series of papers [8.17–20] *Machlin* has undertaken a pair-potential analysis of the structure and stability of binary stoichiometric intermetallic compounds which is based on the following two assumptions:

(i) The interaction between two particles separated by a distance $R_{ll'}$ is given by a Lennard-Jones potential, i.e.

$$\Phi_{ij}(R_{ll'}) = -\alpha_{ij}/R_{ll'}^m + \beta_{ij}/R_{ll'}^n \quad ; \quad i,j = A, B \tag{8.1}$$

in which the values of the exponents are fixed as $m = 4$ and $n = 8$ (see [8.17] for the reasons behind this particular choice). The use of pair-potential model incorporates inherently two important alloy chemical parameters, namely atomic size and coordination.

(ii) the electrochemical factor (difference in electronegativity between the components) is incorporated by allowing the constant of attraction α and the constant of repulsion β to vary upon alloying. The amount of this variation is estimated in the following way: It is well known that after summing up the pair interaction and minimizing with respect to the nearest-neighbour distance, the binding energy per particle of a monatomic system may be expressed as [8.21]

$$E = -\frac{1}{2}\left(\frac{n-m}{n}\right)\frac{\alpha S_m}{d^m} = -\frac{1}{2}\left(\frac{n-m}{m}\right)\frac{\beta S_n}{d^n} \quad , \tag{8.2}$$

where d is the equilibrium nearest-neighbour distance in the given structure and S_n, S_m are lattice sums defined by

$$S_m = \sum_l (R_l/d)^{-m} \quad . \tag{8.3}$$

Machlin further points out that within each row of the periodic table, E varies nearly as d^{-4} [8.17] – thus, according to (8.2) it seems that α is constant for the elements of a given period and that β must vary as d^4. The physical interpretation is that the constant of repulsion varies with the extension of the repulsive core of the atom.

This is where the electrochemical factor comes in: As a consequence of a difference in electronegativity, a transfer of electrons will occur from the more electropositive to the more electronegative element. Such an electron transfer will clearly alter the atomic radii, hence the interatomic distances and further β. Numerically, this relation is established using *Gordy*'s [8.22,23] model for the electronegativity (which assumes the electronegativity to be proportional to a screened nuclear charge and inversely proportional to the interatomic spacing) with the result

$$R_A^* = R_A/[1 - a(X_A - X_B)/(X_A - 0.5)R_A]$$

$$R_B^* = R_B \Big/ \left[1 + a\left(\frac{c_A}{c_B}\right)(X_A - X_B)/(X_B - 0.5)R_B \right] \tag{8.4}$$

for the variation with concentration of the A and B component radii on alloying. X_A and X_B are the electronegativities of A and B and a is a constant; quantities with an asterisk correspond to the A-B compound, those without to the pure A and B metals. If we assume separations $d_{AA}^* \propto 2R_A^*$, $d_{BB}^* \propto 2R_B^*$ and $d_{AB}^* \propto R_A^* + R_B^*$, we obtain the following relation for the altered pair-potential parameter β_{ij}^*:

$$\beta_{ij}^* = \beta_{ij}(d_{ij}^*/d_{ij}) \quad . \tag{8.5}$$

Equations (8.4,5) together describe the influence of the electronegativity difference upon the interatomic pair potential in binary alloys. According to (8.4) when $X_1 > X_2$ one finds $R_1^* > R_1$ i.e. the effective radius of the electronegative atom increases while that of the electropositive atom decreases. Equation (8.5) shows that a decrease of the atomic diameter is reflected in a decrease of the constant of repulsion.

This means that Machlin's model is essentially a simplified scheme for describing the chemical compression effect, which was discussed in Sect. 2.4.1 on the basis of first-principles pseudopotential calculations. In Machlin's original model the amount of expansion and compression is symmetric (weighted by the concentrations). In a later version [8.20] the model was slightly modified to account for the fact that the compression of the electropositive atom is always much stronger than the expansion of the electronegative one (see again Sect. 2.4.1). In spite of its simplicity, this model allows some important predictions: (i) It predicts quantitatively the increasing deviations from Vegard's law with an increasing difference in electronegativity between the constituents [Fig. 8.4(a)]. (ii) A large negative volume of formation should be accompanied by an increasingly exothermic heat of formation (the relation should be approximately $\Delta E/\overline{E} \cong 4/3 \Delta V/\overline{V}$) – this is demonstrated in Fig. 8.4(b) for some examples of simple-metal Laves phases. (iii) The electronegativity effect strongly modifies the predictions made on the basis of atomic size arguments. For example, it follows easily from the pair-potential model that for a zero electronegativity difference ($\Delta X = 0$), stable C15 (MgCu$_2$-type) phases (with $\Delta H < 0$) should be formed only for a radius ratio R_A/R_B in the range $1.07 < R_A/R_B < 1.35$. For a non-zero ΔX, if the initial radius ratio was larger than the ideal, then the decrease in the effective radius of the electropositive component (A) tends to stabilize the compound. For an initial radius ratio less than ideal, a destabilization is predicted. This asymmetric effect is superimposed on an overall lowering of ΔH proportional to ΔV. The net result of these effects on the radius-ratio limits for absolute stability of C15 phases is to leave the lower one relatively unaffected while markedly increasing the upper one [Fig. 8.4(c)] – this is

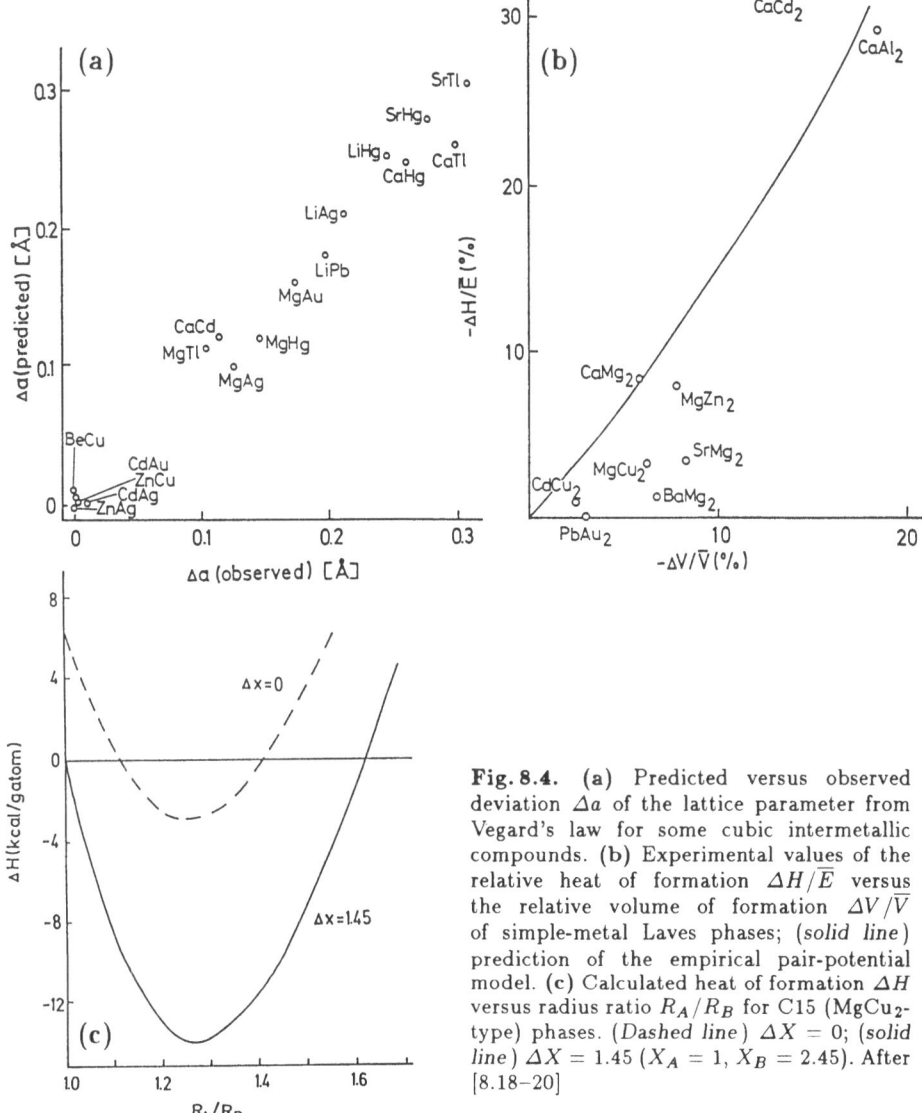

Fig. 8.4. (a) Predicted versus observed deviation Δa of the lattice parameter from Vegard's law for some cubic intermetallic compounds. (b) Experimental values of the relative heat of formation $\Delta H/\overline{E}$ versus the relative volume of formation $\Delta V/\overline{V}$ of simple-metal Laves phases; (*solid line*) prediction of the empirical pair-potential model. (c) Calculated heat of formation ΔH versus radius ratio R_A/R_B for C15 (MgCu$_2$-type) phases. (*Dashed line*) $\Delta X = 0$; (*solid line*) $\Delta X = 1.45$ ($X_A = 1$, $X_B = 2.45$). After [8.18–20]

just what is indeed observed for this structure [8.1]. Other applications include problems such as the relative stability of the A15 vs L1$_2$ and the C15 vs C32 (AlB$_2$-type) structures.

Evidently the merit of the empirical pair-potential model is to help to elucidate the relation between atomic size and electronegativity, and how the interplay of these two parameters influences the stability of binary intermetallic compounds. On the other hand it cannot be overlooked that in its present state the model is in many ways quite unrealistic: it completely ignores the large volume force and also the volume dependence of the pair potential. Furthermore the Lennard-Jones form of the pair potential is perhaps not an ideal choice. It should be relatively easy to correct both points. Recently, simple analytic approximations to realistic metallic pair potentials have become available [8.24, 25]; the electron-gas contribution to the heat of formation is also well known (see Sect. 6.1.5). What is left to be done is to find a simple recipe to describe alloying effects on the potential parameters (the pseudopotential first of all), in the spirit of (8.4–5). The search for such a simple model should of course be guided by the results of ab initio calculations.

8.3 Classification of Intermetallic Phases According to Building Principles and Properties

Before we discuss the detailed results of microscopic calculations, we should reflect on which classes of intermetallic compounds we may legitimately expect to be amenable to a perturbative approach. Very recently *Pearson* [8.25] has attempted to classify the crystal structures of intermetallic phases according to their building principles and properties. He proposes to distinguish the following five classes:

(i) *Geometrical packings.* These include most structures based on the close packing of spheres and on tetrahedral close packings. They are characterized by a negative volume of formation and the closest interatomic distances are distinctly shorter than those expected from *Pauling's* [8.27] empirical equation $R(1) - R(n) = 0.3 \log n$ (where n is the valence divided by the coordination number and R is in Ångstrom). This equation relates bond radii R to the bond strength measured in terms of the number of valence electrons per ligand.

(ii) *Phases where the band-structure energy appears to be an unusually large fraction of the total energy.* Examples are the Hume-Rothery phases formed by the noble metals and the *B*-group metals and the Zintl phases formed by *A*- and *B*-group metals. Characteristic properties include structural changes at well-defined values of the valence electron concentration, and the fact that extrapolation of the volume vs concentration relation (which is linear over a wide range) to 100 % of the electropositive component gives an apparent volume (the "increment", see Sect. 6.1.2) which is much smaller than that determined from the crystal structure of the pure element.

(iii) *Valency compounds.* The compositions of these compounds are such that normal valency rules are satisfied. At stoichiometric compositions these phases are usually insulating or semiconducting; nonstoichiometry may result in a metallic behaviour.

(iv) *Framework structures.* All atoms in the structure form a "framework" (i.e. a packing) in which the nearest neighbour distances conform with Pauling's rule, no rules for valency or valence electron concentration are satisfied.

(v) *Hybrid-framework structures with geometrical packings.* In this case atoms on certain sites form a framework (in the sense just defined) and the remaining atoms are accommodated in the large voids according to geometrical packing principles.

It is clear that class (i) contains the best candidates for a discussion in terms of refined pseudopotentials. Class (ii) comprises the difficult cases where the bonding effects might be strong enough to exceed the limits of applicability of a perturbation approach. Class (iii) is definitely outside its scope since it is impossible to describe ionic or covalent bonding in a formalism based on response theory. The situation as to the remaining two classes of intermetallic compounds is still largely unexplored.

8.4 Topologically Close-Packed Intermetallic Compounds (Frank-Kasper Phases)

It has been pointed out by *Frank* and *Kasper* [8.28] that a large number of relatively complex structures found in intermetallic compounds may be understood in terms of the geometrical requirements of sphere packing.[1] Examples are the Laves phases (C14-$MgZn_2$ type, C15-$MgCu_2$ type, C36-$MgNi_2$ type – more than 350 binary alloys are known to assume one of these three Laves structures) and the μ, χ, and σ phases. Their topological and geometrical properties lead one to expect that a pair-potential approach will be sufficient to explain their bonding properties (the very elevated degree of space filling and the high coordination numbers suggest the absence of pronounced directional bonding effects). Nevertheless only a very few attempts at microscopic calculations have been undertaken up to now.

8.4.1 Heat and Volume of Formation of Laves Phases

Perhaps the simplest case for an ab initio study of the phase stability of Laves phases are the three hexagonal (C14) compounds formed by alloys of two alkali metals. The structures of Na_2K, Na_2Cs, and K_2Cs have been studied in detail [8.29], and it has also been shown that an occasionally postulated compound Na_2Rb definitely does not exist. The phase diagrams are shown in Fig. 8.5. They point to an only moderate stability of these compounds: Na_2K and Na_2Cs melt incongruently at temperatures that are lower than the average melting points of the components; K_2Cs decomposes

---

[1] The Frank-Kasper phases are crystalline phases, with rather large unit cells, in which most atoms have 12-particle icosahedral coordination shells (the icosahedron is the most stable geometry for a cluster of 13 particles interacting by central pair potentials, but it is impossible to fill three-dimensional space with a network with only icosahedral coordination). This local icosahedral order is interrupted by particles with high coordination numbers 14, 15, and 16.

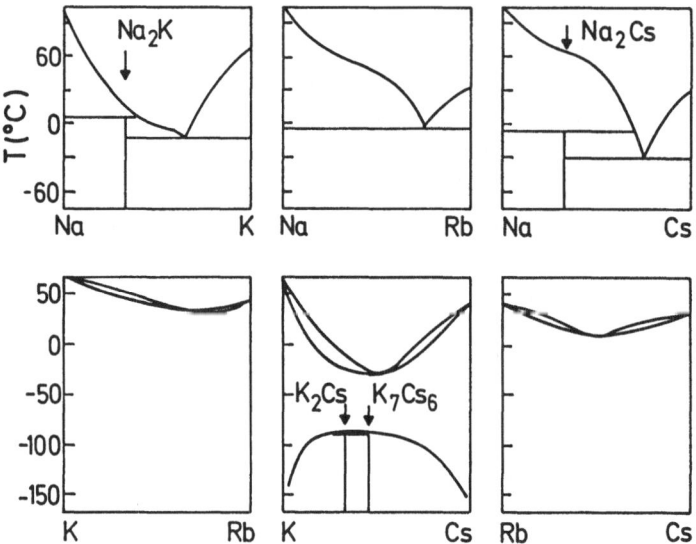

Fig. 8.5. Phase diagrams of binary alkali-metal alloys (compiled after [8.30])

into a bcc solid solution at $T \sim 180$ K. The volume and the heat for formation of all possible alkali-metal Laves phases (except Li_2Rb and Li_2Cs) have been calculated by *Hafner* using an optimized first-principles pseudopotential [8.31] and a model-potential calculation for Na_2K has been performed by *Zhdanov* and *Teordorovich* [8.32]. The results are compiled in Table 8.1. A negative heat of formation is predicted for Na_2K, K_2Cs and Rb_2Cs, which, however, does not necessarily mean that these compounds are pre-

Table 8.1. Heat and volume of formation of binary alkali-metal Laves phases with the C14 ($MgZn_2$-type) structure [a] (after [8.31])

Compound		ΔH [cal/g-atom]		$\Delta V/V$ [%]	
		theor.	exp.	theor.	exp.
Li_2Na		680		0.1	
Li_2K		3600		−7.5	
Na_2K	(ideal)	−63		−1.5	
	(min)	−125	−145	−1.5	−1.5
Na_2Rb		1100		−5.4	
Na_2Cs		2070		−11.6	−7.8
K_2Rb		30		3.0	
K_2Cs	(ideal)	−125		−1.3	
	(min)	−180		−1.3	
Rb_2Cs		−130		0.6	

[a] Calculated using the parameters of the ideal structure unless stated otherwise.
For Na_2K and K_2Cs all structural parameters have been varied to minimize the total energy.

dicted to be stable relative to the corresponding solid solutions. For Na_2K it follows from the rather high ΔH of the solid solution, that the compound will be stable up to the observed melting temperature. For the other two compounds the transition temperature (compound $\rightarrow$ disordered alloy) may be estimated from the equality of the free enthalpies of formation (assuming for simplicity an ideal entropy of formation for the disordered alloy and the experimental value $\Delta S\,(Na_2K) = 1.3\,cal/g\text{-}atom\,deg$ for all Laves compounds). Using the calculated enthalpy of mixing of the solid solutions with an ideal lattice structure (see Sects. 7.1.1 and 7.3), one obtains $T_c = 612\,K$ (K_2Cs) and $T_c = 174\,K\,(Rb_2Cs)$. Using the experimental ΔH's for the solid solutions one finds $T_c = 240\,K\,(K_2Cs)$ and $T_c = 98\,K\,(Rb_2Cs)$; experimentally [8.29] $T_c \sim 180\,K\,(K_2Cs)$. The important thing is not a quantitative agreement with experiment (which can hardly be expected to be better), but the correct trend: Rb_2Cs is predicted to be stable only at temperatures much lower than the decomposition temperature of K_2Cs. Experimentally, Rb-Cs alloys have been investigated down to $T = 90\,K$, no indication of compound formation has been found in that range. Thus it appears that the theory is capable of predicting quantitatively meaningful values of ΔH for Na_2K, K_2Cs and Rb_2Cs, but it fails to explain the existence of the Laves phase Na_2Cs – predicting large positive enthalpies of formation for both Na_2Rb and Na_2Cs. This failure is probably associated with the fact that Cs behaves as a transition metal under moderately elevated pressure and might be expected to behave similarly when alloyed with another element with a higher electron density.

Compared to the random solid solutions the calculated excess volumes of formation are more negative for systems with large differences in the atomic size of the components. Based on earlier work of *Pauling* [8.33] and of *C.B. Shoemaker* and *D.P. Shoemaker* [8.34], *Simon* [8.35] has pointed out that the lattice constants and interatomic spacings of the Laves phases and those of close-packed intermetallic compounds of stoichiometry AB_n in general follow quite closely the relation

$$d_{ij} = \frac{f_{ij}}{n+1}(R_A + nR_B) \quad ; \quad i, j = A, B \tag{8.6}$$

expressing the additivity of the atomic radii (d is a lattice constant or interatomic distance, f_{ij} is some constant depending on the particular crystal structure). Equation (8.6) is the appropriate generalization of Vegard's law for disordered solid solutions to the case of an intermetallic compound. The physical significance of this relation is most evident in a *Pearson* [8.1] nearest-neighbour diagram (Fig. 8.6). Such a diagram is a plot of the structural strain parameter S versus the radius ratio, where

$$S = (2R_A - d_{AA})/2R_B \tag{8.7}$$

is the difference between the actual interatomic spacing d_{AA} and the dis-

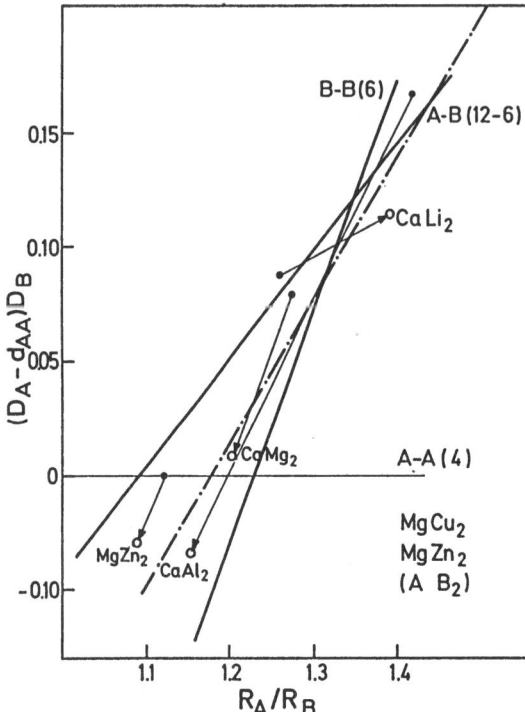

Fig. 8.6. Pearson nearest-neighbour diagram for the Laves phases. The solid straight lines correspond to the strain-free atomic contacts indicated, the bold dot-dashed line to the additivity of the atomic radii as expressed by (8.6). The dots characterize real compounds, represented by their Goldschmidt radii. The positions of KNa$_2$, CsK$_2$, CaLi$_2$, CaMg$_2$ and CaAl$_2$ are specially emphasized, open symbols are effective atomic radii deduced from the interatomic potentials in the compound (see text)

tance $2R_A$ for a simple strain-free AA contact. One finds that there is no unique radius ratio for which hard spheres can maintain all three contacts AA, BB and AB exactly. The ideal $(R_A/R_B) = (3/2)^{1/2} = 1.225$ is usually defined as corresponding to the centroid position of the triangle formed by the "strain-free" lines (incidentally it is the same as that corresponding to the point of intersection of the AA and BB lines). The line defined by the Simon law (8.6) passes through the triangle spanned by the AA, BB, and AB lines, and the real compounds are scattered along this line. Very recently *Churcher* and *Heine* [8.36] have given an explanation of this law in terms of a pair-potential model: the effect of a change in the radius ratio on the lattice constant is calculated from the virial $\sum_i F_i R_i$ (where F_i is the force at a distance R_i from a central atom whose radius has been changed). The essential point is that the A and B sites have to be elastically equivalent in the sense that a change δR of the atomic radius on either site produces the same overall change in the lattice constant. This condition imposes certain restrictions on the interatomic force law; in the approximation of purely harmonic interactions,

$$\Phi_{ii}(R) = \lambda_{ii}(R - 2R_i)^2,$$

$$\Phi_{ij}(R) = \lambda_{ij}(R - R_i - R_j)^2 \quad (i \neq j) \tag{8.8}$$

one finds that the force constant λ_{ij} has to be proportional to d_{ij}^{-m} where the exponent m lies in the range $3 < m < 4$, depending slightly on the crystal structure[2] [see also (8.5) in the preceding section but note that the two relations are not completely equivalent], i.e.

$$\lambda_{ij} = \lambda_0 d_{ij}^{-m} = \lambda_0 (f_{ij} d)^{-m} \quad , \tag{8.9}$$

where λ_0 is a constant and the f_{ij} are defined in (8.6). Again this relation connects atomic size and stiffness of the interatomic forces and demonstrates, as is natural, that large atoms consist of less tightly bound electrons and hence are softer. However, it seems unlikely that this is the complete story. The relations found by Churcher and Heine merely determine the conditions under which the relatively large individual AA, BB and AB strains compensate to zero along the Pauling-Simon line (8.6), and it appears unlikely that a lattice with large internal strains will be energetically favourable. An example is the Laves phases Na_2Cs, for which the microscopic calculation reproduces the Pauling-Simon law quite well, but predicts a positive ΔH.

In that connection it is instructive to look at heterovalent Laves phases, and in particular the series $CaLi_2$, $CaMg_2$ and $CaAl_2$. In spite of a nearly ideal radius ratio $R_{Ca}/R_{Li} = 1.26$ (in terms of the Goldschmidt radii of the pure components), $CaLi_2$ (C14 structure) is rather unstable and decomposes peritectically at $T \sim 500\,\mathrm{K}$; $CaMg_2$ $(R_{Ca}/R_{Mg} = 1.24)$ has a modest negative ΔH and the compound melts congruently with a melting temperature which is about the concentration-weighted average of the melting temperatures of the pure metals. Finally, $CaAl_2$ $(R_{Ca}/R_{Al} \sim 1.39)$ has a large negative ΔH $(-13\,400\,\mathrm{cal/g\text{-}atom})$ and a melting point which is considerably higher than the average. Thus the relative stability of these three compounds goes directly against a radius-ratio rule expressed in terms of the pure metal radii.

The case of Ca-Al alloys has already been discussed as an example of the renormalization of the interatomic potentials by chemical effects. The chemical compression of the more electropositive atom (Ca) changes the potentials in such a way that the interatomic distances fit exactly into the minima of the pair potential (see Sect. 2.4.1 and Fig. 2.8). The case of Ca-Li alloys serves to illustrate that the same effect can also lead to a destabilization of a given structure. In this case the more electropositive ion

[2] For the MgCu$_2$ and CaF$_2$ structures, *Churcher* and *Heine* [8.36] have determined m to be $m = 3.9$ and $m = 3.8$ respectively.

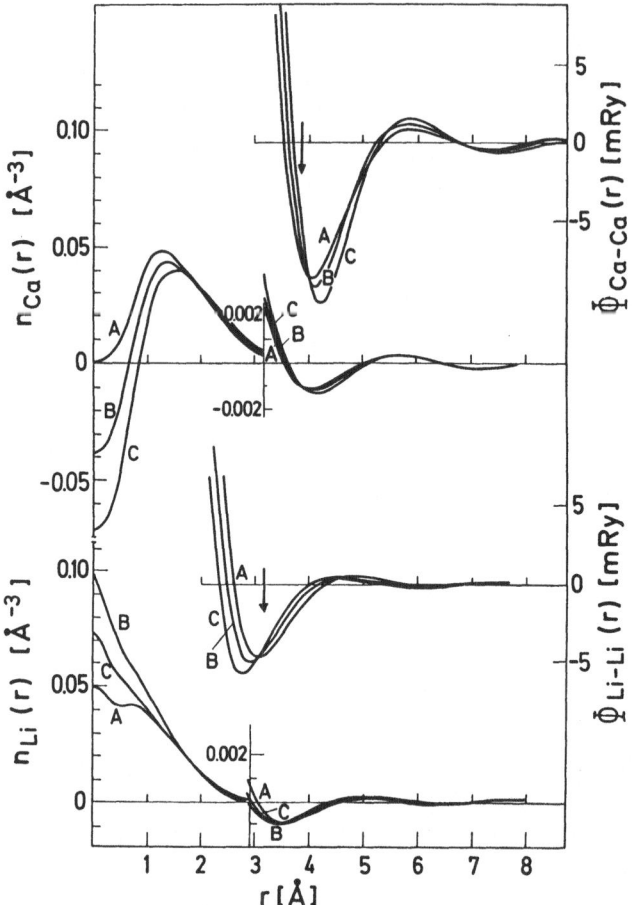

Fig. 8.7. Screening charge densities $n_i(r)$ and interatomic potentials for Ca and Li, calculated for the pure metals (*curve A*) and for alloys of the composition Ca_2Li (*curve B*) and $CaLi_2$ (*curve C*). The vertical arrows mark the interatomic distances $d_{Ca\text{-}Ca}$ and $d_{Li\text{-}Li}$ in the hexagonal Laves phase $CaLi_2$ (after [8.46])

is Li and consequently $\Phi_{Li\text{-}Li}$ is compressed and $\Phi_{Ca\text{-}Ca}$ slightly expanded. As a result it is found that the interatomic distances of the C14-$CaLi_2$ – which fit very nicely into the minima of the pair potentials of the pure metals – are not well adapted to the form of the potentials in the alloy (Fig. 8.7). This suggests that an effective radius ratio be defined in terms of the positions of the minima in $\Phi_{ij}(R)$, calculated at the composition of the alloy. If we redraw the Pearson plot with these effective concentration-dependent radii, we find that the points representing the actual compounds are shifted essentially parallel to the Pauling-Simon line (see Fig. 8.6), the stable compounds ($CaAl_2$, $CaMg_2$) now fall right in the middle of the "low-strain" triangle and the less stable $CaLi_2$ is shifted in the direction of larger strains.

Radwan and *Taut* [8.37, 38] have used a linearly screened optimized model potential to calculate the band structure and the soft x-ray spectra of $CaAl_2$. The calculated total density of states is surprsingly free-electron like, with small structures which correlate rather well with the observed soft x-ray spectra.

Thus it appears that the microscopic calculations give quite a reliable picture of Laves phase formation. What seems now to be highly desirable is to look for a simple, yet quantitative model which will enable us to understand the chemical compression effect.

8.4.2 Structural Stability of Laves Phases

It is well known that the crystal structures based on the close-packing of equal-sized spheres may be regarded as a stacking sequence of hexagonal layers which are the closest packing in two dimensions. There is a double choice in stacking each layer on the last one. They may be designated as ∇ and Δ and any repeating sequence defines a member of an infinite series of close-packed structures, of which the face-centred cubic and the hexagonal close-packed lattices are only the simplest examples of many.

In complete analogy to this, the close-packed intermetallic compounds may be considered as different stacking sequences of one or more different sandwiches of layers. For the AB_2 Laves phases such a sandwich is made up of a Kagomé net[3] of B atoms, a triangular net of A atoms centering its hexagons, then a triangular net of B atoms centering half of the Kagomé triangles (say those in orientation Δ) and by a third triangular layer of A atoms centering the remaining triangles. A second Kagomé layer of B atoms with its ∇ triangles over the Δ triangles of the first one begins the next sandwich which may be similar to the first one, or alternatively have its triangular B net over the ∇ triangles of the Kagomé net. The first stacking is called Δ, the second ∇ and an infinite series of structures is described by the different sequences of Δ and ∇. As in the case of the monatomic close-packed structures, all these structures have the same nearest neighbour configuration: the larger A atoms are surrounded by four other A atoms and twelve B atoms, while the B atoms have six B and six A as nearest neighbours. The majority atoms form a tetrahedral network, with the A atoms filling the large holes. As an example, a sketch of the C14 structure is shown in Fig. 8.8.

Repeated Δ stackings make the cubic Laves-phases C15, the repeated sequence $\nabla\Delta$ defines the C14 lattice and the sequence $\Delta\Delta\nabla\nabla$ the C36 structure. In addition to these three fundamental structures, several other

---

[3] A Kagomé consists of hexagonal rings of atoms, each side of the hexagon being caped with a triangle. See Fig. 8.8. The Kagomé nets are situated at $z = -1/4$, $z = 1/4$ and $z = 3/4$.

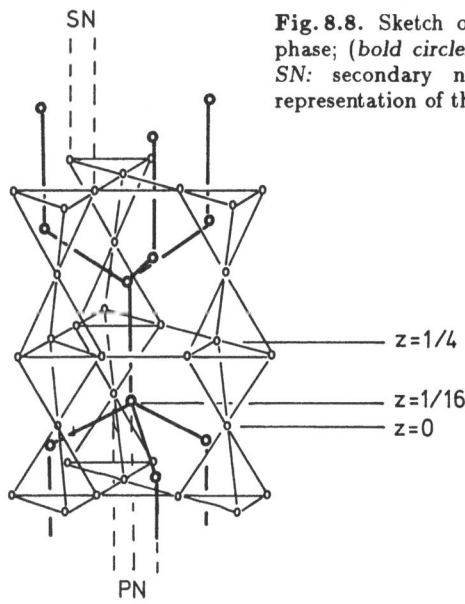

SN

PN

z = 1/4

z = 1/16
z = 0

Fig. 8.8. Sketch of the C14 structure of an AB_2 Laves-phase; (*bold circles*) A atoms, (○) B atoms. *PN:* primary; *SN:* secondary network (according to *Pearson's* [8.1] representation of the Laves phases)

stacking variants were found in pseudobinary AB_2-AC_2 alloys [8.39–42]: 6-, 8-, 9-, 10-, 14-, 16-, and 21-layer structures were found in systems like $MgCu_2$-$MgZn_2$, $MgCu_2$-$MgAl_2$, $MgZn_2$-$MgLi_2$ etc.

A considerable amount of effort has gone into explaining the occurrence of the different stacking variants. *Laves* and *Witte* [8.39] showed that for those quasi-binary alloys containing only simple metals there exists a Hume-Rothery-type rule for the valence electron concentration, attributed to Fermi-surface-Brillouin-zone interactions. *Elliott* and *Rostoker* [8.43] and *Dwight* [8.44] attempted to establish the importance of the electronic factor for transition-metal Laves phases as well. *Edwards* [8.45] noted a qualitative correlation between the crystal structure of the pure B metal and the occurrence of C15 or C14.

A number of microscopic calculations of structural energy differences between C14, C15 and C36 polytypes have been performed by *Hafner* [8.31, 46]. *Rennert* and *Radwan* [8.47, 48] examined in addition two possible six-layer stackings for the example of $MgZn_2$. The results are compiled in Table 8.2. Hafner's calculations are based on optimized first-principles pseudopotentials and predict the correct structure in each case; the calculations of Rennert and Radwan are based on different model potentials and fail to predict the correct C14 structure of $MgZn_2$ – only after introducing an artificial (in the sense that it does not follow from the calculations) electron transfer from Mg to Zn does the calculation yield the correct structure. Let us note that the alloys treated here follow Edward's correlation: if the B metal is hexagonally close packed (Li, Na, Mg, Zn), the Laves phase assumes the C14 structure (KNa_2, $CsNa_2$, $CaLi_2$, $CaMg_2$, $MgZn_2$); Al is fcc and the

Table 8.2. Structural energy difference relative to the energetically most favourable Laves-phase stacking variant (after [8.31, 46]) except for the results marked a and b, which are from [8.47, 48]. The symbols 6^1 and 6^2 stand for two other stacking variants consisting of six layers [8.47, 48]

Structures	ΔE [mRy/ion]				
	C14	C15	C36	6^1	6^2
Compound					
KNa_2	0	0.30	0.20		
KCs_2	0	0.40	0.27		
$CsNa_2$	0	0.50	0.40		
$CaLi_2$	0	0.15	0.46		
$CaMg_2$	0	0.10	1.80		
$CaAl_2$	0.80	0	5.30		
$MgZn_2$	0	0.70	2.20		
$MgZn_2$ [a]	0.60	0	0.14	0.04	0.30
$MgZn_2$ [b]	0	0.72		0.52	0.32

[a] Without an imposed charge transfer.
[b] With an imposed charge transfer.

compound $CaAl_2$ is of the C15 type. The only exception is CsK_2 where the B metal (K) has no close-packed structure (but note that all microscopic calculations predict K to be hcp at very low temperatures).

If we try to understand the physical mechanism behind these structural trends, we can again work either in reciprocal or in real space. However, there is now an important difference from the case of pure metals or substitutional solid solutions: In that case the stable stacking variant among the close-packed polytypes was determined by electronic effects alone since the electrostatic part of the structural energy differences is negligible (see Appendix C for the numerical values of the Madelung coefficients). This is not the case for the binary close-packed compounds. Here the differences in the Madelung coefficients (see Table 8.3 and Appendix C) are about a factor of ten larger than for the monatomic close-packed structures. Another point is that in the monatomic case, the minimum of the electrostatic energy is found for a structure very close to the ideal one, whereas for the hexagonal Laves phases the electrostatic energy has its minimum in a heavily distorted structure: for a homovalent ($Z_A = Z_B$) C14 lattice, the minimum is found at $c/a = 1.52$, $x = -0.18$, $z = 0.070$ to be compared with $c/a = 1.633$, $x = -0.1667$, $z = 0.0625$ for an ideal structure. The parameter x fixes the position of the B atoms in the Kagomé net and z measures the distance of the triangular net of A atoms from the Kagomé plane (for details and similar results for other structures see Table 8.3).

If only the ideal structures are compared, the total electrostatic energy and that of the A and B sublattices are lower for the C15 than for the C36 and the C14 structures and the A-B coordination is electrostatically more favourable in the hexagonal stacking variants.

Table 8.3. Madelung coefficients α_{ij}, $i,j = A, B$ $\alpha = \alpha_{AA} + \alpha_{AB} + \alpha_{BB}$ for some topologically close-packed binary crystal structures. The coefficients are given for the ideal structure and for the structure which minimizes the electrostatic energy $E_{es} = (Z_A^2 \alpha_{AA} + Z_A Z_B \alpha_{AB} + Z_B^2 \alpha_{BB})/R_a$ for the case of a homovalent compound ($Z_A = Z_B$). After [8.31]

Structure		α_{AA}	α_{AB}	α_{BB}	α
C15-MgCu$_2$-type[a]					
ideal		−0.386168	−0.478229	−0.907916	−1.772313
C14 MgZn$_2$-type					
ideal	$c/a = 1.633$ $x = -1/6$ $z = 1/16$	−0.385602	−0.480707	−0.905207	−1.771515
min.	$c/a = 1.52$ $x = -0.180$ $z = 0.070$	−0.383662	−0.473688	−0.917111	−1.774461
C36-MgNi$_2$-type					
ideal	$c/a = 3.266$ $x = 1/6$ $z = 3/32$	−0.385886	−0.479468	−0.906562	−1.771915
min.	$c/a = 3.22$ $x = 0.158$ $z = 0.093$	−0.385892	−0.479431	−0.906597	−1.771920
Al$_3$Zr$_4$-type					
ideal	$c/a = 1.0$ $z = 1/4$	−0.449350	−0.544329	−0.777983	−1.771662
min.	$c/a = 0.925$ $z = 0.275$	−0.457412	−0.554828	−0.765164	−1.777404
D8$_5$-Co$_7$Mo$_6$-type[b]					
ideal	$c/a = 5.45$ $z_1 = 0.3526$ $z_2 = 0.4552$	−0.544175	−0.647638	−.0578190	−1.770002
obs.	$c/a = 5.38$ $z_1 = 0.3483$ $z_2 = 0.4518$	−0.557958	−0.611739	−0.602004	−1.771701
min.	$c/a = 5.30$ $z_1 = 0.350$ $z_2 = 0.458$	−0.544649	−0.616476	−0.614297	−1.775422
K$_7$Cs$_6$-type[b]					
ideal	$c/a = 3.633$ $z_1 = 0.97125$ $z_2 = 0.8175$	−0.547710	−0.637783	−0.584698	−1.770190
obs.	$c/a = 3.629$ $z_1 = 0.9769$ $z_2 = 0.8208$	−0.553746	−0.623254	−0.594748	−1.771749
min.	$c/a = 3.630$ $z_1 = 0.9740$ $z_2 = 0.8100$	−0.529568	−0.651135	−0.594748	−1.775451

[a] All atomic positions fixed by symmetry.
[b] In the Co$_7$Mo$_6$ and the K$_7$Cs$_6$ structures there are more than two open parameters not fixed by symmetry. Only the parameters z_1 and z_2 determining the shortest distances between the large minority atoms (Mo and Cs) have been varied to minimize the Madelung energy.

The band-structure contribution to the total energy is given by (2.52). In Fig. 8.9 the normalized energy-wave-number characteristics of the KNa_2 and CsK_2 phases are plotted together with the corresponding structural weight functions $N_{ij}(q)$ (defined as the square of the partial structure factor times the multiplicity of the reciprocal lattice vector). The lowering of the symmetry in the hexagonal Laves phases is reflected by the distribution of the structural weight over a larger number of reciprocal lattice vectors. This

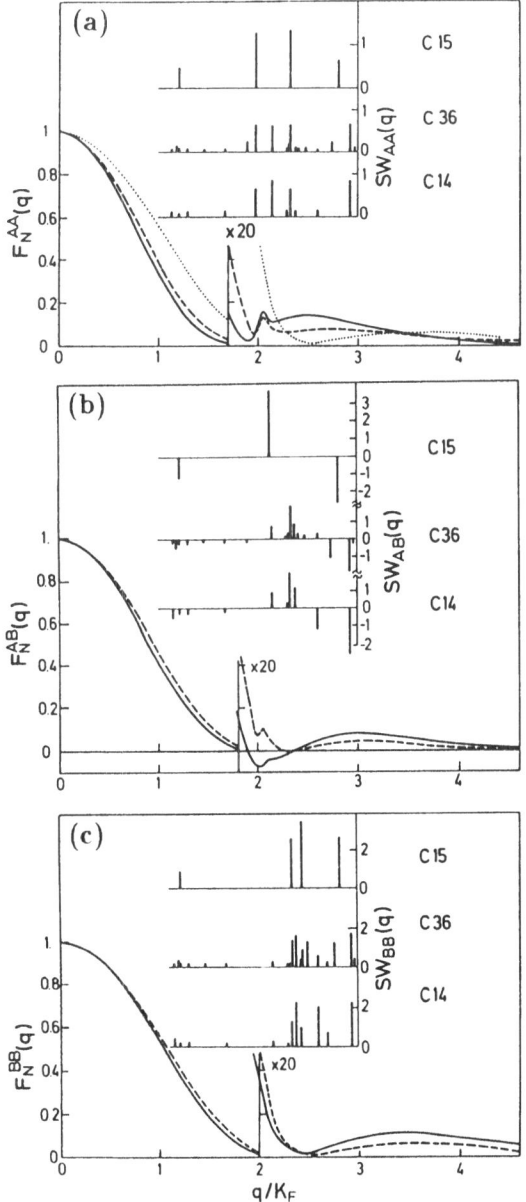

Fig. 8.9a–c. Normalized energy-wave-number characteristics $F_N^{ij}(q)$ and structural weight functions $N_{ij}(q)$ for the three Laves-phase structures of the compounds KNa_2 (——) and CsK_2 (- - -). In part (a) we include also the characteristic for K-K majority atom interactions in CsK_2 (···) to emphasize chemical effects in the indirect ion-electron-ion interaction (compare with the K-K minority interactions in KNa_2!). After [8.31]

results in energetically more favourable contributions to the band-structure energy for the C14 (and also to some extent for the C36 structure) than for the C15 structure. This is the analogue of the q_0 effect discussed in connection with the structural stability of pure metals (see Sect. 3.2).

A deeper understanding can be reached by varying the open parameters of the atomic positions within the unit cell. As has been shown above, the electrostatic energy has its minimum (for exactly equal valences Z_A and Z_B) in a heavily distorted structure. Now, even for KNa$_2$ and CsK$_2$, the ratio of the effective valences Z_i^* changes from about 1.04 for the pure metals to a value of about 1.12 in the alloy (the larger ion sees an increased electron density and has to form a larger orthogonalization hole, see also Sect. 2.4 and Appendix A.1). This relatively small change is sufficient to shift the minimum in the electrostatic energy to $c/a = 1.62$, $x = -0.175$ and $z = 0.0635$, i.e. the electron transfer associated with the orthogonality constraint removes most of the anisotropy present in the Madelung energy of a homovalent C14 compound and suffices to stabilize the wurtzite-like sublattice of the larger minority atoms (c/a and z are now quite close to their ideal values). The tetrahedral network of the majority atoms is still appreciably distorted – this predicted distortion is removed by the inclusion of band-structure effects. Here again chemical effects in the pseudopotential turn out to be important, as can be seen by comparing the energy-wave-number characteristic for K-K interactions in KNa$_2$ and in CsK$_2$ [Fig. 8.9(a)]. The complete set of calculated structural parameters is compared with experiment in Table 8.4.

Table 8.4. Lattice constants, structural parameters and enthalpies of formation for the C14-Laves phases KNa$_2$ and CsK$_2$ after [8.31]

	KNa$_2$ theor.	exp.	CsK$_2$ theor.	exp.
a [Å]	7.62	7.50	9.56	9.07
c [Å]	12.79	12.31	16.26	14.76
c/a	1.68	1.62	1.70	1.63
x	0.0635	0.0625	0.064	0.0625
z	−0.1675	−0.1667	−0.1670	−0.1667
ΔH [cal/g-atom]	−125	−145	−180	<0

In a real-space picture these effects depend on the details of the pair potentials: varying z changes mainly the A-A distances, varying x mainly the B-B distances. Hence a proper prediction of the correct values of these parameters is a stringent test of the form of the pair potentials around the nearest-neighbour distance. The stable stacking sequence of ∇ and Δ layers is determined by the interaction with second and more distant neighbours and this again is analogous to the fcc vs hcp stability problem in the pure metals and in substitutional alloys. In that case it was possible to discuss the stability problem in terms of lattice sums over a single Friedel potential (7.13), whereas in the present case the size difference cannot be neglected and the single potential would have to be replaced by a set of partial pair potentials – this remains to be done. But of course the most important factor will be the wavelength of the oscillatory part of the pair potentials, $\lambda_F = 2\pi/(2k_F)$ which is determined by the valence electron concentration alone (the size difference on the other hand sets the phase difference between the three individual potentials at small R, which goes assymptotically towards zero). The valence electron concentration is of course set mainly by the majority atoms. So we find that the *Laves-Witte* [8.39] and *Edwards* [8.45] correlations are well understood, at least for simple metals.

A notable attempt to interpret the phase stability of transition-metal Laves phases was made by *Johannes* et al. [8.49] and *Haydock* and *Johannes* [8.50]. They consider a common-band model alloy and calculate d-band densities of states for the three Laves phase structures using the recursion method of *Haydock* et al. [8.51]. The three densities of states all show similar highly peaked structures. Cohesive energies are calculated by summing up the one-electron energies. The structural energy differences show a cyclic variation with band filling, which is related to the moments of the densities of states and hence to the layer stacking of the structures. The predicted stable structures are first C15 (for up to an average number of 3.4 d-electrons per atom, e/atom ≤ 3.4), then C14 (e/atom ≤ 4.9), next C36 ($6.5 \leq$ e/atom ≤ 6.9), again C15 (e/atom ≤ 7.7) and finally from e/atom $= 7.7$ onwards C36 – in rather good agreement with experiment. The remaining small differences may be understood in terms of the major factor neglected in this common-band model, i.e. site-diagonal energy differences depending on the group-number difference between the two components.

Thus the interpretation of the structural stability of the transition-metal Laves phases is entirely analogous to *Pettifor*'s [8.52] interpretation of the crystal structures of the pure transition metals (see Sect. 3.3.1) and in direct contrast to the attempts [8.42, 43] to extend the Laves-Witte correlation from the simple to the transition metals.

8.4.3 Other Topologically Close-Packed Compounds

Microscopic calculations for other topologically close-packed intermetallic compounds are extremely scarce – indeed most of the crystal structures are prohibitively complex. A notable exception is *Hafner*'s work [8.31, 53] on intermetallic compounds with A_6B_7 stoichiometry and crystal structures closely related to the μ phases.

These crystal structures consist of the Δ and ∇ sandwiches building the Laves phases, intercalated with an alternate sandwich of layers which we will denote as O. The O sandwich is built up of a central hexagonal (honeycomb)

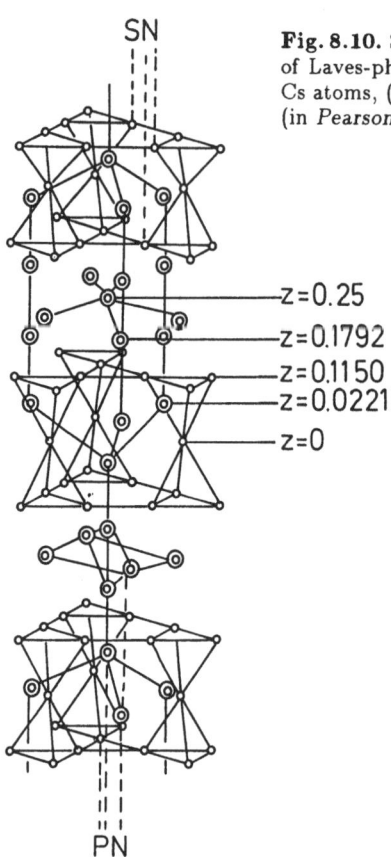

SN

z=0.25
z=0.1792
z=0.1150
z=0.0221
z=0

PN

Fig. 8.10. Sketch of the crystal structure of Cs_6K_7 consisting of Laves-phase-like and Al_3Zr_4-type layers. (*double circles*) Cs atoms, (o) K atoms. *PN:* primary; *SN:* secondary network (in *Pearson's* [8.1] terminology)

layer of A atoms, two triangular close-packed layers of A atoms (centering the hexagons) and two Kagomé nets of B atoms (with the triangles over the central A atoms) in centrosymmetric positions [8.28]. A repeated sequence of O layers defines the Al_3Zr_4 structure [8.54], and any sequence of O, Δ and ∇ layers defines a conceivable structure. For example, repeated ∇O layers make the μ phase ($D8_5$, typified by Fe_7W_6 and Co_7Mo_6) [8.55], repeated ∇OΔO layers build the structure of Cs_6K_7 [8.28, 29] (which is sketched in Fig. 8.10) and a further combination which would yield A_6B_7 stoichiometry is $\nabla\Delta$OO – however, no example of this structure is known [8.28].

The heat and volume of formation have been calculated for all possible alkali-metal alloys with these structures. All possible Al_3Zr_4-type phases have a positive ΔH and among the possible A_6B_7-stacking variants the lowest energy is always found for the ∇OΔO (K_7Cs_6-type) stacking. Figure 8.11 summarizes the calculated heats of formation (together with those of the Laves phases and the substitutional bcc solid solutions) as a function of the nominal radius ratio [8.31]. Negative values for the enthalpy of formation are predicted for Cs_6K_7, Cs_6Rb_7, and Rb_6K_7. Experimentally, only Cs_6K_7 has been found to be stable at temperatures below 178 K. The

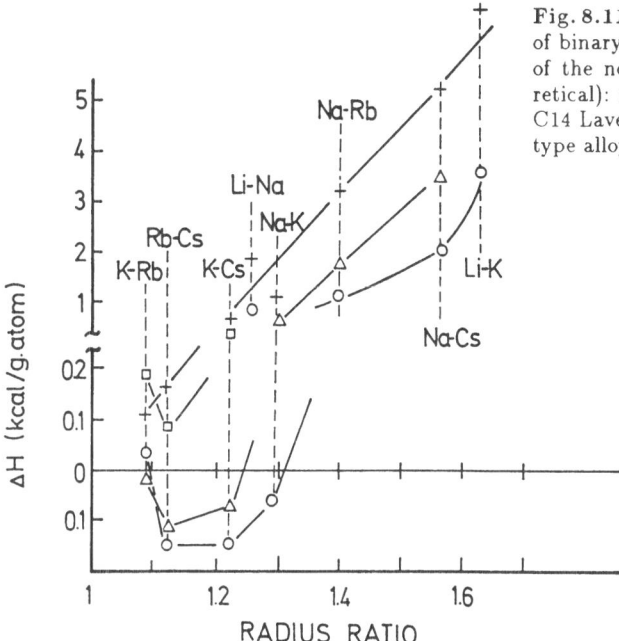

Fig. 8.11. Enthalpies of formation of binary alkali alloys as a function of the nominal radius ratio (theoretical): (+) bcc solid solution; (○) C14 Laves-phase AB_2; (Δ) K_7Cs_6-type alloys; (□) Al_3Zr_4-type alloys

critical temperature for a transition into a disordered solid solution may be estimated in the same manner as for the Laves phases with similarly encouraging results [8.31].

The physical mechanisms stabilizing the Cs_6K_7 structure are closely analogous to those already discussed for the Laves phases, but they are rather difficult to discuss since the structures have many more open parameters than the Laves phases. Here we shall pick out only one of the prominent features of the μ and the Cs_6K_7 phases and their stacking variants, namely the close distances between the large minority atoms in the Al_3Zr_4-like regions of the lattice (see the chains of four Cs atoms in z-direction, Fig. 8.10). In the transition-metal μ phases, these short interatomic spacings are often interpreted as resulting from a covalent contribution to the bonding – an interpretation which is rather unlikely to hold for the simple-metal compound Cs_6K_7. Here again one finds that the electrostatic energy has its minimum (for the homovalent case) in a strongly distorted structure – one in which these Cs-Cs distances are 12.6 % smaller than in the ideal structure. Taking the appropriate effective valences into account removes most of the distortion, and the band-structure contribution pushes these atoms even further apart so that finally they are somewhat larger than in an ideally close-packed structure – in reasonable agreement now with experiment. Thus one finds that the actual Cs_6K_7 structure is the result of a delicate balance between electrostatic and electronic forces.

Again it is instructive to cast a brief glance at the interatomic potentials for Cs_6K_7 (Fig. 8.12). According to the large number of crystallo-

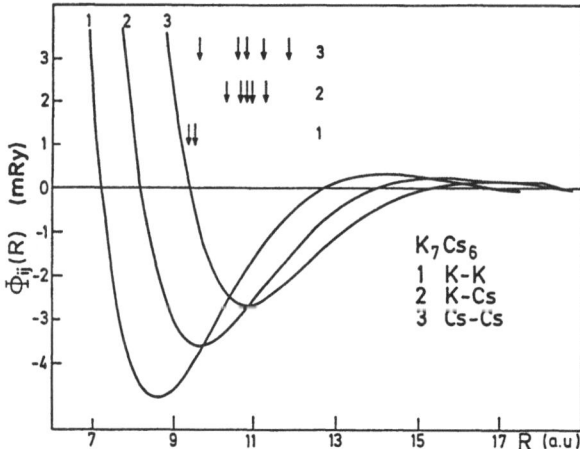

Fig. 8.12. Effective interatomic pair potentials in Cs_6K_7. The arrows indicate the nearest-neighbour distances in the observed Cs_6K_7 structure

graphically inequivalent sites and the deviation from the ideal geometry, the nearest-neighbour distances are now split, but they still fit quite well into the minima of the pair potentials. The argument that the stability of the topologically close-packed compound depends on the nearest-neighbour distances fitting into the minima of the pair potentials seems to have a general validity: it was first found in the case of the Laves phases and further confirmed for the more complex structure of Cs_6K_7. The changes of the pair potential on alloying were found to be important. Additional support for this stability criterion is provided by the Ca-Zn system: in this system a large number of intermetallic compounds with stoichiometries ranging from Ca_3Zn to $CaZn_{13}$ are found, most of which can be considered as being of a topologically close-packed type (for a description of the crystal structures see [8.56–58] and further references therein). The pair potentials shown in Fig. 8.13 show that in each case the interatomic potential is well adapted to the nearest neighbour spacings. It is the variation of $\Phi_{ij}(R)$ with concentration which makes possible the stability of this large number of intermetallic compounds. This variation of the pair potential depends on the rather subtle rearrangement of the valence electrons which is again best visualized in terms of the valence electron density.

8.4.4 Charge-Density Analysis of Bonding

As in the case of pure metals, a study of the valence electron distribution is very helpful in understanding the chemical bond in intermetallic compounds. However, both experimental and theoretical electron density determinations remain very scarce [8.53, 59, 60].

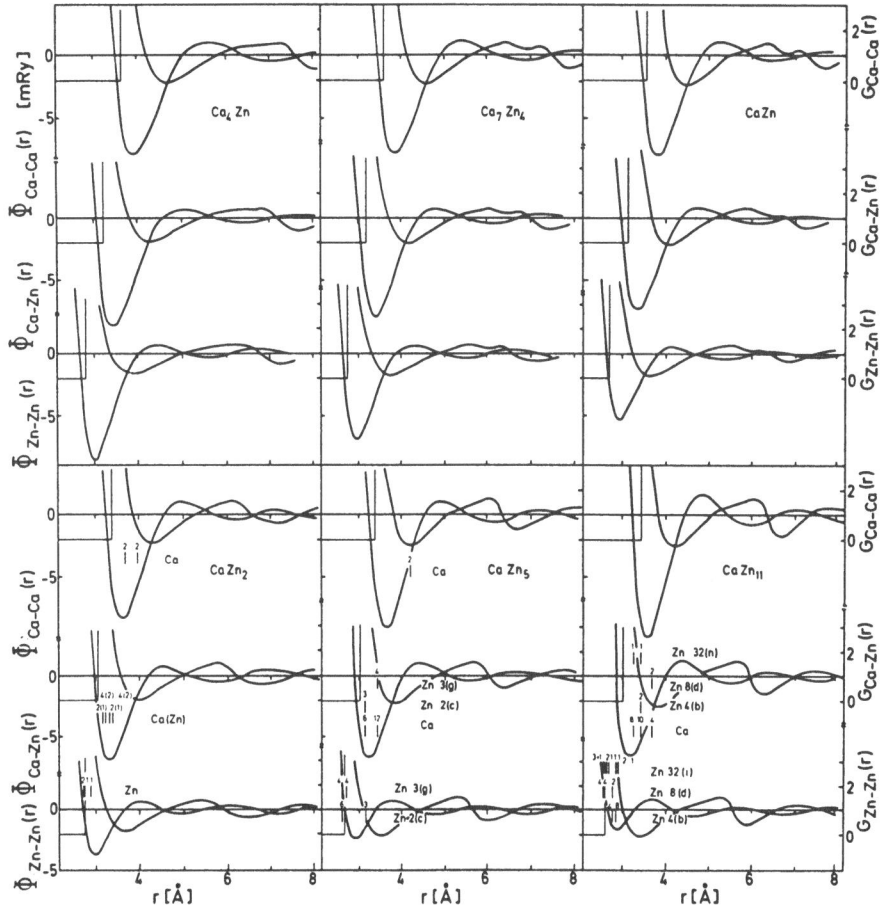

Fig. 8.13. Effective interatomic pair potentials and interatomic spacing for a number of Ca-Zn intermetallic compounds (see text)

The few results obtained so far show that the valence electron densities in binary intermetallic compounds are characteristically different from those found in the pure metals. Even in compounds with rather small chemical bonding effects such as KNa_2 or CsK_2, the electron density is far more inhomogeneous than in the pure metals (cf. Sect. 3.3.3): the interatomic density maximum is more than twice as large as the minimum density (Fig. 8.14)[4]. This is by no means unexpected – it simply corresponds to the picture of free-atom-like pseudoatoms. On the other hand, the electron density in the Na tetrahedra of KNa_2 (Fig. 8.14) is appreciably higher than in pure Na. This is consistent with the results of Sect. 8.4.2. There we found that the tetrahedral network of the minority atoms is stabilized by band-structure

---

[4] The charge densities presented here are based on the same pseudopotentials as the calculations on the thermodynamic excess functions and the structural energy differences discussed in the previous sections.

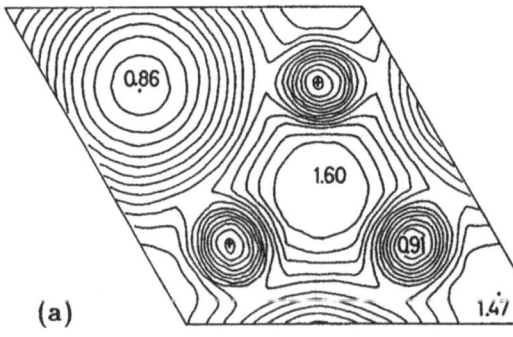

(a)

Fig. 8.14a,b. Pseudo-valence electron density in the hexagonal Laves phase KNa_2, in units of electrons per mean atomic volume, contour lines are drawn in intervals of 0.05 electrons/atomic volume. (a) In the $Z = 1/4$ plane (forming a Kagomé network), occupied by Na atoms at the sites marked $+$. (b) Primary network (see Fig. 8.8), parallel to the c axis (after [8.59])

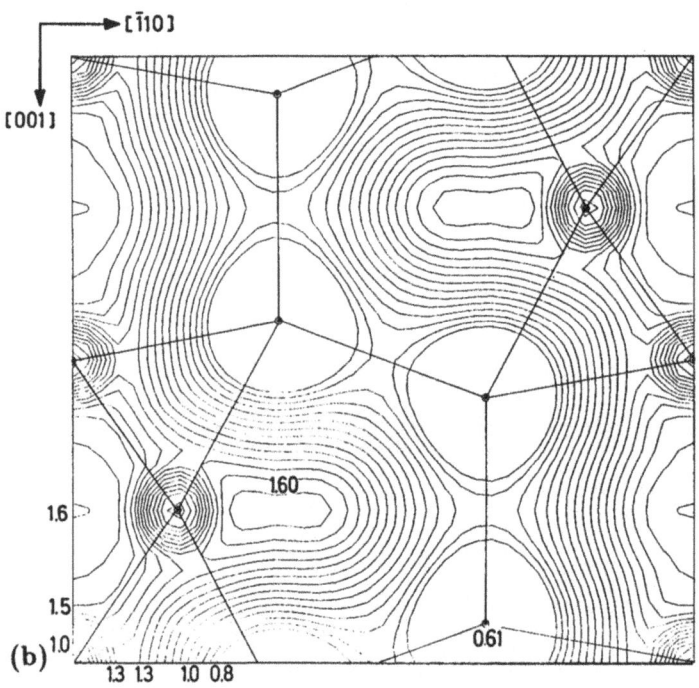

(b)

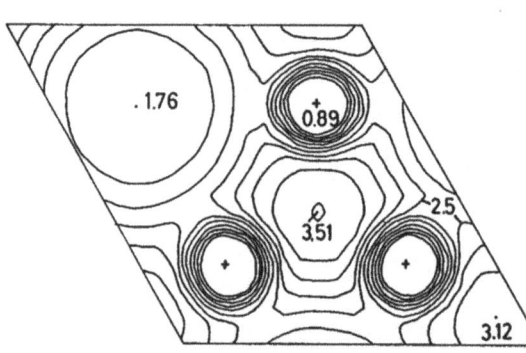

Fig. 8.15. Electron density (pseudo-valence electron density) in the Kagomé net of $MgZn_2$. The contour interval is 0.25 electrons/mean atomic volume. Zn-atoms are represented by crosses (after [8.59])

effects and here we see that this stabilization is realized by increasing the electron density in the tetrahedra. Figure 8.15 shows the electron density in the Kagomé plane of the hexagonal Laves phase $MgZn_2$. The result emphasizes the effects found in KNa_2, i.e. absence of bonding electrons in nearest neighbour bonds and increased electron density at the centres of the double tetrahedra formed by the small majority (B) atoms. Note that in the Laves phases there are many tetrahedra formed by A and B atoms (there are AB_3, A_2B_2 and B_4 aggregates), but a distinct accumulation of electrons occurs only in the $2 \times B_4$ aggregates formed by the majority atoms. For a complete discussion of electron-transfer effects, the orthogonalization effect has to be added to the pseudo-valence electron densities. In KNa_2, CsK_2 and $MgZn_2$ it is always found that the electronic charge around the core of the large A atom is lower than in the pure A metal; it is slightly enhanced around the core of the small B atoms, and a few residual electrons are found in the interstitial region, mainly in the B tetrahedra.

The redistribution of electrons on alloying appears even more distinctly in a plot of the difference electron density defined as

$$\varrho(\mathbf{r}) = \sum_{i(A)} [\varrho_A^{AB}(|\mathbf{r} - \mathbf{R}_i|) - \varrho_A(|\mathbf{r} - \mathbf{R}_i|)]$$

$$+ \sum_{i(B)} [\varrho_B^{AB}(|\mathbf{r} - \mathbf{R}_i|) - \varrho_B(|\mathbf{r} - \mathbf{R}_i|)] \quad , \tag{8.10}$$

where ϱ_i is the pseudo-valence electron density in the pure i-metal and ϱ_i^{AB}, $i = A, B$ is the pseudo-valence electron density in the AB alloy and the sum $\sum_{i(A)}$ sums only over lattice sites occupied by A atoms.

In the heterovalent Laves phases, the effect of the electronic rearrangements is very distinctly expressed. Here again it is instructive to consider the series $CaAl_2$, $CaMg_2$ and $CaLi_2$. The difference electron densities in their Kagomé nets is shown in Fig. 8.16 (note that $CaAl_2$ is cubic whereas the other two compounds are hexagonal). There is a strong accumulation of electrons in the Al_4 tetrahedra, a somewhat weaker and more diffuse one in the Mg_4 tetrahedra, and in the Li_4 tetrahedra the electron density is actually reduced, in striking analogy to the variation of crystal stability in this series of compounds. Thus there is good evidence that the accumulation of electrons in the B_4 tetrahedra really plays a crucial role in the bonding of Laves phases.

Up to now, we know of only a single experimental determination of the electron density in a binary simple-metal compound, by *Ohba* et al. [8.60] for $MgZn_2$ and $MgCu_2$. Their difference electron densities for these two compounds are shown in Fig. 8.17 in the form of contour maps in the Kagomé planes. Although they are not directly comparable to the calculated results (they are defined relative to a free-atom electron configuration), they confirm the main results of the theoretical analysis: (a) only

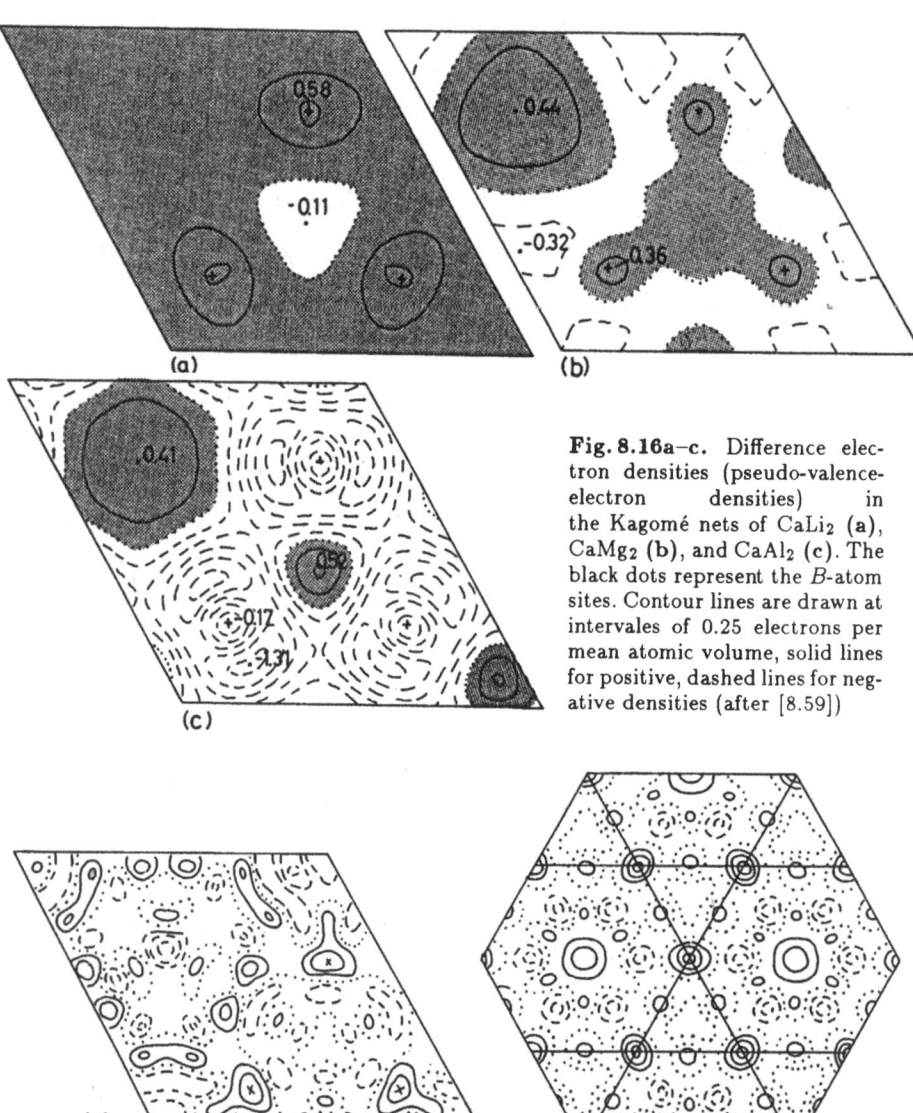

Fig. 8.16a–c. Difference electron densities (pseudo-valence-electron densities) in the Kagomé nets of $CaLi_2$ (a), $CaMg_2$ (b), and $CaAl_2$ (c). The black dots represent the B-atom sites. Contour lines are drawn at intervals of 0.25 electrons per mean atomic volume, solid lines for positive, dashed lines for negative densities (after [8.59])

Fig. 8.17a,b. Difference electron densities (relative to free-atom reference configurations) for the Kagomé planes of $MgZn_2$ (a) and $MgCu_2$ (b), as determined by *Ohba* et al. [8.60] using x-ray diffraction. The contour intervals are $0.05\,e/\text{Å}^3$ ($MgZn_2$) and $0.1\,e/\text{Å}^3$ ($MgCu_2$); solid and dashed lines represent positive and negative densities, respectively

minimal asphericity of the individual atoms, substantiating our picture of a superposition of spherical pseudoatoms; (b) absence of bonding electrons between the central atom and the surrounding atoms in the coordination polyhedra, confirming again the absence of directional bonding effects; (c) increased electron density in the B_4 tetrahedra and (d) electron transfer from Mg to Zn (MgZn$_2$), and from Cu to Mg (MgCu$_2$). Thus we conclude that the microscopic picture for Laves phase formation is corroborated by the experimental charge density analysis.

Charge density calculations have also been presented for Cs$_6$K$_7$ [8.53]. It is found that in the Δ and ∇ layers of the structure, the electron density distribution is very similar to CsK$_2$, whereas in the intermediate Cs-rich layers, it is rather homogeneous and close to pure Cs. Thus from the point of view of the charge density analysis, Cs$_6$K$_7$ appears as Cs-intercalated Cs$_2$K.

8.4.5 Lattice Dynamics of Topologically Close-Packed Compounds

For the stoichiometric intermetallic compounds with a well-defined site occupancy, the configurational part of the entropy of formation vanishes and therefore the entropy of formation depends only on the vibrational properties of the compounds. However, owing to the complexity of the crystal structures, only a very small number of lattice dynamical investigations of intermetallic compounds have so far been undertaken [8.61–65], although thermodynamic investigations suggest that soft phonon modes exist in a number of C14 phases[5].

A generalization of the linear response theory of harmonic lattice dynamics sketched in Sect. 4.5.1 to the case of a binary compound is trivial. The dynamical matrix $D_{\alpha\beta}^{\kappa\kappa'}(q)$ $(\alpha, \beta = 1, 2, 3; \kappa, \kappa' = 1, \ldots r)$ is now a $3r \times 3r$ matrix where r is the number of atoms in the unit cell, each matrix element being given by a lattice sum over the Fourier transform $\hat{\Phi}_{\kappa\kappa'}(q)$ of the corresponding effective interatomic pair potential [see (3.9–12)].

The first phonon dispersion calculation for an intermetallic compound was performed by *Eschrig* et al. [8.61] for MgZn$_2$[6] and showed some very interesting features. In this case there are 12 atoms per unit cell and therefore 36 dispersion branches. Only the low-energy part of the phonon dispersion relations is shown in Fig. 8.18. The most remarkable aspect of these results is the existence of very low-lying transverse optical modes at energies of only ~ 3 meV. Two of these modes (Δ_3, Δ_4) correspond to a librational motion of the Zn double tetrahedra around the c axis. A group-theoretical analysis [8.62] shows that the polarization vectors of these modes are completely determined by symmetry, independent of the interatomic potential. This would suggest that the soft-mode character of these vibrations is a general property of the crystal structure. Unfortunately, the existence of these soft

---

[5] This is inferred mainly from specific heat measurements which show a violation of Debye's T^3 law at comparatively low temperatures between 5 and 20 K in MgZn$_2$ and other hexagonal Laves phases. See [8.61] for details and references.

[6] The calculation is based on the optimized version of the Heine-Abarenkov-Animalu pseudopotential [8.66] and a simple exchange-correlation correction to the RPA screening.

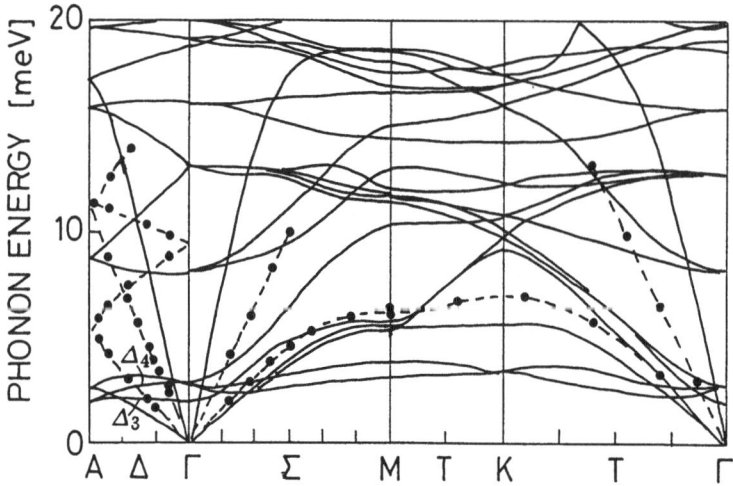

Fig. 8.18. Low-energy part of the phonon dispersion relations in the hexagonal Laves phase (C14) MgZn$_2$. (——) theoretical results of *Eschrig* et al. [8.61, 62]; (•, - - -) experimental results of *Dorner* et al. [8.63]

modes could not be confirmed experimentally [8.63]; the experimental results are, however, still rather incomplete. This first example demonstrates nonetheless, that the phonon frequencies are very sensitive to the interatomic potential and that the approach of neglecting all chemical changes in the pseudopotential is not appropriate.

A second calculation [8.64] has been performed for CaMg$_2$ (which has the same structure). In this case the soft modes are not predicted by the theory – which would suggest that they have to be attributed to the very large mass of the Zn double tetrahedra in MgZn$_2$. For CaMg$_2$ the only comparison with experiment is via a generalized phonon density of states (essentially an integral over the inelastic neutron scattering law, extended over a large region in q space); see Fig. 8.19. In this case the width of the distribution and the position of the main peak are well reproduced.

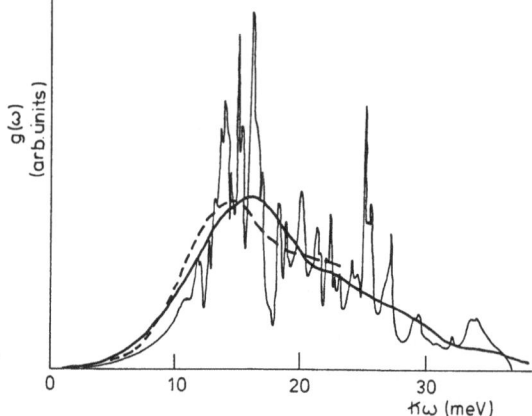

Fig. 8.19. Phonon density of states for the hexagonal Laves phase (C14) CaMg$_2$. (*Thin line*) phonon density of states, calculated from the dispersion relations; (*bold line*) generalized DOS (calculated including effects of instrumental resolution); (*dashed line*) experimental generalized DOS (after [8.64])

The first complete determination of phonon dispersion relations for a tetrahedrally close-packed simple-metal compound was performed by *Reichardt* and *Kobbelt* [8.65] for $CaAl_2$, following earlier inelastic neutron scattering experiments on the isomorphic compounds $LaAl_2$ and YAl_2 [8.67]. In the cubic Laves phase there are 18 independent phonon branches. A Born-von Kármán force-constant analysis shows that the nearest-neighbour interactions dominate the phonon spectrum. The higher-lying optical branches, which all involve deformations of the Al double tetrahedra, turn out to be especially sensitive to the form of the Al-Al potential around the nearest-neighbour distance, therefore the reproduction of these dispersion relations is the most stringent test of any pair potential in a binary compound performed so far. Reichardt and Kobbelt analyzed their data using simple model potentials and found that the pseudopotentials in the alloys are very different from those in the pure metals – with the modified potentials a reasonably accurate fit can be achieved.

8.5 Intermetallic Phases with Large Band-Structure Stabilization

The intermetallic phases where the band-structure energy constitutes an unusually large fraction of the total energy are an important testing ground for the limits of the applicability of perturbative calculations of energy and other properties. Of the intermetallic phases belonging to this group, the disordered Hume-Rothery phases have already been discussed in Sect. 7.2, and we have found that perturbation theory offers a valid explanation of the valence-electron-concentration rules. The Zintl phases have been discussed as ordered superstructures of a mean bcc lattice in Sect. 7.4.2, where we found that three-body forces are necessary for a stabilization of the B32 vs the B2-type superstructure. Here we return briefly to this topic in order to illustrate the appearance of covalent bonding effects in the electron density distribution and in the lattice dynamics of these compounds.

The Zintl phases with the B32 structure (LiAl, LiGa, LiIn, NaIn, NaTl, LiZn and LiCd; see Sect. 7.4.2) occupy a fascinating niche in our knowledge of intermetallic phases. In early studies, the bonding in the Zintl compounds was variously, and sometimes simultaneously, described as ionic, covalent, or metallic [8.68]. Modern band-structure techniques have been applied repeatedly to the calculation of the electronic structure and of the total energy of Zintl phases (see e.g. [8.69,70] and further references therein) and attempts to extend pseudopotential-perturbation methods have been discussed in Chap. 7 [8.71,72]. The most recent band-structure calculations agree in their support for the classical conjecture of a covalent sp^3-type bond between the polyvalent metal atoms: a strong intra-atomic s to p promotion leads to the formation of incomplete sp^3-hybrids. This picture of the chemical bond is most directly visualized in the electron density distributions.

8.5.1 Charge-Density Analysis of Bonding in Zintl Phases

Figure 8.20 shows contour maps of the valence electron densities in the (110) plane of LiAl in the B32 and in the B2 structures, based on self-consistent tight-binding calculations [8.70] and on pseudopotential-perturbation calculations (in that case the pseudo-valence electron density is shown). The most striking feature is the appearance of a "bond charge" along the Al-Al nearest-neighbour bonds in the B32 structure, quite similar in form to, though of course much weaker than, in a tetrahedrally bonded semiconductor [compare with the electron density in the same crystallographic plane of Si as shown in Fig. 3.7(d)]. Along the Li-Li bonds on the other hand, the

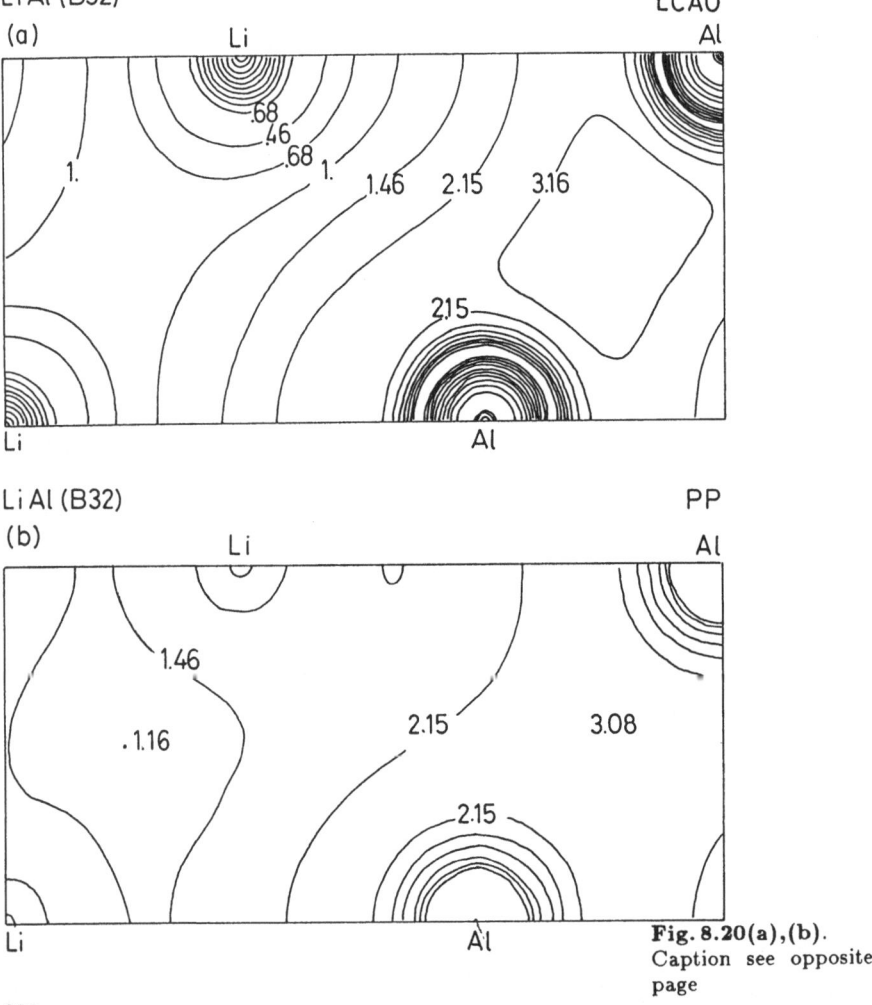

Fig. 8.20(a),(b). Caption see opposite page

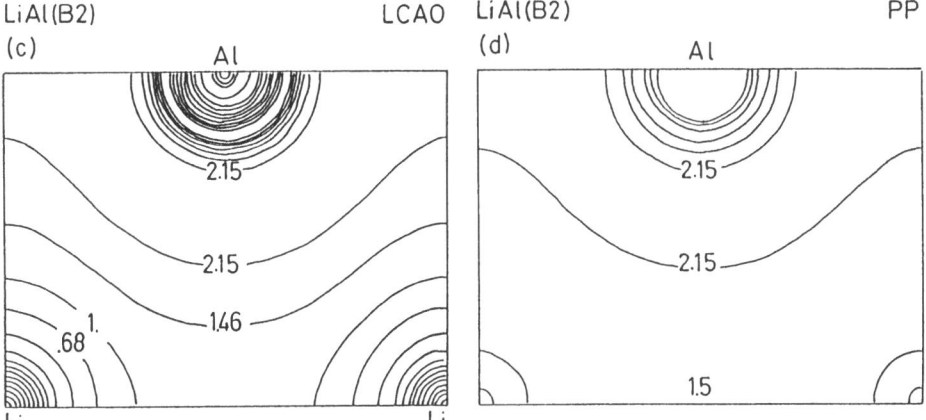

Fig. 8.20(a)–(d). Contour maps of the valence-electron distributions in the (110) plane of LiAl, in the B32 [(a) and (b)] and in the B2 structures [(c) and (d)]. (a) and (c) show the result of self-consistent tight-binding calculations [8.70], (b) and (c) the pseudo-valence electron densities (screening charge densities) derived from a perturbation calculation based on optimized pseudopotentials. Contour lines are drawn at geometrically spaced intervals, and the numbers give the electron density in 10^{-2} electrons/a.u.^{3}

electron density is quite uniform. The Li-Al bonds show an ionic polarization towards the Al site. These results suggest that, at least for the I-III B32-compounds, the bond is generally metallic, with an appreciable covalent contribution to the Al-Al bond. In the B2 structure the bond appears to be purely metallic.

If we compare these results derived from the self-consistent tight-binding calculations with the linear-screening calculations based on optimized pseudopotentials, we find that the perturbative approach yields an adequate representation of the electron density in the B2 structure (keeping in mind that we compare a pseudo-charge density with a real charge density and hence should look only at the interstitial region), but not in the Zintl phase. In the latter, the electron density is lower in the Al-Al bonds, but higher in the Li-Li bonds than predicted by the tight-binding calculations. Thus it appears that the linear-screening approach misses at least part of the characteristic transfer of electrons from the Li-Li to the Al-Al bonds [8.69, 70].

8.5.2 Lattice Dynamics of Zintl Phases

Finally we would like to make a few remarks about the dynamical properties of the Zintl phases. The phonon dispersion curves should give valuable information about the interatomic forces. Such a study is also motivated by the fact that LiAl is one of the rare intermetallic compounds where a complete set of phonon dispersion relations has been measured using inelastic neutron scattering [8.73]. Figure 8.21 compares the experimental results

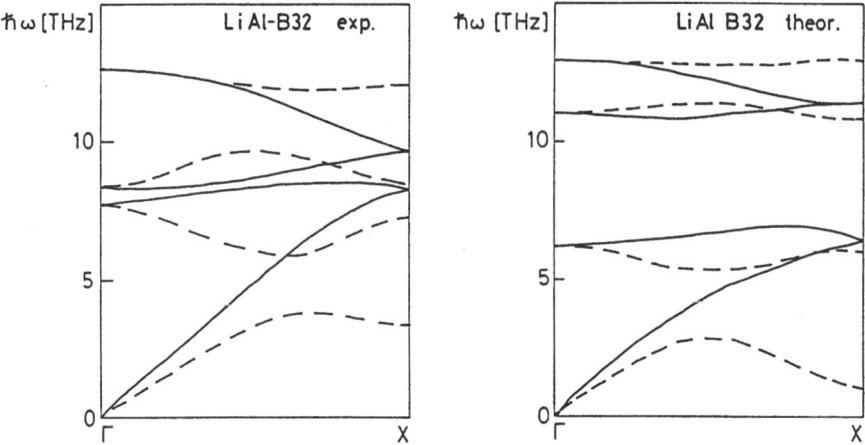

Fig. 8.21a,b. Phonon dispersion curves for B32 LiAl; (a) Empirical force-constant fit to the neutron scattering data [8.73]. (b) Theory. (——) longitudinal modes; (- - -) doubly degenerate transverse modes

with the theoretical prediction derived from interatomic pair forces based on optimized ab initio pseudopotentials. The calculation yields a good fit to the elastic constants and to the highest optical modes, but differs from experiment in two important aspects: (i) the dip in the transverse acoustic branch near the X-point is too deep, and (ii) the nearly degenerate longitudinal optical modes near 8–9 THz cannot be reproduced. The theoretical prediction of a rather soft acoustic zone-boundary mode probably stems from the neglect of covalent bonding forces – a similar situation is found for the flat transverse-acoustic branches in the elemental semiconductors. The reason behind the overestimate for the splitting of the LO modes is not so clear. It could be related to the fact that B32 LiAl is never completely free of defects. Stoichiometric LiAl usually contains both Li vacancies and Li antisite defects (Li atoms on Al sites) at total concentrations of about 3 percent, with appreciable atomic relaxation around the defects. Corrections for the disorder effects will probably reduce the LO splitting.

In any case, conventional second-order perturbation theory represents at best an acceptable starting point for a theoretical description of these strongly bonding compounds. Higher-order corrections, guided by fully self-consistent electronic structure calculations are urgently called for.

Recent Results

The results of the second-order pseudopotential calculations for the simple-metal Laves phases have been confirmed by fully self-consistent linearized-muffin-tin-orbital (LMTO) calculations of the electronic structure [8.74]. *Christensen* [8.75] has extended his LMTO calculations of the electronic structure and of the total energies to ternary Zintl phases, with particular emphasis on the trend in the sequence Si $\rightarrow$ LiAlSi $\rightarrow$ LiAl from strong to weaker covalent bonding and to bonding with a more metallic character.

9. Liquid Alloys

Formally the generalization of liquid state theory from monatomic to binary or multicomponent fluids is trivial: instead of a single integral equation relating the pair-correlation function $g(R)$ to the pair potential $\Phi(R)$, we have now a set of three coupled integral equations relating the partial pair-correlation functions $g_{ij}(R)$ to the pair potentials $\Phi_{ij}(R)$. The pair-correlation functions may be chosen either as the component-related correlation functions $g_{AA}(R)$, $g_{BB}(R)$, and $g_{AB}(R)$, describing the spatial correlation of AA, BB, and AB pairs, or the functions $g_{NN}(R)$, $g_{cc}(R)$, and $G_{Nc}(R)$ introduced by *Bhatia* and *Thornton* [9.1]

$$g_{NN}(R) = c_A^2 g_{AA}(R) + c_B^2 g_{BB}(R) + 2c_A c_B g_{AB}(R)$$

$$g_{cc}(R) = c_A c_B [g_{AA}(R) + g_{BB}(R) - 2g_{AB}(R)]$$

$$g_{Nc}(R) = c_A g_{AA}(R) - c_B g_{BB}(R) + (c_B - c_A) g_{AB}(R) \qquad (9.1)$$

describing respectively number-density fluctuations, concentration fluctuations and the density-concentration correlation in direct space. In reciprocal space, these correlations are described by the partial static structure factors which depending on the convention adopted, are defined either as the Ashcroft-Langreth [9.2] structure factors

$$S_{ij}(q) = \delta_{ij} + (n_i n_j)^{1/2} \int [g_{ij}(R) - 1] e^{i\mathbf{q}\mathbf{R}} d^3 \mathbf{R} \qquad (9.2a)$$

or the Faber-Ziman [9.3,4] structure factors

$$a_{ij}(q) = 1 + n \int [g_{ij}(R) - 1] e^{i\mathbf{q}\mathbf{R}q} d^3 \mathbf{R} \qquad (9.2b)$$

or the Bhatia-Thornton [9.1] structure factors

$$S_{NN}(q) = 1 + n \int [g_{NN}(R) - 1] e^{i\mathbf{q}\mathbf{R}} d^3 \mathbf{R}$$

$$S_{cc}(q) = c_A c_B [1 + n \int g_{cc}(R) e^{i\mathbf{q}\mathbf{R}} d^3 \mathbf{R}]$$

$$S_{Nc}(q) = c_A c_B n \int g_{Nc}(R) e^{i\boldsymbol{q}\boldsymbol{R}} d^3 \boldsymbol{R} \quad . \tag{9.3}$$

The Bhatia-Thornton functions are especially useful for desplaying the short-range order in a liquid alloy: $g_{NN}(R)$ and $S_{NN}(q)$ describe the topological properties of the spatial arrangement of the mean atoms (their topological short-range order – TSRO) as in a one-component fluid. A nice example is the variation of the TSRO in Al-Si alloys [9.5]. The functions $g_{cc}(R)$ and $S_{cc}(q)$ on the other hand describe the local fluctuations in the concentration and hence the chemical short-range order (CSRO) in the liquid alloy. Figure 9.1 shows their typical form for a nearly ideal substitutional alloy (Na-K), for a case with a strong tendency towards heterocoordination (Li-Pb), and finally at a point close to the phase separation curve in Li-Na. Note that $g_{cc}(R)$ is positive for a preferred coordination by like atoms and negative for a preferred coordination by unlike atoms, see (9.1). The fact that the liquid

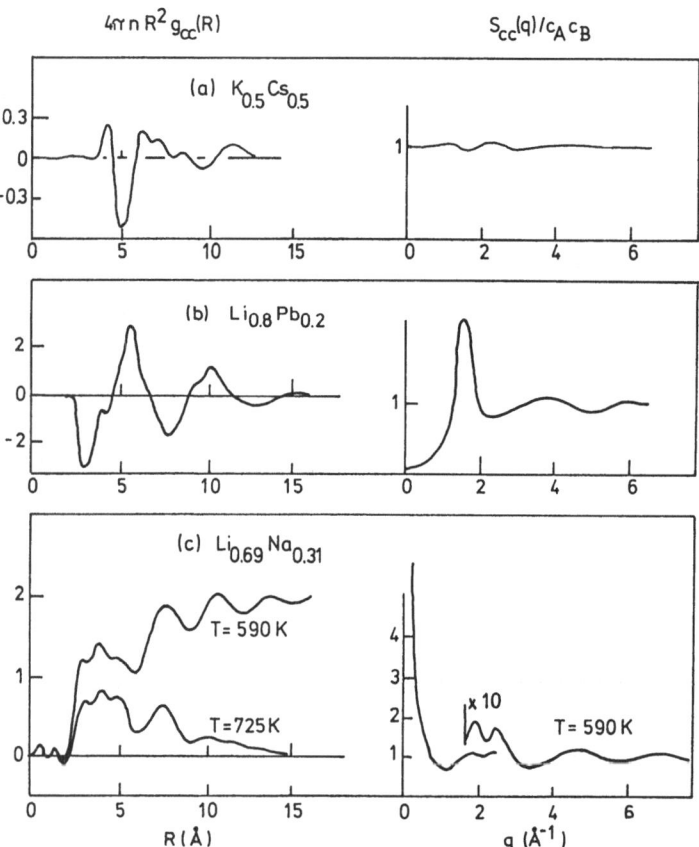

Fig. 9.1a–c. Concentration fluctuation correlation function $g_{cc}(R)$ and static structure factor $S_{cc}(q)$ for **(a)** a Na-K alloy with a nearly ideal mixing behaviour [9.7], **(b)** a Li$_4$Pb alloy with a strong tendency to heterocoordination [9.6] and **(c)** close to the phase-separation curve in Li-Na alloys [9.8]

structure is now described by a set of three partial structure factors, also causes important experimental complications. Each diffraction experiment measures now only a weighted mean value of the partial static structure factors, i.e.

$$
\begin{aligned}
S(q) &= [c_A f_A(q)^2 S_{AA}(q) + c_B f_B(q)^2 S_{BB}(q) \\
&\quad + 2\sqrt{c_A c_B} f_A(q) f_B(q) S_{AB}(q)]/\overline{f^2} \\
&= [\overline{f(q)}^2 S_{NN}(q) + \Delta f(q)^2 S_{cc}(q) + 2f(q)\Delta f(q) S_{Nc}(q)]/\overline{f^2} \quad (9.4)
\end{aligned}
$$

with $\overline{f} = c_A f_A + c_B f_B$ and $\Delta f = f_B - f_A$. The $f_i(q)$ are the atomic scattering form factors for neutrons, x-rays or electrons. This means that *three independent* diffraction experiments are necessary for a complete determination of the partial structure factors – only a very few such determinations have been performed until now. For a recent discussion of diffraction experiments on binary liquid alloys, see [9.9, 10].

Deviations from an ideal mixing behaviour which show up in the atomic structure also appear in the thermodynamic excess functions. The low-q limit of $S_{NN}(q)$ is related to the compressibility and to the excess volume of formation; that of $S_{cc}(q)$ is inversely proportional to the seond derivative of the free enthalpy of formation [9.1]. Therefore the prime aim of a theory of liquid alloys must again be to calculate the partial pair-correlation functions from the interatomic potentials, and hence the thermodynamic excess functions. Again this can be achieved by computer simulation, by solving a set of coupled integral equations, or by thermodynamic perturbation theory. We begin by discussing the results available from computer simulations.

9.1 Computer Simulations of Binary Liquid Alloys

As in the investigation of simple monatomic liquids, the aim of computer simulations of liquid mixtures is twofold: (i) to test the integral equation or thermodynamic perturbation results for a given set of interatomic potentials against the "exact" computer simulation data and (ii) to check the reliability of the interatomic potentials. However, it turns out that even with present-day computers, there are still limitations to the computer simulation of liquid alloys; with the range of interaction as given by typical metallic pair potentials, each particle interacts with at least 60 to 100 neighbours, so that models with 1000–4000 particles are about the largest which can be handled. This means that the statistics of the partial pair correlations is of a rather modest quality, especially for the minority component. As a consequence it seems that the long-wavelength part $(q \lesssim 0.6\,\text{Å}^{-1})$ of the partial structure factors is inaccessible to present-day computer experiments.

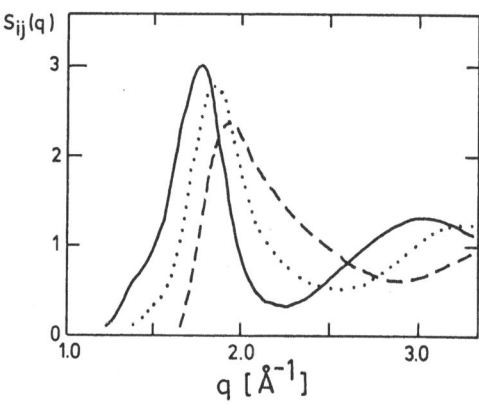

Fig. 9.2. The partial static structure factors $S_{ij}(q)$ for an equiatomic liquid Na-K alloy close to the melting temperature: (——) $S_{KK}(q)$, ($\cdots$) $S_{Na\text{-}K}(q)$, (- - -) $S_{NaNa}(q)$. Molecular dynamics calculations of *Jacucci* et al. [9.7]

The first application of molecular dynamics to a liquid alloy was presented by *Jacucci* et al. [9.11] for Na-K. Their calculation is based on the DRT model potentials [9.12]. The resulting partial structure factors are shown in Fig. 9.2. The most striking feature is the asymmetry of the first peak in the partial static structure factors, especially in that of the smaller atoms. Not surprisingly it turns out that these features are rather well described by the Percus-Yevick structure factors for a mixture of hard spheres of different sizes (analytical solutions have been given by *Ashcroft* and *Langreth* [9.2] and by *Enderby* and *North* [9.4]). Thus the differences in the partial structure factors are to be attributed to the difference in the atomic sizes and the distribution of the smaller atoms is somewhat more diffuse than that of large atoms. We have also learnt that a mixture of hard spheres, treated in the Percus-Yevick approximation is a rather good model for liquid alloys with a nearly ideal mixing behaviour.

Molecular dynamics calculations for Na-K [9.7] and for Na-Cs [9.13] alloys have been performed by the Groningen group. Their calculations are based on interatomic potentials derived from Heine-Abarenkov-type pseudopotentials (see Sect. 2.1.2). They also performed detailed x-ray and neutron diffraction studies of these alloys. For both alloys they achieve a rather good agreement between theory and experiment, thus confirming the validity of the model potential approach for these alloys. For the case of Na-K they confirm the result of Jacucci et al. that a mixture of hard spheres is a good first approach to the structure of the alloy, but they also point out that a HS model has, as in the monatomic case, a tendency to overestimate the amplitude of the oscillations in $S_{ij}(q)$ at large q. The results for NaCs (Fig. 9.3) show a rather pronounced deviation from an ideal mixing behaviour. The peaks in the pair-correlation functions of the like atoms, $g_{NaNa}(R)$ and $g_{CsCs}(R)$ are more pronounced than in the correlation function of the unlike neighbours. This expresses a tendency towards like-atom cluster formation which is attributable to a stronger interaction between like than between unlike atoms and differs from the behaviour of a hard-sphere mixture.

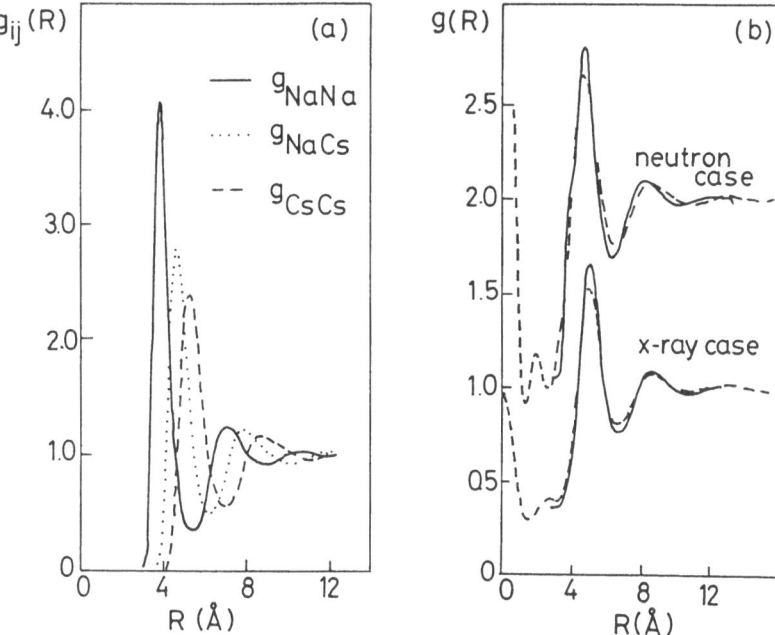

Fig. 9.3. (a) Partial pair correlation functions $g_{ij}(R)$ for a liquid $Na_{60}Cs_{40}$ alloy at $T = 100°\,C$, obtained by a molecular dynamics computer simulation; (b) X-ray and neutron weighted total pair correlation function $g(R) = \{c_A^2 f_A^2 g_{AA}(R) + c_B^2 f_B^2 g_{BB}(R) + 2c_A c_B f_A f_B g_{AB}(R)\}/\bar{f}^2$. (——) molecular dynamics; (- - -) experiment. After [9.13]

Computer simulations for alloys with strong chemical interactions are rather scarce. This is mainly due to the fact that, except for an exploratory investigation based on ab initio pseudopotentials (see Chap. 2), attempts to construct interatomic potentials for such systems have met with only moderate success. *Copestake* et al. [9.14] suggested that hard spheres with positive and negative screened electric charges might be a suitable reference system for describing alloys with chemical short-range order effects. For this system an analytic solution of the mean spherical approximation and a numerical solution of the hypernetted chain approximation are available and a solution of the hypernetted chain equation is also possible for "soft" repulsive interactions. These solutions have been tested by a molecular dynamics simulation of *Jacucci* et al. [9.15]. They find a good agreement between the MD data and the hypernetted chain results and the partial structure factors of this model also provide a rather good description of the experimental concentration fluctuation structure factor $S_{cc}(q)$ of alloys with strong chemical interactions and strong ordering effects such as Li_4Pb. A series of molecular dynamics simulations for liquid alloys with either nearly ideal mixing behaviour (Ca-Mg), weak or moderate chemical short-range order (Mg-Zn, Ca-Al, Li-Mg), or concentration-dependent changes in the topological short-range order (Al-Ge) has been presented just recently [9.16].

9.2 Thermodynamic Variational Calculations

The thermodynamic variational method based on the use of the Gibbs-Bogoljubov inequality (cf. Sect. 4.3.1) is readily generalized to the case of a binary liquid mixture. The crucial step in applying this method is the selection of an appropriate reference system. The model for the liquid mixture should satisfy two essential conditions: (i) Analytic solutions for the thermodynamic functions and the partial static structure factors, or at least sufficiently accurate analytical parametrizations of these quantities should be available. (ii) The model should be sufficiently realistic to include the essential deviations from a nearly ideal mixing behaviour, for example chemical short-range ordering, clustering or incipient segregation. A mixture of hard spheres of different diameters satisfies point (i) − analytic solutions of the Percus-Yevick equation are available [9.2,4] [see Sect. 4.2 and Appendix D.3.1; the generalization of (4.1–8) to the case of a multicomponent system is trivial]. However, it is clear that a mixture of hard spheres is a good model only for systems with a nearly ideal mixing behaviour − deviations from this behaviour are included only insofar as they are attributable to size effects. *Waisman* [9.17] has shown that under two restrictions the mean-spherical approximation (MSA) can be solved for charged hard spheres interacting through screened Coulomb (Yukawa) potentials. The two conditions demand that all spheres have the same diameter and secondly that the system is electrically neutral, i.e. $c_A Q_A + c_B Q_B = 0$ where Q_A, Q_B are the charges on the spheres. This is just the model system proposed by *Copestake* et al. [9.14]. It provides a convenient and rather realistic reference system for alloys which show either a tendency to heterocoordination or to clustering of equal atoms, provided that size effects are less important than the true "chemical" effects.[1] The one-component-plasma (OCP) system, which is another possible reference system for metals and alloys with rather "soft" interactions, has not yet been generalized to the multi-component case. However, *Chaturvedi* et al. [9.18] have pointed out that for homovalent alloys, the bare ionic mixture reduces to the OCP. In that case, any deviation from the ideal mixing behaviour must come from the perturbation acting on the reference system (in this case the perturbation would be essentially just the non-Coulombic part of the effective interatomic interactions). In the following we shall describe the application of the thermodynamic variational technique to various liquid alloys.

---

[1] Imposing the same restrictions, the MSA has also been solved analytically for a system with unscreened Coulomb interactions [9.17b]. For a metallic alloy this would of course be less appropriate as a reference system. The MSA has also been solved for charged hard spheres of different diameters [9.19]. However, the solution is not yet in a form which would make it a useful reference system in a variational calculation (the solution requires the numerical solution of a system of 13 coupled nonlinear algebraic equations!).

9.2.1 Systems with a Nearly Ideal Mixing Behaviour

The thermodynamic variational technique with a binary hard-sphere reference system was applied first by *Umar* et al. [9.26] to liquid Na-K and by *Hafner* to a number of liquid simple-metal alloys involving alkali, alkaline-earth metals and Al [9.20–22]. The variational parameters are the diameters σ_A, σ_B of the spheres representing the two components, i.e. the variational condition is now given by

$$\left(\frac{\partial \overline{F}}{\partial \sigma_i}\right)_{T,V} = 0 \quad ; \quad i = A, B \quad . \tag{9.5}$$

Here $\overline{F}$ is an upper bound to the exact free energy of the system in the spirit of the Gibbs-Bogoljubov inequality,

$$
\begin{aligned}
\overline{F} &= F_{\text{HS}}(\sigma_i, \sigma_j) + E_{\text{v}} + 2\pi n \sum_{i,j=A,B} (c_i c_j)^{1/2} \\
&\quad \times \int \Phi_{ij}(R) g_{ij}^{\text{HS}}(R; \sigma_i, \sigma_j) R^2 dR \\
&= F_{\text{HS}}(\sigma_i, \sigma_j) + E_0 + E_1 + \frac{1}{2\pi^2 n} \sum_{i,j=A,B} \int \hat{\Phi}_{ij}(q) S_{ij}^{\text{HS}}(q) q^2 dq \tag{9.6}
\end{aligned}
$$

in either real- or reciprocal-space formulation [note that $\hat{\Phi}_{ij}(q) = (8\pi Z_i Z_j / V_a q^2) + 2F_{ij}(q)$].

A typical result for the partial structure factors $S_{ij}(q)$ and the x-ray scattering intensity $I(q) = f(q)^2 S(q)$ of liquid Na-K alloys is shown in Fig. 9.4. As the neutron scattering lengths of Na and K are nearly identi-

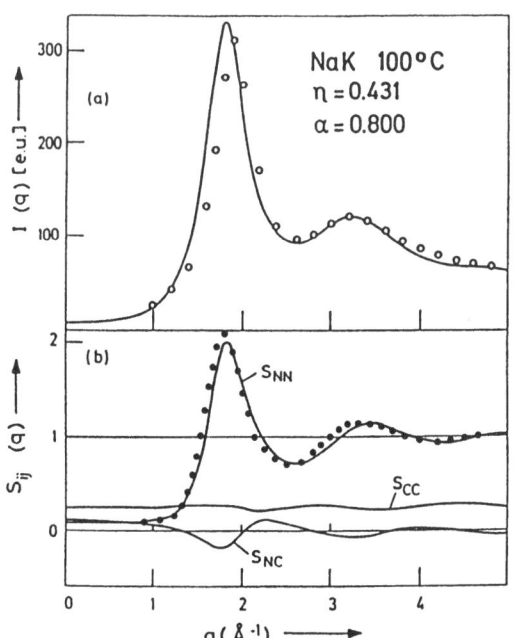

Fig. 9.4. (a) X-ray scattering intensity $I(q) = f^2(q)S(q)$ for liquid Na-K alloys (in electronic units); (—) variational calculation; (o) experiment [9.7, 23]. (b) Partial Bhatia-Thornton structure factors; (—) variational calculation; (o and ●) experiment [9.7, 23] (see text). After [9.21]

cal, $(\Delta f \sim 0)$ a neutron diffraction experiment measures essentially $S_{NN}(q)$. The calculations are based on ab initio pseudopotentials and reproduce the experimental results rather well, except for a small phase shift of the large-q oscillations. Here the situation is completely analogous to the case of the pure alkali metals. It is remarkable that the concentration fluctuation structure factor $S_{cc}(q)$ is very close to the limiting value $S_{cc}^{id}(q) = c_A c_B$ corresponding to an ideal random mixture.

Deviations from a random mixture appear when the size ratio $\alpha = \sigma_B/\sigma_A$ (by definition we have $\sigma_B < \sigma_A$) deviates more strongly from unity. This happens for example in Ca-Li alloys. In this case the concentration fluctuation correlation function $g_{cc}(R)$ displays first a positive peak which shows the nearest neighbour Li pairs, then a deep minimum demonstrating the presence of unlike-atom pairs and finally a second peak representing pairs of the larger Ca atoms, see Fig. 9.5. In this case a HS-variational calculation still yields a good fit to the experiment[2], indicating that the observed effects can indeed be attributed to the size difference alone.

As in the monatomic case, there are again different routes to the thermodynamic functions: via the compressibility or virial equations of state (cf. Sect. 4.2 and Appendix D). For the hard-sphere mixture analytical solutions have been given for both of these [9.24]. *Mansoori* et al. [9.25] have shown that a semiempirical equation of state, which takes a weighted average over

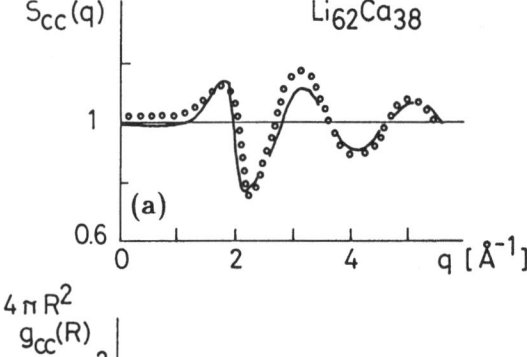

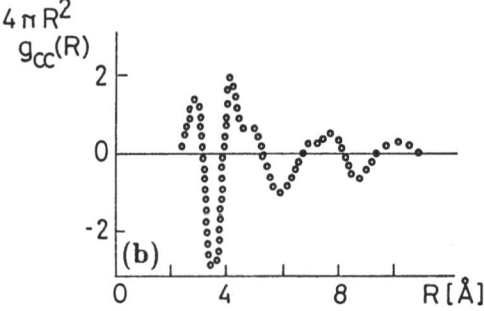

Fig. 9.5. (a) Concentration fluctuation structure factor $S_{cc}(q)$ for the "zero alloy" $Ca_{0.32}$ $Li_{0.68}$ at temperatures close to the liquidus. (b) Concentration fluctuation correlation function $g_{cc}(R)$. (———) hard-sphere variational calculation [9.22]; (o) experiment [9.6]

---

[2] $S_{cc}(q)$ can be measured directly by a neutron diffraction experiment on a "zero alloy". This is an alloy for which the average neutron scattering length $\bar{f} = c_A f_A + c_B f_B$ is zero and $\Delta f \neq 0$. See (9.4) and [9.6, 9, 10].

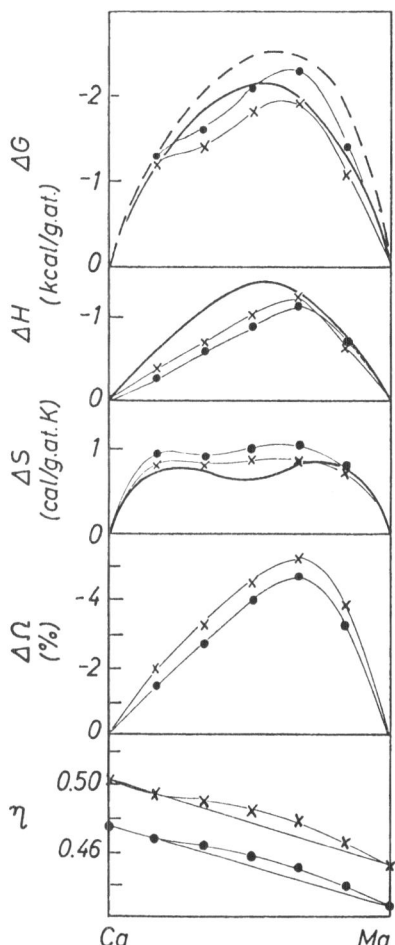

Fig. 9.6. Free energy ΔG, enthalpy ΔH, entropy ΔS and volume ΔV of formation for liquid Ca-Mg alloys. Circles and crosses: thermodynamic variational calculation at two different temperatures [($\bullet$) $T = 900$ K, ($\times$) $T = 700$ K]; (—— and - - -) experiment

the compressibility and virial equations in the ratio $2:1$, yields a very good approximation to the computer-simulation results. It is convenient to use their equation of state in the variational calculations.

A typical result for variationally determined thermodynamic excess functions is given in Fig. 9.6 for the case of liquid Ca-Mg alloys [9.22] and further results are summarized in Table 10.1. To obtain the volume of formation ΔV, the variational conditions (9.5) must be applied at constant temperature as a function of volume. The minimum in the free energy determines the equilibrium volume and hence the volume of formation of the alloy. The theory (the calculations are again based on ab initio pseudopotentials of the type discussed in Sects. 2.1.1 and 2.4.1) results in remarkable agreement with experiment. The most noticeable feature is perhaps the pronounced deviation of the entropy of formation from the ideal entropy of mixing. This is in fact quite easily understood. The hard-sphere contribution to the entropy is given by [9.26]

$$S_{HS} = S_{gas} + S_c + S_\eta + S_\sigma \quad , \tag{9.7}$$

where S_{gas} is the ideal gas entropy, S_c the ideal entropy of mixing, and

$$S_\eta = k_B[3 - 2/(1 - \eta) - 1/(1 - \eta)^2] \tag{9.8}$$

depends only on the average packing fraction η [equation (9.8) gives the Mansoori-Carnahan-Starling-Leland [9.25] form of S_η][3], while S_σ is proportional to the product of the concentrations and the square of the difference in the hard-core diameters, $S_\sigma \propto c_A c_B (\sigma_A - \sigma_B)^2$. The explicit expression for the size-mismatch term is quite lengthy and may be found in Appendix D. The contributions from S_η and S_σ dominate the entropy of formation. S_σ is a small, positive contribution expressing the increased disorder due to the size mismatch. S_η is a negative and monotonically decreasing function of the packing fraction η. Therefore, we obtain a negative contribution from S_η to the entropy of formation for $\Delta V < 0$. Thus we find that a decrease in the entropy of formation is directly related to a negative volume of formation. This correlation between ΔV and the excess entropy of formation $\Delta S^E = \Delta S - \Delta S^{id}$ has been further explored by *Young* and coworkers [9.27–29].

Further applications of the Gibbs-Bogoljubov variational technique include again alkali-alkali alloys [9.30, 31], alloys of aluminium with other polyvalent metals such as Mg, Zn [9.32], and Sn [9.33] and liquid brasses (Cu-Zn, Cu-Al, Cu-Sn) [9.34]. Table 9.1 shows that even in the rather difficult cases of the polyvalent alloys and of the liquid brasses, reasonable agree-

Table 9.1. Volume, enthalpy and entropy of formation for liquid equiatomic AB alloys

Alloy	T [°C]	ΔV [%]		ΔH [cal/g-atom]		ΔS [cal/g-atom K]	
		theor.	exp.	theor.	exp.	theor.	exp.
NaK[a]	100	−1.9	−1.4	470	174	1.03	1.35
NaCs[a]	100	−8.4	−5.4	1640	218	0.33	1.20
KRb[a]	100	−0.19	−0.19	160	30	1.22	
KCs[a]	100	−0.87		200	28	1.08	
RbCs[a]	100	−1.20		−39	−32	1.12	
AlMg[a]	665	−4.4	−3.0	−2170	−806	1.20	1.18
AlSn[b]	800			−1260	−980		
CuZn[c]	1083			−573	−1423	1.39	1.37
CuAl[c]	1083			−1490		1.42	
CuSn[c]	1083			−1193	−475	1.86	1.92

[a] After [9.21], using ab initio pseudopotentials.
[b] After [9.33], and
[c] After [9.34], using model potentials and the experimental volumes of the alloys

---

[3] The same expression applies to a monatomic liquid.

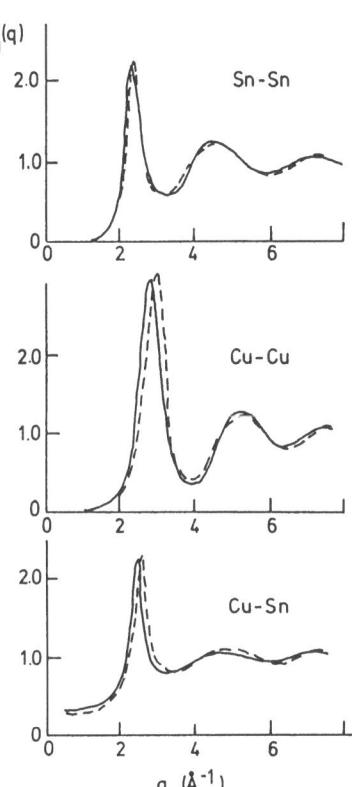

Fig. 9.7. Partial static structure factors $a_{ij}(q)$ for liquid Cu_6Sn_5 alloy. (——) thermodynamic variational calculation [9.34]; (- - -) experiment (neutron diffraction with isotopic substitution) [9.35]

ment is obtained. The most serious discrepancy occurs for Na-Cs where all calculations tend to overestimate the enthalpy of formation. This disagreement, and also the failure to predict the stability of the crystalline compound Na_2Cs (see Chap. 8) are probably due to the neglect of s-d electron transfer effects.

Figure 9.7 shows a comparison of calculated structure factors, $a_{ij}(q)$ $(i, j = Cu, Sn)$ with experimental results for the liquid alloy Cu_6Sn_5. The good agreement between theory and experiment, both for this quantity and also for the thermodynamic exess functions, suggests that the hard-sphere model gives an adequate description of this class of alloys. However, there are several details in the measured partial structure factors [heights of the peaks, position of the peak in $S_{Cu-Sn}(q)$] which suggest that the distribution of the two atomic species in the alloy is not completely random, but that there exists a weak chemical short-range order in the sense of a preferred heterocoordination. These effects are of course not properly taken into account in a binary hard-sphere model.

The simplest way to introduce corrections to the hard-sphere model is again based on the random-phase approximation (RPA). For a binary liquid this means that the partial direct correlation functions $c_{ij}(R)$ are approximated by

$$c_{ij}(R) = c_{ij}^0(R) - \beta\Phi_{ij}(R) \quad , \tag{9.9}$$

where the $c_{ij}^0(R)$ are the direct correlation functions of a reference fluid (for example a hard-sphere mixture) and the $\Phi_{ij}(R)$'s are the pair potentials representing the perturbation. In fact it turns out that for describing ordering effects, it is sufficient in a first approximation to consider only the direct correlation function $c_{cc}(R)$ describing the concentration fluctuations (9.1) and the potential $\Phi_{cc}(R)$ coupling to the concentration fluctuations (2.58). Furthermore, as the RPA has only asymptotic, i.e. low-q, validity, it is convenient to proceed in q-space

$$\hat{c}_{cc}(q) = \hat{c}_{cc}^0(q) - \beta\hat{\Phi}_{cc}(q) \quad . \tag{9.10}$$

Equation (9.10) has to be used in the Ornstein-Zernike relations for the binary system. In terms of the direct correlation functions $\hat{c}_{NN}(q)$, $\hat{c}_{Nc}(q)$, and $\hat{c}_{cc}(q)$ for number density and concentration fluctuations the OZ relations are given in q-space by [see Appendix D, (D.31 ff)]

$$n\hat{h}_{NN}(q) = S_{NN}(q) - 1 = \frac{1 - n\hat{c}_{cc}(q)}{D(q)} - 1$$

$$n\hat{h}_{cc}(q) = \frac{S_{cc}(q) - c_A c_B}{c_A c_B} = \frac{1 - n\hat{c}_{NN}(q)}{D(q)} - 1$$

$$n\hat{h}_{Nc}(q) = \frac{S_{Nc}(q)}{c_A c_B} = \frac{n\hat{c}_{Nc}(q)}{D(q)} \quad \text{with} \tag{9.11}$$

$$D(q) = [1 - n\hat{c}_{NN}(q)][1 - n\hat{c}_{cc}(q)] - n^2 c_A c_B \hat{c}_{Nc}(q)^2 \quad . \tag{9.12}$$

It follows that for $q \to 0$ the concentration fluctuation structure factor is given by

$$S_{cc}(0) = \frac{c_A c_B}{1 - n\hat{c}_{cc}^0(0) + n\beta\hat{\Phi}_{cc}(0) - n^2 c_A c_B \hat{c}_{Nc}^0(0)^2/[1 - n\hat{c}_{NN}^0(0)]} \quad . \tag{9.13}$$

Thus we find that with the approximations introduced above, the RPA yields a result familiar from conformal solution theory [9.36, 37] with the "interchange energy" expressed in terms of Φ_{cc}.

On the other hand it is known that the long-wavelength limit of $S_{cc}(q)$ is related to the second derivative of the Gibbs free energy of formation with respect to the concentration via [9.1]

$$\left(\frac{\partial^2 \Delta G}{\partial c_B^2}\right)_{p,T} = \frac{Nk_BT}{S_{cc}(0)} \quad . \tag{9.14}$$

Thus an incipient phase separation $(\partial^2 \Delta G/\partial c_B^2 \sim 0)$ is related to a divergence in $S_{cc}(0)$. In general a value of $S_{cc}(0)$ greater than ideal [and hence $\hat{\Phi}_{cc}(0)<0$] expresses a tendency to form like-atom clusters, while $S_{cc}(0) < S_{cc}^{id}(0)$ [and thus $\hat{\Phi}_{cc}(0)>0$] reflects a tendency to heterocoordination.

Equations (9.9–14) have been used by *Bhatia* et al. [9.37] to investigate concentration fluctuations in liquid alkali alloys. In this study, the pair potentials $\Phi_{ij}(R)$ were assumed to be of a Yukawa form, i.e.

$$\Phi_{ij}(R) = \varepsilon_{ij} \frac{\sigma_{ij}}{R} e^{-\kappa_{ij}(r-\sigma_{ij})} \tag{9.15}$$

[since only the $q = 0$ Fourier component, $\hat{\Phi}_{cc}(0)$ of Φ_{cc} enters (9.14) via $S_{cc}(0)$, the functional form of the potential is not that important]. The screening constant was set to $\kappa_{ij} = 1/\sigma_{ij}$ and with $\varepsilon_{AA} = \varepsilon_{BB} = 0$ this leaves a single parameter ε_{AB} measuring the non-additivity of the pair interactions at hard contact. In Fig. 9.8, the $S_{cc}(0)$ curves for Na-K and Na-Cs are shown, together with the experimental results from thermodynamic measurements. It turns out that a rather small ordering potential (of the order of a few tenths of a millirydberg) is sufficient to promote a substantial tendency towards formation of like-atom clusters. In both cases it was found sufficient to use a concentration-independent ordering energy, thus the asymmetry of the $S_{cc}(0)$ curves – which reflects a larger tendency

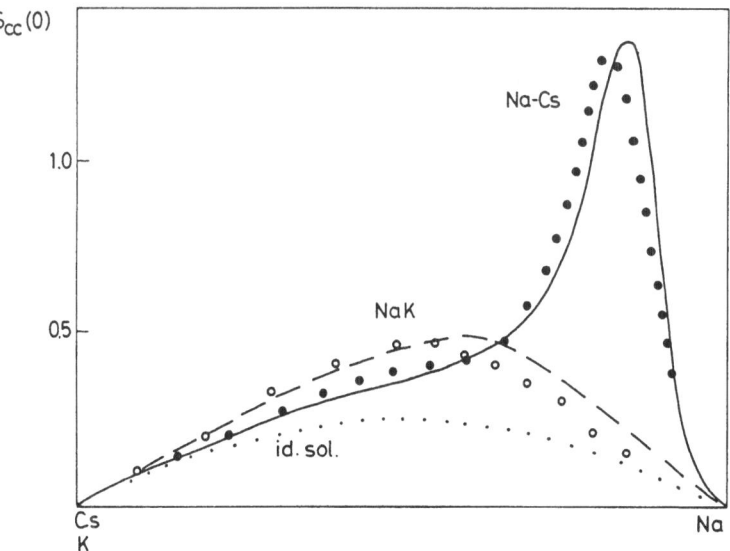

Fig. 9.8. $S_{cc}(0)$ for Na-K [(- - -) theory, (o) experiment] and Na-Cs [(——) theory, (•) experiment] alloys. $\varepsilon_{AB} = 0.0415\, k_B T$ for Na-K and $\varepsilon_{AB} = 0.0693\, k_B T$ for Na-Cs. The dotted line shows $S_{cc}(0) = c_A c_B$ for an alloy with an ideal mixing behaviour. The larger $S_{cc}(0)$, the more pronounced is the tendency to a formation of like-atom clusters. See text; after [9.37]

towards clustering for majority concentrations of the smaller atoms – is a consequence of the size difference. This simple theory also describes $S_{cc}(q)$ quite well for finite q – this is because $S_{cc}(q)$ goes rapidly to its assymptotic value $c_A c_B$ and there are no large oscillations. This is in contrast to the case of chemically ordered alloys $[S_{cc}(0) \ll c_A c_B]$ where one finds large oscillations in $S_{cc}(q)$. Such alloys are not amenable to a simplified treatment using the RPA.

9.2.2 Liquid Alloys with Strong Chemical Short-Range Order

The main structural, thermodynamic, electrical and magnetic properties of binary liquid alloys with a strongly non-ideal mixing behaviour have been of longstanding interest. Liquid alloys of the alkali or alkaline-earth metals with polyvalent metals (Pb, Tl, In, etc.), semiconductors or semimetals (Si, Ge, Bi, Sb, ...), or with the noble metals, have been studied very thoroughly. Their thermodynamic excess functions, static structure factors, electrical transport properties, magnetic susceptibilities etc. are well known and exhibit anomalies at the same characteristic concentration (for recent reviews see [9.6, 9.38, 39]. However, it must be admitted that the theoretical interpretation is still at a rather empirical level. The most popular interpretation of the thermodynamic and structural data assumes the existence of chemical complexes (or "associates"). For a set of assumed interaction parameters, the thermodynamic functions of the pseudoternary mixture $A_x B_y$ associates, A ions and B ions, are calculated via the mass-action law [9.40–42]. The interpretation of structural data is based on essentially equivalent assumptions [9.43]. It is illustrative of the difficulties with the association model that any agreement with experimental diffraction data depends primarily on the form factor of the complex $A_x B_y$.

A more fundamental interpretation should really start from an electronic theory of the interatomic forces. However, this poses two problems: (i) Simple model potentials are not appropriate for calculating the interatomic forces in such complex systems. A more successful approach based on optimized ab initio pseudopotentials has been discussed in Chap. 2. (ii) The calculation of the structure requires an appropriate (in the sense discussed at the beginning of Sect. 9.2) reference system which already incorporates the main ordering effects.

It was first proposed by *Copestake* et al. [9.14, 44, 45] that ordering phenomena in molten salts, liquid semiconductors and even in liquid metallic alloys might be modelled by a system of hard spheres interacting by Coulomb or screened Coulomb (Yukawa) potentials. The particularly attractive feature of this model is that for equal diameters of the two species of spheres and when the condition of charge neutrality is satisfied ($\overline{Q} = c_A Q_A + c_B Q_B = 0$), an analytical solution for the partial structure factors and for the thermodynamic functions may be found within the mean spherical approximation. In this case, the MSA is defined by the Ornstein-Zernike relations (see Sect. 4.2 for a comparison with the monatomic case)

$$h_{ij}(R) = c_{ij}(R) + n \sum_k c_k \int c_{ik}(|\boldsymbol{R} - \boldsymbol{R'}|)h_{kj}(R')d^3\boldsymbol{R'} \quad , \qquad (9.16)$$

in conjunction with the closure relations

$$\begin{aligned} c_{ij}(R) &= -\beta\Phi_{ij}(R) \\ &= -2\beta Q_i Q_j \exp[-\kappa(R - \sigma_{ij})]/R \quad , \quad R > \sigma_{ij} \qquad (9.17a) \\ h_{ij}(R) &= -1 \quad R < \sigma_{ij} \quad . \qquad (9.17b) \end{aligned}$$

If the above-mentioned conditions of equal diameter and charge neutrality are satisfied, one finds that after transforming to the Bhatia-Thornton functions, the system of three coupled integral equations reduces to two independent equations [9.47]

$$h_{\mathrm{NN}}(R) = c_{\mathrm{NN}}(R) + n \int c_{\mathrm{NN}}(|\boldsymbol{R} - \boldsymbol{R'}|)h_{\mathrm{NN}}(R')d^3\boldsymbol{R} \quad , \qquad (9.18a)$$

$$c_{\mathrm{NN}}(R) = 0 \quad R > \sigma \qquad (9.18b)$$

$$h_{\mathrm{NN}}(R) = -1 \quad R < \sigma \qquad (9.18c)$$

for the number-density fluctuations, and

$$h_{\mathrm{cc}}(R) = c_{\mathrm{cc}}(R) + n \int c_{\mathrm{cc}}(|\boldsymbol{R} - \boldsymbol{R'}|)h_{\mathrm{cc}}(R')d^3\boldsymbol{R'} \qquad (9.19a)$$

$$\begin{aligned} c_{\mathrm{cc}}(R) &= -2\beta c_A c_B \Delta Q^2 \exp[-\kappa(R - \sigma)]/R \\ &= -\beta\varepsilon\sigma \exp[-\kappa(R - \sigma)]/R \quad , \quad R > \sigma \qquad (9.19b) \end{aligned}$$

$$h_{\mathrm{cc}}(R) = 0 \quad R < \sigma \quad , \qquad (9.19c)$$

where $\Delta Q = Q_A - Q_B$, for the concentration fluctuations. The parameter ε defined in (9.19b) measures the strength of the ordering potential at hard contact ($R = \sigma$). The third equation has a trivial solution

$$\Phi_{\mathrm{Nc}}(R) = h_{\mathrm{Nc}}(R) = c_{\mathrm{Nc}}(R) = 0 \quad , \qquad (9.20)$$

i.e. in this case the fluctuations in number density and concentration are uncorrelated.

Equation (9.18) is identical to the Percus-Yevick integral equation for hard spheres and therefore we have analytical solutions for the direct correlation function $c_{\mathrm{NN}}(R)$ [9.46]. Equation (9.19) has been solved analytically by *Waisman* [9.47] and explicit results for the correlation function $c_{cc}(R)$, the static structure factor $S_{cc}(q)$ and the thermodynamic functions from [9.48] are given in Appendix D.3.3.

Copestake et al. [9.14] have pointed out that this is a convenient model for the observed structure of Li_4Pb. *Hafner* et al. [9.48–50] demonstrated that the model may be used to rationalize all the thermodynamic and structural information available for a large number of Li- and Na-based alloys with polyvalent and nobel metals. They showed that to a first approximation the heat of formation is given as the sum of an electron-gas term and an ordering contribution,

$$\Delta H = \Delta H_{eg} + \Delta H_{ord} \tag{9.21}$$

with ΔH_{eg} given by (6.20) and the ordering contribution expressed in terms of the hard-sphere-Yukawa (HSY) model. Equation (9.21) may be used to fit the parameters of this model (essentially the strength ε of the ordering interaction at hard contact), and the predictions of the model may be checked agaist the entropy and against structural data. The entropy of formation is found to consist, besides the ideal mixing entropy ΔS_{id}, of a hard-sphere contribution ΔS_{HS}, which takes into account the number density fluctuations (ΔS_{HS} is important in the case of a large volume of formation), an ordering contribution (accounting for the concentration fluctuations) and a small electronic contribution,

$$\Delta S = \Delta S_{HS} + \Delta S_{ord} + \Delta S_{el} + \Delta S_{id} \quad . \tag{9.22}$$

Figure 9.9 shows the entropy of formation of liquid alloys of Li and Na with Sn and Pb and illustrated in Fig. 9.10 are the interplay of packing and ordering effects. In the Li-based alloys there is a single minimum in the entropy of formation close to the "stoichiometric" composition Li_4X (X =Pb,Sn). In the Na-based alloys on the other hand we find, besides the minimum at the composition Na_4Pb, a pronounced dip near the composition NaPb. In Na-Sn the positions of the minimum and of the dip are reversed [9.51, 52]. Similar anomalies at *two* different compositions are found in other thermodynamic functions, e.g. in the excess specific heat. The structural manifestations of these thermodynamic anomalies have been investigated by *Albas* et al. [9.52b] and by *Tamaki* et al. [9.39, 52a]. Within the spirit of the "association" model these specific heat anomalies have been interpreted as arising from the existence of two different types of associates with compositions Na_4Pb and NaPb (and analogously for Na-Sn). Evidently this implies a type of bonding which is much more localized and more complicated then one based on simple volume and pair forces. The present model offers another explanation which appears to be more natural and also simpler: In the Li-based alloys, the maximum volume contraction happens to coincide with

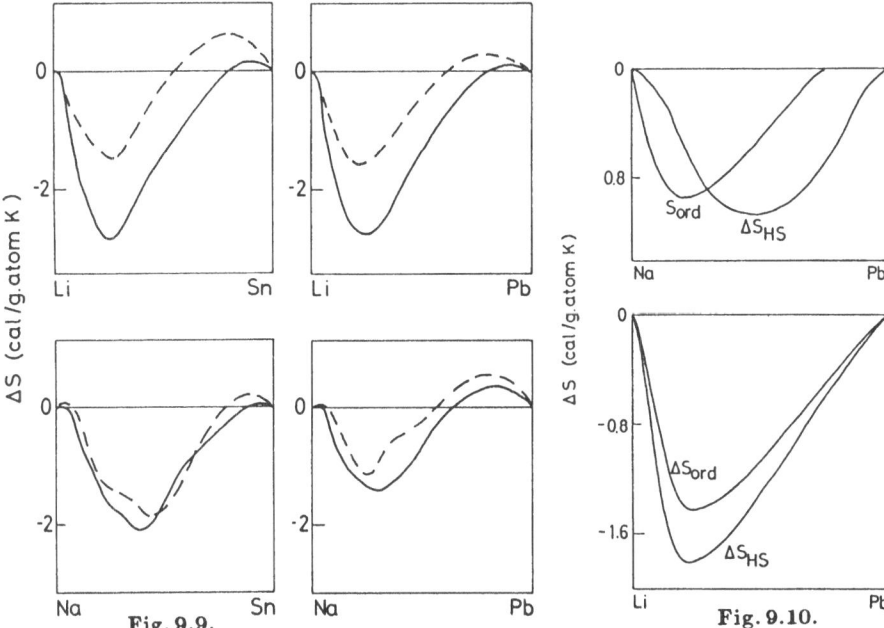

Fig. 9.9. Entropy of formation of liquid Li-Sn, Li-Pb, Na-Sn, and Na-Pb alloys at temperatures close to the liquidus. (——) hard-sphere-Yukawa model fitted to the heat of formation (see text); (- - -) experiment. After [9.48–50]

Fig. 9.10. Packing contribution ΔS_{HS} and ordering contribution ΔS_{ord} to the entropy of formation of liquid Li-Pb and Na-Pb alloys. After [9.48, 49]

the "stoichiometric" composition Li_4X (remember that both the magnitude and the functional form of the volume of formation are dominated by elastic effects, see Sect. 6.1.2) and therefore the hard-sphere and the ordering contribution cooperate to form a single deep minimum in the entropy of formation. In the Na-based alloys the contraction is maximal close to the equiatomic composition whereas the ordering effects still peak close to the "stoichiometric" composition Na_4X (Fig. 9.10). Note that in the Na-based alloys the atomic volume of the monovalent component is larger, whereas in the Li-based alloys it is smaller than that of the polyvalent component. The model can be tested further by comparing its structural predictions with experiment. In Fig. 9.11 we show the concentration fluctuation structure factors $S_{cc}(q)$, for the "zero alloys" (see footnote 2) of the systems Li-Ag, Li-Mg and Li-Pb, as calculated using the HSY model and as determined by elastic neutron scattering. The good agreement shows that the HSY model indeed offers a reasonably accurate and consistent description of the thermodynamic and structural manifestations of the ordering effects.

Remembering that pseudopotential theory seems to give reasonably reliable ordering potentials for systems such as Li-Pb (see Fig. 2.15), it is of course tempting to use the HSY system as a reference system in a Gibbs-

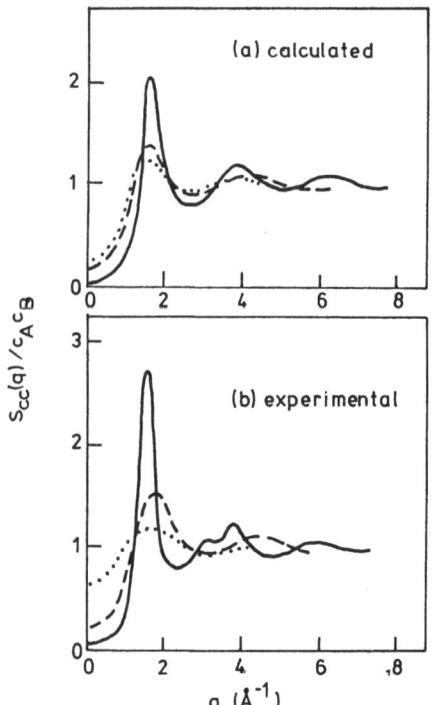

Fig. 9.11a,b. Concentration fluctuation structure factors $S_{cc}(q)$ for the zero alloys $Li_{0.8}Pb_{0.2}$ (——), $Li_{0.72}Ag_{0.28}$ (- - -) and $Li_{0.7}Mg_{0.3}$ at temperatures close to the melting point: **(a)** calculated and **(b)** experimental. After [9.48]

Bogoljubov variational calculation. In this case an upper bound to the exact free energy of the liquid alloy is given by

$$\overline{F} = \frac{3}{2}k_B T + E_v + 2\pi n \left[\int_0^\infty g_{NN}^{HS}(R_j, \sigma)\Phi_{NN}(R)R^2 dR \right.$$

$$\left. + \int_0^\infty g_{cc}^{HSY}(R_j, \sigma, \varepsilon\kappa)\Phi_{cc}(R)R^2 dR \right]$$

$$- T[\Delta S_{id} + \Delta S_{el} + \Delta S_{HS}(\sigma) + \Delta S_{ord}(\sigma, \varepsilon, \kappa)] \qquad (9.23)$$

with the optimal values of the parameters σ, ε, κ being determined by the isothermic-isochoric variational conditions. However, it turns out that if all three parameters are treated as free variational parameters, no minimum in the free energy exists [4] [9.53]. The reason for this failure seems to be the neglect of the cross-correlation between the number-density and the concentration fluctuations, which is a peculiarity of the MSA[5]. The necessary

---

[4] To be more precise: a solution of the unconstrained minimum condition exists for weakly ordering systems such as Mg-Zn, Mg-Cd, or Li-Mg, but not for strongly ordering systems such as Li-Pb. In all cases where an unconstrained solution exists, the condition (9.24a) is automatically respected.

[5] The solution of the hypernetted-chain approximation for the HSY model shows indeed a non-zero $S_{Nc}(q)$ [9.14, 15], but it is not sufficiently simple to be useful in a variational calculation.

coupling may be established by imposing the condition

$$\Phi_{cc}(\sigma) = \varepsilon \quad , \tag{9.24a}$$

i.e. one requires that the strength of the ordering potential at hard contact is exactly the same for both the exact and the reference-system interactions. The hard-sphere diameter σ and the screening constant κ are determined by the variational conditions

$$\left(\frac{\partial \overline{F}}{\partial \sigma}\right)_{T,V,\kappa,\varepsilon} = 0 \tag{9.24b}$$

$$\left(\frac{\partial \overline{F}}{\partial \kappa}\right)_{T,V,\sigma,\varepsilon} = 0 \quad . \tag{9.24c}$$

Figure 9.12 shows the result of such a variational calculation for liquid Na-Pb alloys. The construction of the interatomic potentials is based here upon the same pseudopotentials and screening functions as were used for

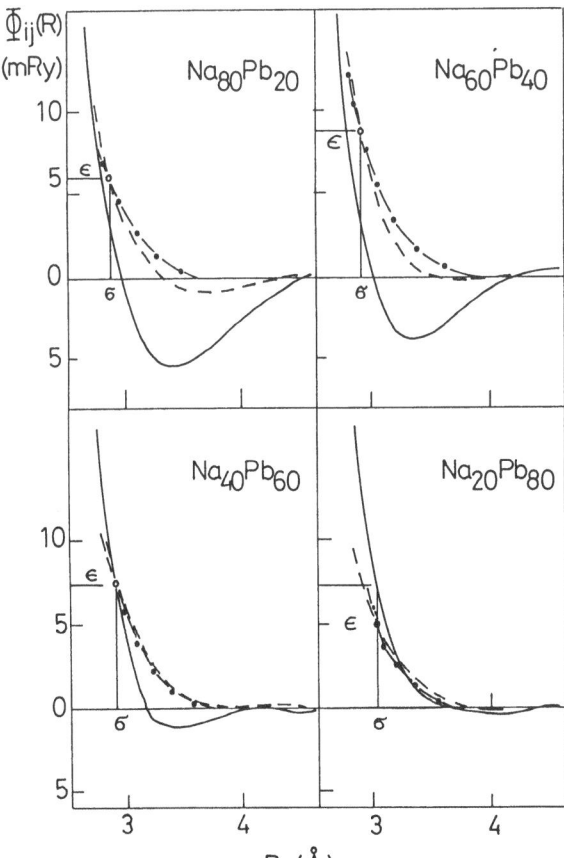

Fig. 9.12. Mean pair potential $\Phi_{NN}(R)$ (——) and ordering potential $\Phi_{cc}(R)$ (- -) for a series of Na-Pb alloys, calculated using ab initio pseudopotential theory [9.49]. The chain curves represent the reference HSY potential determined by applying the restricted variational conditions (9.24)

259

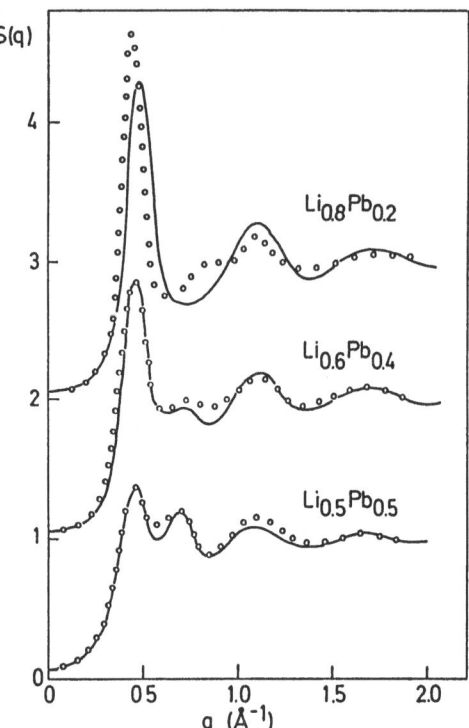

S(q)

Li$_{0.8}$Pb$_{0.2}$

Li$_{0.6}$Pb$_{0.4}$

Li$_{0.5}$Pb$_{0.5}$

q (Å^{-1})

Fig. 9.13. Composite (neutron-weighted) static structure factor $S(q)$ for a series of Li-Pb alloys of different compositions for temperatures close to the liquidus line. (—) as derived from the variational condition; (○) experimental results of *Ruppersberg* et al. [9.54]. After [9.53]

Li-Pb (Fig. 2.15). There we also discussed the physical mechanisms which determine the form of the pair potentials. The pseudopotential calculation confirms the essential aspects of the simple HSY model: the ordering potential is of a strongly damped form and it has the correct order of magnitude. Figure 9.13 displays the composite (neutron-weighted) static structure factors of a series of Li-Pb alloys of different compositions for temperatures close to the liquidus line. This is a good illustration of both the success and the limitations of the HSY reference system combined with effective pair potentials derived from pseudopotentials. The trend in the CSRO is properly described, as can be seen from the variation of the prepeak[6] with concentration. On the other hand it is evident that a hard-core interaction is too crude a model to describe certain fine structures arising from a topological short range order. This is most evident for the Li-rich alloys.

The relative success of the HSY model suggests that we interpret the physical mechanism creating the CSRO as charge ordering; electrons being transferred to the more electronegative element. The amount of charge transferred may be estimated directly from the ordering contribution to the

---

[6] The term "prepeak" refers to a peak in the composite $S(q)$ which appears to the left of the first peak in $S_{NN}(q)$. The position Q_P of the first peak in $S_{NN}(q)$ is related to the mean nearest-neighbour distance via $d \approx 2\pi/Q_P$. Thus the appearance of a prepeak indicates a modulation of the structure on a scale larger than d.

Table 9.2. Strength of the ordering potential at contact ε and charge transfer Q in equiatomic liquid Li alloys, as estimated from the HSY-model. After [9.50]

Alloys	ε [mRy]	Q
Li-Mg	6.7	0.127
Li-Cd	16.2	0.200
Li-Ga	7.0	0.129
Li-In	17.1	0.207
Li-Tl	17.1	0.212
Li-Sn	15.9	0.202
Li-Pb	13.1	0.186
Li-Bi	24.0	0.254

heat of formation (9.21). These estimates are given in Table 9.2 for a series of equiatomic Li-based alloys. We find that only about 0.1 to 0.25 electrons per atom are transferred – this compares reasonably well with the values that might be obtained by integrating up the charge density distributions in comparable heterovalent intermetallic compounds (cf. Sects. 8.3 and 8.4). However, it must be pointed out that there is a certain degree of arbitrariness in these numbers: instead of being described by (9.19b), the ordering potential might equally well have been formulated as (remembering that in the MSA $c_{\mathrm{cc}} = -\beta \Phi_{\mathrm{cc}}$ for $R > \sigma$)

$$\Phi_{\mathrm{cc}}(R) = 2c_A c_B \Delta \overline{Q}^2 \exp(-\kappa R)/R \quad , \tag{9.25}$$

this changes nothing in the model, except that the parameter $\Delta \overline{Q}^2$ is now larger by a factor $\exp(\kappa \sigma)$. This enhancement factor can be quite large, depending upon the value of the screening constant. Keeping this cautionary note in mind, the charge-ordering picture and the analogy to ionic melts can be used as a first approximate model for the CSRO in molten alloys.

The HSY model has also been used by *Holzhey* et al. [9.55, 56] in a first attempt at a self-consistent treatment of the electronic and the atomic structures going beyond the linear response approximation. The local electronic properties (local density of states, energy shifts etc.) of the two components are expressed in a tight-binding picture. The net charge transfer of electrons can be calculated by integrating up the local densities of states. This sets the parameters of the HSY model defining the atomic structure. When these are known, the electronic structure has to be recalculated[7] and the entire scheme has to be iterated to self-consistency. It turns out that there is a strong correlation between the charge transfer and a minimum in the density of states at the Fermi level; both depend mainly on the ratio $\Delta E/t$ between the difference ΔE in the atomic energies and the transfer integral

---

[7] To calculate the DOS for a given set of tight-binding parameters and a given atomic structure, the Cayley tree method of *Falicov* and co-workers [9.57, 58] has been used.

t. The charge transfer and the minimum in the DOS are more pronounced at large values of $\Delta E/t$. At higher temperatures, the charge transfer is reduced, and the minimum in the DOS is flattened. This model may even be extended to alloys such as Cs-Au, which show a genuine metal-insulator transition at the equiatomic composition [9.55, 58b]. Thus we are beginning to develop a consistent qualitative picture of the interrelations between the interesting thermodynamic, structural, and electronic properties of these liquid alloys[8].

9.2.3 Liquid Alloys with a Miscibility Gap

The properties of binary metal alloys with a miscibility gap are of particular interest because of the pronounced fluctuation effects above the critical point for phase separation. Anomalies have been found in the x-ray and neutron scattering intensities, the specific heat, and in the electrical resistivity (see [9.8, 60] and further references therein). The schematic Gibbs free energy isotherms in a liquid alloy exhibiting a miscibility gap are shown in Fig. 9.14. The phase separation curve is determined there by means of a

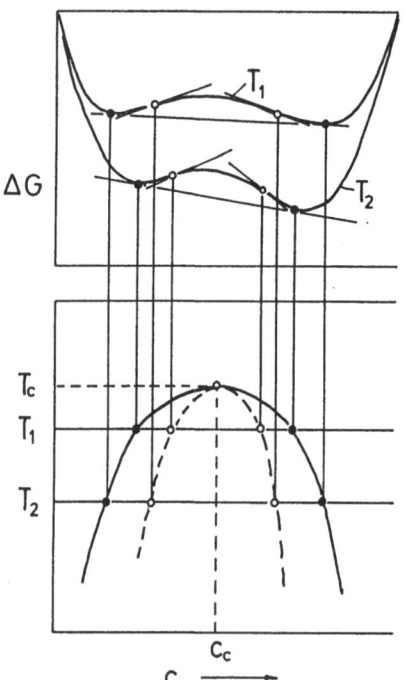

Fig. 9.14. (a) Schematic free energy isotherms in a binary liquid metal alloy exhibiting a miscibility gap. A line is drawn tangent to an isotherm at two points (●), the points of tangency define the solubility limits at that temperature. The turning points of the isotherms (the open circles, defined by a zero curvature, $\partial^2 \Delta G/\partial c^2 = 0$) determine the point at which the critical fluctuations in a supercooled alloy diverge. This is the spinodal curve. (b) Phase separation curve (——) and spinodal line (- - -) corresponding to the isotherm shown in (a)

---

[8] In view of the preceding remarks on the arbitrariness of the numerical value of ΔQ, it appears to be unfortunate that the electronic and atomic structures are coupled just via ΔQ. Very recently *Hafner* and *Pasturel* [9.59] have shown that the coupling might better be achieved in a variational approach based on the Gibbs-Bogoljubov inequality – see also Sect. 9.4.

common-tangent construction applied to the isotherms. For temperatures lower than the critical temperature there is a concentration regime in which the isotherm is concave downward, so that a straight line may be drawn which is tangential to the curve at two points. The abscissae of the points of tangency $c_A(T)$ and $c_B(T)$ define the solubility limits at a given temperature. Except for the critical point, the fluctuation effects are finite along the phase separation curve, however they diverge along the so-called *spinodal* line. This is a line determined by the condition that the curvature $\partial^2(\Delta G)/\partial c^2$ of the Gibbs free energy isotherms vanishes and it marks the absolute limit of stability of the alloy. In the region bounded by the phase separation curve and the spinodal line the alloy is metastable against long-wavelength concentration fluctuations (9.14); below this it is unstable. The main challenge is to give a consistent description of the phase separation curve and of the critical divergences along the spinodal line.

The simplest approach is again based on a hard-sphere variational calculation [9.61]. The free energy is obtained variationally, using hard-sphere structure factors to evaluate the ion-ion interaction energy and the entropy. The phase separation curve is then determined by applying the common-tangent construction. This approach can never yield divergent concentration fluctuations because hard-sphere structure factors, at least in the Percus-Yevick approximation, exhibit no divergences at any temperature or concentration. Hence it always violates (9.14). This method has been applied by *Stroud* [9.61] to Li-Na alloys using simple empirical model potentials. The calculation overestimates the critical temperature, but reproduces the observed asymmetric form of the phase separation curve rather well, and in particular approximately matches the critical concentration of about 65 at. % Li. *Stevenson* [9.62] used a very similar technique (but including third-order contributions to the total energy) to investigate metallic (i.e. pressure-ionized) hydrogen-helium fluid mixtures. He found that the hydrogen-helium mixture decomposes into hydrogen- and helium-rich phases at temperatures below 8000 K and at pressures ranging between 4 and 40 Mbar (these values are relevant to conditions in the interior of Jupiter).

Stroud [9.60] has formulated a mean-field theory for the phase separation problem. The mean-field approximation consists in the assumption that each ion reacts like a free particle to the mean field acting on it – in fact it is just the same assumption as that which enters the random phase approximation (RPA). Indeed as for Ising or Heisenberg ferromagnets, the RPA and the mean-field approximation turn out to be equivalent for the phase separation problem. The RPA applied to a binary system yields the following expressions for the partial structure factors [cf. also (9.10–12)]

$$S_{AA}(q) = [1 + \beta n c_B \hat{\Phi}_{BB}(q)]/D(\beta, c_A, q) \tag{9.26a}$$

$$S_{BB}(q) = [1 + \beta n c_A \hat{\Phi}_{AA}(q)]/D(\beta, c_A, q) \tag{9.26b}$$

$$S_{AB}(q) = -\beta n (c_A c_B)^{1/2} \hat{\Phi}_{AB}(q) / D(\beta, c_A, q) \tag{9.26c}$$

$$D(\beta, c_A, q) = [1 + \beta n c_A \hat{\Phi}_{AA}(q)][1 + \beta n c_B \hat{\Phi}_{BB}(q)] \\ - \beta^2 n^2 c_A c_B \hat{\Phi}^2_{AB}(q) \quad . \tag{9.26d}$$

The spinodal line is characterized by the long-wavelength divergence of the partial structure factors, $S_{ij}(0) \to \infty$ as $T \to T_c$. Hence it is determined by the solution of the equation

$$D(\beta_s, c_A, 0) = 0 \tag{9.27}$$

($\beta_s = 1/k_B T_s$, T_s is the temperature of the spinodal at the concentration c_A). The solution of (9.27) is given approximately by the expression

$$k_B T_s(c_A) = c_A c_B n \frac{\hat{\Phi}^2_{AB}(0) - \hat{\Phi}_{AA}(0) \hat{\Phi}_{BB}(0)}{c_A \hat{\Phi}_{AA}(0) + c_B \hat{\Phi}_{BB}(0)} \quad . \tag{9.28}$$

Table 9.3. Calculated critical concentrations c_c and temperatures T_c for liquid alloys, compared with experiment. After [9.60]

Alloy	c_c		T_c	
	calc.	exp.	calc.	exp.
$Li_c Na_{1-c}$	0.62	0.65	650	580
$Ga_c Hg_{1-c}$	0.44	0.50	406	475
$Ga_c Bi_{1-c}$	0.62	0.70	157	535
$Zn_c Bi_{1-c}$	0.75	0.83	438	878

Equation (9.28) may be used to give a simple estimate of the critical temperature T_c and the critical concentration c_c [which are given by the maximum of the $T_s(c)$ curve]. Table 9.3 compares several such estimates, based on the empty-core model potential, with the core radii fitted to the zero-pressure condition [see (3.4)]. The concentration dependence is determined mainly by the denominator in (9.28) (the average pair interaction) and the asymmetry depends on the electron/atom ratio. In heterovalent alloys, the spinodal tends to be skewed towards the element with the lower valence, and in homovalent alloys towards the element with the lower core radius[9]. The value

[9] The pronounced asymmetry of the coexistence curves in binary metallic mixtures is probably associated with the strong state-dependence of the effective interatomic potentials – cf. our remark on the liquid-vapour coexistence curve in Sect. 4.5. There seems to be good reason to expect a similar singularity in the diameter of the coexistence curve, i.e. an increasing amplitude with increasing difference in the valencies. However, the existing material has not yet been analyzed in sufficient detail (the diameter of the coexistence curve in binary liquid mixtures is discussed in [9.63]).

of the critical temperature is rather sensitive to the difference between the
A-B interaction and the geometrical mean of the A-A and B-B interactions
and is also strongly influenced by small changes in the pseudopotential. In
$\hat{\Phi}_{ij}(0)$, the pair potentials are weighted equally over all distances even inside
the repulsive core at points which are actually never sampled. In a more realistic theory one would therefore wish $\hat{\Phi}_{ij}(0)$ to be replaced by a quantity
in which the pair potentials are weighted with a function expressing the
probability of finding another particle at that distance.

Away from the critical point, the behaviour of the system is characterized by mean-field critical exponents which are known to differ from those
found experimentally.

The common-tangent construction applied to a free energy based on
the usual RPA expression [see (4.26–28) and below] would yield a phase
separation curve which does not meet the spinodal at the critical point.
This is not simply a shortcoming of the RPA, but a result of the general
lack of thermodynamic self-consistency inherent in any approximate liquid-
state theory (see Sect. 4.2). A method which would enable us to calculate
both the phase-separation and the spinodal curves from the same equation
of state has yet to be devised – see Appendix D for a discussion of techniques
intended to achieve thermodynamic consistency.

The main shortcoming of the RPA, at least in the form (9.26), is that
it fails to predict the first peak in the partial static structure factors. One
might be tempted to use the RPA not for the full potential as done by
Stroud, but only for the perturbation to a reference potential, as in (9.10).
Such an approach, with an OCP reference potential has indeed been attempted by *Chaturvedi* et al. [9.18] for Li-Na alloys. But even with this
modified approach, the validity of the theory is still limited to the long-
wavelength limit[10] and the problem of thermodynamic consistency is not
resolved.

Future progress could be achieved in two directions: (i) By performing
variational calculations with a more appropriate reference system, i.e. one
which itself has a critical point. Again the HSY system might be a very
suitable candidate. For $\varepsilon<0$ the MSA equation for the HSY system (9.19)
always has a solution. For $\varepsilon>0$ on the other hand, one finds [see (D.55)]
that a solution exists only if the parameter $\theta \equiv \beta\varepsilon\eta$ is smaller than a critical value θ_c whose exact value depends on the screening constant κ. In fact,
a critical divergence in $S_{cc}(0)$ already occurs at smaller values of θ. A first
result of a HSY-variational calculation for $S_{cc}(q)$ and $g_{cc}(R)$ for the Li-Na
system [9.64] is shown in Fig. 9.15. We find that the calculation reproduces
both the critical divergence of $S_{cc}(q)$ and the characteristic temperature
dependence of the correlation length of the concentration fluctuations described by the large-R behaviour of $g_{cc}(R)$. Except for the core effect and

---

[10] But remember our comments on the often claimed "exactness" of the RPA in the
$q = 0$ limit (Sect. 4.3.3).

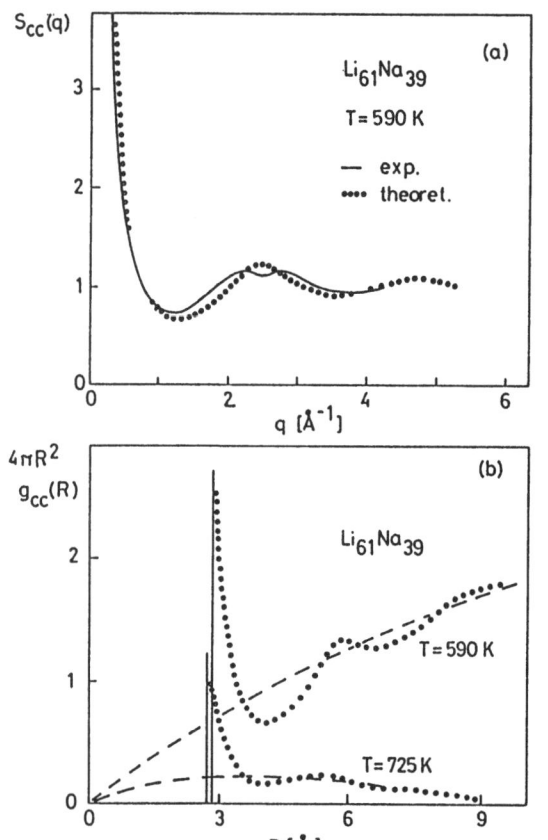

Fig. 9.15a,b. Structure factor $S_{cc}(q)$ (a), and concentration fluctuation correlation function $g_{cc}(R)$ (b) for a $Li_{0.61}Na_{0.39}$ alloy. (—) experiment [9.8]; (- - -) Ornstein-Zernike behaviour for the correlation function. The structure factor refers to a temperature close to the critical point, the correlation function has been drawn at two temperatures in order to demonstrate the decay of the correlation length of the critical concentration fluctuation away from the critical point. Hafner, unpublished results

the oscillations at small distances, the correlation function closely follows the classical Ornstein-Zernike behaviour expressed by [9.65]

$$4\pi R^2 g^{OZ}(R) = \frac{S_{cc}(0)}{c_A c_B} R \frac{\exp(-R/\xi)}{\xi^{-2}} \tag{9.29}$$

where ξ is the correlation length. The expression (9.29) is maximum at $R = \xi$; Figure 9.15 shows very nicely the temperature-dependent decay of the correlation length away from the critical point. As the HSY model yields also a reasonable $S_{NN}(q)$ for all q, this seems to be the most promising structural model for a binary liquid near phase separation (provided that the size ratio plays no dominant role). However, the problem of thermodynamic consistency remains to be solved.

(ii) The thermodynamic perturbation theory should be developed beyond the level of variational calculations and the RPA. This problem will be discussed in some detail in the next section. A first step in that direction has been taken by *Henderson* and *Ashcroft* [9.66]. They applied a binary version of their mean-density approximation [MDA, see (4.36, 37) and accompanying comments] to liquid Li-Na alloys. The MDA brings some numerical improvements over the RPA, but essentially shares the same limitations.

Clustering and phase separation are also characteristic features of the phase diagrams of liquid metals and salts [e.g. $K_c(KCl)_{1-c}$]. Very recently *Chabrier* and *Hansen* [9.67] have shown that second-order pseudopotential theory may be used to describe the structure and the miscibility curves of these systems. Here the conduction electrons not bound by the halide ions screen a mixture of positive and negative ions. The structure is modelled in terms of a mixture of point ions with Coulomb and repulsive Born-mayer forces (the latter are necessary to prevent the collapse of pairs of ions with charges of opposite sign) treated numerically in the HNC approximation [9.67]. This is perhaps the most ambitious, but still very successful attempt to extend this simple perturbation theory.

9.3 Thermodynamic Perturbation Theory

The way in which a thermodynamic perturbation theory for binary liquid mixtures might be developed beyond the level of a simple variational calculation is clearly marked out by the well-known results for monatomic liquids (see Sects. 4.3.2, 3). However, the generalization to the binary case poses some specific problems which up to now have prevented its widespread use. These will be discussed in the following.

The starting point is again a decomposition of the pair potentials $\Phi_{ij}(R)$ into a steeply repulsive short-range term $\Phi_{ij,0}(R)$ and a longer-range part $\Phi_{ij,1}(R)$ according to [for the corresponding separation in the monatomic case, see (4.6) and Fig. 4.2]

$$\Phi_{ij}(R) = \Phi_{ij,0}(R) + \Phi_{ij,1}(R) \qquad \text{and} \qquad \text{9.30a)}$$

$$\Phi_{ij,0}(R) = \begin{cases} \Phi_{ij}(R) - \Phi_{ij}(R_{ij}) & R < R_{ij} \\ 0 & R > R_{ij} \end{cases} \qquad (9.30b)$$

$$\Phi_{ij,1}(R) = \begin{cases} \Phi_{ij}(R_{ij}) & R < R_{ij} \\ \Phi_{ij}(R) & R > R_{ij} \end{cases} , \qquad (9.30c)$$

where R_{ij} is the distance at which the potential $\Phi_{ij}(R)$ has its first minimum.

9.3.1 Repulsive Forces: Weeks-Chandler-Andersen Theory

The functional Taylor-expansion for the free energy of a system with a soft repulsive interaction proposed by *Weeks, Chandler* and *Andersen* (WCA) [9.68–70] has been generalized by *Sung* and *Chandler* [9.71] to the case of binary liquid mixtures. The expansion now reads

$$F = F_\sigma - \frac{n}{2}k_B T \sum_{i,j=A,B} c_i c_j \int B_{\sigma_{ij}}(R) d^3 R \tag{9.31}$$

with the "blip-functions" $B_{\sigma_{ij}}(R)$

$$B_{\sigma_{ij}}(R) = y_{\sigma_{ij}}(R)\{\exp[-\beta\Phi_{ij,0}(R)] - \exp(-\beta\Phi_{\sigma_{ij}})\} \tag{9.32}$$

$$y_{\sigma_{ij}}(R) = g_{\sigma_{ij}}(R)\exp(\beta\Phi_{\sigma_{ij}}) \tag{9.33}$$

measuring the departure of the actual pair potentials $\Phi_{ij}(R)$ from a set of reference HS potentials $\Phi_{\sigma_{ij}}$ with diameters $\sigma = \{\sigma_{AA}, \sigma_{BB}, \sigma_{AB}\}$. The function $g_{\sigma_{ij}}(R)$ is the partial pair correlation of this reference system and F_σ is the free energy of a mixture of hard spheres. In the liquid-metal case, the HS-parameter σ is chosen such that the first term in the functional Taylor series (9.31) vanishes. This suggests a generalization of the WCA condition (4.12) in the form

$$\sum_{i,j=A,B} c_i c_j \int B_{\sigma_{ij}}(R) d^3 R = 0 \tag{9.34}$$

or in the form of the substantially stronger condition

$$\Sigma_{ij} \equiv \int B_{\sigma_{ij}}(R) d^3 R = 0 \quad i,j = A,B \quad . \tag{9.35}$$

Condition (9.34) has the disadvantage that there are numerous sets of parameters $\sigma = \{\sigma_{AA}, \sigma_{BB}, \sigma_{AB}\}$ which satisfy (9.34), but yield rather poor pair-correlation function predictions. Condition (9.35) can be satisfied only if we have a reference system with non-additive parameters [i.e. with $\sigma_{AB} \neq (\sigma_{AA} + \sigma_{BB})/2$]. The analytic solution of the Percus-Yevick equations as well as the semi-empirical results of *Verlet* and *Weis* [9.72] are only valid for the additive case. Solutions for non-additive HS mixtures are available only for a restricted range of parameters and in non-analytic form [9.73], hence they do not constitute an appropriate reference system. As a sort of compromise, *Sandler* [9.74] proposed to stick to the additive case, i.e. $\sigma_{AB} = (\sigma_{AA} + \sigma_{BB})/2$, and to determine σ_{AA} and σ_{BB} such as to fulfil (9.34) and to minimize one of the two functions

$$\Sigma_1 = \sum_{i,j=A,B} c_i c_j |\Sigma_{ij}| \quad \text{or} \tag{9.36}$$

$$\Sigma_2 = \sum_{i,j=A,B} |\Sigma_{ij}| \quad . \tag{9.37}$$

Lee and *Levesque* [9.75] have shown that with condition (9.34) alone, the blip-function expansion reproduces the total thermodynamic properties of a Lennard-Jones mixture to within 1 % of the computer-simulation results. *Sandler* [9.74] demonstrated that upon imposing one of the constraints (9.36, 37) one also obtains reasonably accurate partial pair-correlation functions. However, in these calculations the Lennard-Jones potentials

$$\Phi_{ij}^{LJ}(R) = 4\varepsilon_{ij}[(d_{ij}/R)^{12} - (d_{ij}/R)^6] \tag{9.38a}$$

were subject to the conditions

$$\varepsilon_{AB} = (\varepsilon_{AA}\varepsilon_{BB})^{1/2} \tag{9.38b}$$

and

$$d_{AB} = (d_{AA} + d_{BB})/2 \quad , \tag{9.38c}$$

i.e. they were constrained to additivity both with respect to the effective diameters and the depths of the individual potentials. This is not always the case for metallic alloys.

Kahl and *Hafner* [9.76] have shown that for an alloy with a nearly ideal mixing behaviour such as K-Rb, Σ_{AB} is already very close to zero when σ_{AA} and σ_{BB} are determined by imposing the conditions $\Sigma_{AA} = \Sigma_{BB} = 0$ i.e. (9.35). However, even for a weakly non-ideal alloy such as K-Cs, the three conditions (9.35) cannot be satisfied simultaneously. Although the effect on the relevant partial pair-correlation function is not very pronounced, the violation of one of the blip-function conditions shows up in a drastic deviation of the corresponding partial static structure factor $S_{ij}(q)$,

$$S_{ij}(q) = S_{\sigma_{ij}}(q) + n(c_i c_j)^{1/2}\hat{B}_{\sigma_{ij}}(q) \tag{9.39a}$$

at small q (Fig. 9.16). Imposing (9.36) or (9.37) does not improve the situation – discrepancies now appear in all three partial structure factors, and the calculated compressiblity of the alloy becomes quite unrealistic. For strongly non-ideal systems this problem would become extremely severe. As a way out of the dilemma, *Kahl* and *Hafner* [9.76] proposed to determine σ_{AA} and σ_{BB} from the conditions $\Sigma_{AA} = \Sigma_{BB} = 0$ and then vary R_{AB} for a given $\sigma_{AB} = (\sigma_{AA} + \sigma_{BB})/2$ such that $\Sigma_{AB} = 0$. This means that the separation between unlike neighbours is modified in such a way that the set of repulsive potentials $\Phi_{ij,0}(R)$ is effectively additive. This procedure is justified for two good reasons: (i) If the blip-function treatment of the repulsive part of the potential is supplemented by an optimized-random-phase approximation for the long-range part, then the self-consistency achieved by the iterative algorithm [see (4.30 ff) and below] makes the precise way in which the initial separation is achieved ultimately irrelevant. (ii) The partial structure fac-

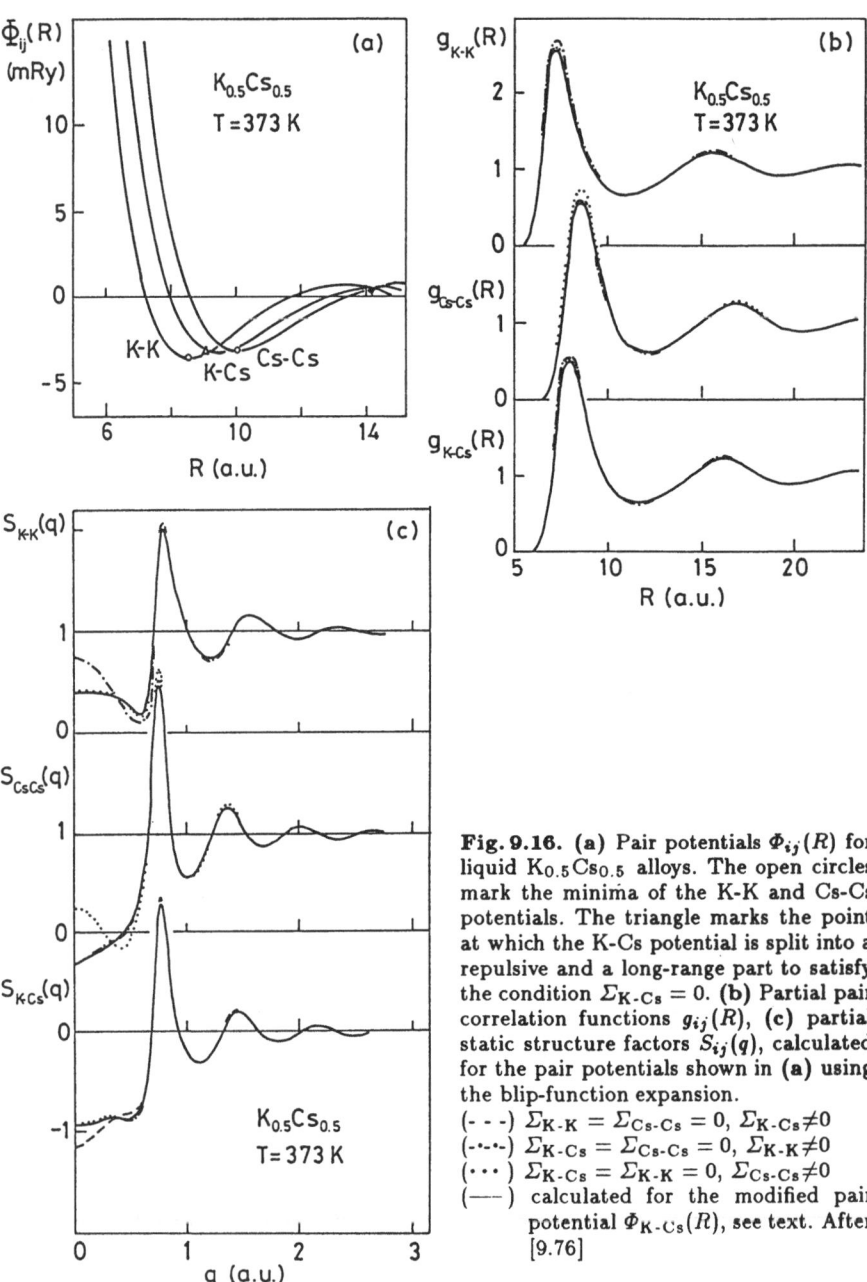

Fig. 9.16. (a) Pair potentials $\Phi_{ij}(R)$ for liquid $K_{0.5}Cs_{0.5}$ alloys. The open circles mark the minima of the K-K and Cs-Cs potentials. The triangle marks the point at which the K-Cs potential is split into a repulsive and a long-range part to satisfy the condition $\Sigma_{K\text{-}Cs} = 0$. (b) Partial pair correlation functions $g_{ij}(R)$, (c) partial static structure factors $S_{ij}(q)$, calculated for the pair potentials shown in (a) using the blip-function expansion.

(- - -) $\Sigma_{K\text{-}K} = \Sigma_{Cs\text{-}Cs} = 0$, $\Sigma_{K\text{-}Cs} \neq 0$
(-·-·-) $\Sigma_{K\text{-}Cs} = \Sigma_{Cs\text{-}Cs} = 0$, $\Sigma_{K\text{-}K} \neq 0$
(· · ·) $\Sigma_{K\text{-}Cs} = \Sigma_{K\text{-}K} = 0$, $\Sigma_{Cs\text{-}Cs} \neq 0$
(———) calculated for the modified pair potential $\Phi_{K\text{-}Cs}(R)$, see text. After [9.76]

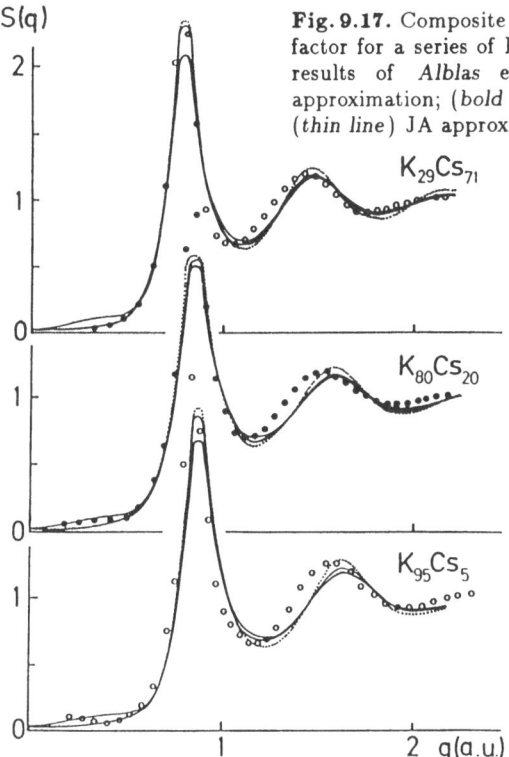

S(q)

Fig. 9.17. Composite (neutron-weighted) static structure factor for a series of K-Cs alloys. (*Circles*) experimental results of *Alblas* et al. [9.76]; (*dotted line*) HS approximation; (*bold line*) WCA approximation (9.39a); (*thin line*) JA approximation (9.39b). After [9.76]

$K_{29}Cs_{71}$

$K_{80}Cs_{20}$

$K_{95}Cs_5$

tors $S_{ij}(q)$ are now realistic even in the long wavelength regime[11], and when added up to the composite neutron or x-ray structure factor (Fig. 9.17) they compare well with the diffraction experiments of *Albas* et al. [9.77].

What was said earlier by way of comment on the WCA structure factors of the pure metals, applies also to the case of alloys: compared to a hard-sphere calculation, the WCA blip-function expansion as modified by Kahl and Hafner substantially improves the large-q behaviour of the structure factor by imposing a damping of the oscillations. However, in the region of the second peak, a substantial phase difference between theory and experiment persists. Again this can only be corrected by taking into account the oscillatory part of the effective interatomic potentials.

---

[11] If we use (9.39a) to calculate the partial structure factors, we find the same unphysical hump near $q = \pi/\sigma_{ij}$ as in the pure metal case. It may be eliminated by adopting the expression proposed by *Jacob* and *Andersen* [9.78] and generalized to the binary case by *Telo da Gama* and *Evans* [9.79], i.e.

$$\hat{c}_{ij}(q) = \hat{c}_{\sigma_{ij}}(q) + \hat{B}_{\sigma_{ij}}(q) \qquad (9.39b)$$

and correspondingly for the static structure factors. However, one has to remember that (9.39b) has only low-q validity, whereas (9.39a) remains valid for large q.

9.3.2 Long-Range Forces: Optimized-Random-Phase Approximation

For the pure liquid metals we have found that in most cases the structure is dominated by the short-range repulsive forces. The structure determined by excluded volume effects is not easily modified by the much weaker attractive forces. Only for the molten group IV and V elements (which are semiconductors or insulators in the crystalline state) and for some "anomalous" metals such as Ga, In or Hg do the attractive forces induce a *qualitative* modification of the hardsphere like structure, the final structure resulting from a competition between the geometrical requirements of sphere packing and the long-range attractive forces originating from electronic effects.

However, in a mixture, it is possible for the attractive interactions to produce structural effects which do not compete with the repulsive forces. As a simple example consider a mixture consisting of hard spheres of diameter σ together with square-well atoms of core diameter σ with attractive well of range $1.5\,\sigma$. The attractions between the square wells tend to make the square-well atoms cluster together, and this can occur without any interference with the excluded-volume correlations produced by the hard cores. Consequently the attractive interactions are able to produce important structural effects.

In fact we have seen in the last sections that even a rudimentary description of deviations from an ideal mixing behaviour is impossible without the attractive forces. A mean-field approach such as the RPA is of limited value because of its failure to describe the short-range correlations. The HSY model is more realistic in this respect, but it can be solved only with the introduction of a complete decoupling of the density and concentration fluctuations.

Therefore it appears to be highly desirable to extend a theory such as the optimized-random-phase approximation – which we had found to be very successful for the pure metals – to the case of binary alloys. The formal generalization is straightforward (in fact it is to be found in the early papers of *Andersen* et al. [9.80–83]). The correlation functions (total and direct) are again decomposed into a reference part and a residual part

$$c_{ij}(R) = c_{ij,0}(R) + c_{ij,1}(R) \tag{9.40}$$

$$h_{ij}(R) = h_{ij,0}(R) + h_{ij,1}(R) \quad . \tag{9.41}$$

A system of residual Ornstein-Zernike equations is easily derived, the closure relations for these equations being given by

$$c_{ij,1}(R) = -\beta \Phi_{ij,1}(R) \quad R > \sigma_{ij} \tag{9.42a}$$

$$h_{ij,1}(R) = -1 \quad\quad R < \sigma_{ij} \tag{9.42b}$$

or equivalently

$$c_{ij,1}(R) = -\beta\Phi^*_{ij,1}(R) = \begin{array}{ll} -\beta\Phi_{ij,1}(R) & R>\sigma_{ij} \\ -\beta\psi_{ij}(R) & R<\sigma_{ij} \end{array} \qquad (9.43)$$

with the optimized potential $\Phi^*_{ij,1}(R)$ (cf. Sect. 4.3.3). This defines a set of three coupled integral equations relating the total correlation functions $h_{ij,1}(R)$ and the optimized potentials $\Phi^*_{ij,1}(R)$. Applying the optimization condition, (9.42b) yields an integral equation for the optimized potential in the range $R<\sigma_{ij}$ [i.e. for $\psi_{ij}(R)$]. Again this is conveniently reformulated in the form of a variational condition which asks for the RPA contribution to the free energy

$$F_{\text{ORPA}} = \frac{1}{(2\pi)^3} \int \left\{ n\beta \text{Tr} \sum_k c_k S_{\sigma_{ik}}(q)\hat{\Phi}^*_{kj}(q) \right.$$

$$\left. - \ln \det \left[\delta_{ij} + n\beta \sum_k c_k S_{\sigma_{ik}}(q)\hat{\Phi}^*_{kj}(q) \right] \right\} d^3q \qquad (9.44)$$

to be stationary with respect to changes in the optimized potentials $\Phi^*_{ij}(R)$ for $R<\sigma_{ij}$, i.e.

$$\frac{\delta F_{\text{ORPA}}[\Phi^*_{ij}]}{\delta\Phi^*_{ij}(R)} = 0 \quad R<\sigma_{ij} \quad , \quad i,j = A, B \quad . \qquad (9.45)$$

For continuous potentials, the repulsive part must be expressed in terms of effective hard-core interactions by means of an intermediate blip-function expansion. The final result for the free energy and the static structure factors is a direct generalization of the pure metal results (4.28 ff).

Despite the conceptual simplicity, the actual application of the binary ORPA represents a large amount of heavy computation – especially in the case of soft alloy potentials where the self-consistency requirements cannot be ignored[12]. This is the reason why applications are rather scarce. Except for the original work of *Andersen* and *Chandler* [9.83] on a primitive model for an electrolyte, we know only of the work of *Sung* et al. [9.84] on a mixture of hard spheres and square-well atoms. Very recently *Kahl* and *Hafner* [9.85] presented the first ORPA-calculation for liquid metallic alloys. Their results for the K-Cs and Al-Ge systems are shown in Fig. 9.18. For K-Cs, the inclusion of the attractive interaction clearly improves the agreement with experiment. The effect is most pronounced near the first peak and around the second peak in $S(q)$. The attractive forces induce a marked change in

---

[12] The new optimized potential $\Phi^*_{ij}(R)$ replaces $\Phi_{ij}(R_{ij})$ in (9.30b,c). Since (9.30a) has to be preserved, this defines a new $\Phi_{ij,0}(R)$ and so on. The calculations have to be repeated until self-consistency is achieved.

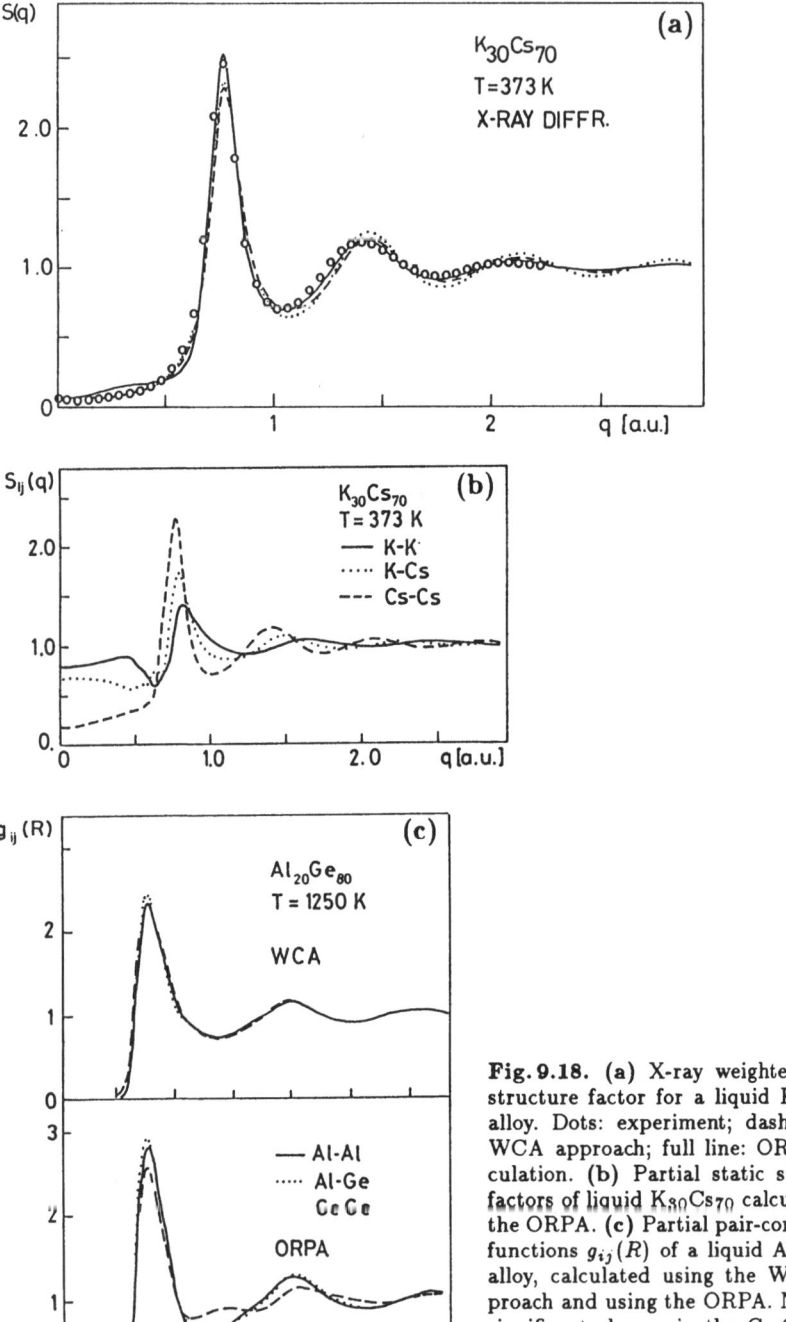

Fig. 9.18. (a) X-ray weighted static structure factor for a liquid $K_{30}Cs_{70}$ alloy. Dots: experiment; dashed line: WCA approach; full line: ORPA calculation. (b) Partial static structure factors of liquid $K_{30}Cs_{70}$ calculated in the ORPA. (c) Partial pair-correlation functions $g_{ij}(R)$ of a liquid $Al_{20}Ge_{80}$ alloy, calculated using the WCA approach and using the ORPA. Note the significant change in the Ge-Ge pair-correlation due to the inclusion of the long-range forces. After [9.85]

the pair-correlation functions of the Al-Ge alloy; the Ge-Ge correlation is most strongly affected – see Fig. 10.18(c). We find that the ORPA offers a valid explanation of the change in the topological short-range order from close-packed hard-sphere like in the Al-rich regime to a distinctly lower coordination at the Ge-rich end.

Clearly the ORPA seems to be the most promising approach for studying the structure and thermodynamics of binary alloys with a non-ideal mixing behaviour.

9.4 Structure and Thermodynamics of Liquid Transition-Metal Alloys

The structural and thermodynamic properties of liquid transition-metal-based alloys are still a largely unexplored field, both experimentally and theoretically. Only quite recently has the technological interest in amorphous transition-metal alloys (see e.g. [9.9, 10]) motivated a renewed interest in the corresponding liquid alloys, but even now we still lack sufficiently detailed knowledge to enable us to establish definite trends across the transition-metal series.

The easiest aspect to handle is as usual the heat of formation. The success of the Miedema model and Pettifor's microscopic interpretation with the aid of simple rectangular d-bands (see Sect. 6.2), shows that the heat of formation is largely independent of structural effects. The effects of disorder and short-range order in liquid or amorphous alloys are most effectively treated in a scheme based on a tight-binding or linearized-muffin-tin orbital representation, combined with the recursion technique or the cluster-Bethe-lattice method for the calculation of the electronic density of states [9.86–89]. The cohesive energy or the heat of formation can be calculated from the density of states, but what we need is an analysis which consistently relates the electronic to the atomic structure.

Only a first attempt has been made to calculate a possible CSRO in liquid transition metals from a microscopic basis. *Hafner* and *Pasturel* [9.59] realized that a variational calculation of the free energy does not necessitate an explicit knowledge of the interatomic forces; all one really needs is an expression of the free energy in terms of some reference pair-correlation functions. For a transition-metal alloy, such a total energy expression may be constructed in a tight-binding framework using the cluster-Bethe-lattice method in the Cayley-tree approximation [9.57, 58]. The electronic contribution to the ground-state energy is calculated using the component-related Green's functions

$$G^i(z) = (z - \varepsilon_i - \Delta_i)^{-1} \quad , \quad i = A, B \tag{9.46}$$

with the self-energy Δ_i given by the solution of

$$\Delta_A = \frac{nc_A \int t_{AA}^2(R)g_{AA}(R)d^3\boldsymbol{R}}{z - \varepsilon_A - \Delta_A} + \frac{nc_B \int t_{AB}^2(R)g_{AB}(R)d^3\boldsymbol{R}}{z - \varepsilon_B - \Delta_B} \tag{9.47a}$$

$$\Delta_B = \frac{nc_A \int t_{AB}^2(R)g_{AB}(R)d^3\boldsymbol{R}}{z - \varepsilon_A - \Delta_A} + \frac{nc_B \int t_{BB}^2(R)g_{BB}(R)d^3\boldsymbol{R}}{z - \varepsilon_B - \Delta_B} \tag{9.47b}$$

The ε_i are the atomic energy eigenvalues of the d-levels, the $t_{ij}(R)$'s are the d-d hopping matrix elements (in an isotropic approximation, see Sect. 3.1.1). This total energy expression is then used in a variational equation of the type (9.23) based on a hard-sphere Yukawa reference system. This upper bound to the exact free energy is then minimized with respect to the HSY parameters. A typical result for the DOS and partial structure factors of a Ni-Ti alloy is shown in Fig. 9.19. The electronic density of states shows

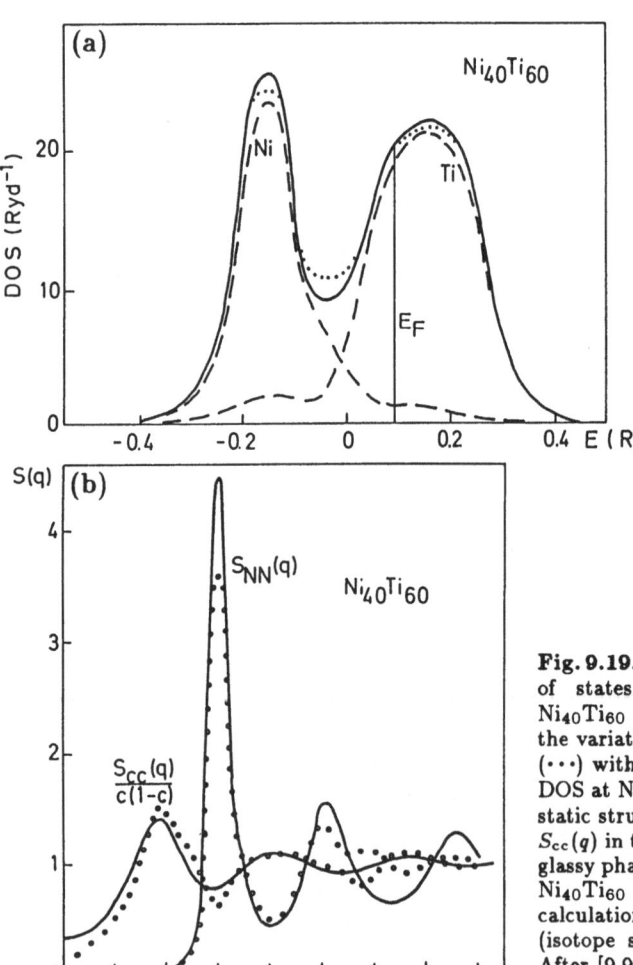

Fig. 9.19. (a) Electronic density of states of supercooled liquid $Ni_{40}Ti_{60}$ alloys. (——) including the variationally calculated CSRO; ($\cdots$) without CSRO; (- - -) partial DOS at Ni and Ti sites. (b) Partial static structure factors $S_{NN}(q)$ and $S_{cc}(q)$ in the supercooled liquid ($\approx$ glassy phase, see text and Chap. 11) $Ni_{40}Ti_{60}$ alloys. (——) variational calculation; (o) neutron scattering (isotope substitution experiment). After [9.90]

a two-peaked structure. The CSRO renders the splitting of the bands even more pronounced: the CSRO tends to keep like atoms apart and the reduced overlap necessarily narrows the bands. If the splitting of the bands is sufficiently large, CSRO tends to be energetically favourable for an approximately half-filled band. This parallels the results of Gautier et al. for the case of crystalline alloys. The variationally determined parameters of the reference system also yield a rather accurate description of the partial structure factors (Fig. 9.19b) (actually this calculation has been performed for a supercooled liquid alloy and is compared to the experimental results for an amorphous alloy – we shall come back to this point in Sect. 11.1.2).

Considerable attention has been paid to liquid alloys of transition metals with polyvalent simple metals or metalloids. These alloys are characterized by exceptionally large negative values of the enthalpy and enthalpy of mixing. A simple approach to these systems has been presented by the Grenoble group [9.91–93]. It is based on the plausible assumption that the main effect of the addition of sp-electrons is to fill the d-band of the transition metal. The enthalpy of mixing is thus written as

$$\Delta H = (1 - c)\Delta E_2 + c\Delta E_1 \quad , \tag{9.48}$$

where c stands for the concentration of the simple metal or metalloid. The change in energy arising from the filling of the d-band is given by ΔE_2 and can be calculated in a rigid-band fashion; ΔE_1 is the contribution from the change in the number of sp-electrons and is estimated in a free-electron approximation [9.92,93]. For the entropy of mixing, it is assumed that the ideal mixing and the electronic contributions dominate; the structural effects contribute a packing term ΔS_η and a size-mismatch term ΔS_σ. Both can be calculated in a hard-sphere approximation. Thus one has

$$\Delta S = \Delta S_{id} + \Delta S_{el} + \Delta S_\eta + \Delta S_\sigma. \tag{9.49}$$

The electronic entropy is related to the density of states at the Fermi level via

$$S_{el} = \tfrac{1}{3}\pi^2 N(E_F)k_B^2 T \quad . \tag{9.50}$$

As pointed out by *Meyer* et al. [9.94] the change of S_{el} gives a non-negligible contribution to the entropy of formation. Note that transition metals have a large electronic entropy $S_{el}/k_B \sim 2$ owing to the large d-electron density of states at the Fermi level and that S_{el} varies rapidly with the d-band occupation. The packing contribution S_η is important only in those cases where there is a non-negligible volume of formation; S_σ is negligible in almost all cases.

Figure 9.20 shows a typical result for liquid Ni-Si. It is found that the enthalpy of formation is rather insensitive to the precise form of the density

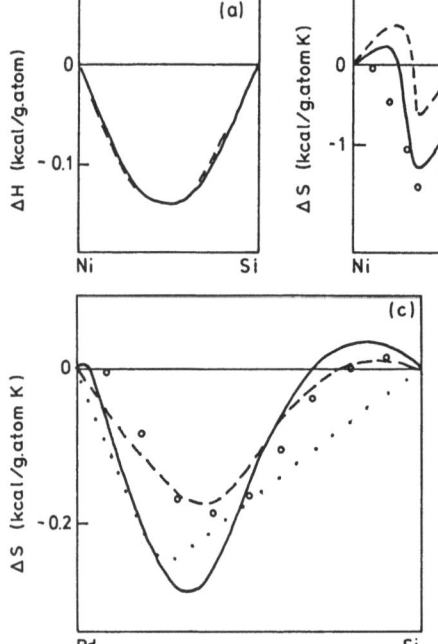

Fig. 9.20a–c. Enthalpy (a) and entropy of formation (b) of liquid Ni-Si alloys. (o) experiment. (- - - and ——) calculated according to (9.46–48) with an approximation using two (- - -), or four (——) moments for the d-band density of states. (c) Contributions to the entropy of formation of liquid Pd-Si alloys; (···) electronic; (- - -) packing contribution; (——) total ΔS. After [9.91–93]

of states as can be seen by comparing the results obtained at different levels of the continued fraction expansion. On the other hand the entropy of mixing is more sensitive to the band structure. In some cases such as in liquid Pd-Si alloys, the electronic and the packing effects peak at different concentrations [9.95]. This gives a very nice explanation of the complicated form of the entropy of formation in this system [Fig. 9.20(c)]. In general the rather good agreement with experiment – which has been documented for (Fe, Co, Ni, Pd)-(Si, Al) alloys – demonstrates the preponderance of electronic over structural effects.

9.5 Collective Excitations in Liquid Alloys

The analysis of the atomic motions in liquid metals which was presented in Sect. 4.7 can be easily generalized to the case of a binary mixture. Instead of a single dynamical structure factor $S(q,\omega)$ [or, equivalently a time-dependent correlation function $g(R,t)$] one now has to consider a set of three partial dynamical structure factors $S_{ij}(q,\omega)$, $i,j = A,B$ (or partial correlation functions) describing the motions of A-A, B-B, and A-B pairs with the inelastic scattering law now being a weighted sum of the three partials. Again it turns out that in some instances a more physical picture is obtained if one considers the dynamical analogues of the number-density and concentration fluctuation structure factors as defined by *Bhatia* and

Thornton [9.1], (see also [9.90])

$$S_{NN}(q,\omega) = \frac{1}{2\pi N} \int d^3 R \int d\omega \langle n(R,t)n(0,0)\rangle \exp(iqR - i\omega t)$$

$$S_{cc}(q,\omega) = \frac{N}{\pi} \int d^3 R \int d\omega \langle c(R,t)c(0,0)\rangle \exp(iqR - i\omega t)$$

$$S_{Nc}(q,\omega) = \frac{1}{2\pi} \int d^3 R \int d\omega [\langle n(R,t)c(0,0)\rangle$$
$$+ \langle c(R,t)n(0,0)\rangle] \exp(iqR - i\omega t) \quad . \tag{9.51}$$

For the pure liquid metals it was found that a dispersion law for collective excitations could be defined in terms of the positions of the inelastic peaks in $S(q,\omega)$ and that the thermal properties of the liquid metals can be described in terms of these collective coordinates.

The dynamical properties of ordered intermetallic compounds are characterized by the existence of optical modes in which one species moves in phase opposition to the other. If an alloy shows a sufficiently high degree of chemical short range order we may expect it to be capable of sustaining collective excitations of the same type which should show up as inelastic peaks in $S_{cc}(q,\omega)$. In molten salts such modes have indeed been found in optical experiments and computer simulations [9.96].

Relatively little is known about the dynamical properties of liquid metallic alloys. *Jacucci* and *McDonald* [9.97] have investigated Na-K alloys and very recently *Jacucci* et al. [9.15, 16] treated liquid Li-Pb alloys using molecular dynamics. The calculations for Na-K alloys are based on pseudopotential-derived interatomic forces and those for Li-Pb on a soft-sphere-Yukawa potential proposed by *Copestake* et al. [9.48]. The density-fluctuation spectra of the alloy and of the pure metals have the same hydrodynamic-like structure with a well resolved Brillouin peak and a central Rayleigh peak. The difference is in the relative intensities of the peaks: whereas in the pure metal the inelastic and quasi-elastic peaks are of comparable height and width, in the alloy the central elastic peak is much more intense. In a nearly ideal mixture such as Na-K, the presence of an inelastic peak in $S_{NN}(q,\omega)$ demonstrates the existence of well-defined propagating density fluctuations. However, no collective excitations of the form of oscillations in the local concentration have been detected in $S_{cc}(q,\omega)$ (**Fig. 9.21**). In principle, the width of the central peak in $S_{NN}(q,\omega)$ should be controlled by thermal conduction and the width in $S_{cc}(q,\omega)$ by interdiffusion. According to the computer simulation, the low-frequency part of $S_{NN}(q,\omega)$ is almost identical to $S_{cc}(q,\omega)$ and therefore the dominance of the Rayleigh peak must again be ascribed to the effects of interdiffusion.

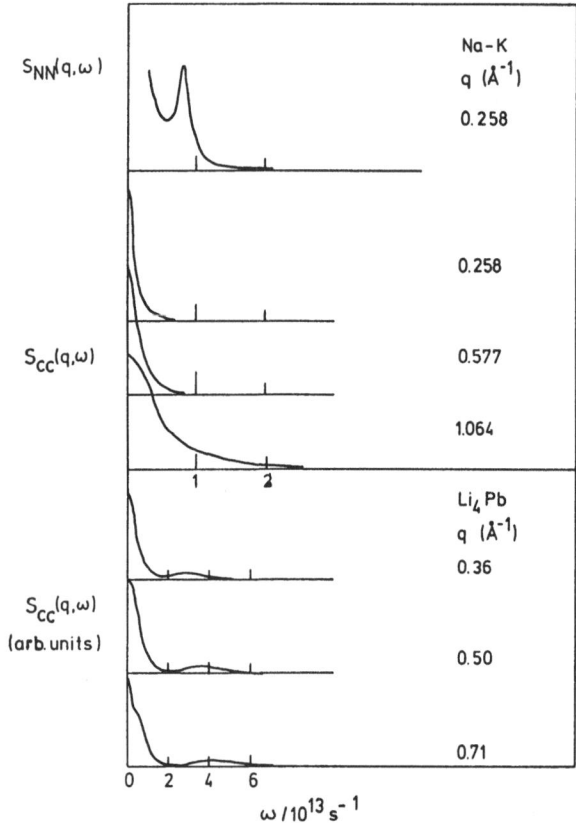

Fig. 9.21. (a) $S_{NN}(q, \omega)$ for liquid Na-K alloys at $q = 0.258\,\text{Å}^{-1}$. **(b)** $S_{cc}(q, \omega)$ for liquid Na-K alloys at different momentum transfers. **(c)** $S_{cc}(q, \omega)$ for liquid Li_4Pb alloys at different momentum transfers. After [9.97] and [9.15]

In the strongly ordering Li-Pb system on the other hand, a propagating concentration fluctuation mode is clearly observable in $S_{cc}(q, \omega)$ for not too large momentum transfers. In liquid Na-K, the diffusive peak has a purely Lorentzian shape. In Li_4Pb, it shows a shoulder which *Jacucci* et al. [9.15, 16] interpret as the signature of a restoring force tending to reestablish a high degree of local chemical order.

Here again it would be interesting to examine the interrelations between the thermodynamic properties and the characteristics of collective excitations – but this subject has yet to be investigated.

Recent Results

The critical point behaviour of the phase-separating hard sphere-Yukawa (HSY) model (see Sect. 9.2.3) has been discussed by *Cummings* and *Stell* [9.98] and *Hafner* and *Jank* [9.99]. It was shown that, depending on the range of the pair potential, the critical exponents show a crossover from mean-spherical values for strong to moderate screening to mean-field values for weak screening [9.99].

Kahl and *Pastore* [9.100] have applied the coupled integral equations with HNC, PY, and mixed closure to calculate the structure of liquid K-Cs alloys.

Bosse et al. [9.101] have analyzed the molecular-dynamics data for the dynamical properties of liquid Li_4Pb alloys in terms of a memory function approch and propose a "fast sound" mechanism for the dispersion of the propagating concentration-fluctuation modes.

10. Alloy Phase Diagrams

With the presentation of a theory of liquid alloys, we have now assembled all the necessary tools for the ab initio construction of an alloy phase diagram.

10.1 First Principles Calculations of Alloy Phase Diagrams

Here we can again proceed either by a direct calculation of the Gibbs free energies of all the competing phases, or by generalizing the order-parameter theory for the solid $\leftrightarrow$ liquid transition to binary alloys. We begin by discussing the "total-free-energy" approach. The basic thermodynamic considerations have already been presented in Sect. 1.2.2. At a given temperature T and pressure p we have to draw the $\Delta G(c,T)$ curves of all the competing solid and liquid phases. The stable phases and their limits of stability are then found through the familiar common tangent construction (the common tangent expressing the equality of the chemical potentials). An example is shown in Fig. 10.1 for the case of Ca-Mg alloys. This diagram contains the Gibbs free energies of the liquid and the terminal solid solutions (with a fcc crystal structure in the Ca-rich regime and a hcp one on the Mg-rich side – in agreement with experiment these are found to be the lowest energy solutions) and of intermetallic compounds with stoichiometry $CaMg_2$ and Ca_2Mg. For the composition $CaMg_2$ the three Laves phase structures (hexagonal $MgZn_2$ and $MgNi_2$ and cubic $MgCu_2$ types) and for the composition Ca_2Mg the Mg_2Cu- and Mg_2Ni-type crystal structures, have been considered. This shows that any such calculation is necessarily limited, as it is based on a finite set of more or less arbitrarily chosen structures and compositions. Of course, the selection of the compound phases to be considered in the calculation has to be made in the light of the general alloy chemical principles. In the present case of a weak chemical interaction between the constituent metals, intermetallic structures were selected which are known to exist in systems with similar size ratios. For a proper calculation of the phase diagram, it is equally important to predict the instability of the Mg_2Cu-type compound Ca_2Mg relative to both the liquid and the supercooled liquid phase and to the crystalline $(Ca-CaMg_2)$ two-phase

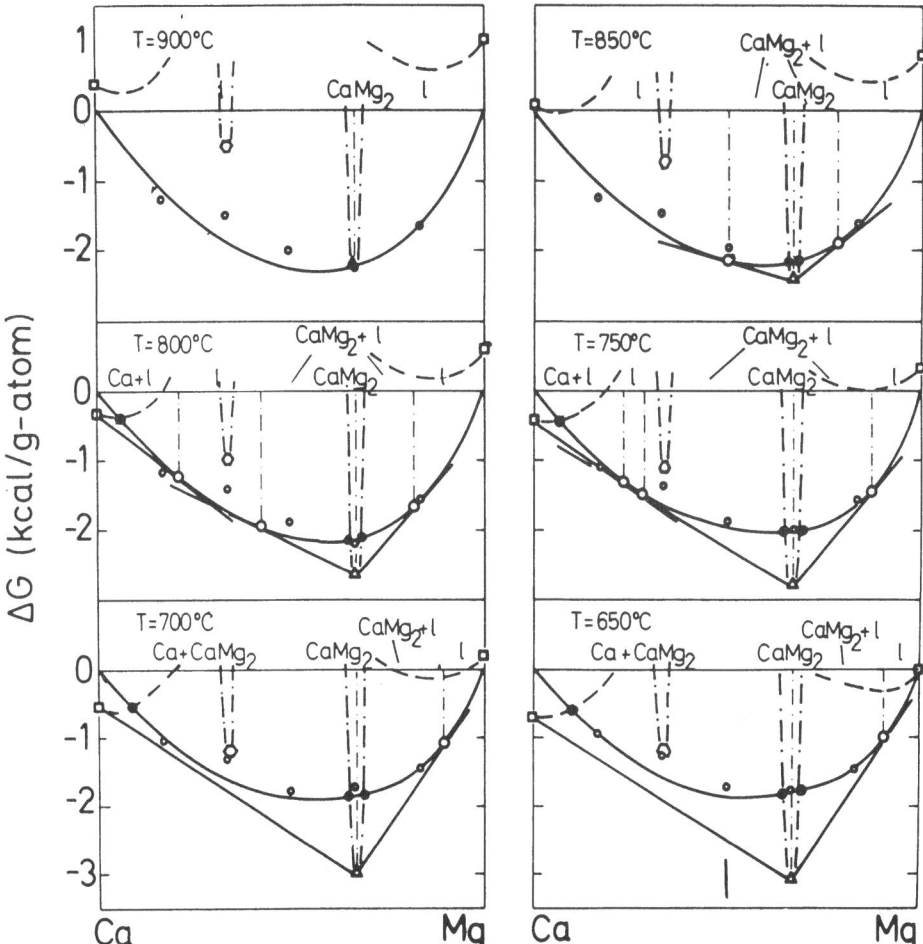

Fig. 10.1. Free energy of formation $\Delta G(c, T)$ relative to the pure liquid metals for Ca-Mg alloys at different temperatures. (□) pure crystalline metals; (- - -) terminal solid solutions; (—— and ○) liquid solutions; (△) hexagonal ($MgZn_2$-type) Laves-phase $CaMg_2$; (*hexagons*) Mg_2Cu-type compound Ca_2Mg. The dot-dashed curves represent ΔG for slightly off-stoichiometric compounds. The common-tangent construction for the determination of the equilibrium phase diagram is shown. The intersections of a $\Delta G(c, T)$ curve for a solid alloy x with the $\Delta G(c, T)$ curve of the liquid alloy determine the $T_0^x(c)$ curves. Below the $T_0^x(c)$ curve a thermodynamic driving potential exists for a partitionless crystallization of the phase x. See text. After [10.1] and [10.2]

mixture, and also the stability of the $MgZn_2$-type compound $CaMg_2$. Table 10.1 summarizes the available experimental thermodynamic information and compares it with the calculations.

The resulting phase diagram for this compound-forming system is compared with experiment in Fig. 10.2 together with analogous phase diagram constructions for a system with unlimited liquid and solid solubility (Mg-Cd), a eutectic system (Cd-Zn), and a system with a melting extremum (K-

Table 10.1 (a) Heat and entropy of formation of liquid Ca-Mg alloys at $T = 900°$ C   (b) Heat of formation of crystalline CaMg$_2$ at $T = 0\,$K

c_{Mg}	ΔH [cal/g-atom]		$\Delta S/k_B$		ΔH [cal/g-atom]	
	theor.	exp.	theor.	exp.	theor.	exp.
0.16	-1295	-1300	0.90	0.71	-3600	-3200
0.33	-1474	-1600	0.85	0.67		
0.50	-2021	-2100	0.98	0.56		
0.67	-2315	-2300	1.00	0.79	After [10.3]	
0.83	-1605	-1650	0.80	0.75		

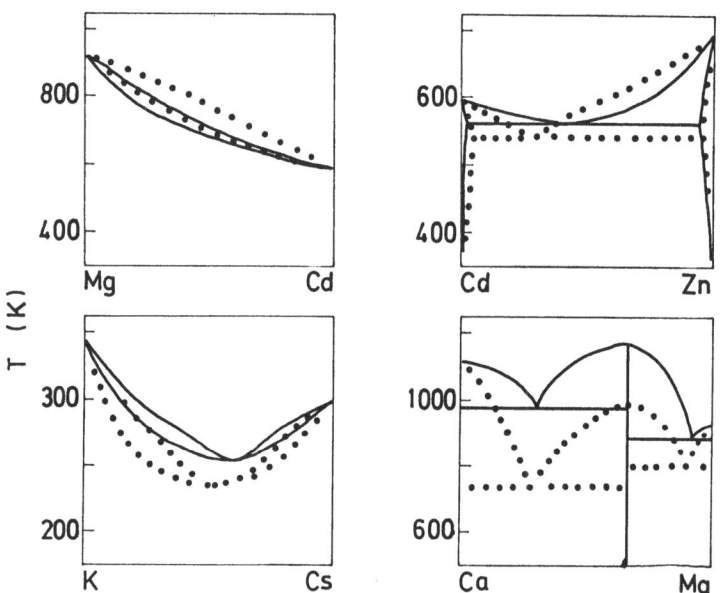

Fig. 10.2. Alloy phase diagrams as calculated from first-principles (——) and compared with experiment ($\bullet$). After [10.4] (Mg-Cd), [10.6] (Cd-Zn), [10.5] (K-Cs), and [10.1, 2] (Ca-Mg)

Cs) (for further examples see e.g. [10.4, 5, 7]). It is clear that at the present stage, the ab initio calculation of alloy phase diagrams is still restricted to the simplest possible systems, i.e. homovalent alloys with relatively small size differences. This is only a reminder as to how ambitious a task we are undertaking: a prediction of the Gibbs free energies of formation of all competing phases within a maximum error of 20 % is a prerequisite for a successful phase diagram calculation.

What makes a phase diagram calculation interesting is that it is not a mere number-crunching exercise, but it gives (a) access to thermodynamic information not accessible to experiment and (b) a deeper insight into the mechanisms of phase formation. An example to illustrate the first point is the determination of the $T_0^x(c)$ curves. $T_0^x(c)$ is defined as the temperature at which the liquid and the solid phase, x, of composition c have the same free energy, below this temperature a thermodynamic driving potential exists for a completely partitionless solidification into a single phase x. As stressed by *Massalski* [10.8], the concept of the T_0^x temperatures is important for discussing the thermodynamic aspects of the formation of metastable phases – we shall return to this point in the following chapter.

A deeper understanding of phase formation hinges primarily on our insight into the origin of the crystalline and liquid phases and their inter-relations. For monatomic systems much progress has been made through the recent developments based on the density-functional theory of freezing (see Chap. 5). This theory can be generalized to binary systems, but given the state of the art of the theory of liquid alloys, it will certainly be some time before it can be applied to realistic models of alloys. However, the thermodynamic equations relating the shape of the phase boundaries to the chemical short-range order in the competing phases have been explored [10.9–13] and can serve as a useful guide for future investigations.

10.2 Chemical Short-Range Order and Alloy Phase Diagrams

It has been shown by *Bhatia* and *March* [10.9–11] that at fixed temperature, pressure and composition the following set of differential equations holds for the phase boundaries of phases α and β in the (c, T)-plane (c_α is the concentration of the B atoms in phase α)[1]

$$\frac{dT}{dc_\alpha} = -(c_\alpha - c_\beta)\frac{RT^2}{S_{cc}^\alpha(0)L^\beta} \tag{10.1a}$$

$$\frac{dT}{dc_\beta} = -(c_\alpha - c_\beta)\frac{RT^2}{S_{cc}^\beta(0)L^\alpha} \quad . \tag{10.1b}$$

Here

$$L^\alpha = (1 - c_\alpha)L_A + c_\alpha L_B \tag{10.2a}$$

$$L^\beta = (1 - c_\beta)L_A + c_\beta L_B \tag{10.2b}$$

are generalized latent heats of transformation, being determined predominantly by the latent heat of melting of the pure A and B metals, corrected

---

[1] For a recent discussion of this topic, see also *Geertsma* [10.13]. The following discussion follows his work quite closely.

for the temperature difference between the melting temperature T_A and T,

$$\frac{L_A}{T} = \frac{L_{0A}(T_A)}{T} + \int_{T_A}^{T} [c_{p_A}^{\alpha}(T,0) - c_{p_A}^{\beta}(T,0)]\frac{dT}{T}$$

$$+ \frac{\partial}{\partial T} RT \ln \left[\frac{a_A^{\beta}(T,c_{\beta})}{a_A^{\alpha}(T,c_{\alpha})}\right] . \tag{10.3}$$

In (10.3) the first term is the latent heat of melting of pure A under standard conditions, the integral over the difference in the partial heat capacities c_{p_A} in phases α and β corrects for the difference in entropy of pure A in phases α and β, the last term accounts for the composition dependence of the chemical potential $\mu_A(T,c)$ [$a_A^{\alpha}(T,c_{\alpha})$ is the activity of the component A in phase α]. Note that for an ideal solution, (10.1a,b) reduce to the well-known equations for the freezing point depression.

From (10.1) it follows that the ratio $R(T)$ of the slope of the phase boundary lines of a two-phase region at a given temperature is expressible in terms of the concentration fluctuation structure factor $S_{cc}^{\alpha}(0)$ and the generalized latent heat L^{β} [10.13]

$$R(T) = \frac{(dT/dc_{\alpha})}{(dT/dc_{\beta})} = \frac{S_{cc}^{\beta}(0)L^{\alpha}}{S_{cc}^{\alpha}(0)L^{\beta}}. \tag{10.4}$$

Relations (10.1) to (10.4) are of importance for a critical evaluation of experimentally determined or calculated phase diagrams.

10.2.1 Melting Extrema

In the following we will identify phase α with the solid $(\alpha \to s)$ and phase β with the melt $(\beta \to l)$. At a melting extremum the solidus and the liquidus touch at T_0, c_0 (see Fig. 10.3a), and thus

$$\frac{dT}{dc_l} = \frac{dT}{dc_s} = 0 \quad , \quad c_l = c_s = c_0, \quad T = T_0 . \tag{10.5}$$

Furthermore, at the extremum we find $L^l = L^s = L^0$ [see (10.2) and (10.3)], so that by differentiating (10.1),

$$\left(\frac{d^2T}{dc_l^2} \Big/ \frac{d^2T}{dc_s^2}\right) = \left(\frac{S_{cc}^s(0)}{S_{cc}^l(0)}\right)^2 . \tag{10.6}$$

The latent heat L^0 is usually positive and in that case a maximum in the phase diagram occurs for $d^2T/dc_l^2 > 0$ and a minimum for $d^2T/dc_s^2 < 0$. This implies [via (10.6)] that for a melting maximum $S_{cc}^s(0) < S_{cc}^l(0)$ [or, recalling (9.14), $\partial^2 \Delta G^s/\partial c_s^2 > \partial^2 \Delta G^l/\partial c_l^2$], while for a melting minimum $S_{cc}^s(0) > S_{cc}^l(0)$ [or $\partial \Delta G^s/\partial c_s^2 < \partial^2 \Delta G^l/\partial c_l^2$]. The inequalities for the Gibbs

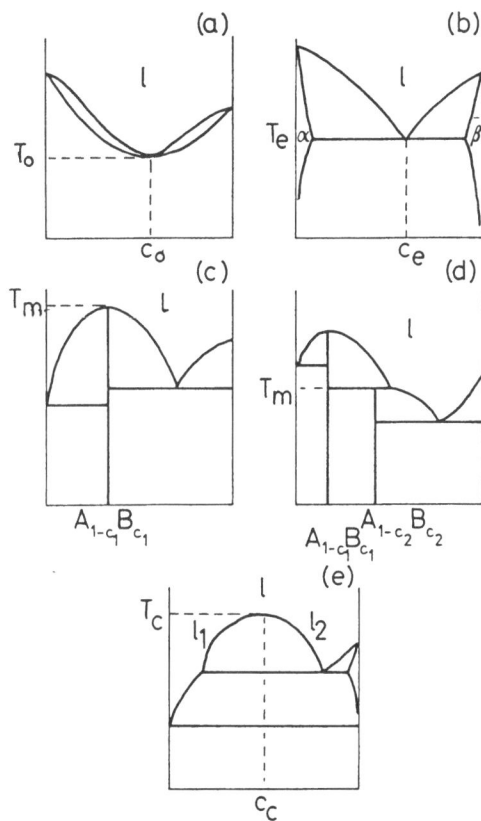

Fig. 10.3a–e. Simple phase diagrams with (**a**) a melting minimum, (**b**) a eutectic point, (**c**) a congruently, (**d**) an incongruently melting compound, and (**e**) phase separation in the liquid state

free energies of formation are immediately evident from the graphical phase diagram construction. The phase diagram has a melting maximum (minimum) if the curvature of the ΔG curve is larger (smaller) for the solid than for the liquid phase. For a small region around the melting extremum, one finds that

$$\frac{c_0 - c_s}{c_0 - c_l} = \pm \frac{S_{cc}^s(0)}{S_{cc}^l(0)} \quad . \tag{10.7}$$

Thus we find that the existence of a melting extremum has clear structural implications: if a melting minimum exists in a system with a tendency towards clustering [i.e. $S_{cc}(0) > S_{cc}^{id}(0)$], then this tendency must be expressed more in the solid than in the liquid phase. If we consider a system with a tendency towards hetero-coordination [i.e. $S_{cc}(0) < S_{cc}^{id}(0)$], then the concentration fluctuation must be more pronounced in the liquid phase for a melting minimum to exist. For the K-Cs system shown in Fig. 10.2 [here $S_{cc}(0) > S_{cc}^{id}$] this means that the calculation underestimates the concentration fluctuations in the solid solution. The liquid is adequately modelled by a hard-sphere mixture, (see Sect. 9.1) – to get a deeper melting minimum in agreement with experiment, one must allow for a short-range order in and/or a static distortion of the solid phase.

10.2.2 Eutectic Diagrams

In a eutectic system we consider the melt in equilibrium with a solid phase α on the A-rich side, and with a solid phase β on the B-rich side[2]. Thus we have two pairs of differential equations describing the solidus and liquidus lines for the $l \to \alpha$ and $l \to \beta$ phase equilibria (see Fig. 10.3b), the eutectic point being determined by the intersection of the liquidus lines. If the solid solubility is very small, each pair of differential equations degenerates into a single equation for the liquidus ($c_\alpha = 0$)

$$\frac{dT}{dc_l} = -c_l \frac{RT^2}{S^l_{cc}(0)L^A} \tag{10.8}$$

for the $l \to \alpha$ and similarly for the $l \to \beta$ equilibrium. This shows that a steep liquidus (and hence a deep eutectic) depends on (a) large concentration fluctuations in the liquid [small $S^l_{cc}(0)$] and (b) a small latent heat. The first point explains why our calculation of the Ca-Mg phase diagram underestimates the depth of the eutectic: a mixture of hard spheres as a reference system excludes all concentration fluctuations other than those arising from size differences (note that *Sommer*'s [10.14] empirical analysis of the thermodynamic data points to a small but distinct tendency to heterocoordination). We also slightly underestimate the heat of formation of the liquid and overestimate that of the solid compound and hence the computed latent heat is somewhat too large. The way to an improved phase diagram calculation proceeds via an improved modelling of the liquid state.

In analogy to (10.4), the slope of the liquidus lines at constant T is now given by

$$R(T) = \frac{(c_{e>} - c_\beta)}{(c_{e<} - c_\alpha)} \frac{S^{l<}_{cc}(0)L^\alpha}{S^{l>}_{cc}(0)L^\beta} \quad ; \tag{10.9}$$

(the subscript e> refers to a concentration of the melt that is larger than the eutetic composition c_e). At the eutectic composition, the concentration fluctuation term drops out, and $R(T)$ is related to the latent heats L^α, L^β alone

$$R_e(T) = \frac{(c_e - c_\beta)}{(c_e - c_\alpha)} \frac{L^\alpha}{L^\beta} \quad . \tag{10.10}$$

10.2.3 Compound Formation

For a congruently melting compound with a very narrow existence range (see Fig. 10.3c), we again have a single equation for the slope of the liquidus

---

[2] A and B can be either a pure metal or a stoichiometric intermetallic compound.

line

$$\frac{dT}{dc_l} = -(c_l - c_c)\frac{RT^2}{S_{cc}^l(0,T)L_c(T)} \quad . \tag{10.11}$$

If the latent heat $L_c(T)$ of the compound is only weakly temperature dependent, then the form of the liquidus line is determined primarily by the temperature and concentration dependence of the concentration fluctuations.

For an incongruently melting compound $A_{1-c_2}B_{c_2}$ which decomposes peritectically at a temperature T_m into a solid compound $A_{1-c_1}B_{c_1}$ and a liquid with composition c_l (see Fig. 10.3d) the ratio of the slopes of the liquidi above and below T_m is found to depend only on the latent heats of the solid compounds

$$R_i(T_m) = \frac{(c_l - c_1)}{(c_l - c_2)}\frac{L_{c_2}}{L_{c_1}} \quad . \tag{10.12}$$

10.2.4 Phase Separation in the Liquid State

At the critical point for a phase separation in the liquid state, one has $dT/dc_{l_1} = dT/dc_{l_2} = 0$ where c_{l_1} and c_{l_2} refer to the phase boundaries at the left and at the right of the extremum of the miscibility gap (see Fig. 10.3e). This case has already been discussed in some detail in Sect. 9.2.3. More generally, a flat liquidus curve $(dT/dc \simeq 0)$ over a larger concentration region is usually due to a large $S_{cc}(0)$ and hence indicates a tendency to phase separation in the liquid [10.10]. A strong change in slope and curvature of the liquidus curve relates to a strong concentration dependence of the ordering potential. This was pointed out by *Geertsma* [10.13] on the basis of an analysis of the Rb-Tl and Cs-Tl phase diagrams, and is in agreement with EMF data of *Tamaki* et al. [10.15] on Na-based alloys with group III metals, and also agrees with preliminary results on pseudopotential-derived interatomic forces [10.16].

10.3 Molecular Theory of the Freezing of Liquid Alloys

The density-functional approach to the freezing transition described in Sect. 5.2 has been generalized by *Barrat* et al. [10.17,18] to binary mixtures. They used Baus's [10.19] real-space version of the theory based on the grand canonical ensemble and applied it to mixtures of hard spheres of different diameters. Formally, the generalization of the theory represents no great difficulty; equation (5.12) is now replaced by

$$\ln\left(\frac{n_{i,S}(\boldsymbol{R}_1)}{n_{i,L}}\right) = \int d^3R_2 \sum_{j=A,B} c_{ij}^{(2)}(\boldsymbol{R}_1,\boldsymbol{R}_2)\Delta n_{j,S}(\boldsymbol{R}_2) \quad , \quad i = A, B \quad , \tag{10.13}$$

and correspondingly for the grand potential difference $\Delta\Omega$ [see (5.13)]. For a random substitutional alloy a real-space expansion of the partial number density $n_{i,S}(\boldsymbol{R})$ of the solid phase

$$n_{i,S}(\boldsymbol{R}) = c_i \left(\frac{\alpha_i}{\pi}\right)^{3/2} \sum_l \exp[-\alpha_i(\boldsymbol{R}-\boldsymbol{R}_l)^2] \tag{10.14}$$

is certainly the simplest possible approach [cf. (5.16)]. In this case the solution of the phase transition problem reduces to a search for non-trivial (i.e. $\alpha_A, \alpha_B \neq 0$) minima in the free energy of the alloy.

Barrat et al. deal with hard-sphere mixtures in the Percus-Yevick approximation [see Chap. 9 and Appendix D.3]. Since the temperature cancels out for hard spheres, the free energy depends only on the packing fraction η, the size ratio $\alpha = \sigma_B/\sigma_A$ ($\sigma_B < \sigma_A$), and the concentration c_A. At fixed values of these thermodynamic variables, a single minimum corresponding to the fluid phase ($\alpha_A = \alpha_B = 0$) is found at low packing fractions. Above a bifurcation value of $\eta = 0.5$ the free energy of the mixture becomes bistable, the second minimum at $\alpha_A, \alpha_B \neq 0$ corresponding to a mechanically stable crystalline (fcc) phase. This is the case for monatomic systems and for mixtures with a size ratio sufficiently close to unity ($0.85 < \alpha < 1$). At lower values of α it becomes increasingly difficult to find a mechanically stable solid phase. Note that the critical value of the radius ratio predicted by the density-functional approach is precisely that of the empirical Hume-Rothery rule!

The coexistence of the fluid and solid phases at a given temperature requires the equality of the pressures $p_S = p_L$ and of the chemical potentials $\mu_S = \mu_L$. Barrat et al. calculate the free energies of the competing phases by adding p_S/n_S (or p_L/n_L) to the free energies (note that this requires a knowledge of the equation of state of both phases) and determine the concentrations of the coexisting phases from the by now familiar common-tangent construction.

For size ratios in the range $0.94 < \alpha < 1$ a spindle-type phase diagram is found, for $0.92 < \alpha < 0.94$ one finds a phase diagram with a melting minimum and for $\alpha < 0.92$ a eutectic phase diagram (see Fig. 10.4). In the range $0.85 < \alpha < 0.92$ the solubility shrinks gradually to zero, but this happens in an asymmetric way: the solid solubility of the smaller spheres is always much larger in a crystal of the larger spheres than vice-versa. All this is in very satisfactory agreement with the general behaviour of real alloys with a near-ideal mixing behaviour and of rare-gas mixtures.

In a second paper *Barrat* [10.18] applied the density-functional approach to the freezing of a symmetric mixture of charged hard spheres (i.e. $\sigma_A = \sigma_B$ and $c_A Q_A + c_B Q_B = 0$; this is the Coulomb limit of the HSY reference system described in Sect. 9.2.2). In this case the two coupled equations (10.13) decouple into two independent equations, one for the mean number

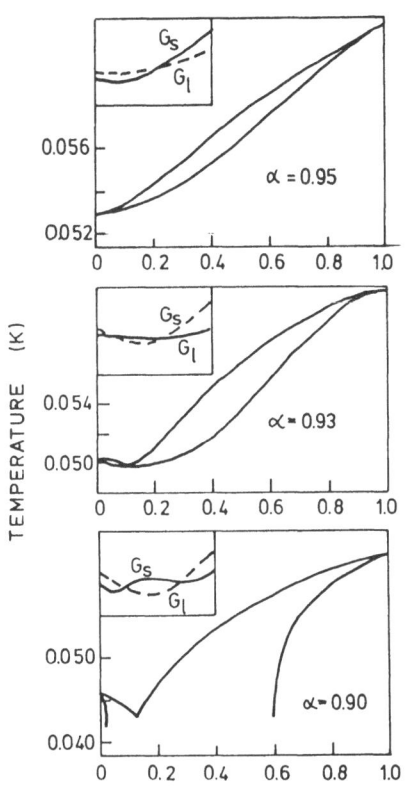

Fig. 10.4. Fluid-solid phase diagram for hard-sphere mixtures with different size ratios $\alpha = \sigma_B/\sigma_A$ under normal pressure, as calculated by *Barrat* et al. [10.14] using density functional theory. The inset gives a schematic representation of the concentration dependence of the Gibbs free energies of the solid and the liquid at a given temperature and pressure

density $n_{N,s}(R) = c_A n_{A,s}(R) + c_{B,s} n_{B,s}(R)$ and one for the charge density $n_{c,s}(R) = n_{A,s}(R) - n_{B,s}(R)$; the direct correlation function $c_{Nc}(R)$ is identically zero in this model. The equation for $n_{N,s}$ is solved in the real-space approach (10.14), whereas the long-range Coulomb term in $c_{cc}(R)$ requires a Fourier-space calculation analogous to (5.14). At low pressures, a 50/50 alloy freezes into an ordered CsCl structure and at higher pressures a phase transition to a random fcc structure takes place.

Thus the results of Barrat et al. provide a very encouraging starting point for investigations of microscopic aspects of the freezing of binary alloys. However, it should not be overlooked that an application to a real alloy has not been presented as yet.

Recent Results

Mbaye et al. [10.20] have presented first-principles calculations of semiconductor alloy phase diagrams, based on selfconsistent electronic band structure calculations.

Podloucky et al. [10.21] have proposed a combined quantum-statistical and electronic structure approach for the calculation of the solid-state part of alloy phase diagrams and an application to Li-Al alloys.

11. Beyond the Phase Diagram: The Formation and Properties of Metastable Phases

In recent years the investigation of metastable intermetallic phases (crystalline [11.1], quasi-crystalline[1] [11.2] and amorphous [11.3]) has attracted considerable attention, both because of the potential technological importance of some of these materials and because of the fundamental scientific interest in the influence of disorder on the properties of solids. The methods discussed in this book may also serve for an investigation of the formation and of the properties of these metastable phases. From the point of view of a calculation of the electronic ground-state energy there is no essential difference between a stable and a metastable crystalline compound, but new aspects arise in the study of non-crystalline solids – these will be briefly addressed below.

11.1 Amorphous Alloys – Metallic Glasses

Metastable amorphous or glassy[2] metallic alloys can now be produced in at least three different ways: (i) by quenching from the high-temperature melt, (ii) by co-deposition from the vapour phase, and according to a recent discovery [11.4] (iii) by solid state reaction. The ability of a given metal or alloy to form an amorphous phase will depend on thermodynamic and kinetic criteria. The main thermodynamic criterion is the thermodynamic driving potential for crystallization, i.e. the difference in the free energy of the metastable amorphous and of a stable or metastable crystalline phase (or two-phase mixture). The kinetic criteria describe the ease with which the particles rearrange to form a new phase. The relative importance of

---

[1] The very recent discovery of quasi-crystalline materials is one of the most exciting results of modern condensed matter physics. A quasi-crystalline solid has a long-range orientational order, but no translational periodicity (the density is a quasi-periodic function). The point-group symmetry is icosahedral (which makes it incompatible with translational symmetry) and results in a diffraction pattern with sharp diffraction spots with five or tenfold symmetry in certain directions.

[2] Conventionally the term "glass" is reserved for an amorphous material produced by quenching from the melt. However, since there is no essential difference in the properties of amorphous materials produced by melt-quenching, vapour-deposition, or solid state reaction, we shall treat "amorphous alloys" and "metallic glasses" as synonymous expressions for the same class of materials.

the thermodynamic and kinetic criteria will depend considerably upon the method of production [11.5]. The following section discusses the possiblity of predicting the thermodynamic and perhaps even some of the kinetic criteria.

11.1.1 Glass-Forming Ability

If we compare the thermodynamic properties of the crystalline, liquid and amorphous phases of a material we find that at low temperatures the glass has an internal energy, entropy, volume and specific heat which are usually a few percent higher than the corresponding crystalline values. On raising the temperature, we find that at a certain temperature T_g the slopes of the energy, entropy and volume vs temperature curves change abruptly and that there is a jump discontinuity in the specific heat and in the thermal expansion coefficient (see Fig. 11.1). These now have values comparable to those extrapolated from data on the supercooled melt. Thus T_g – the glass transition temperature – describes the point at which the glass becomes unstable against the supercooled liquid state. The supercooled melt above T_g exists only in a narrow temperature interval (at most up to 20 K), then it crystallizes at a temperature T_c, either directly into the stable phase or via some intermediate metastable states (sometimes T_c may even be lower than T_g, in that case no glass transition is observed).

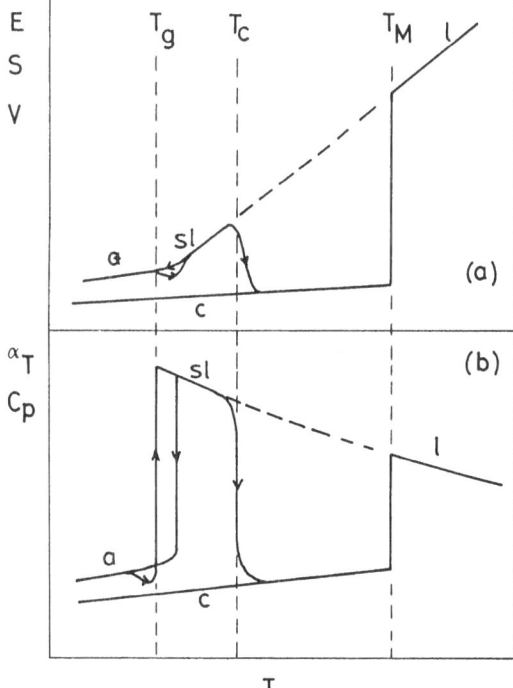

Fig. 11.1a,b. Temperature dependence of the internal energy E, the entropy S and the specific volume V (**a**), and of the specific heat c_V and the thermal expansion coefficient α_p (**b**) in the crystalline, amorphous, and liquid states. T_M is the melting temperature, T_g the glass-transition temperature and T_c the crystallization temperature (schematically)

Thus upon cooling the primary requirement for glass formation is that crystallization from the melt is suppressed until the melt has been cooled below T_g (or T_c, if $T_g > T_c$). It is immediately clear that the narrower the interval between the liquidus temperature T_l and the glass temperature T_g, the easier it will be to bring a system into an amorphous state. This qualitative statement can be quantified within the Davies-Uhlmann theory of crystallization [11.6]: assuming homogeneous nucleation and crystal growth, time-temperature-transformation (TTT) diagrams for the crystallization of a certain fraction of the material may be constructed and the critical cooling rate estimated, with the ratio $T_{rg} = T_g/T_l$ – the reduced glass temperature – as the most important parameter. The higher T_{rg}, the better is the expected glass-forming ability.

The liquidus line $T_l(c)$ summarizes the information on the equilibrium thermodynamics and may be calculated from first principles as discussed in the previous chapter. An important point in the present context is that we find a minimum in the Gibbs free energy G_l of the liquid phase relative to the free energy G_s of the competing solid phase (or two-phase mixture) at the eutectic minimum – this usually being the region of easy glass formation. This argument was extended by *Sommer* [11.7] who used a more empirical analysis of the available thermodynamic data to show that in a variety of metallic glasses there is indeed a minimum in the driving potential for crystallization of the supercooled liquid in the glass-forming concentration region.

A further important thermodynamic aspect of the glass-forming ability is connected with the concept of the $T_0^x(c)$ temperatures (see Sect. 10.1). Within the regions limited by the $T_0^x(c)$ curves of the stable and metastable crystalline phases, a composition-invariant crystallization is thermodynamically possible and we can expect it to take precedence over glass formation during rapid solidification. Thus the concentrations at which the $T_0^x(c)$ curves intersect the $T_g(c)$ curve represent boundaries to the glass-forming compositions. In the case of the Ca-Mg system we find that although the eutectic minimum is not particularly deep, the T_0^x curves of the solid compounds never cross (see Fig. 11.2); this is a very important point for a good glass-forming ability (for further discussion of this point see [11.9]).

A theoretical estimate of T_g requires a precise understanding of the thermodynamic character of the glass transition. Unlike any equilibrium phase transition, the glass transition is dominated by kinetic effects and represents a transition from a ergodic to a non-ergodic behaviour. Below T_g certain diffusive and relaxational degrees of freedom of the liquid are blocked; this results in the sudden reduction of the specific heat, thermal expansion coefficient, and compressibility. Although attempts have been made to extend the formulation of the Gibbs ensemble to the amorphous state by a consideration of ensemble theory in a restricted phase space [11.10], a full statistical-mechanical treatment of the glass transition is still lacking. In the

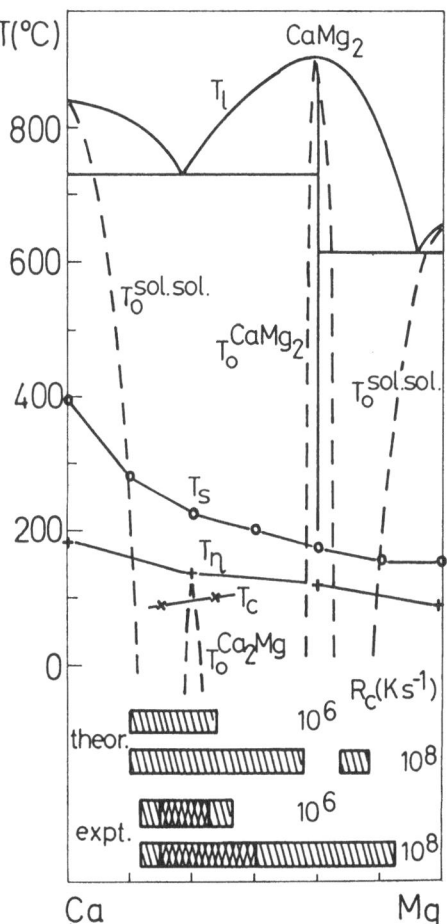

Fig. 11.2. Concentration dependence of T_l, T_0^x, T_s and T_η for Ca-Mg alloys, together with the calculated and observed glass-forming composition ranges for cooling rates of $R_c = 10^6\,\mathrm{Ks^{-1}}$ and $R_c = 10^8\,\mathrm{Ks^{-1}}$ (the cross-hatched area indicates that complete glass formation has been found, a hatched area stands for only partial glass formation). After [11.8], experimental results after [11.7]

absence of such a theory, an estimate of T_g must be based on a quantification of the difference between liquid and glass. The most simple and commonly accepted definition of a glass is to consider the glassy state as an extension of the liquid state for which the viscosity η exceeds 10^{13} Poise. The strong decrease of the fluidity $\Phi = \eta^{-1}$ near the glass transition must be accompanied by a corresponding decrease in the configurational entropy. It was first pointed out by *Kauzmann* [11.11] that extrapolations of the thermodynamic properties of the melt suggest that the difference in entropy between the liquid and the crystal vanishes at a temperature T_s somewhat below the observed T_g. Kauzmann assumed that T_g would approach T_s in the limit of low cooling or heating rates. T_s may be derived straightforwardly in a thermodynamic variational calculation of the crystalline and amorphous phases [11.8]. This is shown in Fig. 11.3 for some Ca-Mg alloys. That this is a reasonable estimate of T_g is confirmed by analyses of experimental thermodynamic data for many systems [11.7, 12] and by computer simulations on hard-sphere systems [11.13].

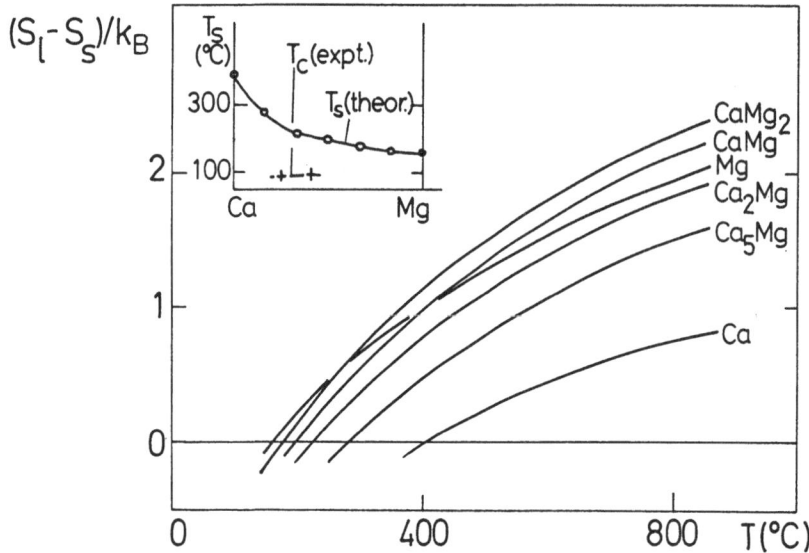

Fig. 11.3. Entropy difference between the liquid phase and the crystalline phase, for pure Ca and Mg, the Laves phase $CaMg_2$ and two-phase mixtures $(Ca+CaMg_2)$ of various compositions. The concentration dependence of the vanishing excess-entropy temperatures T_s is shown in the inset together with the experimentally determined crystallization temperatures T_c. After [11.8]

The temperature dependence of the viscosity can also be used for an estimate of T_g. Within the free-volume model [11.14] it is given by the simple Doolittle equation $\eta = \eta_0 \exp(V/\overline{V}_f)$ where V is the atomic volume and $\overline{V}_f$ the average free volume per atom of the liquid. $\overline{V}_f$ as a function of temperature can be derived from the thermodynamic variational approach which uses a hard-sphere refrence system: $\overline{V}_f$ is set equal to the difference between the atomic volume, and the atomic volume V_{cp} at random close packing, i.e.. $\overline{V}_f = V - V_{cp}$. This is shown in Fig. 11.4 for some Ca-Mg alloys. The viscosity prefactor η_0 may be derived from a calculation of the bulk viscosity near the melting point. Finally the temperature T_η defined

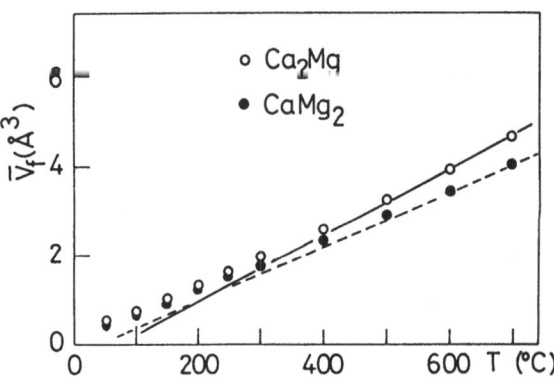

Fig. 11.4. Temperature dependence of the average free volume $\overline{V}_f$ for Ca_2Mg and $CaMg_2$. The straight lines mark the linear temperature dependence of $\overline{V}_f$ observed at higher temperatures [11.8]

by $\eta(T_\eta) = 10^{13}$ Poise is the estimate for the glass temperature in the free volume-model (for details see [11.8]). The variation of T_η and T_s as a function of concentration is also displayed in Fig. 11.2. We find a reasonable consistency of T_s and T_η (which is of course necessary since both the decrease of the entropy of the liquid and the increase in its viscosity are related to the decrease of the average free volume) and a good agreement with the experimentally observed crystallization temperature.

The estimated T_g is only weakly composition dependent and hence the composition dependence of the reduced glass temperature $T_{rg} = T_g/T_l$, is dominated by the variation of T_l. *Davies* et al. [11.6, 15] have shown that the critical cooling rate R_c as a function of the reduced glass temperature falls on a single master curve for many glass-forming systems: R_c is a monotonically and steeply decreasing function of T_{rg}. Using Davies' curve for $R_c = R_c(T_{rg})$ one can obtain the concentration dependence of the critical cooling rates which show a clear maximum with $R_c \lesssim 10^8 \, \text{Ks}^{-1}$ at the composition of the Laves phase and a minimum of $R_c \lesssim 10^6 \, \text{Ks}^{-1}$ in the region of the Ca-rich eutectic. This may be used to determine the concentrations at which glass formation may be expected at a given cooling rate: The limits of the glass-forming region are given either by the concentration at which the estimated cooling rate exceeds the specified value or by the concentrations at which the $T_g(c)$ curve intersects a $T_0^x(c)$ curve. These estimates are shown in Fig. 11.2 together with the experimentally determined glass-forming ranges.

The important thing is not so much that the prediction agrees well with experiment, but that these considerations demonstrate that at least for alloys with relatively weak chemical interactions (and hence rather temperature-insensitive thermodynamic excess functions) the energetics of the competing alloy phases is the determining factor for glass formation. We may expect this to hold also for alloys with more pronounced chemical interactions and hence more strongly temperature- and concentration-dependent thermodynamic functions. In that case we would merely expect a more strongly concentration-dependent T_g.

If an alloy that in equilibrium contains a multi-phase structure is co-evaporated at sufficiently low substrate temperatures, the low mobility of the fully intermixed atoms constrains the film to a single-phase structure. Therefore the atomic structure should reflect the energetically most favourable single-phase structure at the temperature of the substrate. Similarly, if the alloy is formed by homogenization of elemental layered composites at temperatures below T_g, it is clear that the system must lower its energy on forming an amorphous single-phase structure. Therefore in both cases the free energy versus composition diagrams contain the most relevant information on the formation of the glassy phase. The kinetic criteria now concern the formation (nucleation and growth) of a crystalline phase from the amorphous phase rather then from the melt and will be distinctly different from those for liquid quenching. Generally they will be of much lesser importance. For a recent comparative study of thermodynamic and kinetic criteria for liquid quenching, vapour quenching, and solid state reaction, see e.g. [11.5].

11.1.2 Atomic Structure of Metallic Glasses

Our discussions of the glass-forming ability have been based on the relative thermodynamic properties of the supercooled melt and of the crystalline phases – so far without any regard to the actual structure of the amorphous phase. In the liquid phase, equilibrium statistical mechanics can be used to derive certain approximate integral equations relating the pair-correlation functions to the pair potentials. In the amorphous phase we are far from equilibrium and we have to use the various algorithms for building computer-generated structural models. Quite generally any such algorithm consists of two distinct steps: (i) construction of an initial "starting" structure, (ii) refinement of this model by energy minimization, molecular dynamics or Monte Carlo techniques. (A detailed mathematical description of these modelling algorithms goes beyond the scope of this book.)

Recently there has been a lot of discussion as to whether step (i) should be executed using the classical dense-random-packing of hard-sphere (DRPHS) algorithms [11.16], or by a random packing of larger "molecular units" or "associates" [11.17]. However, whether this difference is important at all, depends on the way in which step (ii) is executed. If a static energy minimization routine [11.18] is used, the starting structure and the final relaxed structure are related through a steepest-gradient path on the energy hypersurface in the multidimensional configuration space. So it is mainly the starting structure that determines the final result.

A dynamical relaxation using molecular dynamics [11.19, 20] or Monte Carlo simulation [11.21] would yield in principle the equilibrium configuration corresponding to the chosen set of interatomic potentials and thermodynamic variables so that the starting configuration would be ultimately irrelevant. Hence the relaxation has to be constrained in some way to yield a metastable amorphous phase. In all cases the final result depends in a rather complex way on the starting structure, the interatomic potentials, the compatibility (or incompatibility!) of the two, and on the thermal history during the computer experiment. Here we have to remember that even on very fast computers, the lowest quench rates attainable (10^{14}–10^{12} Ks^{-1}) are still orders of magnitude larger then the highest quench rates ($\gtrsim 10^8$ Ks^{-1}) achievable in a laboratory. An interesting variant of the molecular dynamics quench experiments has recently been proposed by *Stillinger* and *Weber* [11.20]: at several points of a high-temperature molecular dynamics run, the momenta of all particles are set instantaneously to zero and the system is relaxed statically – i.e. its instantaneous configuration is projected onto a nearby potential energy minimum.

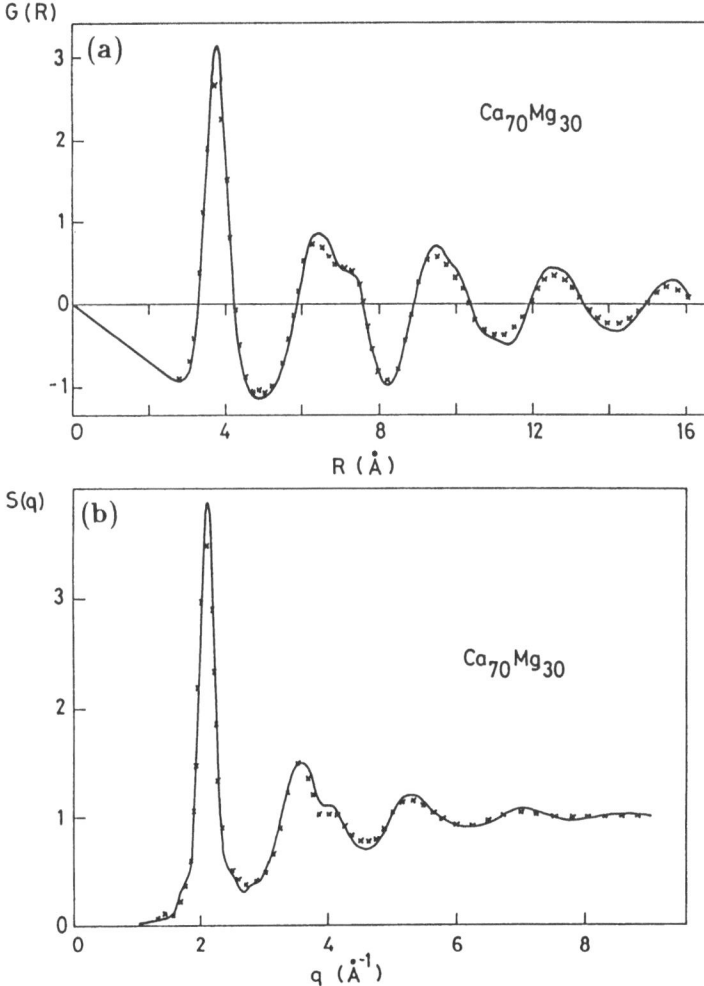

Fig. 11.5a,b. Composite (x-ray and Debye-Waller factor weighted) reduced pair-correlation function $G(R)$ (a) and static structure factor $S(q)$ (b), for a $Ca_{70}Mg_{30}$ glass. (——) theory [11.22, 23]; (×) experiment [11.24]

For simple-metal glasses with weak chemical interactions such as Ca-Mg, a static energy minimization of a DRPHS structure yields results in excellent agreement with experiment (Fig. 11.5). Systems in which we expect a non-negligible chemical short-range order are more difficult to handle and indeed a static relaxation calculation will never produce a chemical short-range order that was not already present in the starting structure (it only searches for the local energy minimum related to the initial structure via a steepest-gradient path). Attempts have been made to equilibrate the system with respect to CSRO in the liquid phase using Monte Carlo [11.25] or molecular dynamics techniques [11.23] – with good success for systems such

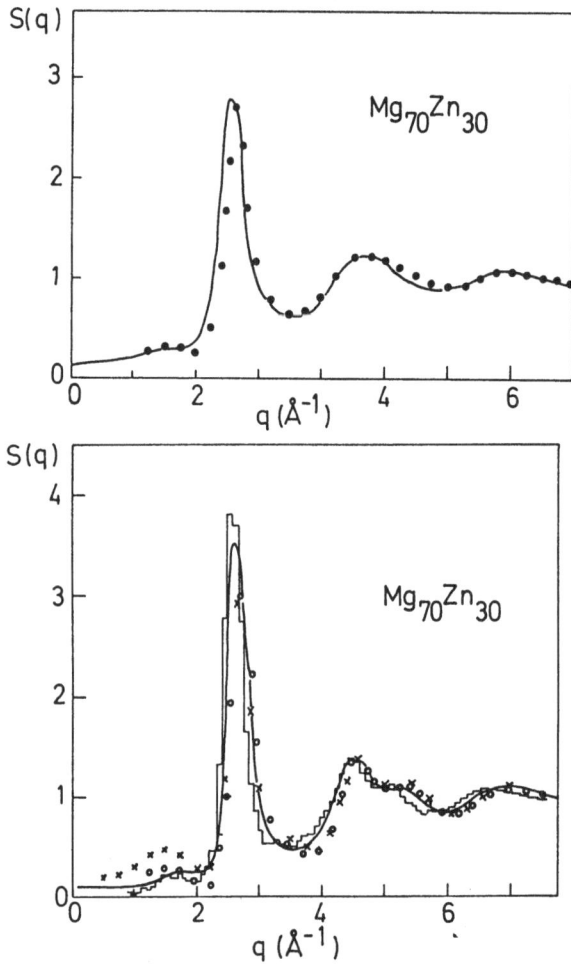

Fig. 11.6. Composite (x-ray weighted) static structure factor $S(q)$ of liquid (*top*, $T = 700$ K) and amorphous (*bottom*, room temperature) $Mg_{70}Zn_{30}$. The small hump at the left hand side of the main peak indicates the presence of a certain degree of chemical short-range order. Note that because of the small difference in the x-ray scattering form factors the concentration-fluctuation term contributes only about 6% of the total scattering. (*Histogram*) molecular-dynamics simulation and quench; (——) thermodynamic variational calculation for liquid and supercooled liquid phase (after [11.23]); (o and ×) experiment [11.26, 27]

as Mg-Zn where we know the pseudopotential-derived ordering potential to be reasonably accurate (Fig. 11.6). These calculations also demonstrate that the short-range order present in the glass is essentially identical to that found in a thermodynamic variational calculation for a supercooled liquid just above the glass transition. The tight-binding calculations for transition-metal alloys that have been briefly sketched in Sect. 9.4 indicate that the same conclusion holds for these materials too, so that we seem to have a reasonable understanding of the structure of amorphous simple-metal and transition-metal alloys and its electronic origin.

A more difficult problem is presented by the transition-metal-metalloid glasses of the type (Fe,Ni)-(P,B), Pd-Si etc. Here there is good experimental evidence [11.17] that besides a pronounced CSRO these glasses possess a certain short- and medium-range topological order similar to that in cementite-type intermetallic compounds. *Gaskell* [11.17] proposed the use of the fundamental steric unit of these crystal structures (trigonal prisms of transition-metal atoms centred by metalloids) as the building block of a random-packing model. According to *Fujiwara* et al. [11.28, 29] it is possible to base the model on a chemically ordered binary DRPHS structure, provided that the interatomic potential has a special form (see Fig. 11.7). A more fundamental approach could perhaps be based on the electronic structure considerations used by *Pettifor* and *Podloucky* [11.30] to successfully explore the crystal structures of the *p-d* compounds.

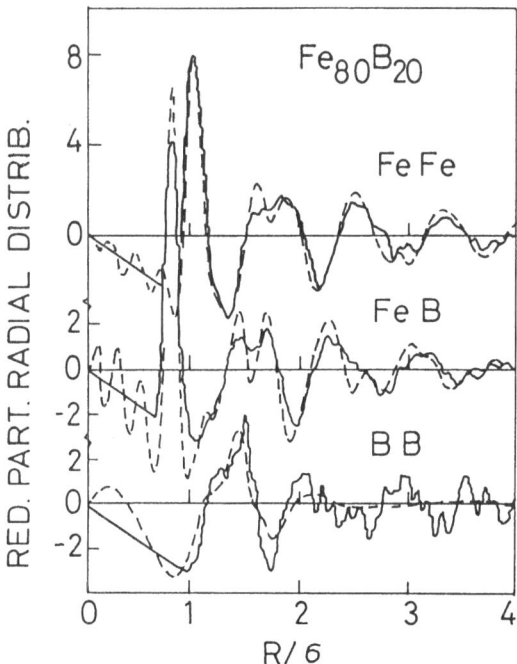

Fig. 11.7. Comparison of the experimental partial radial distribution function for amorphous $Fe_{80}B_{20}$ (- - -) with the model structure. The model is based on truncated Morse potentials. The relative depths of the three potentials are $\varepsilon_{Fe\text{-}Fe}$: $\varepsilon_{Fe\text{-}B}$: $\varepsilon_{B\text{-}B}$ = 1 : 1.07 : 0.047, the B-B potential has its first minimum near $1.3\,\sigma$ (where σ is the mean atomic diameter), so that direct B-B contacts are effectively suppressed. After [11.29]

We would also like to mention some interesting correlations between the crystalline and amorphous structures and the glass-forming ability. These are (i) the relation between the amorphous structure and the liquid structure; (ii) the relation between the amorphous structure and the structure of certain crystalline intermetallic compounds; and (iii) a relation between the atomic and Othe electronic structure of glass-forming alloys.

The first point has already been mentioned: generally we find a distinct similarity between the atomic arrangements in the liquid and in the amorphous states, both in the chemical [11.23–27] and in the topological short-range order [11.31]. This correlation may be explored using the Stillinger-Weber molecular dynamics "quench" procedure [11.20]: during the course of molecular dynamics runs, instantaneous atomic configurations are periodically mapped onto nearby potential energy minima (cf. above). It turns out that the pair-correlation functions and the distribution (in depth) of the potential energy minima are essentially independent of the thermodynamic conditions before the quench and that the mapping produces a significant "image enhancement" with respect to the short-range order present in the system (Fig. 11.8). This observation verifies the existence of a temperature-independent inherent structure for the thermodynamically stable liquid region of the metal or alloy, and we find that this inherent structure is very

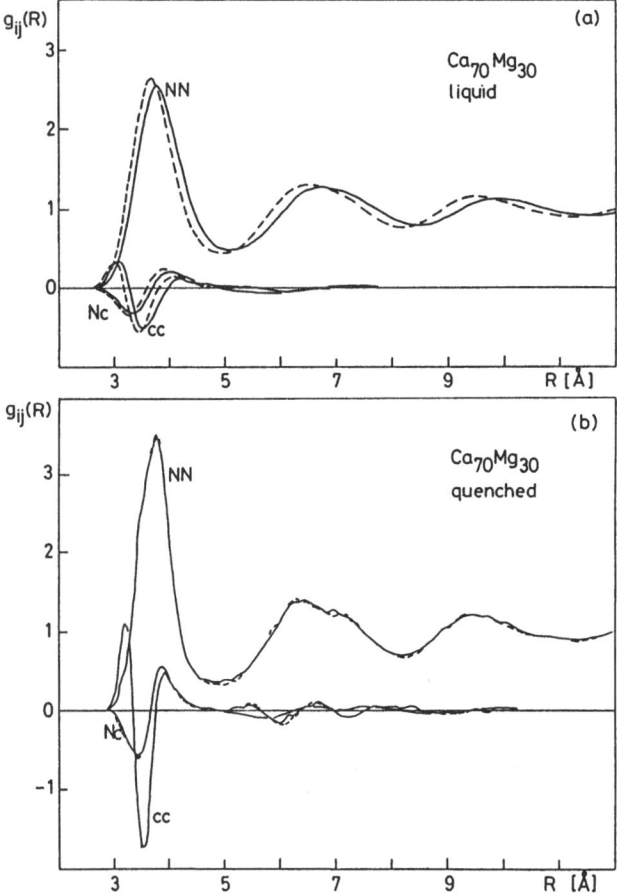

Fig. 11.8a,b. Partial pair-correlation functions $g_{NN}(R)$, $g_{Nc}(R)$ and $g_{cc}(R)$ of $Ca_{70}Mg_{30}$. (a) at $T = 1073$ K; (——) equilibrium density of the liquid; (- - -) at the density of the low-temperature glass. (b) Quenched configurations; full lines: obtained by simultaneous compression from the equilibrium configuration of the liquid; broken lines: isochoric quench from a compressed high-temperature configuration. Unpublished molecular-dynamics calculations of Hafner

close to the structure of an ideal, well relaxed amorphous substance. However, we must bear in mind that a real system produced by a slower quench might be different: in that case the results before the mapping would reflect the incipient non-ergodicity of trapping in special regions of phase space, and hence the thermal history of the sample.

The second point concerns the correlation between the amorphous and the crystalline structures. We have already mentioned that there are strong indications that trigonal prisms might be a fundamental unit in both glassy and crystalline structures of transition-metal-metalloid alloys, but the microscopic mechanism stabilizing this particular geometry is still an open question. For both simple-metal and transition-metal alloys there seems to be a correlation between the formation of topologically close-packed intermetallic compounds and the formation of metallic glasses. This relation may be a complementary one: very stable intermetallic compounds (usually with a Laves or a closely related structure) are found on one side of the phase diagram (where we have a majority concentration of the smaller atoms) with deep eutectic minima and easy glass formation on the other side. Examples are found in the Ca-Mg, Ca-Al, Ca-Cu, Ti-Be, Gd-Co and other systems [11.32, 33]. For other alloys, with smaller size ratios, glasses form in the same composition range as the corresponding Frank-Kasper phase (examples are Nb-Ni, Mn-Si etc.).

A common feature of the crystalline *and* the amorphous structures is that they can both be considered as tetrahedral close packings. This is simply a consequence of the fact that metallic atoms interact by spherically symmetric pair potentials and consequently pack together more or less like spheres, the spheres having their centres on tetrahedral vertices. The first complete coordination shell made up by tetrahedral units would be the icosahedron. In fact it was pointed out many years ago by *Frank* [11.34] that most simple pair potentials lead to a strong preference for icosahedral coordination shells surrounding atoms in both crystalline solids and dense liquids. On a larger scale difficulties arise due to the impossiblity of filling the space with regular tetrahedra. It was shown by *Coxeter* [11.35] that particles with icosahedral coordination shells tile the surface of a four-dimensional sphere, forming a four-dimensional solid called a "polytope". Realistic models are obtained by mapping this four-dimensional object onto the three-dimensional Euclidean space.[3] As pointed out by *Kléman* and *Sadoc* [11.36], defects such as disclinations are necessary to flatten the curved space. Microscopically, the disclination lines correspond to lines of "anomalous" bonds surrounded by four or six tetrahedra instead of the five present in an ideal configuration (see Fig. 11.9). The Frank-Kasper structures [11.38] of intermetallic compounds are an example of ordered arrays of $-72°$ disclination

---

[3] The projection of higher-dimensional regular crystals onto lower-dimensional spaces is also one of the possible techniques for producing "quasi-crystalline structures".

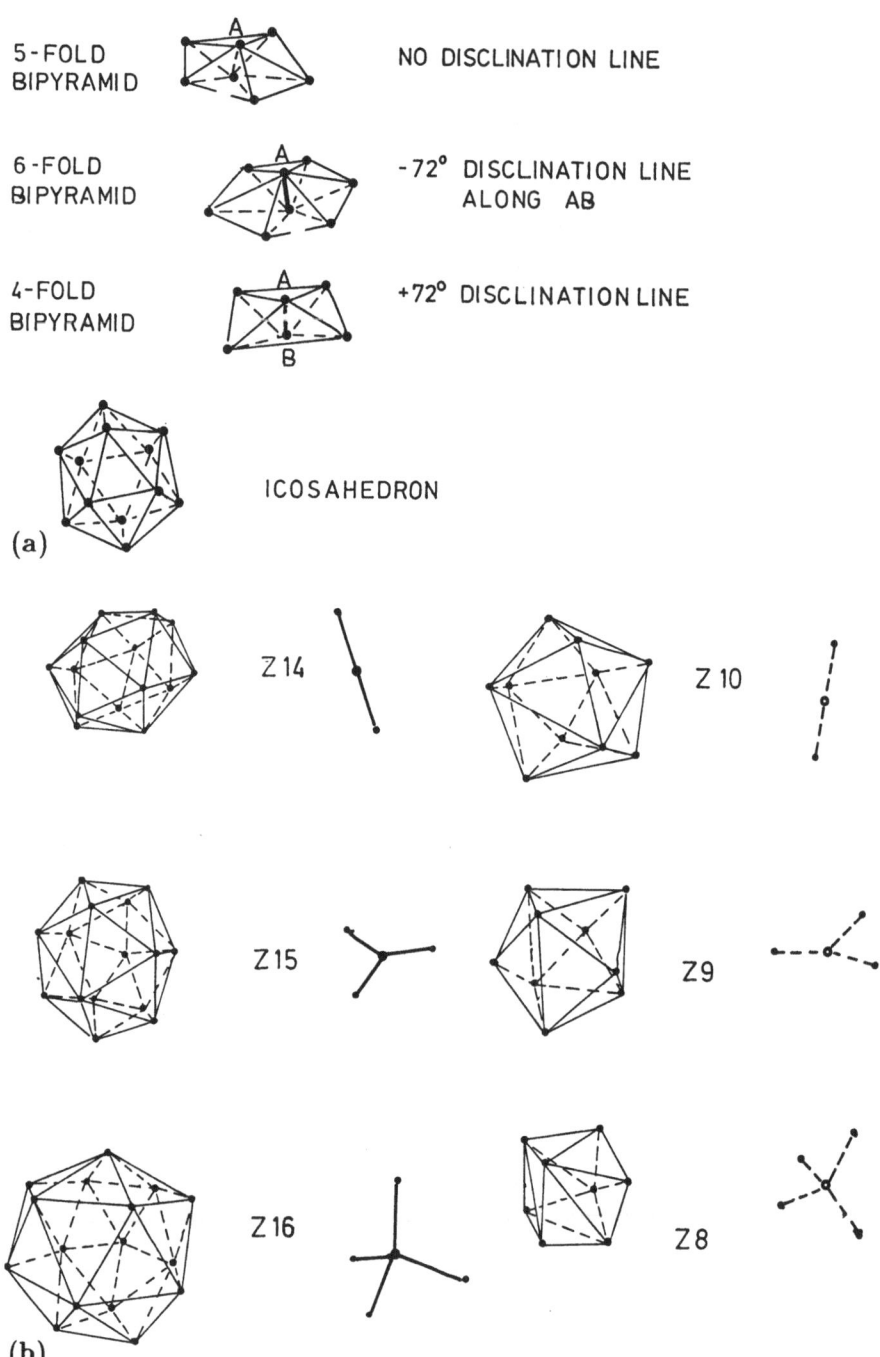

Fig. 11.9. (a) Icosahedron and the microscopic construction for disclinations in a medium composed of tetrahedra. (b) Coordination shells of the canonical Frank-Kasper polyhedra (*left*) and of the canonical Bernal polyhedra (*right*) and their representation as links of ±72° disclination lines. After [11.37]

lines (the curvature mismatch between flat three-dimensional space and the curved space polytope forces an asymmetry into the distribution of positive and negative disclination lines). In these structures the Frank-Kasper polyhedra with coordination numbers $Z = 14, 15$, and 16 alternate with the icosahedron with $Z = 12$. The Frank-Kasper polyhedra may be viewed as links of negative disclination lines and the *Bernal* [11.16, 39] polyhedra with $Z = 9, 10, 11$ as links of positive disclination lines [11.37], see Fig. 11.9(b). Several groups [11.40–42] have proposed to regard the structure of metallic glasses as a disordered, entangled array of disclination lines in an otherwise icosahedral medium. This model is certainly at the same time appealing and realistic. *Sachdev* and *Nelson* [11.42] have shown that the peaks in the static structure factors of metallic glasses may be indexed according to the reciprocal lattice vectors of the curved space icosahedral polytope; *Hafner* [11.43] has shown that the bond-angle distributions of realistic models of metallic glasses point to the presence of a large number of icosahedral sites. Furthermore, the idea provides a rationale for the simultaneous occurrence of tetrahedrally close-packed intermetallic compounds and of glass formation in some many binary systems.

The interest in a possible connection between glass formation and electronic structure was particularly strong in the early days of the research on glassy metallic phases when the glass-forming ability seemed to be rather singular property. A conjecture due to *Nagel* and *Tauc* [11.44] has been widely discussed. They pointed out that in many glassy systems the position Q_p of the first peak of the static structure factor coincides roughly with $2k_F$ (k_F is the Fermi momentum), i.e. $Q_p \simeq 2k_F$. Of course this relation is reminiscent of the Mott-Jones formulation $|Q| = 2k_F$ of the Hume-Rothery rules (Sect. 7.2) where in this case Q is a reciprocal lattice vector, so Nagel and Tauc attempted to extend the classical interpretation of the Hume-Rothery rules to the glassy state. The position of the first peak in the static structure factor, Q_p, acts essentially as a "smeared out" reciprocal lattice vector, and although there are no bands and no Brillouin zones and therefore no band gaps in a glass, the local order (the height of the first peak in $S(q)$ is a rough measure of this order) is strong enough to induce a minimum in the electronic density of states. If the Fermi level falls into this minimum, this would lower the electronic ground-state energy of the glas compared to crystalline structures with anisotropically distributed gaps. The problem is that in its original form, the Nagel-Tauc conjecture has never been really confirmed, except for some recent spectroscopic results [11.45] on noble-metal-polyvalent-metal alloys such as Au-Sn. It has turned out that it is more fruitful to pursue the structural implications of the Nagel-Tauc rule. In a direct-space representation it states that the mean interatomic distance $(d_1 \simeq 2\pi/Q_p)$ is approximately equal to the wavelength of the Friedel oscillations of the interatomic pair potentials $(\lambda_F = 2\pi/2k_F)$, so that the maxima in the pair-distribution function fall into the minima of the pair po-

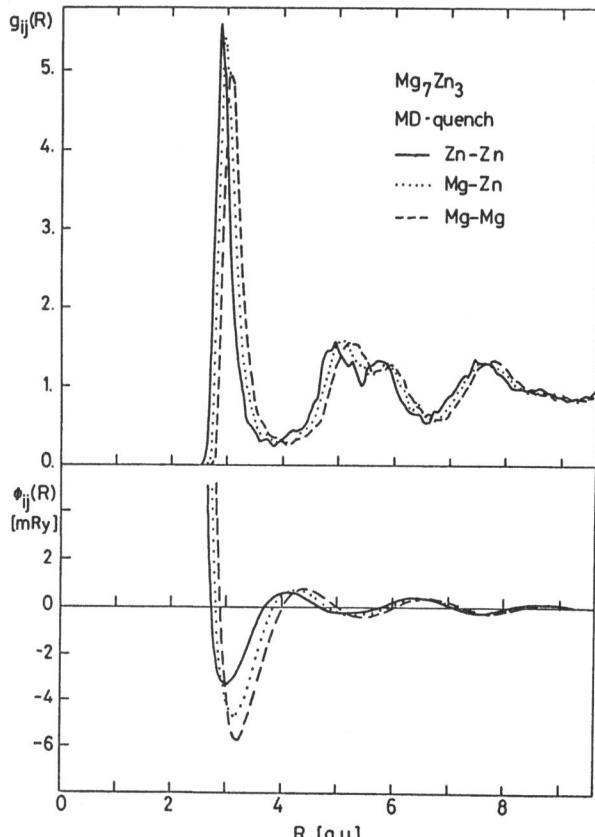

Fig. 11.10. A comparison of the interatomic potentials $\Phi_{ij}(R)$ in a Mg_7Zn_3 alloy and the partial pair correlations $g_{ij}(R)$ of a glassy phase, obtained using the molecular-dynamics quench technique. Note that the maxima of $g_{ij}(R)$ fall into the minima of $\Phi_{ij}(R)$, minimizing the structural contribution to the energy. Similar results have been obtained earlier using static relaxation techniques [11.31, 46]

tential. Indeed it has been shown by model building [11.31, 46, 47] and the analysis of scattering data [11.48, 49] that this correlation exists in many simple-metal glasses and noble-metal-polyvalent-metal glasses (an example is given in Fig. 11.10). In fact the physical background is very similar to that of electronic effects in liquid simple metals discussed in Sect. 4.4. The correlations are particularly striking for the glassy alloys of noble metals with polyvalent metals. Here the structural evidence, supported by extensive electronic structure and transport data suggests that we might consider these alloys as amorphous Hume-Rothery phases [11.49, 50]. The real-space interpretation of the Hume-Rothery rule given in Sect. 7.2 applies directly to this new case as well.

11.1.3 Elementary Excitations in Metallic Glasses

In neither the liquid nor the glassy states, is there any translational symmetry. The glassy state is simpler because it lacks the additional complication of fluidity and therefore it is ideally suited for studying the properties of elementary excitations in disordered materials. As Bloch's theorem is no longer valid, the momentum is no longer a conserved quantity, and the spectral functions $f(q, E)$ of elementary excitations assume a finite width inversely proportional to the lifetime of the excitation. Computationally, the consequence of the absence of translational symmetry is that the diagonalization of the $3N \times 3N$ Hamiltonian matrix (N is the number of particles in the system) can no longer be reduced to a repeated diagonalization of a $3r \times 3r$ matrix in reciprocal space (r being the number of particles in the unit cell of the crystal). This problem can be solved by numerically integrating the equation of motion (i.e. Newton's equation for phonons and the Schrödinger equation for electrons). This can be achieved using an equation of motion [11.51] or molecular dynamics technique, or by using the recursion method [11.52] for the diagonalization of the large real-space Hamiltonian matrix (see also Sect. 1.9). To comment on any of the technical details of these methods would certainly go beyond the scope of this work. Our aim here is only to sketch very briefly the properties of elementary excitations in amorphous solids and their consequences for the thermodynamic and structural properties of glasses.

As examples, we show in Fig. 11.11 the partial spectral functions for longitudinal "phonons" (remember that this term has now to be used with some caution) in the binary glass Mg_7Zn_3 [11.53] and the spectral functions for electrons in a model of amorphous Ni. The phonon calculation is based on pseudopotential-derived interatomic forces, a static cluster relaxation to determine the static structure of the glass, and the recursion technique for the calculation of the vibrational spectral functions. The dynamical structure factor measured in an inelastic neutron scattering experiment is a weighted mean of the partial spectral functions (for details see [11.53]) multiplied by the Bose occupation function. In this case theory and experiment are in very good agreement. The electronic structure calculation [11.54] is based on a dense-random-packing model of the structure and a tight-binding approach which considers both s- and d-electrons. This somewhat academic example has been chosen because it shows the characteristic features very clearly − for examples of electronic structure calculations in real metallic glasses, see e.g. [11.28, 55]. The important results are: (i) Elementary excitations are well defined (i.e. their spectral functions have distinct peaks) not only in the low-q limit, but up to $q \sim Q_p$. For larger momentum transfers the spectral functions approach the incoherent limit, i.e. they become very similar to the corresponding density of states. (ii) The peak positions of the spectral functions define an approximate dispersion relation. In the case of phonons, we find a dispersion maximum near $Q_p/2$ and a dispersion

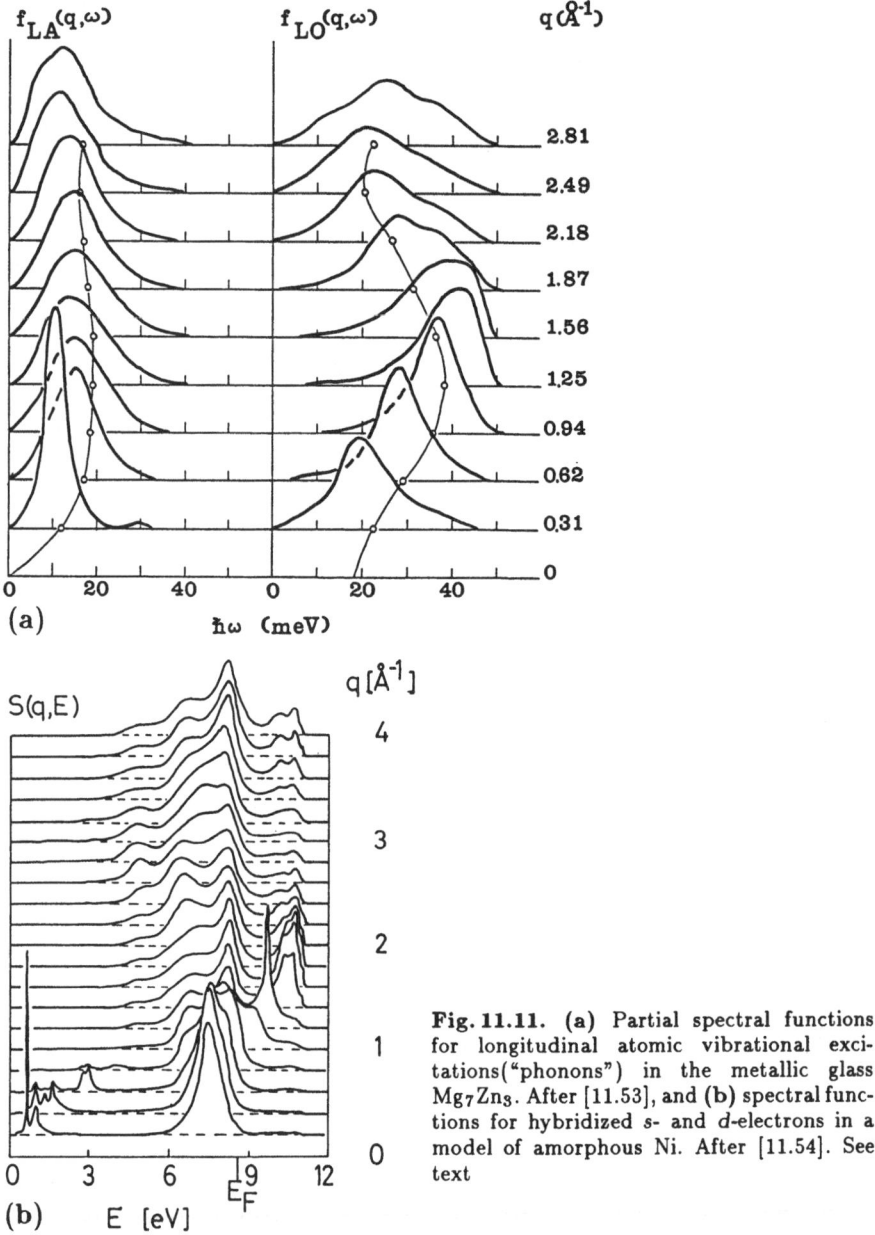

Fig. 11.11. (a) Partial spectral functions for longitudinal atomic vibrational excitations("phonons") in the metallic glass Mg_7Zn_3. After [11.53], and (b) spectral functions for hybridized s- and d-electrons in a model of amorphous Ni. After [11.54]. See text

minimum near Q_p. A similar dispersion minimum is found in the electronic s-band, but in the present case it is somewhat obscured by hybridization with the d-band which itself shows very little dispersion. (iii) The origin of the dispersion minimum is best described as "diffuse umklapp scattering" [11.56], in complete analogy to the crystalline case. The only difference is that the sum over discrete Bragg reflections is replaced by an integral over a continuous static structure factor $S(q)$. Thus the first peak of the static

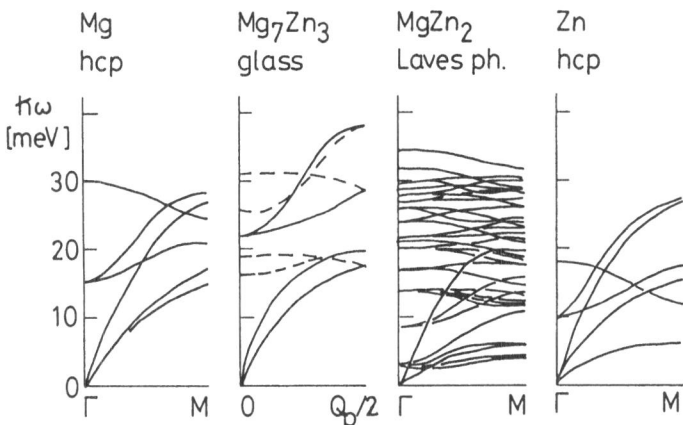

Mg
hcp

Mg₇Zn₃
glass

MgZn₂
Laves ph.

Zn
hcp

Fig. 11.12. Dispersion relations for phonons in crystalline hexagonal close-packed Mg and Zn, in the Laves phase MgZn₂, and in the glass Mg₇Zn₃ (the dispersion relations have been folded pack into the first "pseudo-Brillouin zone", see text). After [11.53]

structure factor really acts like a smeared-out reciprocal lattice vector and we can consider the momentum range $0<q<Q_p/2$ as something like a "first pseudo-Brillouin zone". If we fold the dispersion relations back into the first pseudo-Brillouin zone, we find an astonishing similarity between the glass, the intermetallic compound and the pure component metals (see Fig. 11.12). (iv) In contrast to the crystalline case where the phonon frequency drops to zero at momenta corresponding to reciprocal lattice vectors, the dispersion curve for the glass remains finite at Q_p.

In the vibrational density of states, the finite dispersion minimum contributes an increase of intensity at low frequencies, compared to the corresponding polycrystal. This softening of the low-energy vibrational modes is a general property of many amorphous phases [11.57] and it is reflected in the elastic properties (reduction of the shear constant [11.58]), the thermodynamic properties (increased low-temperature specific heat [11.57,59]), and even in the electron-phonon coupling [11.59,60].

The dispersion minimum in the electronic s-band also has significant manifestations: in this region the group velocity of the electrons becomes negative and this could offer an explanation for the positive Hall coefficient [11.61] and for the positive thermoelectric effect [11.62] found in many amorphous alloys. Its relation to the stability of the amorphous phase (the Nagel-Tauc conjecture) has already been discussed in the previous section.

11.2 Quasi-Crystals

The recent discovery by *Shechtman* et al. [11.2] of aluminum-manganese alloys exhibiting diffraction patterns with both sharp spots and icosahedral symmetry (and thus in striking contradiction to a long-established law of crystallography) was certainly a challenge to theorists. The task was twofold: the first problem consisted in reconciling these two seemingly incompatible facts, the second in explaining the stability of this new type of ordered structure. The answer to the first question was given almost immediately by *Levine* and *Steinhardt* [11.63]. They described a new class of structures based on long-range bond-orientational order, but with quasi-periodic rather then periodic translational order. These structures are generalizations of two-dimensional nonperiodic tilings of a plane [11.64, 65]. One particular "quasi-crystalline structure" studied by Levine and Steinhardt has perfect icosahedral symmetry and consists microscopically of two rhombohedral unit cells (an oblate and a prolate rhombohedron) which are arranged non-periodically, but according to definite mathematical rules. Alternatively, these three-dimensional "Penrose tiles" can also be packed into periodic structures − several complex crystalline intermetallic structures can be described economically in these terms, the rhombohedra being decorated with atoms [11.66]. Both the periodic and the quasi-periodic three-dimensional Penrose tilings can be generated by projections from six-dimensional simple hypercubic lattices [11.67]. A one-dimensional analogue of such a construction is shown in Fig. 11.13. In the meantime many icosahedral phases have been discovered [i-(Al-Mn-Si), i-(Al-Zn-Mg), i-(Al-Cu-Mg), i-(Cu-Cd), i-(Al-Cu-Mg-Li) − see [11.68] for a more complete list and detailed references] and discussed in terms of these ideas.

The second task of explaining the stability of the quasi-crystalline structures was undertaken in a number of investigations [11.69–71]. *Nelson* [11.37, 42, 69] pointed out that the Bragg peaks of an icosahedral crystal can be written as linear combinations of the 12 fundamental lattice vectors pointing to the vertices of an icosahedron. (For a general discussion

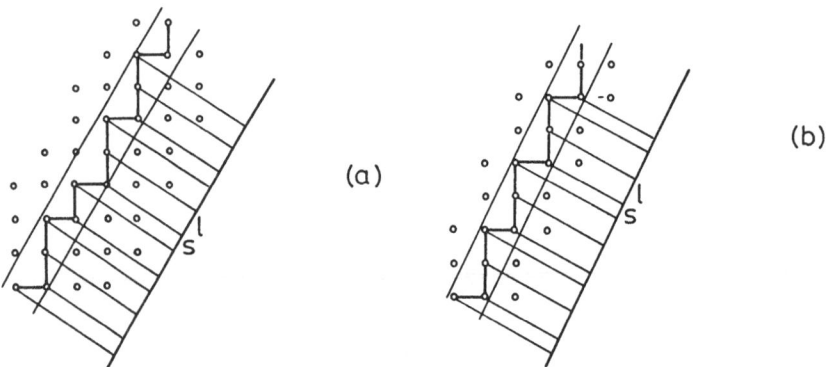

Fig. 11.13a,b. One-dimensional tilings with short and long segments generated by the projection of the edges of a two-dimensional square lattice onto a line with slope t. In **(a)** the slope is $t = \tau = (1 + \sqrt{5})/2$, in **(b)** we have $t = 2/1$. Construction *(a)* generates a one-dimensional analogue of a quasi-periodic Penrose tiling, construction *(b)* a periodic pattern. By choosing t to be a rational continued fraction approximant of τ ($t = 2/1$, $3/2$, $5/3$, ...) one creates structures with a long repeat period which approximate the aperiodic tiling of *(a)* better and better. After [11.66]

(a)

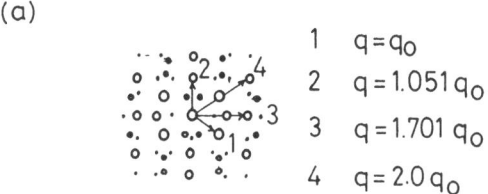

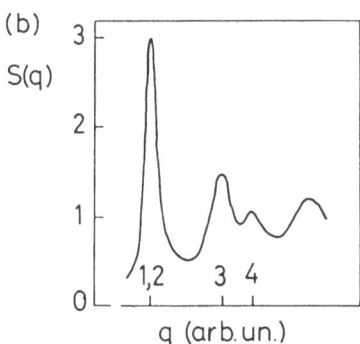

1	$q = q_0$
2	$q = 1.051 q_0$
3	$q = 1.701 q_0$
4	$q = 2.0 q_0$

Fig. 11.14. (a) Bragg peaks in a plane normal to the two-fold symmetry axis of icosahedral Al-Mn. (b) A model amorphous structure factor $S(q)$ indexed according to the Bragg reflections of an icosahedral quasicrystal. After [11.69]

of the problem of indexing icosahedral quasi-crystals, see [11.72].) Figure 11.14 shows 4 elements of this set of 12 which lie in the twofold symmetry plane – they are marked by the large dots. Smaller dots show the positions of "reciprocal lattice vectors" generated by linear combinations of 2, 3 or 4 vectors of this basis set. In terms of the length q_0 of the basis vectors, they are positioned at distances 1.052, 1.701, and 2.0 times q_0 from the centre. Nelson notes that they correlate very well with the positions of the first few peaks in the static structure factors of metallic glasses or quenched-in model liquids. This is just another way of expressing the fact that glasses and supercooled liquids possess a rather high degree of short-range icosahedral bond-orientational order, as had indeed been found in computer experiments on supercooled liquids [11.73]. Thus it is plausible that quasi-crystalline phases can under certain conditions condense out of a supercooled liquid.

This conjecture has been tested using the density-functional theory of freezing presented in Sects. 5.2 and 10.3. For an icosahedral crystal the Fourier expansion for the density of the solid may be written as

$$n_S(\boldsymbol{R}) = n_L(1 + \eta + \sum_n \mu_n \exp(i\boldsymbol{K}_n \boldsymbol{R})) \quad , \tag{11.1}$$

where η is again the fractional density change on solidification; η and the amplitudes μ_n are the order parameters entering the density-functional approach. The reciprocal lattice vectors $\boldsymbol{K}_n$ may be written in terms of the six basis vectors pointing along the directions of the fivefold axes of the icosahedron

$$K_n = \sum_{\alpha=1}^{6} l_{n\alpha} q_\alpha \quad , \tag{11.2}$$

where the l_n are integers and the six basis vectors are given by

$$q_1 = A(0,0,1)$$

$$q_\alpha = A(\sin\gamma\cos\delta_\alpha, \ \sin\gamma\sin\delta_\alpha, \ \cos\gamma) \quad , \quad \alpha = 2,3,\ldots,6 \tag{11.3}$$

with $\delta_\alpha = 2\pi(\alpha - 2)/5$, $\cos\gamma = \sqrt{5}$ [11.69, 70].

Nelson has used model amorphous-metal structure factors as input to the theory and assumed that they are characteristic of the supercooled liquid. He found that near T_g, liquids are metastable with respect to this model icosahedral quasi-crystal. Fcc crystals always have the lowest energy (as has to be expected for a monatomic hard-core system) and the free energy of a bcc phase is always between that of the fcc crystal and the icosahedral quasi-crystal. The DF theory also predicts the Bragg-peak intensities (the μ_n's), and the Fourier transform of the diffraction pattern specifies the decoration of the three-dimensional Penrose tiles with atoms. Based on calculations with relaxed DRPHS structure factors as input, Nelson obtains diffraction patterns that are very similar to those observed experimentally.

As with the DF theory of freezing, these models must be regarded as no more than a first step in the right direction. More realistic calculations will be required to understand the increasing variety of quasi-crystal – they will of course also require improved structural data on liquid alloys as input.

Recent Results

Considerable effort has been devoted very recently to the investigation of the dynamics of supercooled liquids and to the liquid-glass transition [11.74, 75]. For an overview on this subject, see *Götze* [11.76, 77].

Systematic investigations of the chemical and topological short-range order in liquid and glassy alloys using molecular dynamics and potential-energy mapping techniques have been presented [11.78].

A first-principles investigation of the interplay between the atomic and the electronic structures in several metallic glasses has been performed by *Jaswal* and *Hafner* [11.79].

12. Conclusions and Outlook

The theory, or rather the theories, reviewed in this book have achieved for the *sp*-bonded metals and alloys the ambitious goal we stated so boldly in the title. At least in favourable cases, it is now possible to set out with the Hamiltonian of a system of electrons and ions and to compute a phase diagram in good agreement with experiment using methods which are essentially rigorous and developed from first principles. In other cases where this is not yet possible, the theory provides at least a basis for phenomenological models with a minimum number of adjustable parameters.

Of course much work remains to be done. The higher-order effects associated with the transition from metallicity to covalency in the elements (see Sects. 3.2 and 3.3) need further investigation as does charge transfer in alloys with strong chemical interactions (see Sects. 8.4 and 9.2b). The theory of lattice vibrations in both ordered intermetallic compounds and disordered substitutional alloys (Sects. 7.5 and 8.3–4), and the theory of liquid alloys with nonideal mixing behaviour are still in their infancy. With the help of the steadily increasing efficiency of modern computers, future progress appears to be secured.

With this ever-increasing computer performance, non-perturbative methods will become more and more important – for the electronic as well as for the statistical-mechanical part of the theory. Using Hellmann's force theorem and the stress theorem of *Martin* and *Nielsen* [12.1] we can now derive interatomic force constants and hence phonon spectra from self-consistent electronic ground-state energy calculations. However, such a set of force constants leaves us a long way short of the complete description of the interatomic potentials that would be needed as a basis for the statistical-mechanical description of the liquid phase.

The concept of the pseudopotential is fundamental to the subject of the book. The fact that some of the calculations have been done, or will be done in future, more accurately using non-perturbative band-structure methods does not affect the validity of the pseudopotential as a way of thinking about metallic bonding or as a very effective technique for advancing our knowledge. To cite just one example: Several attempts have been made recently to derive empirical interatomic potentials for covalent or partially covalent systems [12.2–5]. They are all based on simple empirical forms of the pair-interaction (e.g. Morse- or Lennard-Jones potentials), with various corrections for the triplet forces. As the assumed form of the two-body forces is quite far from physical reality, these corrections are mostly unphysically large. Here we have found (see the example of liquid and supercooled Ge discussed in Chap. 4) that a physically realistic two-body potential already gives 90 % of the correct answer for the structure factor and correlation functions, and parametrizations of the triplet forces on this basis would be easier and very promising.

This suggests that the relatively "simple" methods discussed in this book will be of continued interest for exploring more and more complex states of metals and alloys and as a means of interpolating between the exact self-consistent total energy calculations.

Appendices

A. Density-Functional Pseudopotentials

Many of the recent advances in the theory of the cohesive properties of materials have been based on the Hohenberg-Kohn-Sham theorem [A.1,2]. This states that the ground-state energy of a many-electron system in an applied external potential is a unique functional $E[\varrho]$ of the electron density $\varrho(\mathbf{r})$. E_{tot} can be expressed (exactly in principle) in terms of the solutions of a set of one-electron self-consistent-field equations which may be derived variationally.

For a metal of atomic number Z_a the Kohn-Sham equations may be written as follows. The one-electron equation to be solved self-consistently is

$$[T + V(\mathbf{r})]\psi_\alpha(\mathbf{r}) = E_\alpha\psi_\alpha(\mathbf{r}) \quad , \tag{A.1}$$

where T is the kinetic-energy operator and the one-electron potential $V(\mathbf{r})$ is given by

$$V(\mathbf{r}) = -\sum_i \frac{2Z_a}{|\mathbf{r} - \mathbf{R}_i|} + \int \frac{2\varrho(\mathbf{r}')}{|\mathbf{r} - \mathbf{r}'|}d^3r' + \mu_{\text{xc}}[\varrho(\mathbf{r})] \quad . \tag{A.2}$$

In terms of the eigenfunctions $\psi_\alpha(\mathbf{r})$ and the eigenvalues E_α of the occupied orbitals, the electron density and total ground state energy E_{tot} are given by

$$\varrho(\mathbf{r}) = \sum_\alpha \psi_\alpha^*(\mathbf{r})\psi_\alpha(\mathbf{r}) \qquad \text{and} \tag{A.3}$$

$$E_{\text{tot}} = \frac{1}{2}\sum_{i,j}' \frac{2Z_a^2}{|\mathbf{R}_i - \mathbf{R}_j|} + \sum_\alpha E_\alpha - \int \frac{\varrho(\mathbf{r})\varrho(\mathbf{r}')}{|\mathbf{r} - \mathbf{r}'|}d^3r\,d^3r'$$
$$+ \int \varrho(\mathbf{r})\{\varepsilon_{\text{xc}}[\varrho(\mathbf{r})] - \mu_{\text{xc}}[\varrho(\mathbf{r})]\}d^3r \quad . \tag{A.4}$$

Many-particle effects are included through an effective exchange and correlation energy $\varepsilon_{\text{xc}}[\varrho]$ and an exchange-correlation potential $\mu_{\text{xc}}[\varrho]$ [A.2] whose exact form is still unknown. When $\varrho(\mathbf{r})$ is sufficiently slowly varying

μ_{xc} is well approximated by the simple functional

$$\mu_{xc}[\varrho(\mathbf{r})] = \frac{\delta\{\varrho(\mathbf{r})\varepsilon_{xc}[\varrho(\mathbf{r})]\}}{\delta\varrho(\mathbf{r})} \quad , \tag{A.5}$$

where $\varepsilon_{xc}[\varrho]$ is the total exchange and correlation energy of a free-electron gas of density ϱ. The functional $\mu_{xc}[\varrho]$ can be separated in an exchange and a correlation potential, where the exchange potential is

$$\mu_x[\varrho(\mathbf{r})] = -4[3\varrho(\mathbf{r})/8\pi]^{1/3} \tag{A.6}$$

and convenient interpolation formulas for the correlation potential may be obtained by fitting the calculated correlation energies and using (A.5), see e.g. [A.3].

The Hohenberg-Kohn-Sham theorem applies to the total electron density. In a pseudopotential-linear-response theory of the cohesive properties and the effective interatomic interactions, one seeks to decompose the one-electron potential $V(\mathbf{r})$ and the electron density $\varrho(\mathbf{r})$ into a linear superposition of individual atomic contributions which are independent of structure, i.e.

$$V(\mathbf{r}) = \sum_i v(\mathbf{r} - \mathbf{R}_i) \tag{A.7}$$

and similarly for $\varrho(\mathbf{r})$. Moreover, one asks for a decomposition of each individual term into a core and a valence contribution; for example we shall want to express $\varrho(\mathbf{r})$ as a sum of core terms together with both homogeneous and spatially varying valence electron terms

$$\varrho(\mathbf{r}) = \sum_i \varrho_c(\mathbf{r} - \mathbf{R}_i) + \varrho_0 + \sum_i \delta\varrho_v(\mathbf{r} - \mathbf{R}_i) \tag{A.8}$$

and analogously for $V(\mathbf{r})$. The nonlinear dependence of the exchange-correlation potential $\mu_{xc}[\varrho]$ on $\varrho(\mathbf{r})$ creates certain difficulties in achieving this decomposition. These can be resolved in different ways, depending on the way in which the pseudopotential is constructed. If we follow the operator approach (see Sect. 2.1.1 and below), starting from an orthogonalized plane-wave expansion of the free-electron-like valence states, we must first achieve the desired decomposition of the total self-consistent potential. If we use the scattering approach (see Sect. 2.1.2 and below), constructing the pseudopotential for an atomic reference configuration, we must derive an ionic pseudopotential by descreening the atomic potential.

The former problem has been treated at length by *Moriarty* [A.4,5]. To arrive at the desired decomposition of $\mu_{xc}[\varrho]$, one starts by rewriting it in the form (using an evident shorthand notation)

$$\mu_{\text{xc}}[\varrho] = \mu_{\text{xc}}\left[\varrho_0 + \sum_i \delta\varrho_v^i\right]$$

$$+ \mu_{\text{xc}}\left[\varrho_0 + \sum_i \delta\varrho_v^i + \sum_i \varrho_c^i\right] - \mu_{\text{xc}}\left[\varrho_0 + \sum_i \delta\varrho_v^i\right] \quad . \qquad \text{(A.9)}$$

The first term is then expanded in powers of $\delta\varrho_v^i$

$$\mu_{\text{xc}}\left[\varrho_0 + \sum_i \delta\varrho_v^i\right] = \mu_{\text{xc}}[\varrho_0] + \frac{d\mu_{\text{xc}}[\varrho_0]}{d\varrho}\sum_i \delta\varrho_v^i + \ldots \quad . \qquad \text{(A.10)}$$

For the second term one proceeds by expanding in the region outside the cores $[\varrho_c^i \ll (\varrho_0 + \sum_i \delta\varrho_v^i)]$ in powers of $\varrho_c^i/(\varrho_0 + \sum_i \delta\varrho_v^i)$, isolating all core-overlap terms and analytically continuing the result over the whole solid by resumming the remaining terms. This yields [A.4]

$$\mu_{\text{xc}}\left[\varrho_0 + \sum_i \delta\varrho_v^i + \sum_i \varrho_c^i\right] - \mu_{\text{xc}}\left[\varrho_0 + \sum_i \delta\varrho_v^i\right]$$

$$= \sum_i \mu_{\text{xc}}\left[\varrho_0 + \sum_i \delta\varrho_v^i + \varrho_c^i\right] - \mu_{\text{xc}}\left[\varrho_0 + \sum_i \delta\varrho_v^i\right] \qquad \text{(A.11)}$$

neglecting all second-order terms. The desired linear dependence of the exchange-correlation potential on structure may then be achieved by setting $\delta\varrho_v^i = 0$ in (A.11), i.e. by approximating the valence electron distribution by its homogeneous average. Combining (A.9) through (A.11) yields finally

$$\mu_{\text{xc}}[\varrho] = \mu_{\text{xc}}[\varrho_0] + \frac{d\mu_{\text{xc}}[\varrho_0]}{d\varrho}\sum_i \delta\varrho_v^i$$

$$+ \sum_i(\mu_{\text{xc}}[\varrho_0 + \varrho_c^i] - \mu_{\text{xc}}[\varrho_0]) \quad , \qquad \text{(A.12)}$$

where the first two terms describe the valence-valence exchange and correlation interaction and the last term defines the core-valence exchange-correlation potential v_{xc}^i. This form of v_{xc}^i had been anticipated [A.6] — although without an adequate derivation — and used in pseudopotential calculations [A.7–10]. With this result the total self-consistent potential $V(\mathbf{r})$ can be written as

$$V(\mathbf{r}) = \sum_i(v_{\text{nuc}}^i + v_c^i + v_{\text{xc}}^i) + v_0 + \mu_{\text{xc}}[\varrho_0]$$

$$+ \sum_i\left(\delta v_v^i + \frac{d\mu_{\text{xc}}[\varrho_0]}{d\varrho}\delta\varrho_v^i\right) \quad , \qquad \text{(A.13)}$$

where a v_x^i stands for the direct Coulomb potential arising from the charge ϱ_x^i, and $v_{\text{nuc}}^i = -2Z_a/|\mathbf{r} - \mathbf{R}_i|$. The last group of terms in (A.13) collects the contributions of the fluctuating part of the valence-electron distribution (i.e.

of the screening-charge distribution in the terminology of Sect. 2.2) to the one-electron potential. We can make the connection with the usual treatment of exchange and correlation within linear response theory by rewriting these terms in Fourier space, i.e.

$$\sum_i \left(\delta v_{\rm v}^i + \frac{d\mu_{\rm xc}[\varrho_0]}{d\varrho}\delta\varrho_{\rm v}^i\right) = \sum_q \frac{8\pi}{q^2} S(q)[1 - G(q)]\delta\varrho_{\rm v}(q)\, {\rm e}^{iqr} \qquad (A.14)$$

with

$$G(q) = -\frac{q^2}{8\pi}\frac{d\mu_{\rm xc}[\varrho_0]}{d\varrho} \qquad (A.15)$$

for the local-field corrections determining the effective electron-electron interaction [see (2.38 ff)]. Equation (A.15) defines precisely the local-density-functional approximation (LDF) to the local field proposed by *Hedin* and *Lundquist* [A.3, 11] and by *Taylor* [A.12]. See Appendix B and *Hafner* and *Heine* [A.13] for a discussion of the practical merit of the LDF approximation to $G(q)$ and that of other approaches involving a full many-body treatment.

A similar expression for the total energy is obtained by treating the exchange-correlation energy in (A.4) in an analogous way, and by expressing the sum over the one-electron eigenvalues as

$$\sum_\alpha E_\alpha = T[\varrho_0] + \varrho_0 V + \sum(T[\varrho_{\rm c}^i] + \varrho_{\rm c}^i V) + \sum_\alpha \delta E_\alpha \quad , \qquad (A.16)$$

where $T[\varrho]$ is the kinetic energy, and the short notation $\varrho V = \int \varrho(r)V(r)d^3r$ has been used. With the further definition of a structure-independent core energy

$$E_{\rm c}^i = T[\varrho_{\rm c}^i] + \varrho_{\rm c}^i(v_{\rm nuc}^i + \tfrac{1}{2}v_{\rm c}^i + \varepsilon_{\rm xc}[\varrho_{\rm c}^i]) \quad , \qquad (A.17)$$

the total energy assumes (using the same notation and neglecting all core-overlap terms)[1] the form

$$E_{\rm tot} = \varrho_0\left[\frac{3}{5}E_{\rm F} + \varepsilon_{\rm xc}(\varrho_0) + \sum_i(\tilde{v}_{\rm c}^i + v_{\rm xc}^i)\right] + E_{\rm es}(Z) + \sum_\alpha \delta E_\alpha$$
$$- \frac{1}{2}\sum_\alpha \delta\varrho_{\rm v}^j\left(\delta v_{\rm v}^i + \frac{d\mu_{\rm xc}[\varrho_0]}{d\varrho}\delta\varrho_{\rm v}^i\right) + \sum_i E_{\rm c}^i \quad , \qquad (A.18)$$

where $\tilde{v}_{\rm c}^i = (Z_{\rm a} - Z/Z_{\rm a})v_{\rm nuc}^i + v_{\rm c}^i$ is a neutral core potential (with the valence charge Z subtracted out) and $E_{\rm es}(Z)$ is the electrostatic energy of point ions of charge Z in a uniform compensating background. To the extent that $E_{\rm c}^i$ in the solid differs insignificantly from its value in the free atom, the binding energy per atom is just given by $E_{\rm bind} = (E_{\rm tot} - \sum_i E_{\rm c}^i)/N$.

---

[1] Core-overlap terms are important only if d-orbitals are present. For a complete treatment involving the overlap terms, see [A.4, 5].

This leaves $\delta\varrho_v$ and $\sum_\alpha \delta E_\alpha$ to be determined as a systematic expansion in terms of the electron-ion pseudopotential w_0. We shall need a first-order calculation of the valence charge density $\delta\varrho_v$ and a second-order calculation for $\sum_\alpha \delta E_\alpha$. The details of this expansion will depend explicitly on the pseudopotential. We turn now to a brief discussion of two important classes of pseudopotentials and we quote the resulting expressions for the total energy.

A.1 Optimized Pseudopotentials –
the Operator Approach

One proceeds here by replacing the true valence orbitals ψ_v (with their familiar nodal behaviour resulting from the orthogonality constraint) by pseudoorbitals φ_v comprising a linear combination of valence and core orbitals [c is a shorthand notation for the quantum numbers $n\,l\,m$ of the core orbitals and for the position R_j of the ion, see (1.35)]; i.e.

$$|\varphi_v\rangle = |\psi_v\rangle + \sum_c b_c|\psi_c\rangle \quad , \tag{A.19}$$

where the coefficients b_c are chosen so that φ_v has certain desired features. The first requirement will be of course that φ_v be nodeless, but it can also be made to satisfy other requirements. One then replaces the true Hamiltonian H for the valence electrons by a pseudo-Hamiltonian H_{ps} which has, however, the same spectrum of valence eigenvalues (see Sect. 2.1). It is easy to show that upon introducing (A.19) into the Schrödinger equation and rearranging terms, the resulting pseudo-Hamiltonian H_{ps} is of the general Austin-Heine-Sham [A.14] family, i.e.

$$H_{ps} = T + V + \sum |\psi_c\rangle\langle F_c| \qquad \text{with} \tag{A.20}$$

$$\langle F_c|\varphi\rangle = (E - E_c)b_c \quad . \tag{A.21}$$

The Phillips-Kleinman [A.15] pseudopotential[2] (2.3) corresponds to the choice $b_c = \langle\psi_c|\varphi\rangle$ or $\langle F_c| = (E - E_c)\langle\psi_c|$. A general pseudopotential of the Austin-Heine-Sham class has a unique eigenstate φ_n with energy E_n. This can be seen by taking the inner product of the pseudo-Schrödinger equation with some core state and introducing some complete set $|\nu\rangle$:

$$(E_c - E_n) \sum_\nu \langle\psi_c|\nu\rangle\langle\nu|\varphi_n\rangle + \sum_\nu \langle F_c|\nu\rangle\langle\nu|\varphi_n\rangle = 0 \tag{A.22}$$

---

[2] In an interesting little paper, *Perdew* and *Vosko* [A.16] have shown (using essentially the *Lödwin* procedure [A.17] for folding down a matrix) that out of any finite set of basis functions consisting of a subset of core orbitals and some arbitrary second subset, the Phillips-Kleinman choice for the coefficients b_c is optimal in the sense of the Rayleigh-Ritz variational principle.

is now an inhomogeneous set of linear equations for the expansion coefficients $\langle \nu|\varphi_n\rangle$ with non-zero and unique solutions, except when $\langle F_c|\nu\rangle = (E_c - E_n)\langle\psi_c|\nu\rangle$, i.e. except for a Phillips-Kleinman pseudopotential. This means that in the case of a Phillips-Kleinman pseudopotential the arbitrariness of the pseudoorbital subsists and must be removed by applying some additional constraint to the pseudo-wave-function.

Several criteria have been proposed to exploit the freedom given in the transformation (A.19). *Cohen* and *Heine* [A.18] and later others [A.19] required the optimum pseudopotential to minimize the kinetic energy [see (2.7)] and hence be spatially smooth; as we have already discussed in Sect. 2.1 this criterion ensures at the same time a maximum cancellation of the attractive electron-ion potential V by the repulsive part of the pseudopotential. *Zunger* and co-workers [A.20] proposed to make φ_n as close as possible to the true orbital (subject to the condition of nodelessness) by requiring

$$\langle\varphi_n|P|\varphi_n\rangle = \text{Minimum} \quad , \tag{A.23}$$

where $P = \sum_c |\psi_c\rangle\langle\psi_c|$ is the core-projection operator. Condition (A.23) asks for the minimum "core admixture" in the pseudopotential φ_n. *Zunger* and *Ratner* [A.21] have made a comparative study of these different criteria. They point out that the "maximum similarity" criterion (A.23) yields a pseudopotential that is strongly repulsive in the core region and has rather large Fourier components around $q \sim 2k_F$ that decay only very slowly with increasing momentum transfer. The latter property is certainly undesirable in a perturbation calculation. A stronger core admixture yields a pseudopotential that is oscillatory in the core region, but possesses a strongly damped Fourier representation.

Harrison [A.22] has applied the Cohen-Heine criterion (2.7) to the Phillips-Kleinman-type pseudopotentials. To solve the variational condition

$$\delta\left(\frac{\langle\varphi|W|\varphi\rangle}{\langle\varphi|\varphi\rangle}\right) = 0$$

we note that the most general variation of the pseudo-wave-function is just an arbitrary linear combination of core states, $\delta\varphi = \sum b_c\psi_c$. This yields directly

$$\langle\psi_c|W|\varphi\rangle = \frac{\langle\varphi|W|\varphi\rangle}{\langle\varphi|\varphi\rangle}\langle\psi_c|\varphi\rangle \quad , \tag{A.24}$$

which is valid for the smoothest pseudo-wave-function. Combining (A.24) and the Phillips-Kleinman equation (which is valid for any φ) leads to

$$W|\varphi\rangle = (1-P)V|\varphi\rangle + \frac{\langle\varphi|(1-P)V|\varphi\rangle}{\langle\varphi|\varphi\rangle - \langle\varphi|P|\varphi\rangle}P|\varphi\rangle \quad . \tag{A.25}$$

Note that this is not a pseudopotential of the Austin-Heine-Sham family because the last term is a nonlinear integral operator. A linearized form may be obtained [Ref. A.22, Chap. 8] by expanding the pseudo-wave-function in terms of plane waves and keeping only a single plane wave in (A.25) (Harrison shows that this does not affect the eigenvalues within second-order perturbation theory), resulting in

$$W|\mathbf{k}\rangle = (1 - P)V|\mathbf{k}\rangle + \frac{\langle \mathbf{k}|(1 - P)V|\mathbf{k}\rangle}{1 - \langle \mathbf{k}|P|\mathbf{k}\rangle} P|\mathbf{k}\rangle \tag{A.26}$$

for the optimized pseudopotential acting on a plane wave $|\mathbf{k}\rangle$. Note that the pseudopotential is a non-Hermitian, nonlocal integral operator. It is easy to show (see [A.22]) that the optimized pseudopotential (A.26) corresponds simply to the Phillips-Kleinman potential with the energy E replaced by the first-order perturbation expression $E \sim k^2 + \langle \mathbf{k}|W|\mathbf{k}\rangle$, i.e.

$$W|\mathbf{k}\rangle = [V + \sum_c (k^2 + \langle \mathbf{k}|W|\mathbf{k}\rangle - E_c)|\psi_c\rangle\langle\psi_c|]|\mathbf{k}\rangle \quad . \tag{A.27}$$

Note also that if the true valence orbital ψ is normalized, then the pseudoorbital φ is not. From (2.5) or (A.19) it follows that (for a projection operator we have $P^2 = P$)

$$\langle\psi|\psi\rangle = \langle\varphi|\varphi\rangle - \langle\varphi|P|\varphi\rangle \quad . \tag{A.28}$$

This must be considered in the calculation of the valence electron distribution $\delta\varrho_v$. We find that it is convenient to separate $\delta\varrho_v$ into orthogonalization-hole and screening contributions

$$\delta\varrho_v = \delta\varrho_{oh} + \delta\varrho_{sc} \quad . \tag{A.29}$$

the orthogonalization charge consists of a homogeneous term to be added to ϱ_0 plus localized charge densities at each ions site

$$\delta\varrho_{oh} = \frac{Z^* - Z}{Z}\varrho_0 + \sum_i \varrho_{oh}^i \quad , \tag{A.30}$$

where $\varrho_{oh}^i = \varrho_{oh}(\mathbf{r} - \mathbf{R}_i)$ and

$$\varrho_{oh}(\mathbf{r}) = -\frac{2V_a}{(2\pi)^3} \int_{k \leq k_F} d^3k \{ [\langle \mathbf{k}|P|\mathbf{k}\rangle\langle\mathbf{k}|\mathbf{r}\rangle + \text{c.c.}$$
$$- \langle\mathbf{r}|P|\mathbf{k}\rangle\langle\mathbf{k}|P|\mathbf{r}\rangle](1 - \langle\mathbf{k}|P|\mathbf{k}\rangle)^{-1} \} \tag{A.31}$$

and the effective valence Z^* is given by

$$Z^* = Z + \frac{2V_a}{(2\pi)^3} \int\limits_{k \leq k_F} d^3k \frac{\langle k|P|k \rangle}{1 - \langle k|P|k \rangle} \quad . \tag{A.32}$$

The expression for the screening-charge distribution $\delta\varrho_{sc}$ is given by (3.17) using (2.35–37) for the screening potential corresponding to a nonlocal pseudopotential. In the total energy expression (A.18) we have to replace ϱ_0 by $\varrho_0^* = (Z^*/Z)\varrho_0$ and use $(\delta\varrho_{oh}^i + \delta\varrho_{sc}^i)$ for $\delta\varrho_v^i$.

The elaboration of the second-order expression for δE_α [see (A.18)] and the completion of the total energy calculation is a rather tedious task; for details the reader is referred to the original papers [A.4, 5, 22, 23]. We note here only the results relevant for the structure-dependent contributions to the total energy: (a) the electrostatic energy has to be calculated with the effective valence Z^*, (b) the energy-wave-number characteristic $F(q)$ takes the form [cf. (2.21, 33–35, 38)]

$$F(q) = \frac{V_a}{(2\pi)^3} \int \frac{n_k - n_{k+q}}{E_k - E_{k+q}} |\langle k + q|w|k \rangle|^2 d^3k$$
$$- \frac{4\pi V_a}{q^2} \{ G(q)\varrho_{oh}(q)^2 + [1 - G(q)]\varrho_{sc}(q)^2 \} \tag{A.33}$$

in essential agreement with the result obtainable by a straightforward application of perturbation theory, but including here a consistent treatment of all exchange-correlation effects within the approximation (A.15).

The optimization procedure can be generalized straightforwardly to binary systems. Note that in this case the pseudopotential will involve two different sets of core states [A.23]:

$$|\varphi_v\rangle = |\psi_v\rangle + \sum_{c\{A\}} |\psi_c^A\rangle + \sum_{c\{B\}} |\psi_c^B\rangle \tag{A.34}$$

and following the same procedure as in the monatomic case, the optimized pseudopotential for the A ions in an A-B alloy is found to be

$$W_A|k\rangle = \left[V_A + \sum_c (k^2 + \langle k|\overline{W}|k \rangle - E_c^A) |\psi_c^A\rangle\langle\psi_c^A| \right] |k\rangle \tag{A.35}$$

with

$$\langle k|\overline{W}|k \rangle = c_A \langle k|W_A|k \rangle + c_B \langle k|W_B|k \rangle \quad . \tag{A.36}$$

The effective valence is given by

$$Z_A^* = Z_A + \frac{2V_a}{(2\pi)^3} \int\limits_{k \leq k_F} d^3k \frac{\langle k|P_A|k \rangle}{1 - \langle k|P|k \rangle} \quad . \tag{A.37}$$

with the operator $P_A = \sum |\psi_c^A\rangle\langle\psi_c^A|$ projecting onto the subspace spanned by A-type core states only and $\langle k|P|k\rangle = c_A\langle k|P_A|k\rangle = c_B\langle k|P_B|k\rangle$. Equations (A.34–36) constitute the formal background to alloy pseudopotentials which were discussed in a more heuristic way in Sect. 2.4. The generalization of the total energy expressions is straightforward, see [A.23].

A.2 Norm-Conserving Pseudopotentials –
the Scattering Approach

In the scattering approach the radial part of the true valence orbital[3] (calculated for some atomic reference state) inside the core region $(r < R_c)$ is replaced by a convenient analytical form [A.24]

$$rR(r) = r^{l+1}f(r) \quad , \quad r < R_c \quad . \tag{A.38}$$

The function $f(r)$ is chosen to give a smooth non-singular potential. It may for example be a polynomial $p(r)$

$$f(r) = \alpha r^4 + \beta r^3 + \gamma r^2 + \delta = p(r) \tag{A.39}$$

or an exponential

$$f(r) = \exp[p(r)] \tag{A.40}$$

[the linear term in $p(r)$ is absent to avoid a singularity in the pseudopotential at $r = 0$]. The coefficients in $p(r)$ are determined by applying the following conditions:

i)   The true orbitals and the pseudoorbitals have the same eigenvalues for some chosen electronic configuration.
ii) $R(r)$ is nodeless and identical to the true valence orbital for $r > R_c$.
iii) the first (and if desired also the second) derivative of the pseudoorbital is matched to the exact values at R_c.
iv)   The pseudo-charge contained in a sphere of radius R_c is identical to the real charge in that sphere.

Conditions (i) and (ii) define a pseudoorbital and condition (iii) ensures that the pseudo- and the true orbitals match continuously and differentiably at R_c. The norm-conservation condition (iv) was introduced by *Hamann* et al. [A.25]. It has been shown that the total charge contained within a sphere is related to the energy derivative of the logarithmic derivative of the orbital at the surface of the sphere [A.26–29]

---

[3]   Alternatively the potential could be modified [A.25], but as the construction of the pseudopotential is achieved by imposing conditions on the wave function, it seems more logical to modify the orbital directly.

$$-2\pi \left[(rR)^2 \frac{d}{dE} \frac{d}{dr} \ln R \right]_{r=R_c} = 4\pi \int_0^{R_c} R^2 r^2 dr \quad . \tag{A.41}$$

Therefore the norm-conservation condition guarantees that not only are the scattering phase shifts (which are related to the logarithmic derivative) of the true potential and the pseudopotential at R_c in agreement for a given energy, but also identical are the linear energy variations of their phase shifts around E. Thus the scattering properties of the pseudopotential and of the full potential have the same energy variation to first order when transferred to another configuration, e.g. from the free atom to an ion in the metal, and the norm-conservation condition is thought to optimize the transferability of the pseudopotential. Conditions (i) to (iv) define the pseudoorbital and the pseudopotential is constructed by inverting the pseudo-Schrödinger equation. This procedure yields a screened pseudopotential and from this the pseudopotential of the ionic core is obtained by a process of unscreening, i.e. by subtracting off the Hartree and the exchange-correlation potentials of the occupied valence states to give

$$W^0(\mathbf{r}) = W(\mathbf{r}) - \int \frac{2\varrho_v(\mathbf{r}')}{|\mathbf{r} - \mathbf{r}'|} d^3\mathbf{r}' - \mu_{xc}[\varrho_v] \quad , \tag{A.42}$$

where ϱ_v is the electron density of the occupied valence states. Note that this valence-unscreening procedure is equivalent to a linearization of the exchange-correlation potential similar to the one discussed at the beginning of this section. The difference is that the remaining parts – and here we think mainly of the valence-core exchange interaction – are treated explicitly in the operator approach, whereas in the scattering approach they are absorbed into the pseudopotential.

The norm-conservation concept was originally introduced by *Topp* and *Hopfield* [A.28] and further developed by a number of workers [A.24, 25, 30–33]. In contrast to the Phillips-Kleinman-type pseudopotentials which are strongly repulsive (or oscillatory) in the core region, the norm-conserving potentials are smooth. This, together with the improved transferability greatly improves their usefulness in band-structure calculations.

For the use in perturbation calculations, things might be different. In Fig. A.1 we compare for the case of sodium the screened pseudopotential matrix element $\langle \mathbf{k} + \mathbf{q} | w | \mathbf{k} \rangle$ as derived from the optimized pseudopotential with that derived from the norm-conserving pseudopotential. We note two things: (i) the norm-conserving pseudopotentials decay more rapidly with increasing wave vector, but the important matrix elements around $q = 2k_F$ are much larger. (ii) The nonlocality is far greater than in the optimized pseudopotential. This would suggest that in a perturbation calculation, the improved transferability is counterbalanced by a much slower convergence.

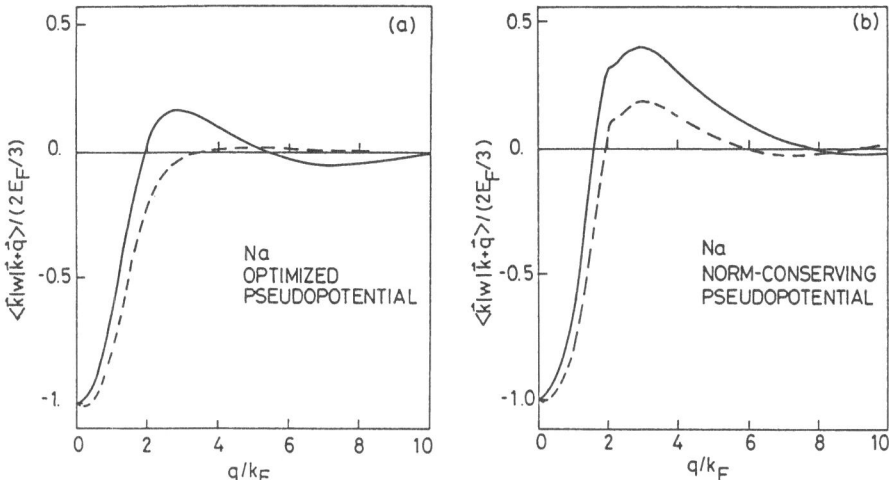

Fig. A.1a,b. Pseudopotential matrix elements $\langle k + q | w | k \rangle$ for sodium (in units of $\frac{1}{2} E_{\mathrm{F}}$); (a) optimized pseudopotential [A.10], (b) norm-conserving pseudopotential of *Bachelet* et al. [A.33] (after [A.29]). (——) backscattering; (- - -) forward scattering, $|k| = k_{\mathrm{F}}$

A further point concerns the choice of the matching radius R_{c}. *Almbladh* and *Morales* have used the norm-conserving pseudopotentials in a phonon-dispersion calculation for Li and Na and they find that the phonon frequencies are quite sensitive to the choice of R_{c}. The dependence of norm-conserving pseudopotentials upon the choice of the matching radius has also been investigated by *Kolar* and *Masek* [A.35] who find that the nonlocality of the pseudopotential increases with decreasing R_{c} and also that the pseudopotential itself shows a pronounced variation. This implies that the use of norm-conserving pseudopotentials in perturbation calculations poses some serious problems – in spite of several admittedly attractive features.

B. Linear Response Theory

Our purpose here is to compile the elementary relations of linear response theory, describing the response of the electron gas to an external perturbing potential w. With a fixed background charge, the perturbing potential w induces a charge density ϱ_{sc}. This induced charge density creates a screening potential w_{sc}. In the Hartree approximation, w_{sc} and ϱ_{sc} are related through the Poisson equation which (in reciprocal space) reads

$$w_{\mathrm{sc}}(q) = \frac{8\pi}{q^2} \varrho_{\mathrm{sc}}(q) \quad . \tag{B.1}$$

A similar relation holds for the total charge density $\tilde{\varrho} = \varrho_0 + \varrho_{\mathrm{sc}}$ and the total potential $\tilde{w} = w + w_{sc}$. Linear response means that the induced charge density is proportional to the perturbing potential. Thus we can define a polarizability (or "non-interacting density response function", by which we mean of the density response function of the non-interacting electron gas) $\chi(q)$ through

$$\varrho_{sc}(q) = \chi(q)\tilde{w}(q) \quad . \tag{B.2}$$

Alternatively, we can introduce the susceptibility (or "interacting density response function") $\Gamma(q)$ through

$$\varrho_{sc}(q) = \Gamma(q)w(q) \quad . \tag{B.3}$$

Combining (B.1) to (B.3) we find the well known relationship between $\chi(q)$ and $\Gamma(q)$

$$\Gamma(q) = \chi(q)\left[1 - \frac{8\pi}{q^2}\chi(q)\right]^{-1} \quad . \tag{B.4}$$

The dielectric function $\varepsilon(q)$ defined by

$$\tilde{w}(q) = w(q) + w_{sc}(q) = w(q)/\varepsilon(q) \tag{B.5}$$

is given in terms of the non-interacting density response function (or polarizability) by

$$\varepsilon(q) = 1 - \frac{8\pi}{q^2}\chi(q) \tag{B.6a}$$

or in terms of the interacting density response function (or susceptibility) by

$$\varepsilon(q) = \left[1 + \frac{8\pi}{q^2}\Gamma(q)\right]^{-1} \quad . \tag{B.6b}$$

In the Hartree approximation, the non-interacting density response function $\chi(q)$ may be calculated very simply by perturbation theory. To first order in a local pseudopotential $w(q)$, the perturbed orbital is given by

$$\varphi_k(r) = \frac{1}{\sqrt{V_a}}\left(e^{ikr} + \sum_q{}' \frac{\tilde{w}(q)}{k^2 - |k-q|^2} e^{i(k+q)r} + \ldots\right) \quad . \tag{B.7}$$

From (B.7) it follows that the charge density, to first order in w, is given by

$$\varrho(r) = \sum_{|k|\leq k_F} |\varphi_k(r)|^2 = \varrho_0 + \frac{2}{V_a}\sum_{|k|\leq k_F}\sum_q{}'\frac{\tilde{w}(q)}{k^2 - |k+q|^2} e^{iqr}$$

$$= \varrho_0 + \sum_q{}'\varrho_{sc}(q) e^{iqr} \quad , \tag{B.8}$$

where $\varrho_0 = Z/V_a$ is the density of the homogeneous electron gas. Comparing (B.8) and (B.2) we find immediately

$$
\begin{aligned}
\chi(q) &= \frac{2}{V_a} \sum_{|\boldsymbol{k}| \leq k_F} (k^2 - |\boldsymbol{k} + \boldsymbol{q}|^2)^{-1} \\
&= -\frac{3}{2} \frac{\varrho_0}{E_F} \left(\frac{1}{2} + \frac{1 - \eta^2}{4\eta} \ln \left| \frac{1 + \eta}{1 - \eta} \right| \right) \quad , \quad \eta = \frac{q}{2k_F}
\end{aligned}
\tag{B.9}
$$

i.e. $\chi(q)$ is given by the Lindhard function [B.1]. If the electronic exchange and correlation interactions are taken into account, then the Coulomb form of the electron-electron interaction in (B.2) and in (B.4) must be replaced by an effective electron-electron interaction $v_{\text{eff}}(q) = [1 - G(q)]8\pi/q^2$ where the "local field factor" $G(q)$ corrects for the exchange and correlation effects. Furthermore, we must now distinguish between the screening of an electron and the screening of an ion (or any other external "test-charge"). The electron dielectric function $\tilde{\varepsilon}(q)$ is defined by

$$
\begin{aligned}
\tilde{w}(q) &= w(q) + w_{\text{sc}}(q) \\
&= w(q) + \frac{8\pi}{q^2}[1 - G(q)]\varrho_{\text{sc}}(q) \\
&\equiv w(q)/\tilde{\varepsilon}(q)
\end{aligned}
\tag{B.10}
$$

i.e. it relates the change in the total Hartree- and exchange and correlation potential to the external potential. From (B.10) and (B.2,3) we find

$$
\begin{aligned}
\tilde{\varepsilon}(q) &= 1 - \frac{8\pi}{q^2}[1 - G(q)]\chi(q) \\
&= \left\{ 1 + \frac{8\pi}{q^2}[1 - G(q)]\Gamma(q) \right\}^{-1} .
\end{aligned}
\tag{B.11}
$$

The test-charge dielectric function (or proton dielectric function) $\varepsilon_p(q)$ takes into account only the change in the electrostatic (Hartree) potential, i.e.

$$
\begin{aligned}
\tilde{w}(q) &= w(q) + \frac{8\pi}{q^2}\varrho_{\text{sc}}(q) \\
&\equiv \frac{w(q)}{\varepsilon_p(q)}
\end{aligned}
\tag{B.12}
$$

so that (B.6b) still holds for $\varepsilon_p(q)$, but (B.6a) has to be modified for the effective interaction, resulting in

$$
\varepsilon_p(q) = \frac{1 - \frac{8\pi}{q^2}[1 - G(q)]\chi(q)}{1 + \frac{8\pi}{q^2}G(q)\chi(q)}
\tag{B.13a}
$$

$$\varepsilon_{\mathrm{p}}(q) = \left[1 + \frac{8\pi}{q^2}\Gamma(q)\right]^{-1} . \tag{B.13b}$$

Thus the relation between the two dielectric functions is given by

$$\varepsilon_{\mathrm{p}}(q) = \tilde{\varepsilon}(q)\left[\tilde{\varepsilon}(q) + \frac{8\pi}{q^2}\chi(q)\right]^{-1} . \tag{B.14}$$

The effective electron-electron interaction $v_{\mathrm{eff}}(q)$ is of course related to the functional for the exchange and correlation energy in the Kohn-Sham density functional theory [B.2], see (A.14–15). In the local density approximation we have [B.3,4] (see (A.5 and 15))

$$G(q) = -\frac{q^2}{8\pi}\frac{d^2[\varrho_0\varepsilon_{\mathrm{xc}}(\varrho_0)]}{d\varrho_0^2} \tag{B.15}$$

i.e. the local field factor $G(q)$ is a quadratic function in the momentum transfer. The prefactor is determined by the second density derivative of the exchange and correlation energy and thus by the exchange and correlation effects on the compressibility of the electron gas. It is a consequence of the general result [B.5]

$$-\left[\Gamma(q)^{-1} + \frac{8\pi}{q^2}\right] = \frac{\delta^2(T[\varrho(r)] + E_{\mathrm{xc}}[\varrho(r)])}{\delta\varrho(r)\delta\varrho(r')} . \tag{B.16}$$

In the limit $q \to 0$ (B.16) reads [using (B.4)]

$$\lim_{q \to 0}\left[\chi(q)^{-1} + G(q)\frac{8\pi}{q^2}\right] = -\frac{d^2 E(\varrho_0)}{d\varrho_0^2} = \varrho_0^{-2}B_{\mathrm{eg}} . \tag{B.17}$$

Equation (B.17) expresses the "compressibility sum rule" for the dielectric function; in the local density approximation (B.15) this sum rule is satisfied by construction. Other sum rules relate the dielectric function to the pair-correlation function of the electrons and to the correlation energy. The connection to the electron-electron pair correlation function is established via the fluctuation-dissipation theorem [B.6,7]

$$S(q) = -\frac{q^2}{8\pi^2\varrho_0}\int_{-\infty}^{\infty}d\omega\,\mathrm{Im}\left\{\frac{1}{\varepsilon(q,\omega)}\right\} , \tag{B.18}$$

where $S(q)$ is the static structure factor of the electron gas [i.e. the Fourier transform of the pair correlation function $g(r)$], and $\varepsilon(q,\omega)$ is the frequency-dependent dielectric function [defined in complete analogy to (B.1) to (B.6) for a frequency-dependent perturbing potential]. The ground-state energy of the interacting electron gas is related to the dielectric function via the coupling-constant integral

$$E_e = E_0 + \sum_q \left[\frac{4\pi \varrho_0}{q^2} + \frac{1}{2\pi} \int_0^2 \frac{d\lambda}{\lambda} \int_{-\infty}^\infty d\omega \operatorname{Im}\left\{ \frac{1}{\varepsilon_\lambda(q,\omega)} \right\} \right]$$

$$= \frac{2.21}{R_s^2} + \frac{2}{\pi R_s^2} \left(\frac{9\pi}{4} \right)^{1/2} \int_0^{R_s} dR_s \int_0^\infty [S(q;R_s) - 1] d(q/k_F) \quad , \quad (B.19)$$

where E_0 is the ground-state energy of the ideal electron gas, and $\varepsilon_\lambda(q,\omega)$ is the dielectric function corresponding to the strength λ of the interaction (remember that $e^2 = 2$ in atomic units). The parameter λ is in fact directly related to the electron density. This allows to re-express the coupling-constant integration as an integral over the electron density parameter R_s. It has been shown [B.6,7] that (B.18) imposes a restriction to the large-q limit of the local field factor

$$\lim_{q \to \infty} G(q) = \tfrac{2}{3}[1 - g(0)] \quad , \quad (B.20)$$

where $g(0)$ is the electron-electron pair-correlation function in the limit $r \to 0$ (which must be positive). However, it is found that, of the conditions (B.17–20), the compressibility sum rule (B.17) is by far the most important for determining the shape of the effective interatomic potential (see e.g. [B.4,8] for an extensive discussion).

Various forms of the local-field correction $G(q)$ have been proposed. The Hubbard [B.9] approximation

$$G(q) = \frac{1}{2} \frac{q^2}{q^2 + k_F^2} \quad (B.21)$$

violates all three consistency relations, but it can be forced to satisfy the compressibility sum rule by multiplying k_F^2 in the denominator by an adjustable factor. In this form the Hubbard correction has been widely used, but nowadays it can no longer be considered as adequate. A more realistic form has been proposed by *Vashishta* and *Singwi* [B.10]

$$G(q) = A\{1 - \exp[-B(q/k_F)^2]\} \quad . \quad (B.22)$$

The parameters A and B depend weakly on the electron density. They have been chosen such as to satisfy the relations (B.17–20). Another realistic form has been derived by *Geldart* and *Taylor* [B.11] using diagramatic techniques. *Ichimaru* and *Utsumi* [B.12,13] have proposed a parametrized form of $G(q)$, based on the quantum Monte Carlo studies of *Ceperley* and

Adler [B.14]. Their $G(q)$ satisfies all consistency relations and represents at the moment the best compromise between accuracy and simplicity.

Finally we comment very briefly on different possible representations of the effective interatomic pair potential, based on the various forms of the response functions introduced above. The standard expression is [see eqs. (2.21–23)], in reciprocal space

$$\hat{\Phi}(q) = \frac{8\pi Z^2}{V_a q^2} + V_a \frac{\chi(q)}{\varepsilon(q)} w(q)^2 \quad .$$

(B.23)

In terms of the ion dielectric function $\varepsilon_p(q)$, the pair potential can be written as

$$\hat{\Phi}(q) = \frac{8\pi Z^2}{V_a q^2} + \frac{V_a q^2}{8\pi} [\varepsilon_p^{-1}(q) - 1] w(q)^2$$

$$= \frac{8\pi Z^2}{V_a q^2} \{ 1 + [\varepsilon_p^{-1}(q) - 1] M(q)^2 \} \quad ,$$

(B.24)

where we have introduced the normalized pseudopotential matrix element $M(q)$ with the Coulomb part factored out [see (3.44)]. *Evans* [B.15] has shown that $\hat{\Phi}(q)$ can also be slightly rearranged to yield

$$\hat{\Phi}(q) = \frac{8\pi Z^2}{V_a q^2} - V_a [1 - G(q)] \frac{8\pi}{q^2} \varrho_{sc}(q)^2 + V_a \chi(q) \tilde{w}(q)^2$$

$$= \hat{\Phi}_n(q) + \hat{\Phi}_{1e}(q) \quad .$$

(B.25)

The first two terms represent a pairwise potential $\hat{\Phi}_n$ between neutral pseudoatoms,

$$\hat{\Phi}_n(q) = \frac{8\pi Z^2}{V_a q^2} - V_a [1 - G(q)] \frac{8\pi}{q^2} \varrho_{sc}(q)^2 \quad .$$

(B.26)

In the absence of exchange and correlation effects [i.e. $G(q) = 0$] this would be just the difference in the electrostatic potentials between a lattice of point ions and a lattice in which the same positive charge has the same density distribution as the electrons. The remaining term

$$\hat{\Phi}_{1e}(q) = V_a \chi(q) \tilde{w}(q)^2$$

(B.27)

is the pair potential arising from the effective one-electron band structure. Evans has shown that the sum over the neutral pseudo-atom pair potential $\Phi_n(R)$ is rapidly convergent and only weakly structure-dependent. The one-electron potential $\Phi_{1e}(R)$ is long-ranged and determines the structural energy differences. Thus we find that for the simple metals the situation is rather similar to that in transition metals where the force theorem [see (3.10)] allows us to express the structural energy differences in terms of the differences of the one-electron energies only.

C. Electrostatic Energies of Crystals and Liquids

In this Appendix we review briefly the calculation of the electrostatic energies of an assembly of point ions in crystalline and liquid structures and we present a collection of Madelung constants for elemental structures and binary compounds together with analytical formulae for the Madelung constants in simple models of liquids and liquid mixtures. We also discuss very briefly some interesting symmetry properties of the electrostatic energies.

C.1 The Madelung Constants of the Elemental Structures

Any straightforward attempt to calculate the electrostatic energy of point ions in a compensating constant background charge by summing the Coulomb interaction over all pairs of ions in real space would run into serious convergence difficulties. In fact such a Coulomb sum is only conditionally convergent. The Ewald-Fuchs method [C.1] solves this problem by dividing the Coulomb interaction into a rapidly decaying part which converges readily in real space and a residual part which gives convergent sums after Fourier transformation. This separation can be achieved in many different ways. The generally accepted choice is that of *Fuchs* [C.1] which yields the following expression for the electrostatic energy (R_a and V_a are the atomic radius and volume)

$$E_{es} = \alpha_M (Z^2/R_a) \tag{C.1}$$

with the Madelung coefficient α_M given by

$$\alpha_M = R_a \left[\frac{4\pi}{V_a} \sum_G {}' \frac{\exp(-G^2/4\eta)}{G^2} |S(G)|^2 + \sum_R {}' \frac{\mathrm{erf}(\sqrt{\eta}R)}{R} \right.$$
$$\left. - \left(\frac{2\sqrt{\eta}}{\sqrt{\pi}} + \frac{\pi}{\eta V_a} \right) \right] \quad , \tag{C.2}$$

where R and G stand for real- and reciprocal-space lattice vectors, $S(G)$ is the structure factor and erfc (x) is the complimentary error function

$$\mathrm{erfc}\,(x) = 2/\pi \int_x^\infty \exp(-x^2) dx \quad . \tag{C.3}$$

Table C.1. Madelung coefficients α_M for the crystal structures of the elements

Structure	Symbol[a]		α_M
fcc	A1	(cF4)	−1.791747
bcc	A2	(cI2)	−1.791858
hcp	A3	(hP2)	−1.791676
diamond	A4	(cF8)	−1.670856
white-thin	A5	(tI4)	−1.773118
In(fct, $c/a = 1.075$)	A6	(tI2)	−1.790670
As	A7	(hR2)	−1.654896
Sb	A7	(hR2)	−1.692771
Bi	A7	(hR2)	−1.716510
Se	A8		−1.646163
Te	A8	(mP4)	−1.677841
Graphite	A9	(hP4)	−0.922843
Hg	A10	(hR1)	−1.789928
α-Ga	A11	(oC4)	−1.728636
β-Ga		(oC8)	−1.774407
simple cubic			−1.760119
simple hex. $[c/a = 1]$			−1.771389
$\quad\quad [c/a = (\sqrt{3}/2)^{1/2}]$			−1.774639
Sm		(hR3)	−1.791719
dhcp		(hP4)	−1.791724
thcp			−1.791724
Cs-IV			−1.772917

[a] We use the conventional Structure report symbols and note in addition the Pearson symbol [C.9] for each structure.

The convergence parameter η is usually chosen to ensure similar rates of convergence for the real- and reciprocal-lattice sums in (C.1).

The Madelung coefficients of the most important crystal structures of metals, semimetals and semiconductors are compiled in Table C.1. They have been calculated using a program capable of handling arbitrary elemental and binary structures − as far as older data are available [C.2–4] they agree with these calculations within all quoted figures.

These values should be compared with the value $\alpha_M = -1.8$ corresponding to a point charge within a compensating spherical charge distribution of radius R_a. The close-packed metallic structures come nearest to this value; an increasingly non-metallic character of the bond is accompanied by a decreasing value of α_M. This is the case e.g. in the series simple cubic $\rightarrow$ Bi $\rightarrow$ Sb $\rightarrow$ As.

Heine and *Weaire* [Ref. C.5, p. 457] and more recently *Ebina* and *Nakamura* [C.6] have pointed out an interesting symmetry of the Ewald-Fuchs formula (C.2): a family of lattices is considered as self-reciprocal when the reciprocal lattice of a member of the family also belongs to the same family. For such structures, the real-space sum (R sum) and the reciprocal-space sum (G sum) in (C.2) are identical in form if the complimentary error function is replaced by its asymptotic expression for large x

$$\text{erfc}\,(x) \simeq \frac{1}{\sqrt{\pi}} \frac{\exp(-x^2)}{x} \quad . \tag{C.4}$$

If the convergence factor η is chosen so as to equalize the rates of convergence $[\eta = \pi/(4Ra)^2]$, then the two series are identical in form and the factors weighting both series are also nearly equal (their ratio is $\pi^{3/2}/6$). The best-known examples of such self-reciprocal structures are the simple hexagonal, the simple tetragonal, the simple rhombohedral and the face-centred tetragonal structures. In each family there is a critical value of the axial ratio (or the rhombohedral angle) for which the reciprocal lattice is identical to the real-space lattice: this is the case for $c/a = (\sqrt{3}/2)^{1/2}$ for the simple hexagonal structure, for $c/a = 1$ (i.e. simple cubic) for the simple tetragonal, for the rhombohedral angle when $\alpha = 90°$ (again simple cubic) and finally when $c/a = 2^{-1/4}$ for the face-centred tetragonal structure. As these are all Bravais lattices [and hence $S(G) = 1$] identical series are obtained for the real- and reciprocal-space contributions to (C.2), and a variation of the axial ratio (or rhombohedral angle) about the critical value will give approximately equal and opposite contributions to the variation of the Madelung coefficient. Hence α_M can be expected to be stationary at the critical value. This is illustrated in Fig. C.1 for the case of the simple hexagonal and the tetragonally distorted diamond lattices and in Fig. 2 (bottom curve) for the rhombohedrally distorted the face-centred cubic lattices. If the symmetry in the Ewald-Fuchs sums were an exact one, these figures would be exactly symmetric and the fcc and bcc lattices would have equal Madelung energies. That the Madelung energy of the simple hexagonal lattice is extremal at $c/a = (\sqrt{3}/2)^{1/2}$ seems to be relevant in the light of a recent investigation of the simple hexagonal high-pressure modification of Si and Ge by *Yin* and *Cohen* [C.7]. According to these authors, the electro-

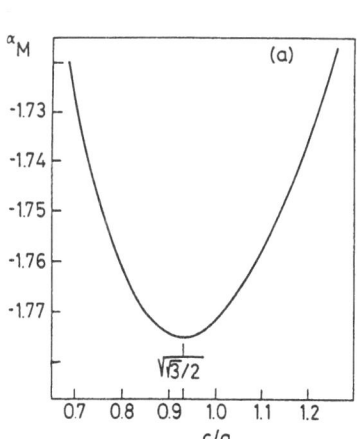
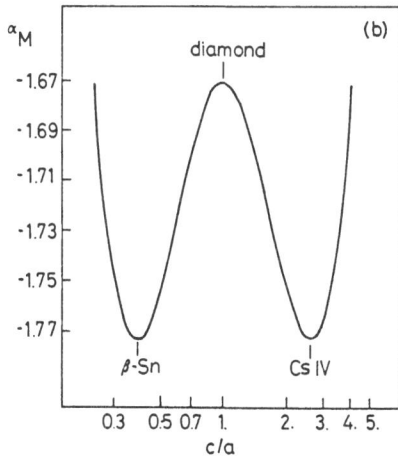

Fig. C.1a,b. Madelung coefficient α_M of a simple hexagonal (a) and of a tetragonally distorted diamond (b) lattice as a function of the axial ratio

static energy plays a dominant role in the determination of the equilibrium axial ratio of the high-pressure form.

The variation of the electrostatic energy of the simple cubic structure with rhombohedral distortion may be described within an even more general framework. Close-packed structures can be classified according to the way in which hexagonal layers of atoms are stacked (cf. Sect. 3.3.2). Denoting hcp by AB and fcc by ABC, there is clearly an infinite number of possible structures of which the hcp, Sm (dhcp), thcp and fcc structures of the lanthanides are just the simplest examples (see Sect. 3.3.2). One can also label a layer according to whether the neighbouring layers are alike or different from one another: the central layer in ABA is in a "hexagonal" environment and denoted by "h", that in ABC is in a cubic environment and denoted by "c". In this notation the series is h...(hcp), hhc...(Sm), hcc...(thcp), and c...(fcc).

For fcc, a variation of the axial ratio of the hexagonal cell corresponds to a rhombohedral distortion and it is well known that a rhombohedral distortion transforms fcc into sc and finally into bcc, the values of the axial ratio being $c/a = \sqrt{6}$ (fcc), $c/a = \sqrt{3/2}$ (sc), $c/a = \sqrt{3}/2$ (bcc). Again we find α_M to be extremal for the self-reciprocal structure (α_M has a local maximum for the sc lattice) and for the fcc and bcc structures which are reciprocal to each other. No such symmetry exists for the other stacking sequences, but as long as the structure has some degree of "cubicity", the α_M vs c/a curve shows a distinct inflection at lower axial ratios (Fig. C.2).

Similar symmetry relations also exist for non-primitive lattices. In that case each point of the reciprocal lattice has a weight $|S(G)|^2$ and a reciprocity relation exists if the weighted G sum can be identified with the R sum of the reciprocal structure. It has been shown that such a relation

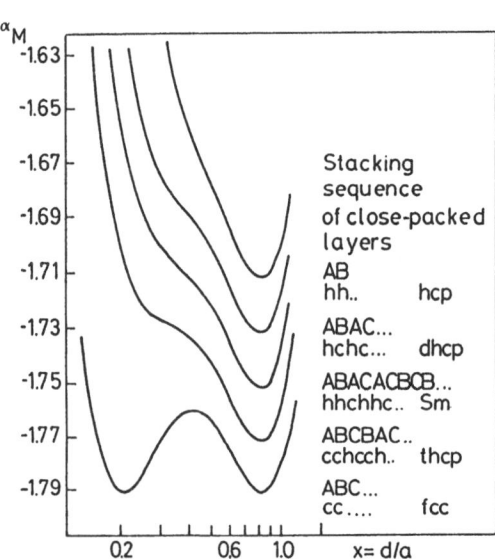

Fig. C.2. Madelung coefficient of closed-packed crystal structures as a function of the interlayer distance d. The curves are shifted by 0.02 with respect to each other, the structures are ordered according to the Madelung coefficient in the ideally close-packed configuration

exists for the family of tetragonally distorted diamond lattices, with the cubic diamond structure as the self-reciprocal member of the family [C.6,8]. The minima in the electrostatic energy occur close to the axial ratios at which the shortest reciprocal lattice and the shortest real lattice vectors cross – contrary to the case of the face-centred tetragonal or the simple rhombohedral lattices these values are not identical. One finds a minimum $\alpha_M = -1.772318$ at $c/a = 0.384$ and $\alpha_M = -1.772917$ at $c/a = 2.603$ (Fig. C.1b). The former value is rather close to the axial ratios of β-Sn $[(c/a) = 0.3857]$ and the metallic high-pressure modifications of Si and Ge $(c/a = 0.55)$. The latter value was first pointed out merely as a mathematical curiosity [C.8], but recently [C.10,11] it was found that this structure is identical to the high-pressure Cs-IV structure. Such symmetry relations for the Madelung constant are certainly not a mere curiosity – they give a first hint as to the character of the chemical bond. A structure with a local minimum in α_M is probably stable under a purely metallic bond, whereas a lattice with a local maximum in α_M requires a large amount of band-structure stabilization and hence an at least partially non-metallic character of the chemical bond.

C.2 The Madelung Constants of Binary Alloys and Intermetallic Compounds

The Ewald-Fuchs formula (C.1,2) is easily generalized to binary phases. We can now represent the electrostatic energy as (R_a is an average atomic radius)

$$E_{es} = \frac{1}{R_a}(Z_A^2 \alpha_{AA} + Z_A Z_B \alpha_{AB} + Z_B^2 \alpha_{BB}) \tag{C.5}$$

with

$$\alpha_{ij} = R_a \left\{ \frac{4\pi}{V_a} \sum_G{}' \frac{\exp(-G^2/4\eta)}{2G^2} [S_i^*(G)S_j(G) + \text{c.c.}] \right.$$
$$\left. + \frac{1}{N} \sum_{\substack{R\{i\} \\ R'\{j\}}}{}' \frac{\text{erfc}(\sqrt{\eta}|R - R'|)}{|R - R'|} + \frac{c_i c_j \pi}{V_a \eta} - 2\sqrt{\frac{\eta}{\pi}} c_i \delta_{ij} \right\} \tag{C.6}$$

expressed in terms of the partial structure factors $S_i(q)$ defined in (2.49) and lattice sums running only over $R\{i\}$ sites occupied by i atoms. Equivalently we might write

$$E_{es} = \frac{1}{R_a}(\overline{Z}^2 \alpha_{NN} + \overline{Z}\Delta Z \alpha_{Nc} + \Delta Z^2 \alpha_{cc}) \quad , \tag{C.7}$$

with $\overline{Z} = (c_A Z_A + c_B Z_B)$, $\Delta Z = Z_A - Z_B$ and α_{NN}, α_{Nc}, and α_{cc} expressed in an obvious way in terms of the structure factors $S(q)$ and $D(q)$; see

(2.51)[1] The latter form is more convenient for disordered substitutional alloys; for Bravais lattices and for non-primitive lattices for which all sites are equivalent, we have $S^*(\boldsymbol{q})D(\boldsymbol{q}) = 0$ [see Sect. 7. and (7.1)] and hence $\alpha_{Nc} = 0$. For a completely random distribution of the two atomic species, one has [see (7.2)] $D^*(\boldsymbol{q})D(\boldsymbol{q}) = c_A c_B$ and it is easy to show [C.12] that in this case $\alpha_{cc} = 0$ so that the electrostatic energy depends only on the mean valence and the Madelung coefficient of the mean lattice.

For stoichiometric intermetallic compounds, the former expression is more convenient. The α_{AA} and α_{BB} represent the electrostatic energies of the A and B sublattices with α_{AB} the interaction between the sublattices. This establishes certain cross correlations. For example, the sublattices of the CsCl structure are simple cubic, therefore we find, correcting for the number of atoms in the unit cell of the lattices: α_M (sc) $= 2\sqrt[3]{2}\,\alpha_{AA}$ (CsCl). The A sublattices in the AB_2 Laves phases are of the diamond type (MgCu$_2$), and of the wurtzite (ZnS) type (MgZn$_2$). Consequently we have α (diamond) $= 3\sqrt[3]{3}\,\alpha_{AA}$(MgCu$_2$) and α (ZnS) $= 3\sqrt[3]{3}\,\alpha_{AA}$(MgZn$_2$), see Tables C.2–5. From (C.5) and (C.7) it follows immediately that

$$\alpha_{NN} = \alpha_{AA} + \alpha_{BB} + \alpha_{AB} \tag{C.8a}$$

$$\alpha_{cc} = c_B^2\alpha_{AA} + c_A^2\alpha_{BB} - c_A c_B\alpha_{AB} \tag{C.8b}$$

$$\alpha_{Nc} = 2c_B\alpha_{AA} - 2c_A\alpha_{BB} + (c_B - c_A)\alpha_{AB} \quad. \tag{C.8c}$$

Table C.2. Madelung coefficients α_{ij}, $\alpha_{NN} = \alpha_{AA} + \alpha_{BB} + \alpha_{AB}$ for binary intermetallic compounds: AB stoichiometry[a]

Structure[a]

Type	Symbol	α_{AA}	α_{AB}	α_{BB}	α_{NN}	Mean lattice
NaCl(B1)	cF8	−0.711055	−0.338008	−0.711055	−1.760119	sc
CsCl(B2)	cP2	−0.698504	−0.394851	−0.698504	−1.791858	bcc
NaTl(B32)	cF16	−0.663078	−0.465703	−0.663078	−1.791858	bcc
CuAuI(L1$_0$)	tP4	−0.668226	−0.455295	−0.668226	−1.791747	fcc
AuCd(B19)	oP4	−0.683789	−0.423962	−0.683789	−1.791540	hcp
ZnS(c)(B3)	cF8	−0.711055	−0.248741	−0.711055	−1.670851	diamond
ZnS(h)(B4)	hP4	−0.707638	−0.253075	−0.707638	−1.668403	
FeB(B27)	oP8	0.607011	0.415119	0.700040	−1.742470	
CrB(B33)	oC8	−0.594007	−0.433223	−0.698174	−1.725408	
NiAs(B8$_1$)	hP4	−0.701168	−0.192341	−0.701168	−1.594677	
HgNa	oC16	−0.637924	−0.445882	−0.678455	−1.762261	

[a] Structures are identified by the structure type and the Pearson [C.9] symbol. For the structures that are often referred to by the Structure Report symbol, this is given in parentheses.

---

[1] Note that (C.5–7) apply to metallic ($\overline{Z}>0$) as well as to ionic ($\overline{Z} = 0$) charge distributions, the background charge compensates only the mean valence.

Table C.3. Madelung coefficients α_{ij}, $\alpha_{NN} = \alpha_{AA} + \alpha_{AB} + \alpha_{BB}$ for binary compounds: AB_2 stoichiometry[a]

Structure[a]

Type	Symbol	α_{AA}	α_{AB}	α_{BB}	α_{NN}	Mean lattice
$MoSi_2(c/a = 3.0)(C11_b)$	tI6	-0.382427	-0.429719	-0.979713	-1.791858	bcc
$MoSi_2(c/a = 3\sqrt[4]{2})$	tI6	-0.353424	-0.487533	-0.950617	-1.791574	fct($c/a = \sqrt[4]{2}$)
$MoSi_2(c/a = 3\sqrt{2})$	tI6	-0.309258	-0.575982	-0.906506	-1.791747	fcc
$MgCu_2(C15)$	cF24	-0.386168	-0.478229	-0.907916	-1.772313	
$MgZn_2(C14)$	hP12	-0.385602	-0.480707	-0.905207	-1.771516	
$MgNi_2(C36)$	hP24	-0.385885	-0.479463	-0.906568	-1.771916	
$CeCu_2$	oI12	-0.401975	-0.427843	-0.933142	-1.762962	
$AlB_2(C32)$	hP3	-0.407386	-0.364018	-0.701985	-1.473388	
$CuAl_2(C16)$	tI12	-0.359952	-0.397462	-1.012332	-1.769746	
CaF_2	cF12	-0.414109	-0.289727	-1.025070	-1.728906	
$CaIn_2$	hP6	-0.406729	-0.235058	-0.837824	-1.479591	

[a] Structures are identified by the structure type and the Pearson [C.9] symbol. For the structures that are often referred to by the Structure Report symbol, this is given in parentheses.

Table C.4. Madelung coefficients α_{ij}, $\alpha_{NN} = \alpha_{AA} + \alpha_{AB} + \alpha_{BB}$ for binary compounds AB_3 stoichiometry[a]

Structure[a]

Type	Symbol	α_{AA}	α_{AB}	α_{BB}	α_{NN}	Mean lattice
$AuCu_3$	cP4	-0.277201	-0.341471	-1.173075	-1.791747	fcc
$TiAl_3$	tI8	-0.280122	-0.335630	-1.175995	-1.791747	fcc
$ZrAl_3$	tI16	-0.275879	-0.344130	-1.171768	-1.791747	fcc
$BiLi_3$	cF16	-0.282183	-0.331564	-1.178112	-1.791858	bcc
$NiSn_3$	hP8	-0.255113	-0.380422	-1.146538	-1.782050	hcp $(c/a) = 1.605$
$SiCr_3(A15)$	cP8	-0.282200	-0.329314	-1.176166	-1.787780	
$CaZn_3$	hP32	-0.205365	-0.172332	-1.060745	-1.438452	

[a] Structures are identified by the structure type and the Pearson [C.9] symbol. For the structures that are often referred to by the Structure Report symbol, this is given in parentheses.

In some sense α_{NN} measures the metallic and α_{cc} the ionic component of the electrostatic energy; for a structure that can be viewed as an ordered superstructure of some mean lattice, α_{NN} is just the Madelung coefficient of the basic structure (see Tables C.2, 4) and $(\alpha_{cc}/c_A c_B)$ is equal to the Madelung coefficient of the ionic superlattice.

For the structures which cannot be considered as superstructures, the Madelung energy is usually slightly lower than for the simple elemental structures. Generally we find a value $\alpha_{NN} \sim -1.77$ for tetrahedrally close-

Table C.5. Madelung coefficients α_{ij}, $\alpha_{NN} = \alpha_{AA} + \alpha_{AB} + \alpha_{BB}$ for binary compounds of various stoichiometries

Structure Type	Symbol	α_{AA}	α_{AB}	α_{BB}	α_{NN}
$BaAl_4$	tI10	−0.20358	−0.372033	−1.159231	−1.734623
$BaLi_4$	hP30	−0.187045	−0.395440	−0.840383	−1.422868
$AuBe_5$	cF24	−0.164334	−0.296603	−1.311370	−1.772307
$CaZn_5$	hP6	−0.161187	−0.200789	−1.220766	−1.582743
$BaCd_{11}$	tI48	−0.064254	−0.230288	−1.442310	−1.736853
$NaZn_{13}$	cF112	−0.052164	−0.231354	−1.390371	−1.673889
Fe_7W_6	hR13	−0.557958	−0.611739	−0.602004	−1.771701
K_7Cs_6	hP26	−0.553746	−0.623254	−0.594748	−1.771749

packed structures, and considerably lower values for the more open structures such as NiAs, AlB_2 or $BaLi_4$. Quite clearly the electrostatic energy makes a significant contribution to structural energy differences for binary compounds.

Little is known about the variation of the electrostatic energy with structural parameters, apart from the investigations of Laves and other closely related phases cited in Sect. 8.3. This investigation shows that there is hardly a single case where the minimum in the electrostatic energy occurs for the most symmetric structure. Another interesting example is the $MoSi_2$ structure (see Table C.3 and Fig. C.3). This is an ordered body-

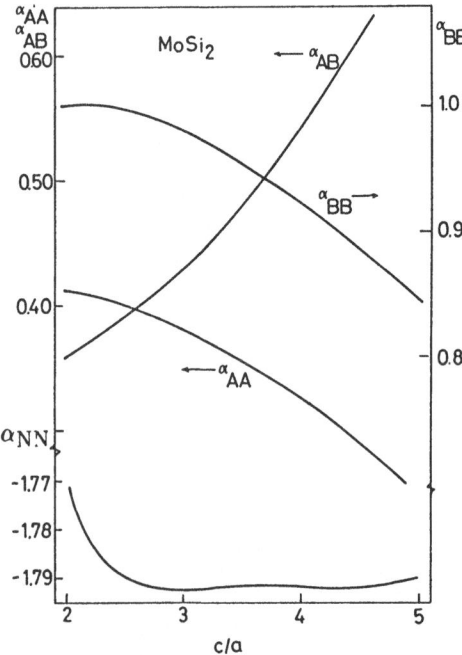

Fig. C.3. Partial Madelung coefficients α_{ij} and $\alpha_{NN} = \alpha_{AA} + \alpha_{AB} + \alpha_{BB}$ for $MoSi_2$-type structures as a function of the axial ratio of the tetragonal cell

centred tetragonal structure which occurs in a large number of intermetal-lic compounds (sp-bonded examples are $MgHg_2$, $CdHg_2$ and Cd_2Hg, they have been discussed by *Inglesfield* [C.13]), with axial ratio varying over a wide range between $c/a = 2.29$ and $c/a = 4.68$. In fact, for $c/a = 3$ the mean lattice is bcc, for $c/a = 2\sqrt{3}$ it is fcc, and for $c/a = 3\sqrt[4]{2}$ it is the self-reciprocal face-centred tetragonal lattice (cf. above). Therefore, in spite of a strong variation of the partial Madelung coefficients α_{ij} with c/a, the mean α_{NN} is approximately constant between $c/a = 2.5$ and $c/a = 5.0$, the actual axial ratio being determined by band-structure effects.

C.3 Electrostatic Energies of Model Liquids and Liquid Mixtures

For a liquid described by the static structure factor $S(q)$, the Madelung coefficient α_M may be expressed as

$$\alpha_M = \frac{2}{\pi} R_a \int_0^\infty [S(q) - 1]dq \quad . \tag{C.9}$$

For model liquids where an analytic solution for the static structure factor is available, the integral (C.9) can also be calculated analytically. The Fourier transform of the right hand side of (C.9) is

$$\int_0^\infty [S(q) - 1]q \sin (qR)dq = 2\pi^2 nR[g(R) - 1] \quad . \tag{C.10}$$

Taking the Laplace transform of (C.10) with

$$G(t) \equiv \int_0^\infty e^{-tx} x g(x)dx \quad , \quad x = R/\sigma \tag{C.11}$$

yields

$$\int_0^\infty [S(q) - 1]\frac{q^2}{(t/\sigma)^2 + q^2}dq = 2\pi^2 \sigma^2 n[G(t) - 1/t^2] \tag{C.12}$$

and after performing the limit $t \to 0$ we find

$$\alpha_M = \frac{2}{\pi} R_a \int_0^\infty [S(q) - 1]dq = 4\pi\sigma^2 n \lim_{t \to 0} [G(t) - 1/t^2] \quad . \tag{C.13}$$

The Laplace transform $G(t)$ of the pair correlation function $g(R)$ may be found by Laplace transforming the integral equation from which $g(R)$ is derived (see Appendix D). For the hard-sphere liquid in the Percus-Yevick approximation α_M was calculated by *Jones* [C.14] with the result (η is the packing fraction)

$$\alpha_M = R_a \frac{2\pi n}{1 + 2\eta} \left[-\sigma^2 + \frac{1}{5}\left(1 - \frac{1}{2}\eta\right)\eta\sigma^2 \right] . \tag{C.14}$$

Umar et al. [C.15] have derived the equivalent result for hard-sphere mixtures (see also [C.16])

$$\alpha_{AA} = R_a c_A^2 \frac{2\pi n}{1 + 2\eta} \left[-\sigma_A^2 + \frac{1}{5}\left(1 - \frac{1}{2}\eta\right)(\eta_A\sigma_A^2 + \eta_B\sigma_B^2) \right.$$
$$\left. + \eta_B\sigma_A(\sigma_B - \sigma_A) + \frac{1}{2}\eta_A\eta_B(\sigma_B - \sigma_A)^2 \right] , \tag{C.15a}$$

$$\alpha_{AB} = R_a c_A c_B \frac{4\pi n}{1 + 2\eta} \left[-\frac{1}{4}(\sigma_A + \sigma_B)^2 + \frac{1}{5}\left(1 - \frac{1}{2}\eta\right)(\eta_A\sigma_A^2 + \eta_B\sigma_B^2) \right.$$
$$\left. + \frac{1}{2}(\eta_B\sigma_A - \eta_A\sigma_B)(\sigma_B - \sigma_A) + \frac{1}{2}\eta_A\eta_B(\sigma_B - \sigma_A)^2 \right] \tag{C.15b}$$

with the partial packing fraction $\eta_i = (\pi/6)n\sigma_i^3$. For the symmetric hard-sphere Yukawa model of mixtures ($\sigma_A = \sigma_B$, $c_A Q_A + c_B Q_B = 0$, see Sect. 9.2.2 κ is the screening constant and ε the strength of the Yukawa potential at hard contact) α_{NN} is given by the one-component hard-sphere result (C.14), while the concentration fluctuation part

$$\alpha_{cc} = R_a \frac{2}{\pi} \int_0^\infty [S_{cc}(q) - c_A c_B]dq = 4\pi n\sigma^2 \lim_{t \to 0} H_{cc}(t)$$

$$H_{cc}(t) \equiv \int_0^\infty e^{-tx} x g_{cc}(x)dx , \quad x = R/\sigma \tag{C.16}$$

may be calculated by Laplace transforming the MSA integral equation [see (9.16–20) and Appendix D] with the final result [C.17]

$$\alpha_{cc} = R_a \frac{2c_A c_B}{\sigma} \frac{[\frac{1}{\omega} - 1 - \frac{1}{ze^z} - \frac{\omega}{2z}(\frac{1}{e^z} + \frac{1}{2ze^z} - \frac{1}{2z})]}{\frac{z}{12\beta\varepsilon\eta\omega} + D(\omega, z)} , \quad z = \kappa\sigma \tag{C.17}$$

$$D(\omega, z) = \left[1 - \frac{2}{z^2} + \frac{2}{ze^z}\left(1 + \frac{1}{z}\right)\right]$$
$$- \frac{\omega}{z}\left[\frac{1}{2e^z} - \frac{1}{z^2 e^z} - \frac{1}{2z}\left(1 - \frac{1}{z}\right) + \frac{1}{2e^z e^{2z}}\left(1 + \frac{1}{z}\right)\right]$$
$$- \frac{1}{\omega}\left(1 + \frac{1}{z}\right) \quad . \tag{C.18}$$

The use of these analytical formulae simplifies the thermodynamic variational calculations based on these reference systems.

D. Liquid State Theory: Integral Equations, Variational Principles and Exactly Soluble Models

The introduction to liquid state theory given in Chapters 4 and 9 was necessarily brief. In this Appendix, we supplement that discussion by (i) a compilation of some elementary relations, (ii) an overview of integral equations and variational principles, and (iii) a collection of results for analytically soluble models.

D.1 Correlation Functions and Equations of State

Within the canonical ensemble, the s-particle distribution function $n^{(s)}(\boldsymbol{R}_1,$ $\ldots, \boldsymbol{R}_s)$ [or the s-particle correlation function $g^{(s)}(\boldsymbol{R}_1, \ldots, \boldsymbol{R}_s) = n^{-s} n^{(s)}(\boldsymbol{R}_1, \ldots, \boldsymbol{R}_s)$ where n is the number density of the ions] of a classical system with the pair-potential Hamiltonian [cf. (2.20)]

$$H = T + E_{\mathrm{v}} + \frac{1}{2N}\sum_{i \neq j}\Phi(R_{ij}, V)$$
$$\equiv U(\boldsymbol{R}_1 \quad, \ldots, \boldsymbol{R}_N) \tag{D.1}$$

is defined by the thermodynamic expectation value (see e.g. [D.1])

$$n^{(s)}(\boldsymbol{R}_1, \ldots \boldsymbol{R}_s) = n^s g^{(s)}(\boldsymbol{R}_1, \ldots \boldsymbol{R}_s)$$
$$= \frac{N!}{(N-s)!}Q_N^{-1}\int \ldots \int \exp[-\beta U(\boldsymbol{R}_1, \ldots \boldsymbol{R}_N)] d^3 \boldsymbol{R}_{s+1} \ldots d^3 \boldsymbol{R}_N \quad, \tag{D.2}$$

where the configurational integral

$$Q_N = \int \ldots \int \exp[-\beta U(\boldsymbol{R}_1, \ldots \boldsymbol{R}_N)] d^3 \boldsymbol{R}_1 \ldots d^3 \boldsymbol{R}_N \tag{D.3}$$

is the structure-sensitive contribution to the canonical N-body partition function Z_N. The thermodynamic functions may be calculated by taking the appropriate expectation value. For a system with pair and volume forces only, we find that the thermodynamic functions are expressible in terms of the pair-correlation function $g^{(2)}(R_1, R_2) = g(R_{12})$ alone (the last equality holds for any homogeneous system). For the internal energy per particle we find immediately

$$E = \frac{3}{2}k_BT + E_v + \frac{n}{2}\int g(R)\Phi(R;V)d^3R. \tag{D.4}$$

the pressure may be calculated via

$$p = -\left(\frac{\partial F}{\partial V}\right)_T = k_BT\left(\frac{\ln Z_N}{\partial V}\right)_T \quad ,$$

but in taking the derivative we have to remember that the pair interactions $\Phi(R;V)$ depend on the volume in a twofold way: explicitly and via the variation of the interatomic distances with volume, cf. (2.31). It follows that the pressure is given by

$$p = nk_BT - \frac{\partial E_v}{\partial V} - \frac{n^2}{2}\int g(R)$$
$$\times \left(\frac{R}{3}\frac{\partial \Phi(R;V)}{\partial R} + V\frac{\partial \Phi(R;V)}{\partial V}\right)d^3R \quad , \tag{D.5}$$

where the second term is characteristic of metallic systems with density-dependent pair potentials. An equation for the isothermal compressibility χ_T may be derived from particle-number fluctuations in the grand-canonical ensemble with the result

$$k_BT\left(\frac{\partial n}{\partial p}\right)_{V,T} = k_BTn\chi_T = 1 + n\int(g(R) - 1)d^3R$$
$$= S(0) \quad . \tag{D.6}$$

Equations (D.4–6) describe three different routes to the thermodynamic properties, they are usually referred to as the energy, pressure (or virial), and compressibility equations. If the *exact* $g(R)$ were known, then the compressibility calculated by differentiating the pressure equation would be the same as that derived from the compressibility equation (D.5) but as we find it necessary to introduce approximations in the equations determining $g(R)$, this poses a problem of *thermodynamic consistency*. For metallic systems we have in addition a problem of *electronic consistency*: the volume derivative of the pair interaction appears in the compressibility calculated via (D.4), but not in that calculated via (D.5). This is entirely analogous to the case of crystalline alloys (see Sect. 2.2).

D.2 Integral Equations and Variational Principles for the Total and Direct Correlation Functions

An integro-differential equation relating $g(R)$ to the pair potential $\Phi(R)$ may be derived by differentiating the defining equation [i.e. (D.2) with $s = 2$] with respect to the coordinates of particle 1. This yields an equation relating $g^{(2)}$ to $\Phi(F)$ and $g^{(3)}$, i.e. the first term of the Born-Green-Yvon hierarchy [see Sect. 1.10.2, (1.71 ff) and Sect. 4.2, (4.1 ff)]. A closed integro-differential equation is obtained by approximating the triplet correlation in terms of pair correlations. The *Kirkwood* [D.2] superposition approximation (4.2) defines the Born-Green equation and *Abe* [D.3] has derived a density expansion of the BGY hierarchy in the superposition approximation by successive resubstitution. An approximate closed form for the resulting infinite series is the hypernetted chain equation (4.4) and a further simplification yields the Percus-Yevick-equation (4.5).

The alternative procedure is via the Ornstein-Zernike equation

$$h(R_{12}) = c(R_{12}) + n \int c(R_{13})h(R_{32})d^3\mathbf{R}_3. \tag{D.7}$$

A closed integro-differential equation results from (D.7) along with a closure relation relating $c(R)$, $h(R)$, and $\Phi(R)$. The derivation of the closure relations is based on cluster expansions for the correlation functions $h(R)$ and $c(R)$ which are most conveniently analyzed diagrammatically. Any detailed discussion of these methods would go far beyond the scope of this book and we present here only a very rough overview, with the aim of giving the reader a feeling for the physical content of the various approximations and their hierarchical order. According to *Rushbrooke* and *Scoins* [D.4], an expression for the Boltzmann term, $\exp[-U(\mathbf{R}_1, \ldots \mathbf{R}_N)/k_BT]$, in terms of products of the Mayer functions, $f(R) = \exp[-\beta\Phi(R)] - 1$, yields an expansion of the direct correlation function $c(R)$ of the form

$$c(R_{12}) = \sum_{n \geq 1} \alpha_{n+1}(R_{12})n^{n-1} \qquad \text{with} \tag{D.8}$$

$$\alpha_{n+1}(R_{12}) = \frac{1}{(n-1)} \int \cdots \int \sum \prod f(R_{ij})d_3^3R_3 \ldots d^3R_{n+1} \tag{D.9}$$

i.e.

$$\alpha_2(R_{12}) = f(R_{12})$$

$$\alpha_3(R_{12}) = \int f(R_{12})f(R_{23})f(R_{31})d^3R_3$$

and so on. Similar expansions may be given for the total correlation function $h(R)$ and for the potential of mean force $\psi(R)$, defined by $g(R) = e^{-\beta\psi(R)}$. These expansions may be expressed diagrammatically, each term $f(R_{ij})$

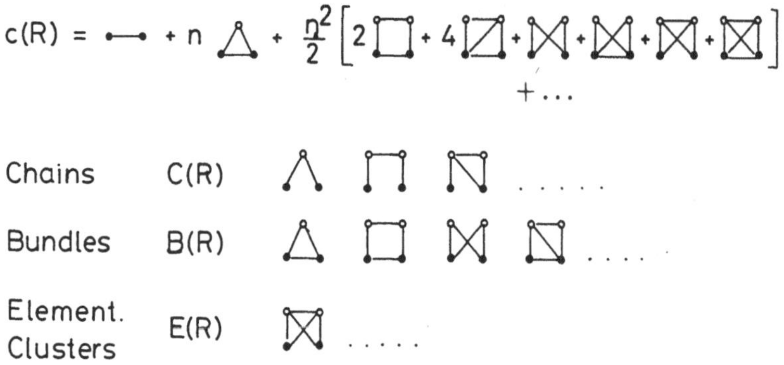

$$c(R) = \bullet\!\!-\!\!\bullet + n \, \triangle \, + \frac{n^2}{2}\left[2\,\square + 4\,\boxslash + \boxtimes + \boxtimes' + \boxtimes + \boxtimes\right]$$

$$+ \dots$$

Chains   C(R) $\wedge$ $\sqcap$ $\boxtimes$ . . . . .

Bundles   B(R) $\triangle$ $\square$ $\boxtimes$ $\boxtimes$ . . . .

Element.
Clusters   E(R) $\boxtimes$ . . . . .

Fig. D.1. Diagrammatic expansion of the direct correlation function $c(R_{12})$ in terms of f-bonds (see text). The closed circles represent the "root points" 1 and 2, the open circles the intermediate points $3, \dots, n+1$ over whose coordinates the cluster integral is extended

being represented by a bond between the points i and j. Figure D.1 shows the diagrammatic expansion for $c(R)$.

The diagrammatic approach consists firstly of identifying various classes of diagrams, and then expressing $h(R)$, $c(R)$ etc. in terms of linear combinations of these diagrams. One distinguishes: (a) Chains, $C(R)$ which contain at least one point (other than the root points 1 and 2) – called a nodal point – which is connected only to two other points so that when the diagram is cut it falls into two separate parts each containing one of the root points 1 and 2. Note that this type of diagram cannot contain a 12-bond. (b) Bundles, $B(R)$ which contain at least two independent routes from 1 to 2, one of which may be a direct 12-bond. (c) Elementary diagrams $E(R)$, i.e. all those which are neither chains nor bundles. Each function $C(R)$, $B(R)$, $E(R)$ stands for the sum over all diagrams of the appropriate class.

The total correlation function is the sum of all connected diagrams, i.e.

$$h(R) = C(R) + B(R) + E(R) \quad . \tag{D.10}$$

From (D.8–9) one can convince oneself that the expansion for $c(R)$ contains no nodes, and thus the direct correlation function is given by

$$c(R) = B(R) + E(R) \quad . \tag{D.11}$$

Furthermore, the indirect interaction $\psi(R) - \Phi(R)$ excludes all 12-interactions, i.e. all bundles, so that

$$\beta[\Phi(R) - \psi(R)] = C(R) + E(R) \quad . \tag{D.12}$$

From (D.10–12) we get the exact relation

$$c(R) = h(R) - \ln g(R) - \beta\Phi(R) + E(R) \quad . \tag{D.13}$$

The hypernetted chain approximation consists of dropping the very difficult elementary graphs, thus giving a closure of the form [see also (4.4)]

$$c_{HNC}(R) = h(R) - \ln g(R) - \beta \Phi(R) \quad . \tag{D.14}$$

This may be expressed in the form $c(R) = h(R) - \ln \{1 + [y(R) - 1]\}$ with $y(R) = g(R) \exp[\beta \Phi(R)]$. Linearizing the logarithm, and introducing a new class $D(R)$ to preserve (D.14) yields

$$c(R) = h(R) - [y(R) - 1] + D(R) \quad . \tag{D.15}$$

The Percus-Yevick approximation consists of setting $D(R) = 0$, i.e. in neglecting an even larger class of diagrams,

$$\begin{aligned} c_{PY}(R) &= g(R) - y(R) \\ &= g(R)\{1 - \exp[\beta \Phi(R)]\} \quad . \end{aligned} \tag{D.16}$$

The mean-spherical approximation cannot be obtained from a cluster expansion in terms of Mayer functions. It is based on the separation of $\Phi(R)$ into a reference part $\Phi_0(R)$ and a perturbation $\Phi_1(R)$ according to the WCA prescription (4.6). *Madden* and *Rice* [D.5] have shown that the MSA can be obtained by using the PY closure for the reference system, i.e.

$$c_0(R) = g_0(R)\{1 - \exp[\beta \Phi_0(R)]\} \tag{D.17a}$$

and describing the effects of the long-range perturbations by the HNC closure in the residual Ornstein-Zernike equation connecting $c_1(R)$, $h_1(R)$ and $\Phi_1(R)$ [see (4.17–21)]

$$c_1(R) = g_1(R) - \beta \Phi_1(R) - \ln [1 - g_1(R)/g_0(R)] \quad . \tag{D.17b}$$

The soft-potential MSA follows from (D.17) with $g(R) \approx g_0(R)$

$$c(R) = g(R)\{1 - \exp[\beta \Phi_0(R)]\} - \beta \Phi_1(R) \quad , \tag{D.18}$$

(a complete diagrammatic derivation is given in [D.6]). The HNC, PY, and MSA integral equations have been the fundamental tools of liquid state theory for a long time and their respective strengths and weaknesses are discussed at length in the specialized literature. Recently, attempts have been made to improve upon these standard fluid integral equations. One has been based upon the discovery that the sum $E(R)$ of the elementary graphs is (within reasonable accuracy) a universal function [D.7] and may be taken from computer-simulation data on e.g. hard-sphere fluids with the effective hard-sphere diameter being determined by a variational criterion [D.8]. This modified hypernetted chain scheme (MHNC) yields very good structural and thermodynamic properties for various monatomic model fluids [D.7], but its extension to mixtures is nontrivial because the effective interactions are not necessarily additive as in a hard-sphere mixture (cf. the discussion in Sect. 9.3). The alternative route towards improving the fluid integral equations is based on the old ideas of mixing the HNC and PY closures [D.9]. This attempt has recently been renewed by *Rogers* and *Young* [D.10] who proposed the following interpolation formula between the two closures (D.14, 15)

$$g(R) = \exp[-\beta\Phi(R)]\exp[h(R) - c(R)] \qquad \text{HNC}$$

$$g(R) = \exp[-\beta\Phi(R)]\{1 + [h(R) - c(R)]\} \qquad \text{PY}$$

$$g_{RY}(R) = \exp[-\beta\Phi(R)]\left(1 + \frac{\exp\{f(R)[h(R) - c(R)]\} - 1}{f(R)}\right) \quad \text{(D.19)}$$

with a "switching" function $f(R)$ that satisfies $f(0) = 0$ and $f(\infty) = 1$, e.g. $f(R) = 1 - \exp(-\alpha R)$. The value of the parameter α is then determined by imposing thermodynamic consistency for the pressure and the compressibility equations of state (D.5,6). This method was motivated by the fact that the HNC and PY pair-distribution functions "bracket" the exact solution for purely repulsive potentials – in that case the method yields excellent results, even for binary mixtures [D.11]. For potentials with attractive interactions such as the Lennard-Jones, however, it has been shown [D.12] that the consistency problem cannot be solved by varying the mixing parameter α in (D.19) between zero and infinity. In that case, *Zerah* and *Hansen* [D.12] proposed to interpolate between the HNC closure (D.14) and the SMSA closure (D.18) in much the same way as Rogers and Young interpolated between HNC and PY, i.e.

$$g(R) = \exp[-\beta\Phi(R)]\exp[h(R) - c(R)] \qquad \text{HNC}$$

$$g(R) = \exp[-\beta\Phi_0(R)][1 + h(R) - c(R) - \beta\Phi_1(R)] \quad \text{SMSA}$$

$$g_{ZH}(R) = \exp[-\beta\Phi_0(R)]$$
$$\times \left(1 + \frac{\exp\{f(R)[h(R) - c(R) - \beta\Phi_1(R)]\} - 1}{f(R)}\right) \quad , \text{(D.20)}$$

where $f(R) = 1 - \exp(-\alpha R)$ is again the same switching function[1]. With the Zerah-Hansen closure (D.20) excellent thermodynamic consistency and accurate structural data may be obtained for both monatomic systems and mixtures with Lennard-Jones interactions. This would suggest that (D.20) very nearly achieves the goal of defining a thermodynamically consistent and universal fluid integral equation.

---

[1] Note that for a purely repulsive potential, i.e. $\Phi_1(R) = 0$, the RY closure is recovered, because in this case the SMSA reduces to the PY closure.

However, the implementation of this technique for metallic systems is nontrivial because of the additional problem of electronic consistency: we cannot simply equate the compressibility given by (D.6) and that derived by differentiating (D.5) because they are different even for the exact $g(R)$ (for a discussion of the compressibility problem with special reference to liquid metals, see [D.13, 14]). We must look for a way to account for the difference.

For the MSA we have shown in Sects. 4.3 and 9.3 that the integral equation is equivalent to a variational problem. This is only one special case: in fact any of the fluid integral equations is equivalent to a special variational condition [D.15, 16]. Defining the functionals

$$R_h[c] = \frac{1}{(2\pi)^3} \int \{n\hat{c}(q) + \ln [1 - n\hat{c}(q)]\} d^3q$$

$$R_c[h] = \frac{1}{(2\pi)^3} \int \{n\hat{h}(q) - \ln [1 + n\hat{h}(q)]\} d^3q \tag{D.21}$$

we find that the functional differential equations

$$\frac{\delta R_h[c]}{\delta c}(q) = -\frac{n^2}{(2\pi)^3}\hat{h}(q) \quad \text{or} \quad \frac{\delta R_h[c]}{\delta c}(R) = -n^2 h(R) \tag{D.22a}$$

$$\frac{\delta R_c[h]}{\delta h}(q) = -\frac{n^2}{(2\pi)^3}\hat{c}(q) \quad \text{or} \quad \frac{\delta R_c[h]}{\delta h}(R) = -n^2 c(R) \tag{D.22b}$$

are equivalent to the Ornstein-Zernike equation and its inverse, in reciprocal and in real space. Based on the relations (D.21, 22) and one of the closure relations it is easy for any of the closures, to find a functional $\mathcal{F}_h^x[c]$ or $\mathcal{F}_c^x[h]$ so that at least one of the variational conditions

$$\frac{\delta \mathcal{F}_h^x[c]}{\delta c}(R) = 0 \quad \text{or} \quad \frac{\delta \mathcal{F}_c^x[h]}{\delta h}(R) = 0 \tag{D.23}$$

is equivalent to the corresponding integral equation. These functionals are given in the HNC approximation by

$$\mathcal{F}_c^{\text{HNC}}[h] = R_c[h] + n^2 \int [\tfrac{1}{2}h^2(R) - g(R) \ln g(R)$$
$$+ h(R) - \beta\Phi(R)h(R)] d^3 R \tag{D.24}$$

and in the Percus-Yevick approximation by

$$\mathcal{F}_h^{\text{PY}}[c] = R_h[c] + n^2 \int \left[\frac{1}{2}\frac{c^2(R)}{1 - \exp[\beta\Phi(R)]} - c(R)\right] d^3 R \tag{D.25a}$$

and

$$\mathcal{F}_c^{PY}[h] = R_c[h] + n^2 \int \{ \tfrac{1}{2} h^2(R)[1 - e^{\beta\Phi(R)}] + h(R) \} d^3 R \qquad \text{(D.25b)}$$

and in the MSA by

$$
\begin{aligned}
\mathcal{F}_h^{SMSA}[c] = R_h[c] + n^2 \int \Bigg[& \frac{1}{2} \frac{c^2(R)}{1 - \exp[\beta\Phi_0(R)]} \\
& + \left(\frac{\beta\Phi_1(R)}{1 - \exp[\beta\Phi_0(R)]} - 1 \right) c(R) \Bigg] d^3 R \quad .
\end{aligned}
\qquad \text{(D.26)}
$$

For the hard-core limit, (D.23) reduces immediately to

$$\mathcal{F}_h^{MSA}[c] = R_h[c] - n^2 \int c(R) d^3 R \quad \text{for} \quad R < \sigma \quad . \qquad \text{(D.27)}$$

In each case the "generating" functionals (D.24–27) have a well-defined physical meaning. In the PY approximation we have [D.16,17]

$$\frac{1}{n} \mathcal{F}_h^{PY}[c] = Z_c^{PY} - 1 = \left(\frac{\beta P}{n} \right)_c - 1 \qquad \text{(D.28)}$$

where the suffix "c" means that the compressibility factor $Z = (\beta P/n)$ has been calculated using the compressibility equation of state (D.6). Equation (D.25a) predicts Z_c^{PY} to be stationary with respect to small variations in $c(R)$. The functional $\mathcal{F}_c^{HNC}$ is found to represent the excess free energy obtained from either the virial or the energy equations within the HNC approximation [D.15,16,18], i.e.

$$\frac{1}{n} \mathcal{F}_c^{HNC}[h] = \frac{1}{N} \beta F \quad . \qquad \text{(D.29)}$$

The variational principle is equivalent here to a minimization of the associated free energy. Within the MSA, the physical role of the generating functional depends on the context of its application; in the HS limit $\mathcal{F}_h^{HSA}[c]$ is just the RPA contribution to the free energy [see (4.25, 26)]. Thus we can now place the simple variational expression (4.27) of the ORPA into a large class of variational criteria.

Finally, the variational expressions allow us to make the connection between the integral-equation methods and the thermodynamic variational methods. For a system depending on some parameter σ (i.e. a hard-sphere diameter), the Gibbs-Bogoljubov inequality (see Sect. 3.5.3) predicts that the free energy is a non-decreasing function of the parameter σ:

$$\frac{\partial F}{\partial \sigma} \geq 0 \quad . \qquad \text{(D.30)}$$

Using the variational principles *Rosenfeld* [D.16] has shown that the excess free energies calculated within PY theory using the compressibility equation or the energy or virial equation in HNC theory, satisfy the exact inequality (D.30). If the generating functional $\mathcal{F}$ represents the excess free energy, this will also be the case for the MSA.

Up to now, the discussion has centred on monatomic liquids. The Ornstein-Zernike equation may be generalized straightforwardly to binary liquids and reads either

$$h_{ij}(R_{12}) = c_{ij}(R_{12}) + \sum_k n_k \int c_{ik}(R_{13})h_{kj}(R_{32})d^3R_3 \qquad \text{(D.31a)}$$

or

$$h_{NN}(R_{12}) = c_{NN}(R_{12}) + n \int c_{NN}(R_{13})h_{NN}(R_{32})d^3R_3$$
$$+ n c_A c_B \int c_{Nc}(R_{13})h_{cN}(R_{32})d^3R_3$$

$$h_{Nc}(R_{12}) = c_{Nc}(R_{12}) + n \int c_{NN}(R_{13})h_{Nc}(R_{32})d^3R_3$$
$$+ n \int c_{Nc}(R_{13})h_{cc}(R_{32})d^3R_3$$

$$h_{cc}(R_{12}) = c_{cc}(R_{12}) + n c_A c_B \int c_{cN}(R_{13})h_{Nc}(R_{32})d^3R_3$$
$$+ n \int c_{cc}(R_{13})h_{cc}(R_{32})d^3R_3 \quad , \qquad \text{(D.31b)}$$

depending on the conventions adopted for the partial correlation functions. The Fourier-transformed OZ equations yield the following expressions for the static structure factors

$$S_{ii}(q) = [1 - n_j \hat{c}_{jj}(q)]/D(q) = 1 + n_i \hat{h}_{ii}(q) \quad , \quad i \neq j$$

$$S_{ij}(q) = (n_i n_j)^{1/2} \hat{c}_{ij}(q)/D(q) = (n_i n_j)^{1/2} \hat{h}_{ij}(q) \quad , \quad i \neq j$$

$$D(q) = [1 - n_A \hat{c}_{AA}(q)][1 - n_B \hat{c}_{BB}(q)] - n_A n_B \hat{c}_{AB}(q)^2$$

and also

$$S_{NN}(q) = \frac{1 - n\hat{c}_{cc}(q)}{\tilde{D}(q)} = \hat{h}_{NN}(q) + 1$$

$$S_{Nc}(q) = \frac{c_A c_B n \hat{c}_{Nc}(q)}{\tilde{D}(q)} = n c_A c_B \hat{h}_{Nc}(q)$$

$$S_{cc}(q) = \frac{c_A c_B (1 - n\hat{c}_{NN})}{\tilde{D}(q)} = c_A c_B [n\hat{h}_{cc}(q) - 1]$$

$$\tilde{D}(q) = [1 - n\hat{c}_{NN}(q)][1 - n\hat{c}_{cc}(q)] - n^2 c_A c_B \hat{c}_{Nc}(q)^2$$
$$= D(q) \quad .$$

The closure relations defining the various approximations are valid for the partial correlation functions and partial pair potentials $\Phi_{ij}(R)$. Together with (D.31) they define a coupled set of three integro-differential equations. The variational principles are also straightforwardly generalized to binary mixtures – explicit formulations are found in [D.15–18].

D.3 Analytical Solutions for Model Liquids and Mixtures

D.3.1 Solution of the PY Equation for the Hard-Sphere Fluid

We give here a very brief summary of *Wertheim*'s [D.19] analytical solution of the PY equation for hard spheres (see also [D.20]). For a hard-sphere liquid the PY closure assumes the form

$$g(R) = 0 \quad R < \sigma$$

$$c(R) = 0 \quad R > \sigma \quad . \tag{D.32}$$

This suggests that we define a new function $\tau(R)$

$$\tau(R) = -c(R) \quad R < \sigma$$
$$\tau(R) = \quad g(R) \quad R > \sigma \tag{D.33}$$

that satisfies the closure automatically. In terms of $\tau(R)$, the OZ equation reads

$$\tau(R) = 1 + n \int_{R' < \sigma} \tau(R') d^3 R'$$
$$- n \int_{\substack{R' < \sigma \\ |R - R'| > \sigma}} \tau(R') \tau(|R - R'|) d^3 R' \quad . \tag{D.34}$$

Equation (D.34) shows that $\tau(R)$ is continuous and continuously differentiable at $R = \sigma$. A Laplace transformation of (D.34) according to

$$F(t) = \sigma^{-2} \int_0^{\sigma} R\tau(R) e^{-tR/\sigma} dR \tag{D.35a}$$

$$G(t) = \sigma^{-2} \int_{\sigma}^{\infty} R\tau(R)\, e^{-tR/\sigma} dR \tag{D.35b}$$

$$K = \sigma^{-3} \int_{0}^{\sigma} \tau(R) R^2 dR \tag{D.35c}$$

yields

$$t[G(t) - F(t)] = (1 - 24\eta K)\frac{1}{t} + 12\eta[F(-t) - F(t)]G(t). \tag{D.36}$$

The investigation of the analytical properties of $F(t)$ and $G(t)$ in the complex t-plane suggests that inside the hard core $c(R)$ is a cubic polynomial in R :

$$-c(R) = \alpha + \beta\left(\frac{R}{\sigma}\right) + \gamma\left(\frac{R}{\sigma}\right)^2 + \delta\left(\frac{R}{\sigma}\right)^3 \quad , \quad R<\sigma \tag{D.37}$$

whose coefficients may be determined from the continuity of $c(R)$ and its derivative at $R = \sigma$ with the result

$$\alpha = (1 + 2\eta)^2/(1 - \eta)^4$$

$$\beta = -6\eta(1 + \eta/2)^2/(1 - \eta)^4$$

$$\gamma = 0$$

$$\delta = \tfrac{1}{2}\eta(1 + 2\eta)^2/(1 - \eta)^4 \quad . \tag{D.38}$$

The pressure and compressibility equations of state are obtained with the result

$$\left(\frac{\beta p}{n}\right)_p = \frac{1 + \eta + \eta^2 - 3\eta^3}{(1 - \eta)^3}$$

$$\left(\frac{\beta p}{n}\right)_c = \frac{1 + \eta + \eta^2}{(1 - \eta)^3} \quad . \tag{D.39}$$

The semiempirical Carnahan-Starling (CS) equation of state (which gives an exact fit to machine calculations [D.21]) is

$$\left(\frac{\beta p}{n}\right)_{CS} = \frac{2}{3}\left(\frac{\beta p}{n}\right)_c + \frac{1}{3}\left(\frac{\beta p}{n}\right)_p \quad . \tag{D.40}$$

The hard-sphere entropies are calculated by integrating the equations of state to give

$$(S_\eta/k_B)_c = \ln(1 - \eta) + \tfrac{3}{2}[1 - 1/(1 - \eta)^2]$$

$$(S_\eta/k_B)_p = -2 \ln (1 - \eta) + 6[1 - 1/(1 - \eta)] \tag{D.41}$$

and the same average as in (D.37) defines the CS entropy. The static structure factor $S(q)$ follows by Fourier transformation of (D.34), see [D.20] (here $u = \sigma q$)

$$\begin{aligned}
\hat{c}(q) = \sigma^3 \{ &\alpha(\sin u/u^3 - \cos u/u^2) \\
&+ \beta[2 \sin u/u^3 - \cos u/u^2 + 2(\cos u - 1)/u^4] \\
&+ \gamma[\sin u(4/u^3 - 24/u^5) + \cos u(12/u^4 - 1/u^2) \\
&+ 24(1 - \cos u)/u^6]\} \quad .
\end{aligned} \tag{D.42}$$

D.3.2 Solution of the PY Equation for Hard-Sphere Mixtures

The analytical solution of the PY equation for binary mixtures of hard spheres with diameters σ_A and σ_B has been given by *Lebowitz* [D.22] and *Lebowitz* and *Rowlinson* [D.23]. The partial direct correlation functions are now given by (they vanish again outside the hard spheres)

$$-c_{AA}(R) = \alpha_A + \beta_A R + \gamma R^3 \quad , \quad R < \sigma_A$$

$$-c_{BB}(R) = \alpha_B + \beta_B R + \gamma R^3 \quad , \quad R < \sigma_B \tag{D.43}$$

$$-c_{AB}(R) = \begin{cases} \alpha_A \quad , & R < \lambda = (\sigma_A - \sigma_B)/2 \\ \alpha_A + [\beta(R - \lambda)^2 + 4\lambda(R - \lambda)^3 + \delta(R - \lambda)^4]/R \quad , \\ & \lambda < R < \sigma_{AB} = (\sigma_A + \sigma_B)/2 \end{cases}$$

The coefficients are given by

$$\alpha_i = \beta \frac{\partial p_c}{\partial n_i}$$

$$\beta_i = -6[\eta_i g_{ii}^2(\sigma_i)/\sigma_i + \eta_j g_{ij}^2(\sigma_{jj})\sigma_{ij}^2/\sigma_{jj}^3] \quad , \quad j \neq i$$

$$\beta = -6[\eta_A g_{AA}(\sigma_A)/\sigma_A^2 + \eta_B g_{BB}(\sigma_B)/\sigma_B^2]\sigma_{AB} g_{AB}(\sigma_{AB})$$

$$d = \tfrac{1}{2}[\eta_A \alpha_A/\sigma_A^3 + \eta_B \alpha_B/\sigma_B^3]$$

$$g_{ii}(\sigma_i) = \{(1 - \eta) + \tfrac{3}{2}(\eta_i/\sigma_i + \eta_j/\sigma_j)\sigma_i\}/(1 - \eta)^2 \quad , \quad j \neq i$$

$$g_{ij}(\sigma_{ij}) = \{\sigma_B g_{AA}(\sigma_A) + \sigma_A g_{BB}(\sigma_B)\}/2\sigma_{AB} \tag{D.44}$$

with the total and partial packing fractions η, η_A, η_B. The equations of state are now given by

$$\left(\frac{\beta p}{n}\right)_c = \frac{1 + \eta + \eta^2 - 3\eta(y_1 + y_2 n)}{(1 - \eta)^3} \tag{D.45a}$$

$$\left(\frac{\beta p}{n}\right)_p = \frac{1 + \eta + \eta^2 - 3\eta(y_1 + y_2\eta + y_3\eta^2)}{(1 - \eta)^3} \tag{D.45b}$$

with

$$y_1 = \frac{1}{2}\sum_{i \neq j} \Delta_{ij}(\sigma_i + \sigma_j)/(\sigma_i\sigma_j)^{1/2}$$

$$y_2 = \frac{1}{2}\sum_{i \neq j} \Delta_{ij}\sum_k (\eta_k/\eta)(\sigma_i\sigma_j)^{1/2}/\sigma_k$$

$$y_3 = \left[\sum_i (\eta_i/\eta)^{2/3} c_i^{1/3}\right]^3$$

$$\Delta_{ij} = [(\eta_i\eta_j)^{1/2}/\eta][(\sigma_i - \sigma_j)^2/\sigma_i\sigma_j](c_ic_j)^{1/2} \quad . \tag{D.46}$$

Again the generalization of the CS relation (D.40) defines an equation of state which is in excellent agreement with simulation data [D.24]. The entropy is now given by a term S_η depending only on the mean packing fraction η [S_η is given by (D.41)] and a term S_σ which depends on the size mismatch, e.g. in the compressibility equation of state

$$\frac{S_\sigma}{\kappa_B} = \frac{\pi c_A c_B n(\sigma_A - \sigma_B)^2[12(\sigma_A + \sigma_B) - \pi n(c_A\sigma_A^4 + c_B\sigma_B^4]}{24(1 - \eta)^2} \tag{D.47}$$

The partial structure factors $S_{ij}(q)$ are defined by the Fourier transformation of (D.43) in conjunction with (D.30). The explicit expressions are quite lengthy, and may be found in [D.25–27].

D.3.3 The Solution of the MSA for Charged Hard-Spheres with Yukawa Interactions

Waisman [D.28] has applied Wertheim's method to the MSA for charged hard spheres with Yukawa interactions and the Coulomb limit has been treated by *Palmer* and *Weeks* [D.29]. In the Yukawa case the interatomic potential of the form

$$\Phi(R) = \begin{cases} \infty & R < \sigma \\ 2Q^2\exp[-\kappa(R - \sigma)]/R & R \geq \sigma \end{cases} \tag{D.48}$$

defines the closure relation

$$c(R) = -\beta\Phi(R) = -2\beta Q^2\exp[-\kappa(R - \sigma)]/R \quad R \geq \sigma$$

$$h(R) = -1 \quad R < \sigma \quad . \tag{D.49}$$

Defining now a function $c_0(R) = c(R) - C \exp[-\kappa(R-\sigma)]/R$ with $C = -2\beta Q^2$, that vanishes outside the core we can follow directly Wertheim's path. The Laplace-transformed OZ equation analogous to (D.36) now reads

$$t[G(t) - F(t)] = (1 - 24\eta K')\frac{1}{t} + Ce^z\frac{t}{t+z}$$
$$+ 12\eta\left\{ G(t)[F(-t) - F(t)] + [G(z) - Ce^z]\frac{t}{t^2 - z^2}\right\} \qquad (D.50)$$

with $z = \kappa\sigma$. The functions $G(t)$ and $F(t)$ are the Laplace transforms of $g(R)$ and $c_0(R)$, and K' is defined analogously to (D.35c), but with the full $c(R)$. The analytical properties of this equation imply again a solution of the form

$$-c(R) = a + b\left(\frac{R}{\sigma}\right) + \frac{1}{2}a\left(\frac{R}{\sigma}\right)^3 + v\frac{1 - \exp(-\kappa R)}{\kappa R}$$
$$+ v^2\frac{\cosh(\kappa R) - 1}{2C\kappa^2\sigma R \exp(\kappa\sigma)}, \quad R < \sigma. \qquad (D.51)$$

Here the coefficients a, b and v have to satisfy the following equations

$$a = 1 - 24\eta K' = 1 - 24\eta\sigma^{-3}\int_0^\infty c(R)R^2 dR \qquad (D.52)$$

$$24\eta y_0^2 = -4b + 2vz - \frac{v^2}{Ce^z}$$

$$24\eta(y_1^2 - 2y_0y_2) = 24\eta a - 2vz^3 + \frac{v^2z^2}{Ce^z}, \quad \text{where}$$

$$y_i = -\left.\frac{d^i}{dx^i}xc_0(x)\right|_{\substack{x=1-\epsilon \\ \lim\ \epsilon \to 0}} \quad ; \quad x = \frac{R}{\sigma}\ . \qquad (D.53)$$

The thermodynamic excess functions may be calculated following the method proposed by *Hoye* and *Stell* [D.30], explicit expressions can be found in [D.31].

D.3.4 The Solution of the MSA for a Symmetric Mixture of Charged Hard Spheres with Yukawa Interactions

Waisman and *Lebowitz* [D.32] have solved the MSA for a mixture of hard spheres with diameters σ_A and σ_B and charges Q_A and Q_B with Coulomb interactions (see also [D.33,34]) and the case of Yukawa interactions has been investigated by *Ginoza* [D.35]. The solutions are rather complex and only for the "symmetric" case $(\sigma_A = \sigma_B,\ c_A Q_A + c_B Q_B = 0)$ are they sufficiently simple to have found application in thermodynamic variational calculations. The closure relations for this case have been formulated in Sect. 9.2.2 and Eqs. (9.16 ff). According to *Waisman* [D.36], $c_{NN}(R)$ is identical to the PY result for a monatomic HS fluid (D.37 ff), $c_{Nc}(R) = 0$ and $c_{cc}(R)$ is given by [compare with (D.51)!]

$$
c_{cc}(R) = \begin{cases} -\beta\varepsilon\omega\left(\dfrac{1 - \exp(-\kappa R)}{\kappa R} + \omega\,\dfrac{\cosh(\kappa R) - 1}{2\kappa^2\sigma R}\right) & R<\sigma \\[2mm] \beta\varepsilon\exp[-\kappa(R-\sigma)]/R & R>\sigma \end{cases} \tag{D.54}
$$

with $\varepsilon = -2c_A c_B(Q_A - Q_B)^2/\sigma$ being the strength of the interaction at hard contact and where the parameter ω is the solution of the quartic equation

$$
\beta\varepsilon\eta = \frac{\omega\{z - \omega/[2\exp(z)]\}}{12\{1 + \omega[1 - \exp(-z)]/2Z\}^4} \tag{D.55}
$$

with $z = \kappa\sigma$. The Fourier transform of the direct correlation function is given by [D.37]

$$
\begin{aligned}
\hat{c}_{cc}(q) = \frac{4\pi\beta\varepsilon}{q}\, A_1 &\left(\frac{1 - \cos(q\sigma)}{q} - \frac{q}{\kappa^2 + q^2}\right) \\
&+ A_2\left(\frac{\exp(-\kappa z)[\kappa\sin(q\sigma) + q\cos(q\sigma)]}{\kappa^2 + q^2}\right) \\
&+ A_3\left(\frac{\kappa\sinh(z)\sin(q\sigma) - q\cosh(z)\cos(q\sigma)}{\kappa^2 + q^2}\right)
\end{aligned} \tag{D.56}
$$

with

$$
A_1 = \frac{\omega}{\kappa}\left(\frac{\omega}{2z\exp(z)} - 1\right)
$$

$$
A_2 = \sigma\exp(z) - \omega/\kappa
$$

$$
A_3 = \omega^2/\kappa z\exp(z) \tag{D.57}
$$

and defines the concentration fluctuation structure factor through $S_{cc}(q) = c_A c_B[1 - n\hat{c}_{cc}(q)]^{-1}$. The ordering contributions to the enthalpy and entropy of formation are given by [D.37]

$$
\Delta H_{ord} = 2\pi n\int g_{cc}(R)\Phi_{cc}(R)R^2 dR = -\varepsilon\omega/2 \quad \text{and} \tag{D.58}
$$

$$\Delta S_{\text{ord.}} = \frac{f(\omega) - f(0)}{2\eta} \quad \text{with} \tag{D.59}$$

$$f(\omega) = -\frac{B_1 + B_2\omega + B_3\omega^2}{72B_4^2(1 + B_4\omega)^3}$$

$$B_1 = z\{2/[\exp(z) - 1] - 1\} \qquad B_2 = -\tfrac{3}{2}[1 - 3\exp(-z)]/2$$
$$B_3 = \exp(-z)[1 - \exp(-z)]/4z \qquad B_4 = [1 - \exp(-z)]/2z \quad . \tag{D.60}$$

References

1.1   W. Hume-Rothery: *Electrons, Atoms, Metals,and Alloys*, (Louis Cassier, London 1948); republ. by (Dover, New York 1973)

1.2   E.P. Wigner, F. Seitz: Phys. Rev. **43**, 804 (1933)

1.3   H. Brooks: Trans. Metall. Soc. AIME **227**, 546 (1963)

1.4.   P. Hohenberg, W. Kohn: Phys. Rev. **136**, B864 (1964)

1.5   W. Kohn, L.J. Sham: Phys. Rev. **140**, A1133 (1965)

1.6   O. Gunnarsson, B.I. Lundqvist: Phys. Rev. B **13**, 4274 (1976)

1.7   J. Harris, R.O. Jones: J. Chem. Phys. **68**, 3316 (1978)

1.8   V.L. Moruzzi, J.F. Janak, A.R. Williams: *Calculated Electronic Properies of Metals*, (Pergamon, New York 1978)

1.9   J.F. Janak, A.R. Williams, Phys. Rev. B14, 4199 (1976)

1.10   H.L. Skriver, O.K. Andersen, B. Johansson: Phys. Rev. Lett. **41**, 41 (1978); ibid. **44**, 1230 (1980)

1.11   J. Friedel: *The Physics of Metals I - Electrons*, ed. by J.M. Ziman (Cambridge Uni. Press, Cambridge 1968) p.494

1.12   D. de Fontaine: Solid State Phys. **34**, 73 (1979)

1.13   W.A. Harrison: *Pseudopotentials in the Theory of Metals* (Benjamin, New York 1966)

1.14   V. Heine, D. Weaire: Solid State Phys. **24**, 247 (1971)

1.15   W.A. Harrison: *Electronic Structure and the Properties of Solids - The Physics of the Chemical Bond* (Freeman, San Francisco 1980)

1.16   J.A. Moriarty: Phys. Rev. Lett. **55**, 1502 (1985)

1.17   F. Ducastelle, F. Gautier: J. Phys. F **6**, 2039 (1976)

1.18   A. Bieber, F. Gautier: Z. Phys. B **57**, 335 (1984)

1.19   M. Born, J.R. Oppenheimer: Ann. Phys. **84**, 457 (1927)

1.20   G.S. Chester: Adv. Phys. **10**, 357 (1958)

1.21   S. Raimes: *The Wave Mechanics of Electrons in Metals* (North-Holland, Amsterdam 1961)

1.22   D. Bohm, D. Pines: Phys. Rev. **85**, 338 (1952); ibid. **92**, 609 (1953)

1.23   P. Nozières, D. Pines: Phys. Rev. **111**, 442 (1958)

1.24   See, e.g., D.M. Ceperley, B.J. Alder: Phys. Rev. Lett. **45**, 566 (1980);
       S.H. Vosko, L. Wilk, M. Nusair, Can. J. Phys. **58**, 1200 (1980)

1.25 S. Hedin, S. Lundqvist: Solid State Phys. **23**, 1 (1969)

1.26 J. Callaway, N.H. March: Solid State Phys. **38**, 136 (1984)

1.27 P. Gombas: *Die statistische Theorie des Atoms und ihre Anwendungen* (Springer, Berlin 1949)

1.28 J.C. Slater: Phys. Rev. **81**, 385 (1951)

1.29 F. Herman, S. Skillman: *Atomic Structure Calculations* (Prentice Hall, Englewood Cliffs, NJ 1963)

1.30 J.B. Mann: *Atomic Structure Calculations: Hartree-Fock Energy Results for Elements from Hydrogen to Lawrencium* (Clearinghouse for Technical Information, Springfield, Va 1967)

1.31 J.C. Slater, G.F. Koster: Phys. Rev. **94**, 1498 (1954)

1.32 J.C. Slater: Phys. Rev. **51**, 846 (1937)

1.33 T. Loucks: *Augmented Plane Wave Method* (Benjamin, New York 1967)

1.34 C. Herring: Phys. Rev. **57**, 1169 (1940)

1.35 H. Hellmann: Acta Physicochim. USSR **1**, 913 (1935)

1.36 H. Hellmann: Acta Physicochim. USSR **4**, 225 (1938)

1.37 P. Gombas: Z. Phys. **111**, 938 (1938)

1.38 E. Antoncik: J. Phys. Chem. Solids **10**, 314 (1959)

1.39 J.C. Phillips: Phys. Rev. **112**, 685 (1958)

1.40 J.C. Phillips, L. Kleinman: Phys. Rev. **116**, 287 (1959); ibid. p.880

1.41 V. Heine: Solid State Phys. **24**, 1 (1971)

1.42 J.M. Ziman: Solid State Phys. **26**, 1 (1971)

1.43 R. Evans: *Electrons in Disordered Metals and at Metallic Surfaces*, ed. by P. Phariseau, B.L. Gyorffy, L. Scheire, NATO Adv. Study Institutes Series B (Plenum, New York 1978) Vol.42, p.415

1.44 D. Pines, P. Nozières: *Quantum Liquids* (Benjamin, New York 1966)

1.45 M.H. Cohen: *Metallic Solid Solutions*, ed. by J. Friedel, A. Guinier, (Benjamin, New York 1962) p.IX-1

1.46 H. Ehrenreich, M.H. Cohen: Phys. Rev. **115**, 786 (1959)

1.47 J. Lindhard: K. Dan. Vidensk. Selsk. Mat.-Fys. Medd. **28**, 8 (1954)

1.48 A. Blandin: *Phase Stability in Metals and Alloys*, ed. by P.S. Rudman, J. Stringer, R.I. Jaffee (McGraw Hill, New York 1965) p.115

1.49 J. Hafner, V. Heine: J. Phys. F **16**, 5779 (1986)

1.50 A.A. Maradudin, E.W. Montroll, G.H. Weiss, I.P. Ipatova: *Theory of the Lattice Dynamics in the Harmonic Approximation*, 2nd ed., Solid State Phys. Suppl. 3 (Academic, New York 1971)

1.51 P. Choquard: *The Anharmonic Crystal* (Benjamin, New York 1967)

1.52	See, e.g., A. Isihara: J. Phys. A **1**, 539 (1968)
1.53	See, e.g., A.S. Besicovich: *Almost Periodic Functions* (Cambridge University Press, London 1932)
1.54	P.J. Steindhardt: *Amorphous Metals and Semiconductors*, ed. by P. Haasen, R.I. Jaffee (Pergamon, London 1986) p.70
1.55	See, e.g., Z.G. Pinsker: *Dynamical Scattering of X-Rays in Crystals*, Springer Ser. Solid-State Sci., Vol.3 (Springer, Berlin, Heidelberg 1978)
1.56	See, e.g., P. Chieux: *Neutron Diffraction*, ed. by H. Dachs, Topics Curr. Phys., Vol. 6 (Springer, Berlin, Heidelberg 1978) p.271
1.57	R. Penrose: Bull. Inst. Math. and Its Appl. **10**, 266 (1974)
1.58	D. R. Nelson, S. Sachdev: *Amorphous Metals and Semiconductors*, ed. by P. Haasen, R.I. Jaffee (Pergamon, London 1986) p.28
1.59	L. Bendersky: Phys. Rev. Lett. **55**, 1461 (1985)
1.60	J.L. Beeby, S.F. Edwards: Proc. R. Soc. London **A274**, 395 (1963); P.L. Leath, B. Goodman: Phys. Rev. **181**, 1062 (1969)
1.61	D.W. Taylor: Phys. Rev. **156**, 1017 (1967)
1.62	P. Soven: Phys. Rev. **156**, 809 (1967)
1.63	M. Mostoller, T. Kaplan: Phys. Rev. B **16**, 2350 (1977)
1.64	F. Ducastelle: J. Phys. F. **2**, 468 (1972)
1.65	L.M. Roth: Phys. Rev. B **9**, 2476 (1974); L. Huisman, D. Nicholson, L. Schwartz, A. Bansil: Phys. Rev. B **24**, 1824 (1981)
1.66	V. Heine: Solid State Phys. **35**, 1 (1980)
1.67	J. Friedel: Adv. Phys. **3**, 446 (1954)
1.68	See the articles by V. Heine, R. Haydock, M.J. Kelly in *Solid State Phys.* **35** (Academic, New York 1980)
1.69	F. Cyrot-Lackmann: J. Physique C1, 67 (1970)
1.70	L.M. Falicov, F. Yndurain: Phys. Rev. B **12**, 5664 (1975)
1.71	D. Weaire, A.R. Williams: J. Phys. C **10**, 1239 (1977)
1.72	B.E. Warren: *X-Ray Diffraction* (Addison-Wesley, Reading, MA 1969) p.229; J.M. Cowley: Phys. Rev. **77**, 669 (1950); ibid. **120**, 1179 (1960)
1.73	W.L. Bragg, E.J. Williams: Proc. R. Soc. London **A145**, 699 (1934)
1.74	A.G. Khatchaturyan: Phys. Status Solidi **60**, 9 (1973)
1.75	J.P. Hansen, I.R. McDonald: *Theory of Simple Liquids* (Wiley, New York 1971)
1.76	M. Born, H.S. Green: Proc. R. Soc. London **A188**, 10 (1946); J. Yvon: *La Théorie Statistique des Fluides et l'Equation d'Etat*, Actualités Scientifiques et Actuelles, Vol.203 (Hermann, Paris 1935)

1.77 J.G. Kirkwood: J. Chem. Phys. **3**, 300 (1935)

1.78 W. Olivares, D.A. McQuarrie: J. Chem. Phys. **65**, 3604 (1976); Y. Rosenfeld: J. Stat. Phys. **37**, 215 (1984); ibid. **42**, 437 (1986)

1.79 L.D. Landau: Phys. Z. Sowjetunion **5**, 172 (1934)

1.80 L.D. Landau, E.M. Lifshitz: *Statistical Physics*, Course on Theoretical Physics, Vol.5 (Pergamon, London 1958) Chap.XIV

1.81 J.G. Kirkwodd, E. Monroe: J. Chem. Phys. **9**, 514 (1941)

1.82 R. Evans: Adv. Phys. **28**, 143 (1979)

1.83 T.V. Ramakrishnan, M. Yussouff: Phys. Rev. B **19**, 2775 (1979)

1.84 A.D.J. Haymet, D.W. Oxtoby: J. Chem. Phys. **74**, 2559 (1981); ibid. **84**, 1769 (1986)

1.85 F. Igloi, G. Kahl, J. Hafner: J. Phys. C **20**, 1803 (1987)

2.1 M. Born, J.R.Oppenheimer: Ann. Phys. **84**, 457 (1927)

2.2 G.S. Chester: Adv. Phys. **10**, 357 (1958)

2.3 F. Ducastelle, F. Gautier: J. Phys. F **6**, 2039 (1976)

2.4 C.M. Varma, W. Weber: Phys. Rev. B **19**, 6142 (1979)

2.5 M.W. Finnis, K. Kear, D.G. Pettifor: Phys. Rev. Lett. **52**, 291 (1984)

2.6 A.E. Carlsson, N.W. Ashcroft, Phys. Rev. B **27**, 2101 (1983)

2.7 A.H. MacDonald, R. Taylor: Can. J. Phys. **62**, 796 (1984)

2.8 L. Dagens: J. Phys. F **16**, 1705 (1986)

2.9 H. Hellmann: Acta Physicochim. USSR, **1**, 913 (1935)

2.10 H. Hellmann: Acta Physicochim. USSR, **4**, 225 (1936)

2.11 P. Gombas: Z. Phys. **111**, 938 (1938)

2.12 J.C. Phillips, L. Kleinman: Phys. Rev. **115**, 287 (1959); ibid p.880

2.13 E. Antoncik: J. Phys. Chem. Solids **10**, 314 (1959) and references therein

2.14 B.J. Austin, V. Heine, L.J. Sham: Phys. Rev. **127**, 276 (1962)

2.15 M.H. Cohen, V.Heine: Phys. Rev. **122**, 1821 (1961)

2.16 W.A. Harrison: *Pseudopotentials in the Theory of Metals* (Benjamin, New York 1966)

2.17 J. Hafner: J. Phys. F **6**, 1243 (1976)

2.18 A. Zunger, M.L. Cohen: Phys. Rev. B **18**, 5449 (1978), and references cited therein

2.19 W.A. Harrison, Phys. Rev. **181**, 1036 (1969)

2.20 J.A. Moriarty: Phys. Rev. B **1**, 1363 (1970); ibid. B **6**, 1239 (1972)

2.21 J.A. Moriarty: Phys. Rev. B **16**, 2537 (1977)

2.22 J.A. Moriarty: Phys. Rev. B **26**, 1754 (1982)

2.23 V. Heine: Phys. Rev. **153**, 673 (1967)

2.24 V. Heine, I.V. Abarenkov: Philos. Mag. **9**, 451 (1964)

2.25 A.O.E. Animalu, V. Heine: Philos. Mag. **12**, 1249 (1965)

2.26   R.W. Shaw: Phys. Rev. **174**, 769 (1968)

2.27   M. Appapillai, A.R. Williams: J. Phys. F **3**, 759 (1973)

2.28   A.R. Williams, M. Appapillai: J. Phys. F **3**, 772 (1973)

2.29   L. Dagens, M. Rasolt, R. Taylor: Phys. Rev. B **11**, 2726 (1975)

2.30   M. Rasolt, R. Taylor: Phys. Rev. B **11**, 2717 (1975)

2.31   A.O.E. Animalu: Phys. Rev. B **8**, 3542 (1973)

2.32   L. Dagens: J. Phys. F **6**, 1801 (1976)

2.33   R.W. Shaw, W.A. Harrison: Phys. Rev. **163**, 604 (1967)

3.34   W.C. Topp, J.J. Hopfield: Phys. Rev. B **7**, 1295 (1974)

2.35   D.R. Hamann, M. Schlüter, C. Chiang: Phys. Rev. Lett. **43**, 1494 (1979)

2.36   A. Zunger: J. Vac. Sci. Technol. **16**, 1337 (1979)

2.37   P.A. Christiansen, Y.S. Lee, K.S. Pitzer: J. Chem. Phys. **71**, 4445 (1979)

2.38   G.P. Kerker: J. Phys. C **13**, L189 (1980)

2.39   G.B. Bachelet, H.S. Greenside, G.A. Baraff, M. Schlüter: Phys. Rev. B **24**, 4745 (1981)

2.40   G.B. Bachelet, D.R. Hamann, M. Schlüter: Phys. Rev. B **26**, 4199 (1982)

2.41   L.I. Schiff: *Quantum Mechanics*, 3rd ed. (McGraw Hill, New York 1968)

2.42   J. Hafner: Z. Phys. B **22**, 351 (1975)

2.43   J. Hafner: Z. Phys. B **24**, 41 (1976)

2.44   N.W. Ashcroft: Phys. Lett. **23**, 48 (1966)

2.45   E.G. Brovman, Yu. Kagan: *Dynamical Properties of Solids*, ed. by A.A. Maradudin, G.K. Horton (North-Holland, Amsterdam 1974) Vol.1, Chap.4

2.46   M.H. Cohen: Phys. Rev. **130**, 1301 (1963)

2.47   J. Lindhard: K. Dan. Vidensk. Selsk. Mat.-Fys. Medd. **28**, 8 (1954)

2.48   P. Lloyd, C.A. Sholl: J. Phys. C **1**, 1620 (1968)

2.49   E.G. Brovman, Yu. Kagan, A. Holas: Sov. Phys. - JETP **34**, 1300 (1972); **35**, 783 (1972)

2.50   J. Hammerberg, N.W. Ashcroft: Phys. Rev. B **9**, 409 (1974)

2.51   Yu. Kagan, V.V. Pushkarev, A. Holas: Sov. Phys. - JETP **46**, 511 (1977)

2.52   V. Heine, D. Weaire: Solid State Phys. **24**, 247 (1970)

2.53   M.W. Finnis: J. Phys. F **4**, 1654 (1974)

2.54   D. Pines: *Elementary Excitations in Solids* (Benjamin, New York 1962)

2.55   Ph. Nozières, D. Pines: Phys. Rev. **111**, 442 (1958)

2.56   E.G. Brovman, Yu. Kagan: Sov. Phys. - JETP **30**, 883 (1970)

2.57   C.J. Pethick: Phys. Rev. B **2**, 1789 (1970)

2.58   E.G. Brovman, Yu. Kagan, A. Holas: Sov. Phys. - Solid State **11**, 733 (1969)

2.59   J.M. Luttinger, J.C. Ward: Phys. Rev. **118**, 1417 (1960)

2.60   Z.D. Popovic, J. Carbotte: J. Phys. F **4**, 1599 (1974)

2.61   G. Solt: Phys. Rev. B **18**, 720 (1978)

2.62   G. Solt, K. Werner: Phys. Rev. B **24**, 817 (1981)

2.63   G. Solt, A.P. Zhernov: Solid State Commun. **23**, 759 (1977)

2.64   J. Hafner, G. Punz: J. Phys. F **13**, 1393 (1983)

2.65   R.E. McLaren: J. Phys. F **6**, 767 (1976)

2.66   C.M. Bertoni, V. Bortolani, C. Calandra, F. Nizzoli: J. Phys. F **4**, 19 (1974)

2.67   L. Hedin, S. Lundquist: Solid State Phys. **23**, 1 (1969)

2.68   L. Hedin, B.I. Lundquist: J. Phys. C **4**, 2064 (1971)

2.69   R. Taylor: J. Phys. F **8**, 1699 (1978)

2.70   D. Ceperley: Phys. Rev. B **18**, 3126 (1978)

2.71   D. Ceperley, B.J. Alder: Phys. Rev. Lett. **45**, 566 (1980)

2.72   S.H. Vosko, L. Wilk, M. Nusair: Can. J. Phys. **58**, 1200 (1980)

2.73   S. Ichimaru: Rev. Mod. Phys. **54**, 1017 (1982)

2.74   P. Vashishta, K.S. Singwi: Phys. Rev. B **6**, 875 (1972)

2.75   F. Toigo, T.O. Woodruff: Phys. Rev. B **1**, 3958 (1970)

2.76   D.J.W. Geldart, R. Taylor: Can. J. Phys. **48**, 167 (1970)

2.77   S. Ichimaru, K. Utsumi: Phys. Rev. B **24**, 7385 (1981)

2.78   C.R. Leavens, A.H. MacDonald, R. Taylor, A. Ferraz, N.H.March: Phys. Chem. Liq. **11**, 115 (1981)

2.79   W.A. Harrison: Phys. Rev. B **7**, 2403 (1973)

2.80   P. Lloyd, J. Oglesby: J. Phys. F **3**, 1683 (1973)

2.81   T.E. Faber: *An Introduction to the Theory of Liquid Metals* (Camb. Uni. Press, Cambridge 1972)

2.82   G. Jacucci, R. Taylor: J. Phys. F **11**, 787 (1981)

2.83   R.W. Shaw, V. Heine: Phys. Rev. B **5**, 1646 (1972)

2.84   J. Hafner, V. Heine: J. Phys. F **13**, 2479 (1983)

2.85   J. Hafner, V. Heine: J. Phys. F **16**, 1429 (1986)

2.86   D.G. Pettifor, M.D. Ward: Solid State Commun. **49**, 291 (1984)

2.87   T.M. Hayes, H. Brooks, A.R. Bienenstock: Phys. Rev. **175**, 699 (1968)

2.88   J. Hafner: Phys. Rev. B **27**, 678 (1983)

2.89   J. Hafner: Phys. Rev. B **21**, 406 (1980)

2.90   W. Biltz, F. Weibke: Z. Anorg. Allg. Chem. **223**, 321 (1935)

2.91   H. Ruppersberg, W. Speicher: Z. Naturfosch. **31a**, 47 (1976)

2.92   J.J. van der Broek: Phys. Lett. A **40**, 219 (1972)

2.93   L. Pauling: *The Nature of the Chemical Bond*, 2nd ed. (Cornell Uni. Press, Ithaca 1952) Chap.12.5

2.94   M. Schlüter, C.M. Varma: Phys. Rev. B **23**, 1633 (1981)

2.95   W.B. Pearson: *The Crystal Chemistry and Physics of Metals and Alloys* (Wiley-Interscience, New York 1972)

2.96   J. Hafner: *Liquid Metals '76*, ed. by R. Evans, D.A. Greenwood (The Institute of Physics, Bristol 1977) p.107

2.97   D. Stroud: Phys. Rev. B 7, 4405 (1973)

2.98   D. Stroud: Phys. Rev. B 8, 1308 (1973)

2.99   J. Hafner, A. Pasturel, P. Hicter: J. Phys. F 14, 1137 (1984)

2.100 P. Beauchamp, R. Taylor, V. Vitek: J. Phys. F 5, 2017 (1975)

2.101 J. Hafner, A. Pasturel, J. Hicter: J. Phys. F 14, 2279 (1984)
       A. Pasturel, J. Hafner: Phys. Rev. B 32, 5009 (1985)

2.102 H. Ruppersberg, H., Reiter: J. Phys. F 12, 1311 (1982)

2.103 A.P. Copestake, R. Evans, H.Ruppersberg, W. Schirmacher: J. Phys. R 13, 1993 (1983)

2.104 W. van der Lugt, W. Geertsma: Proc. 5th Intern. Conf. on Liquid and Amorphous Metals, ed. by C.N.J. Wagner, W.A. Johnson, J. Non Cryst. Solids 61&62, 187 (1984)

2.105 N.Q. Lam, N. van Doan, Y. Adda: J. Phys. F 10, 2359 (1980)

2.106 N.Q. Lam, N. Van Doan, L. Dagens, Y. Adda: J. Phys. F 11, 2231 (1981)

2.107 A. Bieber, F. Gautier: Solid State Commun. 38, 1219 (1981); Z. Physik B 57, 335 (1984)

2.108 A. Bieber, F. Ducastelle, F. Gautier, G. Treglia, P. Turchi: Solid State Commun. 45, 585 (1983)

2.109 J.H. Wills, W.A. Harrison: Phys. Rev. B 25, 5007 (1982)

2.110 A.R. Mackintosh, O.K. Andersen: *Electrons at the Fermi Surface*, ed. by M. Springford (Camb. Uni. Press, London 1980) Sect.5.3

2.111 D.G. Pettifor: Commun. Phys. 1, 141 (1976); J. Phys. F 8, 219 (1978)

2.112 D.G. Pettifor: J. Chem. Phys. 69, 2930 (1978)

2.113 H.C. Skriver: Phys. Rev. Lett. 49, 1768 (1982)

2.114 V. Russier, C. Regnaut, J.P. Badiali: Proc. 6th Intern. Conf. on Liquid and Amorphous Metals, ed. by F. Hensel, E. Lüscher, Z. Phys. Chem. (in press)
       Ch. Hausleitner, J. Hafner: J. Phys. F (in press)

2.115 P. Hohenberg, W. Kohn: Phys. Rev. 136, B 864 (1964)

2.116 W. Kohn, L.J. Sham: Phys. Rev. 140, A 1133 (1965)

2.117 W. Jones, W.H. Young: J. Phys. C 4, 1322 (1971)

2.118 J.R. Chelikowsky: Solid State Commun. 31, 19 (1979)

2.119 J.R. Chelikowsky: Phys. Rev. B 19, 686 (1979)

2.120 M.P. Iniguez, J.A. Alonso: J. Phys. F. 11, 2045 (1981)

2.121 M.P. Iniguez, J.A. Alonso, D.J. Gonzalez, M.L. Zorita: J. Phys. F 12, 1907 (1982)

2.122 A.M. Rosenfeld, M.J. Stott: J. Phys. F 17, 605 (1987)

2.123 J. Hafner, T. Egami, S. Aur, B.C. Giessen: J. Phys. F (in press)

2.124 M. Manninen: Phys. Rev. B 34, 8486 (1986)

2.125 M.S. Daw, M.I. Baskes: Phys. Rev. B 29, 6443 (1984)

2.126 A.E. Carlsson: Phys. Rev. B 35, 4858 (1987)

3.1 V. Heine, D. Weaire: Solid State Phys. **24**, 247 (1970)

3.2 W.A. Harrison: *Electronic Structure and the Properties of Solids - The Physics of the Chemical Bond* (Freeman, San Francisco 1980)

3.3 L.A. Girifalco: Acta Metall. **24**, 759 (1976)

3.4 D.G. Pettifor: *Physical Metallurgy*, ed. by R.W. Cahn, P. Haasen (North-Holland, Amsterdam 1984) Chap.3

3.5 K.A. Gschneidner: Solid State Phys. **16**, 304 (1966)

3.6 G. Solt, J. Kollar: Solid State Commun. **15**, 957 (1974)

3.7 M.W. Finnis: J. Phys. F **4**, 1645 (1974)

3.8 J.R. Chelikowsky: Phys. Rev. B **21**, 3074 (1980)

3.9 J.R. Chelikowsky: Phys. Rev. Lett. **47**, 387 (1981)

3.10 M.T. Yin, M.L. Cohen: Phys. Rev. Lett. **45**, 1004 (1980)

3.11 M.T. Yin, M.L. Cohen: Phys. Rev. B **24**, 6121 (1981); ibid. B **26**, 5668 (1982)

3.12 V.L. Moruzzi, J.F. Williams: *Calculated Electronic Properties of Metals* (Pergamon, New York 1978)

3.13 J.F. Janak, V.L. Moruzzi, A.R. Williams: Phys. Rev. B **12**, 1257 (1975)

3.14 J. Ihm, M.L. Cohen: Phys. Rev. B **23**, 1576 (1981)

3.15 P.K. Lam, M.L. Cohen: Phys. Rev. B **24**, 4224 (1981)

3.16 M.Y. Chou, P.K. Lam, M.L. Cohen: Phys. Rev. B **28**, 4179 (1983)

3.17 F.W. Averill: Phys. Rev. B **6**, 3637 (1972)

3.18 J.A. Moriarty: Phys. Rev. B **26**, 1754 (1982)

3.19 J. Hafner: Z. Phys. B **22**, 351 (1975)

3.20 J. Hafner: Z. Phys. B **24**, 41 (1976)

3.21 J.A. Moriarty: Phys. Rev. B **16**, 2537 (1977)

3.22 A.K. McMahan, J.A. Moriarty: Phys. Rev. B **27**, 3235 (1983)

3.23 J. Hafner: Phys. Status Solidi B **56**, 579 (1973); **57**, 101, 497 (1973)

3.24 J. Hafner, H. Eschrig: Phys. Status Solidi B **72**, 179 (1975)

3.25 J. Hafner: J. Phys. F **6**, 1243 (1976)

3.26 J.P. Jan, H.L. Skriver: J. Phys. F **11**, 805 (1981)

3.27 S.G. Louie, M.L. Cohen: Phys. Rev. B **10**, 3237 (1974)

3.28 J. Friedel: *The Physics of Metals*, Vol.1, ed. by J.M. Ziman (Cambridge University Press, Cambridge 1968) p.494
J. Friedel: Trans. Metall. Soc. AIME **230**, 616 (1964)

3.29 A.R. Mackintosh, O.K. Andersen: *Electrons at the Fermi Surface*, ed. by M. Springford (Cambridge Uni. Press, Cambridge 1980)

3.30 V. Heine: Phys. Rev. **153**, 673 (1967)

3.31 D.G. Pettifor: J. Magn. Magn. Mat. **15-18**, 847 (1980)

3.32 A.R. Williams, C.D. Gelatt, J.F. Janak: *Theory of Alloy Phase Formation*, ed. by L.H. Bennett (The Metallurgical Society of AIME, Warrendale, Pa 1980) p.40

3.33	J.A. Moriarty: Phys. Rev. B 1, 1363 (1970)
3.34	J.D. Upadhyaya, L. Dagens: J. Phys. F 8, L21 (1978)
3.35	H. Wendel; R.M. Martin: Phys. Rev. B 19, 5251 (1979)
3.36	J. Hafner: Phys. Rev. B 10, 4151 (1974)
3.37	J.A. Moriarty: Phys. Rev. B 19, 609 (1979); Solid State Commun. 31, 881 (1979)
3.38	A.R. Williams, M. Appapillai: J. Phys. F. 3, 772 (1973)
3.39	C.S.G. Cousins: J. Phys. C 3, L197 (1970)
3.40	J. Hafner, O. Hittmair: Anz. Österr. Akad. Wiss. Math.-Naturwiss. Kl. 4, 1 (1975)
3.41	A.R. Williams: J. Phys. F 3, 781 (1973)
3.42	J.A. Moriarty: Phys. Rev. B 28, 4818 (1983)
3.43	A.Morita, T. Soma: Solid State Commun. 11, 927 (1972)
3.44	D. Schiferl: Phys. Rev. B 19, 806 (1979)
3.45	Y. Abe, I. Okoshi, A. Morita: J. Phys. Soc. Japan 42, 504 (1977)
3.46	A.B. Maknovetskij, G.L. Krasko: Phys. Status Solidi B 81, 263 (1977)
3.47	R.B. Arkhipov, E.S. Aleskeev, A.I. Barabanov: Sov. Phys. - JETP 36, 323 (1972)
3.48	S.M.M. Rahman, M.M. Rahman: J. Phys. F 11, 2055 (1981)
3.49	L. Kaufman, H. Bernstein: *Computer Calculations of Phase Diagrams* (Academic, New York 1970)
3.50	C.M. Bertoni, V. Bortolani, C. Calandra, F. Nizzoli: J. Phys. 4, 19 (1974)
3.51	D. Weaire: Phys. Status Solidi B 42, 767 (1970)
3.52	B. Predel, D.W. Stein: Acta Metall. 20, 515 (1972)
3.53	L. Kaufman, H. Nesor: *Titanium Science and Technology*, ed. by R.I. Jaffe, H. Burte (Plenum, New York 1973) Vol.2, p.773
3.54	A.P. Miodownik: *Metallurgical Chemistry*, ed. by O. Kubaschewski (HMSO, London 1972) p.233
3.55	J. Lumsden: *Thermodynamics of Alloys*, Institute of Metals Monograph Series 11 (The Institute of Metals, London 1952)
3.56	J. Hafner, V. Heine: J. Phys. F 13, 2479 (1983)
3.57	B.J. Austin, V. Heine: J. Chem. Phys. 45, 928 (1966)
3.58	P. Pykkö: *Adv. Quantum Chemistry*, Vol.11 (Academic, New York 1978) p.389
3.59	M.L. Cohen, V. Heine: Solid State Phys. 24, 35 (1970)
3.60	D.G. Pettifor: M.D. Ward: Solid State Commun. 49, 291 (1984)
3.61	D.G. Pettifor: J. Phys. C 3, 367 (1970)
3.62	D.G. Pettifor: *Metallurgical Chemistry*, ed. by O. Kubaschewski (HMSO, London 1972) p.191
3.63	H.L. Skriver: Phys. Rev. Lett. 49, 1768 (1982)
3.64	B. Johansson, H.L. Skriver, O.K. Andersen: *Physics of Solids under High Pressures*, ed. by J.S. Schilling, R.N. Shelton (North-Holland, Amsterdam 1981) p.245

3.65　A.K. McMahan, D.A. Young: Phys. Lett. **105A**, 129 (1984)

3.66　G.S. Smith, J. Akella: Phys. Lett. **105A**, 132 (1984)

3.67　J.C. Duthie, D.G. Pettifor: Phys. Rev. Lett. **38**, 564 (1977)

3.68　B. Johansson, A. Rosengren: Phys. Rev. B **11**, 2836 (1975)

3.69　P. Coppens, M.B. Hall: *Electron Distributions and the Chemical Bond* (Plenum, New York 1980)

3.70　P. Coppens: *Neutron Diffraction*, ed. by H. Dachs, Topics Curr. Phys., Vol.6 (Springer, Berlin, Heidelberg 1978)

3.71　P. Becker(ed.): *Electron and Magnetization Densities in Molecules and Crystals*, NATO-Advanced Study Institutes, Series B (Plenum, New York 1980)

3.72　W.A. Harrison: *Pseudopotentials in the Theory of Metals* (Benjamin, New York 1966)

3.73　J.P. Walter, C.Y. Fong, M.L. Cohen: Solid State Commun. **12**, 303 (1973)

3.74　C.M. Bertoni, V. Bortolani, C. Calandra, F. Nizzoli: J. Phys. **3**, L244 (1973)

3.75　P.J. Bunyan, J.A. Nelson: J. Phys. F. **7**, 2323 (1977)

3.76　J. Hafner: Solid State Commun: **27**, 263 (1978)

3.77　Y.W. Yang, P. Coppens: Solid State Commun. **15**, 1555 (1974)

3.78　J.R. Chelikowsky, M.L. Cohen: Phys. Rev. Lett. **33**, 1339 (1974)

3.79　H. Olijnyk, W.B. Holzapfel: Phys. Lett. **99A**, 381 (1983)

3.80　H. Olijnyk, W.B. Holzapfel: Phys. Lett. **100A**, 191 (1984)

3.81　D. Glötzel; A.K. McMahan: Phys. Rev. B **20**, 3240 (1979)

3.82　A.P. Frolov, K.P. Radionov: Sov. Phys. - Solid State **16**, 2297 (1975)

3.83　A.W. Olinger, J.W. Shaner: Science **219**, 1071 (1983)

3.84　W. Klement, A. Jayaraman: Prog. Solid State Chem. **3**, 289 (1967)

3.85　J.E. Inglesfield: J. Phys. C **1**, 1337 (1968)

3.86　A.A. Maradudin, E.W. Montroll, G.H. Weiss, I.P. Ipatova: *Theory of Lattice Dynamics in the Harmonic Approximation*, 2nd ed., Solid State Phys., Suppl.3 (Academic, New York 1971)

3.87　P. Brüesch: *Phonons: Theory and Experiments I*, Springer Ser. Solid-State Sci., Vol.34 (Springer, Berlin, Heidelberg 1982)

3.88　G. Grimvall: *Ab - Initio Calculations of Phonon Spectra*, ed. by J.T. Devreese, V. van Doren, P.E. Van Camp (Plenum, New York 1983) p.117

3.89　E.G. Brovman, Yu. Kagan, A. Holas: Sov. Phys. - Solid State **11**, 733 (1969)

3.90　E.G. Brovman, Yu. Kagan, A. Holas: Sov. Phys. - JETP **34**, 1300 (1972)

3.91　C.M. Bertoni, V. Bortolani, C. Calandra, F. Nizzoli: Phys. Rev. Lett. **24**, 1466 (1973)

3.92   C.M. Bertoni, V. Bortolani, C. Calandra, E. Tosatti: Phys. Rev. B 9, 1810 (1974)

3.93   C.M. Bertoni, O. Bisi, C. Calandra, F. Nizzoli: J. Phys. F 5, 419 (1975)

3.94   D.L. Price, S.K. Sinha, R.P. Gupta: Phys. Rev. B 9, 2573 (1974)

3.95   A.M. Altshuler, Yu. Velikov, T.I. Kacherets: Sov. Phys. - Solid State 20, 9 (1978)

3.96   P.E. Van Camp, V.E. Van Doren, J.T. Devreese: *Ab initio Calculations of Phonon Spectra*, ed. by J.T. Devreese, V.E. Van Doren, P.E. Van Camp: (Plenum, New York 1983) p.25

3.97   J.A. Moriarty: Phys. Rev. B 8, 1338 (1973)

3.98   J. Panitz, P.H. Cutler, W.F. King: J. Phys. F 4, L106 (1974)

3.99   D. Stroud, N.W. Ashcroft: Phys. Rev. B 5, 371 (1972)

3.100 D.C. Wallace: *Thermodynamics of Crystals* (Wiley, New York 1972)

3.101 V.G. Vaks, E.V. Zarochentsev, S.P. Kravchuk, V.P. Safronov, A.V. Trefilov: Phys. Status Solidi B 85, 63 (1978)

3.102 T.H.K. Barron, M.L. Klein: *Dynamical Properties of Solids*, ed. by A.A. Maradudin, G.K. Horton, Vol.1, (North-Holland, Amsterdam 1974) p.391

3.103 J. Meyer, G. Dolling, J. Kalus, C. Vettier, J. Paureau: J. Phys. F 6, 1899 (1976)

3.104 J. Hafner: Solid State Commun. 21, 179 (1977)

3.105 Ph. Choquard: *The Anharmonic Crystal* (Benjamin, New York 1967)

3.106 W. Götze, K.H. Michel: *Dynamical Properties of Solids*, ed. by A.A. Maradudin, G.K. Horton, Vol.1 (North-Holland, Amsterdam 1974) p.499

3.107 V.G. Vaks, S.P. Kravchuk, A.V. Trefilov: J. Phys. F 10, 2105 (1980)

3.108 H.R. Glyde, J.P. Hansen, M.L. Klein: Phys. Rev. B 16, 3476 (1977)

3.109 W.J.L. Buyers, G. Dolling, G. Jacucci, M.L. Klein, H.R. Glyde: Phys. Rev. B 20, 4859 (1979)

3.110 S.S. Cohen, M.L. Klein, M.S. Duesbery, R. Taylor: J. Phys. F 6, 337 (1976)

3.111 J.R.D. Copley: Can. J. Phys. 51, 2564 (1973)

3.112 H.R. Glyde, R. Taylor: Phys. Rev. B 5, 1206 (1972)

3.113 W.J.L. Buyers, R.A. Cowley: Phys. Rev. 180, 755 (1969)

3.114 S.H. Taole, H.R. Glyde, R. Taylor: Phys. Rev. B 18, 2643 (1978)

3.115 T.R. Koehler, N.S. Gillis, D.C. Wallace: Phys. Rev. B 1, 4521 (1970)

3.116 V.G. Vaks, E.V. Zarochentsev, S.P. Kravchuk, V.P. Safronov: J. Phys. F 8, 725 (1978)

3.117 R.C. Shukla, R. Taylor: Phys. Rev. B9, 4116 (1974)
3.118 R.A. McDonald, R.D. Mountain, R.C. Shukla: Phys. Rev. B 20, 4012 (1979)
3.119 V.G. Vaks, S.P. Kravchuk, A.V. Trefilov: J. Phys. F 10, 2325 (1980)
3.120 A. Isihara: J. Phys. A 1, 539 (1968)
3.121 T. Lukes, R. Jones: J. Phys. A 1, 29 (1968)
3.122 H. Jones: Phys. Rev. A 8, 3215 (1973)
3.123 M. Hasegawa, M.J. Stott, W.H. Young: J. Phys. F 9, 1 (1979)
3.124 C.H. Leung, M.J. Stott, W.H. Young: J. Phys. F 6, 1039 (1976)
3.125 C. Zener: Elasticity and Inelasticity of Metals (Chicago University Press, Chicago 1948) p.32 and in Phase Stability in Metals and Alloys, ed. by P.S. Rudman, J. Stringer, R.I. Jaffee (McGraw-Hill, New York 1967) p.25
3.126 A.O.E. Animalu: Phys. Rev. 161, 445 (1967)
3.127 J. Hafner: Phys. Rev. B 21, 406 (1980)
3.128 C.S. Barrett, O.R. Trautz: Trans. Am. Inst. Min. Metall. Pet. Eng. 175, 579 (1948)
3.129 C.S. Barrett: Acta Crystallogr. 9, 671 (1956)
3.130 R.Stedman: J. Phys. F 6, 2239 (1976)
3.131 C.M. McCarthy, C.W. Tompson, S.A. Werner: Phys. Rev. B 22, 574 (1980)
3.132 T. Schneider, E. Stoll: Solid State Commun. 8, 1729 (1970)
3.133 R. Pynn, I. Ebbsjö: J. Phys. F 1, 744 (1971)
3.134 G.K. Straub, D.C. Wallace: Phys. Rev. 3, 1234 (1971)
3.135 M.J. Kelly: J. Phys. F 9, 1921 (1979)
3.136 G. Grimvall: Phys. Scr. 12, 168 (1975)
3.137 M.J. Kelly, W.M. Stobbs: Phys. Rev. Lett. 45, 922 (1980)
3.138 D.L. Martin: Proc. R. Soc. London Ser. A 254, 433 (1960)
3.139 S.M.M. Rahman, M.M. Rahman, S.M.M.R. Chowdhury: J. Phys. F 14, 833 (1984)
3.140 J.C. Upadhyaya, S. Wang: J. Phys. F 10, 441 (1980)
3.141 J.L. Pélissier: Physica 128A, 363 (1984)
3.142 A. Jayaraman, W. Klement, G.C. Kennedy: Phys. Rev. 132, 1620 (1963)
3.143 D.M. Straus, N.W. Ashcroft: Phys. Rev. Lett. 38, 415 (1977)
3.144 D.G Pettifor: Solid State Phys. (In press)
3.145 M.Y. Chou, M.L. Cohen: Solid State Commun. 57, 785 (1986)
3.146 K.J. Chang, M.L. Cohen: Phys. Rev. B 30, 5376 (1984); B 31, 7819 (1985); B 34, 8581 (1986)
3.147 O.K. Andersen, O. Jepsen, D. Glötzel: In Highlights in Condensed Matter Theory, ed. by F. Bassani, F. Fumi, M.P. Tosi (North-Holland, Amsterdam 1985)
3.148 N.E. Christensen, V. Heine: Phys. Rev. B 32, 6145 (1985)
3.149 J.A. Moriarty: Phys. Rev. B 34, 6738 (1986)

3.150 M. Heiroth, U. Buchenau, H.R. Schober: Phys. Rev. B **34**, 6681 (1986)

3.151 Y.R. Wang, A.W. Overhauser: Phys. Rev. B **35**, 497, 501 (1987)

3.152 See, e.g., the articles by R.M. Martin, K. Kunc, and O. Nielsen, R.M. Martin in *Electronic Structure, Dynamics, and Quantum Structural Properties of Condensed Matter*, ed. by J.T. Devreese, P. van Camp, NATO Advanced Study Institutes Series B, Vol.121 (Plenum, New York 1985)

4.1 C.A. Croxton: *Liquid State Physics - A Statistical Mechanical Introduction* (Cambridge University Press, Cambridge 1974); J.P. Hansen, I.R. McDonald: *Theory of Simple Liquids* (Wiley, New York 1971)

4.2 B. Alder: *Physics of Simple Liquids*, ed. by H.N.V. Temperley, J.S. Rowlinson (North-Holland, Amsterdam 1968) Chap.4; W.G. Hoover: *Molecular Dynamics*, Lecture Notes Phys., Vol.258 (Springer, Berlin, Heidelberg 1986)

4.3 K. Binder: *Monte Carlo Methods in Statistical Physics*, 2nd ed., Topics in Curr. Phys., Vol.7 (Springer, Berlin, Heidelberg 1986)

4.4 J.M. Hammersley, D.C. Handscomb: *Monte Carlo Methods* (Methuen, London 1964)

4.5 F. Vesely: *Computerexperimente an Flüssigkeitsmodellen* (Physik Verlag, Weinheim 1978); D.W. Heermann: *Computer Simulation Methods in Theoretical Physics* (Springer, Berlin, Heidelberg 1986)

4.6 A. Paskin, A. Rahman: Phys. Rev. Lett. **16**, 300 (1966)

4.7 D. Schiff: Phys. Rev. **186**, 151 (1969)

4.8 R.D. Murphy, M.L. Klein: Phys. Rev. A **8**, 2640 (1973)

4.9 A. Rahman: Phys. Rev. A **9**, 1667 (1974)

4.10 G. Jacucci, M.L. Klein, R. Taylor: Solid State Commun. **19**, 657 (1976)

4.11 R.D. Murphy: Phys. Rev. A **15**, 1188 (1977)

4.12 R.H. Fowler: J. Chem. Phys. **59**, 3453 (1973)

4.13 T. Lee, J. Bischop, W. van der Lugt, W.F. van Gunsteren, Physica C **93**, 59 (1978)

4.14 R.S. Day, F. Sun, P.H. Cutler: Phys. Rev. A **19**, 328 (1979)

4.15 R.D. Mountain: J. Phys. F **8**, 1637 (1978)

4.16 F. Sun, R.S. Day, P.H. Cutler: Phys. Lett. A **68**, 236 (1978)

4.17 E. Michler, H. Hahn, P. Schofield: J. Phys. F **6**, L319 (1976)

4.18 I. Ebbsjö, T. Kinell, I. Waller: J. Phys. C **13**, 1865 (1980)

4.19 G. Jacucci, R. Taylor, S. Tenenbaum, N. van Doan: J. Phys. F **11**, 793 (1981)

4.20 L. Dagens, M. Rasolt, R. Taylor: Phys. Rev. B **11**, 2726 (1975)

4.21 J.M.J. van Leeuwen, J. Groeneveld, J. de Boer: Physica 25, 792 (1954)
4.22 J.K. Percus, G.J. Yevick: Phys. Rev. 110, 1 (1958)
4.23 J. A. Barker, D. Henderson: Rev. Mod. Phys. 48, 587 (1976)
4.24 W.R. Smith, D. Henderson, Y. Tago: J. Chem. Phys. 67, 5308 (1977)
4.25 G. Kahl, J. Hafner: Phys. Chem. Liq. 12, 109 (1982)
4.26 H.C. Andersen, D. Chandler, J.D. Weeks: Adv. Chem. Phys. 34, 105 (1976)
4.27 R. Taylor, R.O. Watts: Solid State Commun. 38, 965 (1981)
4.28 H.R. Leribaux, L.F. Miller: J. Chem. Phys. 61, 3327 (1974)
4.29 L.U. Belan-Gaiko, V.I. Bogdanov, D.L. Fuks: Phys. Status Solidi B 102, 265 (1980)
4.30 A.M. Bratkovsky, V.G. Vaks, A.V. Trefilov: J. Phys. F 12, 611 (1982)
4.31 A.M. Bratkovsky, V.G. Vaks, S.P. Kravchuk, A.V. Trefilov: J. Phys. F 12, 1293 (1982)
4.32 A.M. Bratkovsky, V.G. Vaks: J. Phys. F 13, 2207 (1983)
4.33 Y. Rosenfeld, N.W. Ashcroft: Phys. Rev. A 20, 1208 (1979)
4.34 J.L. Lebowitz, J.K. Percus: Phys. Rev. 144, 251 (1966); E. Waisman: Mol. Phys. 25, 45 (1973)
4.35 W.G. Madden, S.A. Rice: J. Chem. Phys. 72, 4208 (1980)
4.36 N.W. Ashcroft, J. Lekner: Phys. Rev. 165, 83 (1966)
4.37 G.A. Mansoori, F.B. Canfield: J. Chem. Phys. 51, 4958 (1969)
4.38 N.F. Carnahan, K.E. Starling, J. Chem. Phys. 51, 635 (1969)
4.39 H.D. Jones: J. Chem. Phys. 55, 2640 (1971)
4.40 D. Stroud, N.W. Ashcroft: Phys. Rev. B 5, 371 (1972)
4.41 W.H. Young: Liquid Metals '76, ed. by R. Evans, D.A. Greenwood (The Institute of Physics, Bristol 1977) p.1
4.42 J. Hafner: J. Phys. Rev. A 16, 351 (1977)
4.43 M. Shimoji: J. Physique 41, C8-547 (1980)
4.44 J. Hafner: Phys. Rev. B 21, 406 (1980)
4.45 J. Hafner: Liquid Metals '76, ed. by R. Evans, D.A. Greenwood (The Institute of Physics, Bristol 1977) p.107
4.46 R. Kumaravadivel, R. Evans: J. Phys. C 9, 3877 (1977)
4.47 M. Ross, H.E. de Witt, W.B. Hubbard: Phys. Rev. A 24, 1016 (1981)
4.48 W.L. Slattery, G.D. Doolen, H.E. De Witt: Phys. Rev. A 21, 2087 (1980)
4.49 W.L. Slattery, g.D. Doolen, H.E. De Witt: Phys. Rev. A 26, 2255 (1982)
4.50 K.C. Ng: J. Chem. Phys. 61, 2080 (1974)
4.51 F.A. Rogers, D.A. Young, H.E. De Witt, M. Ross: Phys. Rev. A 28, 2990 (1983)
4.52 K.K. Mon, R. Gann, D. Stroud: Phys. Rev. A 24, 2145 (1981)

4.53   J.A. Moriarty, D.A. Young, M. Ross: Phys. Rev. B **30**, 578 (1984)

4.54   M.Ross: J. Chem. Phys. **71**, 1567 (1979)

4.55   E. Waisman, J.L. Lebowitz: J. Chem Phys. **52**, 4307 (1970)

4.56   H.B. Singh, A. Holz: Phys. Rev. A **28**, 1108 (1983)

4.57   R. Hayter, R. Pynn, J.B. Suck: J. Phys. F **13**, L1 (1983); S.K. Lai: Phys. Rev. A **31**, 3886 (1985)

4.58   J.D. Weeks, D. Chandler: Phys. Rev. Lett. **25**, 149 (1970)

4.59   J.D. Weeks, D. Chandler, H.C. Andersen: J. Chem. Phys. **54**, 5237 (1971)

4.60   J.D. Weeks, D. Chandler, H.C. Andersen: J. Chem Phys. **55**, 5422 (1971)

4.61   R.E. Jacobs, H.C. Andersen: Chem. Phys. **10**, 73 (1975)

4.62   M.M. Telo da Gama, R. Evans: Mol. Phys. **41**, 1091 (1980)

4.63   J.H. Wehling, W.M. Shyu, G.D. Gaspari: Phys. Lett. **39A**, 59 (1972)

4.64   M. Hasegawa, M. Watabe: J. Phys. Soc. Jpn. **36**, 1510 (1974)

4.65   M. Hasegawa: J. Phys. F **6**, 649 (1976)

4.66   W.G. Madden: J. Chem. Phys. **75**, 1984 (1981)

4.67   H.C. Andersen, D. Chandler: J. Chem. Phys. **53**, 547 (1970)

4.68   H.C. Andersen, D. Chandler: J: Chem. Phys. **57**, 1918 (1972)

4.69   H.C. Andersen, D. Chandler, J.D. Weeks: J. Chem. Phys. **56**, 3812 (1972)

4.70   L. Verlet, J.J. Weis: Phys. Rev. A **5**, 939 (1972)

4.71   E.W. Grundke, D. Henderson: Mol. Phys. **24**, 269 (1972)

4.72   J. Hafner: Proc. 4th Intern. Conf. on Liquid and Amorphous Metals, ed. by C.N.J. Wagner, W.L. Johnson: J. Non-Cryst. Solids **61&62**, 175 (1984)

4.73   C. Regnaut, J.P. Badiali, M. Dupont: Phys. Lett. A **74**, 245 (1979); J. Physique **41**, C8-603 (1980)

4.74   R. Oberle, H. Beck: Solid State Commun. **32**, 510 (1979)

4.75   C. Regnaut, E. Fusco, J.P. Badiali: Phys. Status Solidi B**12**, 373 (1983)

4.76   J.L. Bretonnet, C. Regnaut: Proc. 5th Int'l. Conf. on Liquid and Amorphous Metals, ed. by C.N.J. Wagner W.L. Johnson, J. Non-Cryst. Solids **61&62**, 147 (1984)

4.77   G. Kahl, J. Hafner: Phys. Rev. A **9**, 3310 (1984)

4.78   G. Kahl, J. Hafner: Solid State Commun. **49**, 1125 (1984)

4.79   J. Hafner, G. Kahl: J. Phys. F **14**, 2259 (1984)

4.80   P.T. Cummings, G. Stell: Mol. Phys. **43**, 1267 (1981)

4.81   P.T. Cummings, P. Egelstaff: J. Phys. F **12**, 233 (1982)

4.82   J.C. Wheeler, D. Chandler: J. Chem. Phys. **55**, 1645 (1971)

4.83   Y. Waseda: *The Structure of Non-Crystalline Materials - Liquids and Amorphous Solids* (McGraw-Hill, New York 1980)

4.84   J.P. Gabathuler, S.Steeb: Z. Naturforsch. **34a**, 1314 (1979)

4.85   M.P. Tosi: *Electron Correlations in Solids, Molecules and Atoms*, ed. by J.T.Devreese, F. Brosens (Plenum, New York 1983) p.235

4.86   D.K. Chaturvedi, G. Senatore, M.P. Tosi: Lett. Nuovo Cimento **30**, 47 (1981);
G.K. Chaturvedi, M. Rovere, G. Senatore, M.P. Tosi: Physica B **111**, 11 (1981)

4.87   G. Pastore, G. Senatore, M.P. Tosi: Physica B111, 283 (1981)

4.88   G. Senatore, M.P. Tosi: Phys. Chem. Liq. **11**, 365 (1982)

4.89   V. Heine, D. Weaire: Phys. Rev. **152**, 603 (1966)

4.90   D.Weaire: J. Phys. C **1**, 210 (1968)

4.91   S. Steeb, P. Lamparter: Proc. 5th Int. Conf. on Liquid and Amorphous Metals, ed. by C.N.J. Wagner, W.L. Johnson, J. Non-Cryst. Solids **61&62**, 237 (1984)

4.92   S. Fishman, M.E. Fisher: Physica A **108**, 1 (1981)

4.93   S.M. Foiles, N.W. Ashcroft: Phys. Rev. A **24**, 424 (1981)

4.94   S.H. Sung, D. Chandler: Phys. Rev. A **9**, 1688 (1974)

4.95   F. Gallerani, G. LoVecchio, A. Parola, L. Reatto: J. Phys. C **16**, 5793 (1983)

4.96   W. Freyland: Comments Solid State Phys. **10**, 1 (1981)

4.97   L.R. Friedman, D.P. Tunstall (Eds.): *The Metal-Nonmetal Transition in Disordered Systems*, 19th Scottish University Summer School in Physics, St. Andrews (1978)

4.98   S. Jüngst, B. Knuth, F. Hensel: Phys. Rev. Lett. **55**, 2160 (1985)

4.99   R.E. Goldstein, N.W. Ashcroft: Phys. Rev. Lett. **55**, 2164 (1985)

4.100   R. Evans, T.J. Sluckin: J. Phys. C **14**, 3137 (1981)

4.101   I.L. McLaughlin, W.H. Young: J. Phys. F **12**, 245 (1982)

4.102   I.L. McLaughlin, W.H. Young: J. Phys. C **15**, 1121 (1982)

4.103   R.L. Henderson, N.W. Ashcroft: Phys. Rev. A **13**, 859 (1976)

4.104   I.L. McLaughlin, W.H. Young: J. Phys. F **14**, 1 (1984)

4.105   J.L. Pelissier: Phys. Lett. **103A**, 133 (1984)

4.106   G. Franz, W. Freyland; W. Gläser, F. Hensel, E. Schneider: J. Physique **41**, C8-194 (1980)

4.107   A. Meyer, M.J: Stott, W.H. Young: Phil. Mag. **33**, 381 (1976);
S. Tamaki, Y. Waseda: J. Phys. F. **6**, L89 (1976)

4.108   S.N. Khanna, F. Cyrot-Lackmann: J. Physique Lett. **40**, L45 (1979)

4.109   T. Itami, M. Shimoji: J. Phys. F. **14**, L15 (1984)

4.110   N.H. March, M.P. Tosi: *Atomic Dynamics of Liquids* (MacMillan, London 1976)

4.111   W. Marshall, S. Lovesey: *Theory of Thermal Neutron Scattering* (Clarendon, Oxford 1971)

4.112   N.H. March, M. Parrinello: *Collective Effects in Solid and Liquids* (Hilger, Bristol 1982)

4.113 J.R.D. Copley, J.M. Rowe: Phys. Rev. Lett. 32, 49 (1974), Phys. Rev. A 9, 1656 (1974)

4.114 W. Gläser, Ch. Morkel: Proc. 5th Int. Conf. on Liquid and Amorphous Metals, ed. by C.N.J. Wagner and W.L. Johnson, J. Non-Cryst. Solids 61&65, 309 (1984); G. Bucher, W. Gläser: Verh. Dtsch. Phys. Ges. (VI) 19, 452 (1984)

4.115 A. Rahman, Phys. Rev. Lett. 32, 52 (1974)

4.116 G. Jacucci, I.R. McDonald: *Liquid and Amorphous Metals*, ed. by E. Lüscher, H. Coufal (Sijthoff and Nordhoff, Alphen van Rijn, Netherlands 1980) p.143

4.117 I. Ebbsjö, T. Kinell, I. Waller: J. Phys. C 11, L501 (1978)

4.118 J. Bosse, W. Götze, M. Lücke: Phys. Rev. A 18, 1176 (1978)

4.119 L. Sjögren: Phys. Rev. A 22, 2866 (1980); ibid. p.2883

4.120 P.A. Egelstaff, W. Gläser: Phys. Rev. A 31, 3802 (1985); P.A. Egelstaff: Proc. 6th Int'l. Conf. on Liquid and Amorphous Metals 1986, Z. Phys. Chem (in press)

4.121 R. Eisenschitz, M.J. Wilford: Proc. Phys. Soc. London 80, 1078 (1962)

4.122 P. Bratby, T. Gaskell, N.H. March: Phys. Chem. Liq. 2, 53 (1970)

4.123 P. Gray, I. Yokoyama, W.H. Young: J. Phys. F 10, 197 (1980)

4.124 I. Okhosi, I. Yokoyama, Y. Waseda, W.H. Young: J. Physique 41, C8-599 (1980)

4.125 P.A. Egelstaff: Rep. Prog. Phys. 29, 333 (1966)

4.126 R. Biswas, D.R. Hamann: In *Computer Based Microscopic Description of the Structure and Properties of Materials*, ed. by J. Broughton, W. Krakov, S.T. Pantelides, MRS Syp. Proc. Vol.63 (MRS, Pittsburgh 1986)

4.127 F. Stillinger, T.A. Weber: Phys. Rev. B 31, 5262 (1985)

4.128 J. Hafner, G. Kahl: Europhys. Lett. (in print)

4.129 R. Car, M. Parinello: Phys. Rev. Lett. 55, 2471 (1985)

4.130 R. Car, M. Parrinello: Solid State Commun. 62, 403 (1987); R. Car, M. Parrinello, W. Andreoni: In *Microclusters*, ed. by S. Sugano, Y. Nishina, S. Ohnishi, Springer Ser. Mat. Sci., Vol.4 (Springer, Berlin, Heidelberg 1987) R. Car, M. Parrinello: Proc. 18th Int'l. Conf. on The Physics of Semiconductors, Stockholm 1986, ed. O. Engström (World Scientific, Singapore 1987)

4.131 G. Pastore, G. Kahl, J.P. Hansen: to be published

4.132 D.H. Li, R.A. Moore, S. Wang: Phys. Lett. A 118, 405 (1986)

4.133 C. Hausleitner, J. Hafner: J. Phys. F (submitted)

4.134 V. Russier, C. Regnault, J.P. Badiali: Proc. 6th Int'l. Conf. Liquid and Amorphous Metals, ed. by F. Hensel, E. Lüscher, Z. Phys. Chem. (in press)

5.1   D. Stroud, N.W. Ashcroft: Phys. Rev. B **5**, 371 (1972)
5.2   W.H. Hartmann: Phys. Rev. Lett. **26**, 1640 (1971)
5.3   H. Jones: Phys. Rev. A **8**, 3215 (1973)
5.4   D.A. Young, M.Ross: Phys. Rev. B **29**, 682 (1984)
5.5   A.M. Bratkovsky, V.G. Vaks, A.V. Trefilov: Phys. Lett. A **103**, 75 (1984)
5.6   J.A. Moriarty, D.A. Young, M. Ross: Phys. Rev. B **30**, 578 (1984)
5.7   A. Angelie, J.L. Pelissier: Physica **121A**, 207 (1983)
5.8   J.L. Pelissier: Physica **121A**, 217 (1983)
5.9   J.L. Pelissier: Physica **126A**, 474 (1984)
5.10 J.L. Pelissier: Phys. Lett. A **103**, 345 (1984)
5.11 J.L. Pelissier: Physica **128A**, 363 (1984)
5.12 J.L. Pelissier: Physica **126A**, 271 (1984)
5.13 B.L. Holian, G.K. Straub, R.E. Swanson, D.C. Wallage: Phys. Rev. B **27**, 2873 (1983);
      A.B. Walker, W. Smith, J.E. Inglesfield: J. Phys. F. **16**, L35 (1986)
5.14 S.M. Stishov: Sov. Phys. - Usp. **17**, 625 (1975)
5.15 W.B. Street, H.J. Raveché, R.D. Mountain: J. Chem. Phys. **61**, 1960 (1974)
5.16 M.J. Mandell, J.P. McTague, A. Rahman: J. Chem. Phys. **66**, 3070 (1977)
5.17 J.P. Hansen, L. Verlet: Phys. Rev. **184**, 150 (1969)
5.18 F.A. Lindemann: Z. Phys. **11**, 609 (1910)
5.19 L. Verlet: Phys. Rev. **165**, 201 (1968)
5.20 M. Born: J. Chem. Phys. **7**, 591 (1939)
5.21 P.M. Platzmann, H. Fukuyama: Solid State Commun. **15**, 677 (1974)
5.22 R.M.J. Cotterill, J.U. Madsen: Nature **288**, 467 (1980)
5.23 See e.g. F.R.N. Nabarro: *Theory of Crystal Dislocations* (Pergamon, Oxford 1967)
5.24 D.J. Thouless, A.M. Kosterlitz: J. Phys. C **6**, 1181 (1973)
5.25 R.M.J. Cotterill, W. Damgaard-Kristensen, E.J. Jensen: Philos. Mag. **30**, 245 (1974)
5.26 V.N. Novikov: Phys. Lett. A **104**, 103 (1984)
5.27 M. Kléman: *Continuum Models of Discrete Systems*, ed. by O. Brulin, R.K.T. Hsieh, Vol.4 (North-Holland, Amsterdam 1981)
5.28 N. Rivier, D.M. Duffy: J. Physique **43** 293 (1982)
5.29 T.V. Ramakrishnan, M. Yussouf: Phys. Rev. B **19**, 2775 (1979)
5.30 N.D. Mermin: Phys. Rev. **137**, A1441 (1965)
5.31 W.F. Saam, C. Ebner: Phys. Rev. A **15**, 2566 (1977)
5.32 R. Evans: Adv. Phys. **28**, 143 (1979)
5.33 P. Hohenberg, W. Kohn: Phys. Rev. **136**, B864 (1964)
5.34 W. Kohn, L.J. Sham: Phys. Rev. **140A**, 1133 (1965)

5.35   A.J.M. Yang, P.D. Fleming, J.H. Gibbs: J. Chem. Phys. **64**, 3732 (1976)

5.36   A.D.J. Haymet, D.W. Oxtoby: J. Chem. Phys. **74**, 2559 (1981); ibid. **84**, 1769 (1986)

5.37   P. Tarazona: Mol. Phys. **52**, 81 (1984)

5.38   G.L. Jones, U. Mohanty: Mol. Phys. **54**, 1241 (1985)

5.39   M. Baus, J.L. Colot: J. Phys. C **18**, L365 (1985)

5.40   F. Igloi, J. Hafner: J. Phys. C **19**, 5799 (1986)

5.41   F. Igloi: J. Phys. C **19**, 6907 (1986)

5.42   A.D.J. Haymet: J. Chem. Phys. **78**, 4641 (1983)

5.43   C. Marshall, B.B. Laird, A.D.J. Haymet: Chem. Phys. Lett. **122**, 320 (1985)

5.44   M. Rovere, M.P. Tosi: J. Phys. C **18**, 3445 (1985)

5.45   A.D.J. Haymet: Phys. Rev. Lett **52**, 1013 (1984)

5.46   N.H. March, M.P. Tosi: Phys. Chem. Liq. **10**, 185 (1980)

5.47   B. D'Aguanno, M. Rovere, M.P. Tosi, N.H. March: Phys. Chem. Liq. **13**, 113 (1983)

5.48   D.W. Oxtoby, A.D.J. Haymet: J. Chem. Phys. **76**, 6262 (1982)

5.49   F. Igloi, G. Kahl, J. Hafner: J. Phys. C **20**, 1803 (1987)

5.50   C. Cerjan, B. Bagchi: Phys. Rev. A **31**, 1647 (1985)

5.51   P. Minchin, M. Watabe, W.H. Young: J. Phys. F **7**, 653 (1977)

6.1   L.H. Bennett, R.E.Watson: *Theory of Alloy Phase Formation*, ed. by L.H. Bennet (The Metallurgical Society of AIME, Warrendale 1980) p.390

6.2   L.A. Girifalco: Acta Metall. **24**, 759 (1976)

6.3   N.W. Ashcroft: *Liquid Metals'76*, ed. by R. Evans, D.A. Greenwood (The Institute of Physics, Bristol 1977), p.39

6.4   L. Vegard: Z. Physik **5**, 17 (1921)

6.5   E-an Zen: Am. Mineral. **41**, 523 (1956)

6.6   W.B. Pearson: *The Crystal Chemistry and Physics of Metals and Alloys* (Wiley-Interscience, New York 1972)

6.7   W.B. Pearson: *Handbook of Lattice Spacings in Metals and Alloys* (Pergamon Oxford 1958);
P. Villars, N. Calvert: *Pearson's Handbook of Crystallographic Data of Intermetallic Phases*, Vols.1-3 (Am. Soc. Metals, Metals Park, Ohio 1985)

6.8   W.A. Kamitakahara, J.R.D. Copley: Phys. Rev. B **18**, 3782 (1978)

6.9   M.J. Hujben, J.Ph. van Hasselt, K. van der Weg, W. van der Lugt: Scr. Metall. **10**, 571 (1976)

6.10   M.J. Hujben, H. Klaucke, J. Hennephof, W. van der Lugt: Scr. Metall. **9**, 653 (1975)

6.11   J. Hafner, G. Punz: J. Phys. F **13**, 1393 (1983)

6.12   J. Hafner: Phys. Rev. A **16**, 351 (1977)

6.13   J. Hafner, V. Heine: J. Phys. F **13**, 2479 (1983)

6.14   J.J. van der Broek: Phys. Lett. A **40**, 219 (1972)

6.15   S. Steeb, S. Wörner: Z. Metallkd. **56**, 771 (1965)

6.16   J. Hafner: J. Phys. F **6**, 1243 (1976)

6.17   C. van der Marel: Dissertation, University of Groningen (1981)

6.18   H. Ruppersberg, W. Speicher: Z. Naturforsch. **31a**, 47 (1976)

6.19   G. Steinleitner, W. Freyland, F. Hensel: Ber. Bunsenges. Phys. Chem. **79**, 1156 (1975)

6.20   W. Biltz, F. Weibke: Z. Anorg. Allg. Chem. **223**, 321 (1935)

6.21   C.H. Hodges, M.J. Stott: Philos. Mag. **26**, 375 (1972)

6.22   J.A. Alonso, L.A. Girifalco: J. Phys. F **8**, 2455 (1978)

6.23   J.A. Alonso, L.A. Girifalco: Phys. Rev. B **19**, 3889 (1979)

6.24   J.A. Alonso, D.J. Gonzalez, M.P. Iniguez: Solid. State Commun. **31**, 9 (1979)

6.25   A. Meyer, I.H. Umar, W.H. Young: Phys. Rev. B **4**, 3287 (1971);
W. Jones, W.H. Young: J. Phys. C **4**, 1322 (1971)

6.26   J.R. Chelikowsky: Phys. Rev. B **21**, 3074 (1980)

6.27   J.R. Chelikowsky: Phys. Rev. Lett. **47**, 347 (1981)

6.28   J. Prater, H. Liebermann, L.A. Girifalco: J. Phys. Chem. Solids **38**, 1307 (1977)

6.29   D.M. Cragg, G.C. Fletcher: J. Phys. F **8**, 87 (1978)

6.30   V.K. Ratti, J.M. Ziman: J. Phys. F **4**, 1684 (1974)

6.31   M. Schlüter, C.M. Varma: Phys. Rev. B **23**, 1633 (1981)

6.32   D.G. Pettifor, C.M. Varma: J. Phys. C **12**, L253 (1979)

6.33   A. Zunger: Phys. Rev. B **17**, 2582 (1978)

6.34   J. Robertson: Solid State Commun. **47**, 899 (1983)

6.35   M. Taut, A.M. Radwan: Phys. Status Solidi B **76**, 605 (1976); ibid **82**, 507 (1977)

6.36   J.A. Tagle, E.T. Arakawa, T.A. Callott: Phys. Rev. B **22**, 2716 (1980)

6.37   S.R. Nagel, U.H. Gubler, C.F. Hague, J. Krieg, R. Lapka, P. Oelhafen, H.J. Güntherodt, J. Evers, A. Weiss, V.L. Moruzzi, A.R. Williams: Phys. Rev. Lett. **49**, 555 (1982)

6.38   M.W. Finnis: J. Phys. F **4**, 1645 (1974)

6.39   D.G. Pettifor: *Atomistics of Fracture*, Proc. NATO Advanced Study Institute, Corsica 1981, ed. by R. Latanision (Plenum, New York 1983)

6.40   R. Hultgren, P.D. Desai, D.T. Hawkins, M. Gleiser, K.K. Kelley: *Selected Values of the Thermodynamic Properties of Binary Alloys* (Am. Soc. Metals, Metals Park, Ohio 1973)

6.41   J. Hafner, A. Pasturel, P. Hicter: Z. Metallkd. **76**, 432 (1985)

6.42   J. Hafner: Phys. Rev. B **21**, 406 (1980)

6.43   J. Hafner: Phys. Rev. B **15**, 617 (1977)

6.44 D.G. Pettifor: Solid State Phys. (in press)

6.45 A.R. Miedema, P.F. de Chatel, F.R. de Boer: Physica 100B, 1 (1980)

6.46 L. Pauling: *The Nature of the Chemical Bond*, 2nd ed. (Cornell Uni. Press, Ithaca, NY 1952)

6.47 A.R. Williams, C.D. Gelatt, V.L. Moruzzi: Phys. Rev. Lett. **44**, 429 (1980)

6.48 D.G. Pettifor: Phys. Rev. Lett. **42**, 846 (1979); in *Physical Metallurgy*, ed. by R.W. Cahn, P. Haasen (North-Holland, Amsterdam 1984) Chap.3

6.49 J. Friedel: Trans. Metall. Soc. AIME **230**, 616 (1964)

6.50 J. Friedel: In *Physics of Metals*, Vol. 1, ed. by J.M. Ziman (Cambridge University Press, Cambridge 1968)

6.51 M. Cyrot, F. Cyrot-Lackmann: J. Phys. F **6**, 2257 (1976)

6.52 V. Heine: Phys. Rev. **153**, 673 (1967);
O.K. Andersen: Solid State Commun. **13**, 501 (1973)

6.53 R.E. Watson, L.H. Bennett: Phys. Rev. Lett **43**, 1130 (1979)

6.54 R.E. Watson, L.H. Bennett: CALPHAD **5**, 25 (1981)

7.1 J.E. Inglesfield: J. Phys. C **2**, 1285 (1969)

7.2 T.M. Hayes, H. Brooks, A.R. Bienenstock: Phys. Rev. **175**, 699 (1968)

7.3 G.L. Krasko: Sov. Phys. - JETP Lett. **13**, 155 (1971)

7.4 W. Hume-Rothery: *The Metallic State* (Oxford Uni. Press, Oxford 1931) p.328

7.5 W. Hume-Rothery, R.E. Smallmann, C.W. Haworth: *The Structure of Metals and Alloys*, 5th ed. (The Institute of Metals, London 1969) pp.47-62 and 124-136

7.6 L.S. Darken, R.W. Gurry: *Physical Chemistry of Metals* (McGraw-Hill, New York 1953) pp.74-92

7.7 N.F. Mott, J. Jones: *The Theory of the Porperties of Metals and Alloys* (Oxford Uni. Press, Oxford 1936)

7.8 J. Hafner: Phys. Rev. B **15**, 617 (1977)

7.9 J.E. Inglesfield: J. Phys. C **4**, 1003 (1971)

7.10 C.H. Leung: J. Phys. F **9**, 179 (1979)

7.11 S. Tanigawa, M. Doyama: J. Phys. F **3**, 977 (1973); Phys. Status Solidi B **56**, 665 (1973)

7.12 T. Soma, H. Matsua, Y. Kohbu: Phys. Status Solidi B **107**, 761 (1981)

7.13 M.P. Iniguez, J.A. Alonso: J. Phys. F **11**, 2045 (1981)

7.14 T. Yokokawa, O.J. Kleppa: J. Chem. Phys. **40**, 46 (1964)

7.15 G. Solt, A.P. Zhernov: Solid State Commun. **23**, 759 (1977)

7.16 J. Hafner, G. Punz: J. Phys. F **13**, 1393 (1983)

7.17 J. Hafner: Phys. Rev. B **21**, 406 (1980)

7.18	C.H. Leung, M.J. Stott, W.H. Young: J. Phys. F **6**, 1039 (1976)
7.19	D.L. Fuks, M.F. Zhorovkov, V.E. Panin: Phys. Status Solidi B **70**, 793 (1975)
7.20	A.I. Landa, V.E. Panin, M.F. Zhorovkov: Phys. Status Solidi B **108**, 113 (1981)
7.21	D.J. Gonzalez, J.A. Alonso: Physica 12 B, 73 (1981)
7.22	D.J. Gonzalez, M.L. Zorita, J.A. Alonso, M.P. Iniguez: Phys. Status Solidi B **119**, 589 (1983)
7.23	M.W. Finnis, V. Heine: J. Phys. F **4**, 960 (1974); M.W. Finnis: J. Phys. F **4**, 969 (1974)
7.24	D.J. Gunton, G.A. Saunders: Solid State Commun. **12**, 569 (1973)
7.25	J. Hafner, V. Heine: J. Phys. F **13**, 2479 (1983)
7.26	B. Predel, W. Schermann: Acta Metall. **19**, 81 (1971)
7.27	W. Gordy, W.J.O. Thomas: J. Chem. Phys. **24**, 439 (1956)
7.28	V.I. Bublik, S.S. Gorelik, A.A. Zaitsev, A.Y. Polyakov: Phys. Status Solidi B **65**, K79 (1974)
7.29	R.A. Logan, J.M. Rowell, F.A. Trumbore: Phys. Rev. **136B**, 1751 (1964)
7.30	S.P. Singh, W.H. Young: J. Phys. F **2**, 672 (1972)
7.31	J. Hafner: J. Phys. F **6**, 1243 (1976)
7.32	P. Beauchamp, R. Taylor, V. Vitek: J. Phys. F **5**, 2017 (1975)
7.33	Ya.A. Belenkij, Z.A. Gurskij, G.L. Krasko: Sov. Phys. - Solid State **15**, 2326 (1974)
7.34	C.S. Barrett: Acta Cryst. **9**, 671 (1956); C.S. Barrett, O.R. Trautz: Trans. Am. Inst. Min. Metal. Pet. Eng. **175**, 579 (1948)
7.35	G. Fichtner: Diplomarbeit, TU Wien (1974); G. Fichtner, J. Hafner: unpublished
7.36	F.H. Herbstein, B.L. Averbach: Acta Metall. **4**, 407 (1956)
7.37	J.L. Murray: *Alloy Phase Diagrams*, ed. by L.H. Bennett, T.B. Massalski, B.C. Giessen (North-Holland, New York 1983) p.223
7.38	B.C. Giessen: *Advances in X-Ray Analysis*, ed. by C.S. Barrett, G.R. Mallett, J.B. Newkirk, Vol.12 (Plenum, New York 1969) p.23
7.39	B.C. Giessen: *Alloy Phase Diagrams*, ed. by L.H. Bennett, T.B. Massalski, B.C. Giessen (North-Holland, New York 1983) p.263 and further references cited therein
7.40	U. Mizutani, T.B. Massalski: Prog. Mater. Sci. **22**, 152 (1978)
7.41	A.P. Blandin: *Alloying Behaviour and Effects of Concentrated Solid Solutions*, ed. by T.B. Massalski (Gordon and Breach, New York 1963)
7.42	A.P. Blandin: *Phase Stability in Metals and Alloys*, ed. by P.S. Rudman, J. Stringer, R.I. Jaffee (McGraw-Hill, New York 1967) p.115

7.43   V. Heine: *The Physics of Metals*, Vol.1, ed. by J.M. Ziman (Cambridge, Uni. Press, Cambridge 1969) p.1

7.44   D. Stroud, N.W. Ashcroft: J. Phys. F **1**, 113 (1979)

7.45   R. Evans, P. Lloyd, S.M.M. Rahman: J. Phys. F **9**, 1939 (1979)

7.46   S.M.M. Rahman: J. Phys. F **11**, 1191 (1981)

7.47   M. Kogachi: J. Phys. Chem. Solids **34**, 67 (1973); ibid **35**, 109 (1974)

7.48   S.D. Borisova, M.F. Zhorovkov, Yu.I. Paskal: Sov. Phys. - Solid State **27**, 400 (1985)

7.49   C.S.G. Cousins: J. Phys. F **4**, 1 (1974)

7.50   Ch.W. Krause, J.W. Morris: Acta Metall **22**, 767 (1974)

7.51   D. de Fontaine: Solid State Phys. **34**, 231 (1979)

7.52   S. Froyen, C. Herring: J. Appl. Phys. **52**, 7165 (1981)

7.53   G. Solt: Phys. Rev. B **18**, 720 (1978)

7.54   G. Solt, K. Werner: Phys. Rev. B **24**, 817 (1981)

7.55   H. Kanzaki: J. Phys. Chem. Solids **2**, 24 (1957)

7.56   F. Toigo, T.O. Woodruff: Phys. Rev. B **1**, 3958 (1970)

7.57   K. Werner, W. Schmatz, G.S. Bauer, E. Seitz, H.J. Fenzl, A.Baratoff: J. Phys. F **8**, L207 (1978)

7.58   J.E. Inglesfield: J. Phys. C **1**, 1337 (1968)

7.59   M. Kogachi, Y. Matsuo: J. Phys. Chem. Solids **32**, 2393 (1973)

7.60   M. Kogachi, K. Katada: J. Phys. Chem. Solids **36**, 501 (1975)

7.61   Y. Matsuo, S. Minimagawa, K. Katada: Acta Metall. **25**, 1179 (1977)

7.62   J.M. Cowley: Phys. Rev. **77**, 669 (1950)

7.63   J.M. Cowley: Phys. Rev. **120**, 1648 (1960)

7.64   W.L. Bragg, E.J. Williams: Proc. R. Soc. London A **145**, 699 (1934)

7.65   L.D. Landau, E.M. Lifshitz: *Statistical Physics*, Course on Theoretical Phys., Vol.5 (Pergamon, Oxford 1980) Sect.139

7.66   A.G. Khachaturyan: Phys. Status Solidi B **60**, 9 (1973)

7.67   G.L. Krasko, A.B. Maknovetskij: Phys. Status Solidi B **65**, 869 (1974); ibid **66**, 349 (1974)

7.68   A.B. Maknovetskij, G.L. Krasko: Phys. Status Solidi B **80**, 341 (1977)

7.69   Z.A. Gurskij, W.I. Baranitskij: Preprint ITF-82-3P, Ukrainian Academy of Sciences (Kiev 1982)

7.70   S.M.M. Rahman: J. Phys. F **11**, 1011 (1981)

7.71   P.L. Fuks, M.F. Zhorovkov, V.E. Panin: Fiz. Met. Metalloved. **43**, 1058 (1977)

7.72   V. Heine, R.D. Jones: J. Phys. C **2**, 719 (1969)

7.73   M.B. McNeil, W.B. Pearson, L.H. Bennett, R.E. Watson: J. Phys. C **6**, 1 (1973)

7.74   W.B. Pearson: *The Crystal Chemistry and Physics of Metals and Alloys* (Wiley-Interscience, New York 1972)

7.75   F.A. Khvaya, A.A. Knatsnelson, V.M. Silonov, M.M. Khrush-
        chov: Phys. Status Solidi B **82**, 701 (1977)
7.76   F.A. Khvaya, A.A. Knatsnelson, V.M. Silonov: Phys. Status
        Solidi B **88**, 477 (1978)
7.77   F. Ducastelle, F. Gautier: J. Phys. F **6**, 2039 (1976)
7.78   A. Bieber, F. Gautier: Solid State Commun. **38**, 1219 (1981)
7.79   A. Bieber, F. Ducastelle, F. Gautier, G. Treglia, P. Turchi:
        Solid State Commun. **45**, 585 (1983)
7.80   A. Bieber, F. Gautier: Z. Physik B **57**, 335 (1984)
7.81   D. de Fontaine: *Alloy Phase Diagrams*, ed. by L.H. Bennett,
        T.B. Massalski, B.C. Giessen (North-Holland, New York 1983)
        p.149
7.82   D.W. Taylor: Phys. Rev. **156**, 1017 (1967)
7.83   P. Soven: Phys. Rev. **156**, 809 (1967)
7.84   P. Soven: Phys. Rev. **178**, 1136 (1969)
7.85   N. Wakabayashi, R.M. Nicklow, H.G. Smith: Phys. Rev. B **4**,
        2558 (1971)
7.86   T. Kaplan, M. Mostoller: Phys. Rev. B **9**, 353 (1974)
7.87   W.A. Kamitakahara, J.R.D. Copley: Phys. Rev. B **18**, 3782
        (1978)
7.88   T. Kaplan, M. Mostoller: Phys. Rev. B **9**, 1783 (1974)
7.89   M. Mostoller, T. Kaplan: Phys. Rev. B **16**, 2350 (1977)
7.90   G. Grünewald: J. Phys. F **6**, 999 (1976)
7.91   G. Grünewald, K. Scharnberg: Proc. Intern. Conf. on Lattice
        Dynamics, Paris 1977, ed. by M. Balkanski (Flammarion, Paris
        1978) p.443
7.92   G. Grünewald, R. Schopohl: J. Phys. F **9**, 1047 (1979)
7.93   G. Jacucci, M.L. Klein, R. Taylor: Phys. Rev. B **18**, 3782
        (1978)
7.94   J. Hafner, G. Punz: Phys. Rev. B **30**, 7336 (1984); Z. Physik B
        61, 231 (1985)
7.95   R. Haydock, V. Heine, M.J. Kelly: J. Phys. C **5**, 2845 (1972);
        ibid. **8**, 2591 (1975)
7.96   See the review articles of V. Heine, R. Haydock, M.J. Kelly:
        Solid State Physics **35** (1980)
7.97   D.M. Straus, N.W. Ashcroft, H. Beck: Phys. Rev. B **15**, 1914
        (1977)
7.98   G. Punz: Dissertation, Techn. University of Vienna (1986)
        unpublished;
        G. Punz, J. Hafner: Z. Physik B **65**, 465 (1987)
7.99   G.S. Barrett, O.R. Trautz: Trans. Am. Inst. Min. Metall., Pet.
        Eng. **175**, 579 (1948)
7.100   M. Hasegawa, W.H. Young: J. Phys. F **8**, 1105 (1978)
7.101   W.H. Young: *Liquid Metals '76*, ed. by R. Evans, D.A. Green-
        wood, Inst. Phys. Conf. Ser. Vol.30 (The Institute of Physics,
        Bristol 1977)

7.102   D.D. Johnson, D.M. Nicholson, F.J. Pinski, B.L. Györffy, G.M. Stocks: Phys. Rev. Lett. **56** 2088 (1986)

7.103   G.M. Stocks, M. Boring, D.M. Nicholson, F.J. Pinski, D.D. Johnson, J.S. Faulkner, B.L. Györffy: In *Noble Metal Alloys*, ed. by T.B. Massalski (AIME, Warrendale 1986)

8.1   W.B. Pearson: *The Crystal Chemistry and Physics of Metals and Alloys* (Wiley-Interscience, New York 1972)

8.2   W.B. Pearson: *Handbook of Lattice Spacings in Metals and Alloys* (Pergamon, Oxford 1958)

8.3   P. Villars, N. Calvert: *Pearson's Handbook of Crystallographic Data of Intermetallic Compounds*, Vols.1-3 (American Society for Metals, Metals Park 1985)

8.4   J. St. John, A.N. Bloch: Phys. Rev. Lett. **33**, 1095 (1974)

8.5   A. Zunger, M.L. Cohen: Phys. Rev. Lett. **41**, 53 (1978)

8.6   J.C. Phillips: Solid State Commun. **22**, 549 (1977)

8.7   A. Zunger: Phys. Rev. B **22**, 5839 (1980)

8.8   J.K. Burdett, G.D. Price, S.L. Price: Phys. Rev. B **24**, 2903 (1981)

8.9   D.G. Pettifor: Solid State Commun. **51**, 37 (1984); J. Phys. C **19**, 285 (1986)

8.10   A. Bieber, F. Gautier: Solid State Commun. **38**, 1219 (1981)

8.11   E.S. Machlin, B. Loh: Phys. Rev. Lett. **45**, 1642 (1980)

8.12   J.K. Burdett: J. Solid State Chem. **45**, 399 (1982)

8.13   P. Villars, K. Girgis, F. Hulliger: J. Solid State Chem. **42**, 89 (1982)

8.14   R.E. Watson, L.H. Bennett: Phys. Rev. B **18**, 6439 (1978)

8.15   D.G. Pettifor: *Physical Metallurgy*, ed. by R.W. Cahn, P. Haasen (North-Holland, Amsterdam 1984) Chap.3

8.16   D. Pettifor, R. Podloucky: Phys. Rev. Lett. **53**, 1080 (1984); J. Phys. C **19**, 315 (1986)

8.17   E.S. Machlin: Acta Metall. **22**, 95; 109; 367; 1433 (1974)

8.18   E.S. Machlin: Acta Metall. **24**, 543 (1976)

8.19   E.S. Machlin: CALPHAD **1**, 361 (1977)

8.20   E.S. Machlin: *Theory of Alloy Phase Formation*, ed. by L.H. Bennett (The Metallurgical Society of AIME, Warrendale 1980) p.127

8.21   M. Born: Proc. Cambridge Philos. Soc. **36**, 160 (1940)

8.22   W. Gordy: W.J.O. Thomas: J. Chem. Phys. **24**, 439 (1956)

8.23   W. Gordy: Phys. Rev. **69**, 604 (1946)

8.24   J. Hafner, V. Heine: J. Phys. F **16**, 1429 (1986)

8.25   D.G. Pettifor, M.D. Ward: Solid State Commun. **49**, 291 (1984)

8.26   W.B. Pearson: J. Less Common Met. **109**, L3 (1985)

8.27   L. Pauling: J. Am. Chem. Soc. **69**, 542 (1947)

8.28 F.C. Frank, J.S. Kasper: Acta Crystallogr. 11, 184 (1958); ibid 12, 483 (1959)

8.29 A. Simon, G. Ebbinghaus: Z. Naturforsch. 29b, 618 (1974)

8.30 A. Simon, W. Brämer, B. Hillenköter, H.J. Kullmann: Z. Anorg. Allg. Chem. 419, 253 (1976)

8.31 J. Hafner: Phys. Rev. B 15, 617 (1977)

8.32 V.A. Zhdanov, S.B. Teordorovich: Izv. Vuz. Fiz. (USSR) 3, 147 (1978)

8.33 L. Pauling: Acta Crystallogr. 10, 374 (1957)

8.34 C.B. Shoemaker, D.P. Shoemaker: Developments in the Structural Chemistry of Alloy Phases, ed. by B.C. Giessen (Plenum, London 1969) p.107

8.35 A. Simon: Angew. Chem. 95, 94 (1983)

8.36 C.D. Churcher, V. Heine: Acta Crystallogr. A 40, 291 (1984)

8.37 A.M. Radwan, M. Taut: Phys. Status Solidi B 76, 605 (1976)

8.38 M. Taut, A.M. Radwan: Phys. Status Solidi B 82, 507 (1977)

8.39 F. Laves, H. Witte: Metallwirtsch., Metallwiss., Metalltech. 15, 840 (1936)

8.40 Y. Komura: Acta Crystallogr. 15, 770 (1962)

8.41 Y. Komura, Y. Kitano: Acta Crystallogr. B33, 2496 (1977)

8.42 Y. Komura, K. Tokunaga: Acta Crystallogr. B36, 1548 (1980)

8.43 R.P. Elliott, W. Rostoker: Trans. Am. Soc. Met. 50, 617 (1958)

8.44 A.E. Dwight: Trans. Am. Soc. Met. 53, 479 (1961)

8.45 A.R. Edwards: Metall. Trans. 3, 1365 (1972)

8.46 J. Hafner: Phys. Rev. B 21, 406 (1980)

8.47 P. Rennert, A.M. Radwan: Phys. Status Solidi B 77, 615 (1976)

8.48 P. Rennert, A.M. Radwan: Phys. Status Solidi B 79, 167 (1977)

8.49 R.L. Johannes, R. Haydock, V. Heine: Phys. Rev. Lett. 36, 372 (1976)

8.50 R. Haydock, R.L. Johannes: J. Phys. F 5, 2055 (1975)

8.51 R. Haydock, V. Heine, M.J. Kelly: J. Phys. C 5, 2845 (1972); ibid 8, 2591 (1975)

8.52 D.G. Pettifor: Metallurgical Chemistry, ed. by O. Kubaschewski (HMSO, London 1972) p.191

8.53 J. Hafner: Phys. Rev. B 19, 5094 (1979)

8.54 C.G. Wilson; D.K. Thomas, F.J. Spooner: Acta Crystalllogr. 13, 56 (1960)

8.55 J.B. Forsyth, L.M. d'Alte da Veiga: Acta Crystallogr. 15, 543 (1962)

8.56 A. Iandelli, A. Palenzona: J. Less Common Met. 12, 333 (1967)

8.57 M.L. Fornasini, F. Merlo: J. Less Common Met. 79, 111 (1981)

8.58 A.F. Messing, M.D. Adams, R.K. Stemmenberg: Trans Am. Soc. Met. 56, 345 (1963)

8.59 J. Hafner: Phys. Blätter 35, 598 (1979); J. Phys. F 15, 1879 (1985)

8.60 T. Ohba, Y. Kitano, Y. Komura: Acta Crystallogr. C **40**, 1 (1984)

8.61 H. Eschrig, K. Feldmann, K. Henning, L. Weiß: *Neutron Inelastic Scattering* (IAEA, Wien 1972) p.157

8.62 H. Eschrig, P. Urwank, H. Wonn: Phys. Status Solidi B **49**, 807 (1972)

8.63 B. Dorner, K. Feldmann, K. Henning, P. Urwank, L. Weiß: Phys. Status Solidi B **63**, K135 (1974)

8.64 H. Eschrig, K. Feldmann, K. Henning, W. Matz, P. Paufler: Phys. Status Solidi B **79**, 283 (1977)

8.65 W. Reichardt, Ch. Kobbelt: Primärbericht, Kerforschungszentrum Karlsruhe, Inst. für Nukleare Festkörperphysik (1984) unpublished

8.66 R.W. Shaw: Phys. Rev. **174**, 769 (1968)

8.67 W. Reichardt: unpublished

8.68 P.W. Anderson: *Concepts in Solids* (Benjamin, New York 1963) p.6;
L. Brewer: *High Strength Materials*, ed. by V. Zackay (Wiley, New York 1964);
N. Engel: *Developments in the Structural Chemistry of Alloy Phases*, ed. by B.C. Giessen (Plenum, New York 1969) p.25;
W. Hückel: *Structural Chemistry of Inorganic Compounds* (Elsevier, Amsterdam 1951) p.1529

8.69 N.E. Christensen: Phys. Rev. B **32**, 207 (1985)

8.70 J. Hafner, W. Weber: Phys. Rev. B **33**, 754 (1986)

8.71 G.L. Krasko, A.B. Maknovetskii: Phys. Status Solidi B **65**, 869 (1974); ibid. B **65**, 349 (1974)
A.B. Maknovetskii, G.L. Krasko: Phys. Status Solids B **80**, 341 (1977); ibid. B **81**, 263 (1977)

8.72 V. Heine, R.O. Jones: J. Phys. C **2**, 719 (1969)

8.73 T.O. Brun, J.E. Robinson, S. Susman, D.F.R. Mildner, R. Dejus, K. Sköld: Solid State Ionics **9**, 485 (1983)

8.74 S.S. Jaswal, J. Hafner: to be submitted

8.75 N.E. Christensen: Phys. Rev. B **32**, 6490 (1985)

9.1 A.B. Bhatia, D.E. Thornton: Phys. Rev. B **2**, 3004 (1970)

9.2 N.W. Ashcroft, D.C. Langreth: Phys. Rev. **156**, 685 (1967)

9.3 T.E. Faber: *An Introduction to the Theory of Liquid Metals* (Cambridge Univ. Press, Cambridge 1972)

9.4 J.E. Enderby, D.M. North: J. Phys. Chem. Liq. **1**, 1 (1968)

9.5 J.P. Gabathuler, S. Steeb, P. Lamparter: Z. Naturforsch. **34a**, 1305 (1979)

9.6 P. Chieux, H. Ruppersberg: J. Physique **41**, C8-321 (1980)

9.7 B.P. Alblas, W. van der Lugt, H.J.L. van der Valk, J.Th. de Hosson, C. van Dijk: Physica **101B**, 177 (1980)

9.8 H. Ruppersberg, W. Knoll: Z. Naturforsch. **32a**, 1374 (1977)

9.9 S. Steeb, P. Lamparter: Proc. 5th Int'l. Conf. on Liquid and Amorphous Metals, ed. by C.N.J. Wagner, W.A. Johnson, J. Non-Cryst. Solids **61&62**, 237 (1984)

9.10 C.N.J. Wagner: In *Proc. 5th Int'l. Conf. on Rapidly Quenched Metals*, ed. by S. Steeb, H. Warlimont (North-Holland, Amsterdam 1985)

9.11 G. Jacucci, I.R. McDonald, R. Taylor: J. Phys. F **8**, L121 (1978)

9.12 L. Dagens, M. Rasolt, R. Taylor: Phys. Rev. B **11**, 2726 (1975)

9.13 M.J. Hujben, W. van der Lugt, W.A.M. Reinert, J.Th. de Hosson: Physica **97B**, 338 (1979)

9.14 A.P. Copestake, R. Evans, H. Ruppersberg, W. Schirmacher: J. Phys. F **13**, 1993 (1983)

9.15 G. Jacucci, M. Ronchetti, W. Schirmacher: *Condensed Matter Research Using Neutrons: Today and Tomorrow*, ed. by S.W. Lovesey, R. Scherm (Plenum, New York 1984), p.139;
G. Jacucci, M. Ronchetti, W. Schirmacher: Proc. 3rd Intern. Conf. on the Structure of Non-Cryst. Materials, ed. by Ch. Janot (Les Editions de Physique, Paris 1985) p.385

9.16 J. Hafner: Proc. 6th Int'l. Conf. on Liquid and Amorphous Metals, ed. by F. Hensel, E. Lüscher, Z. Phys. Chem. (in press)

9.17 E. Waisman: Mol. Phys. **25**, 45 (1973);
E. Waisman, J.L. Lebowitz: J. Chem. Phys. **56**, 3086 (1972); ibid p. 3093

9.18 G.K. Chaturvedi, M. Rovere, G. Senatore, M.P. Tosi: Physica B **111**, 11 (1981)

9.19 M.C. Abramo, C. Caccamo, G. Pizzimenti: J. Chem. Phys. **78**, 357 (1983)

9.20 J. Hafner: *Liquid Metals '76*, ed. by R. Evans and D.A. Greenwood (The Inst. of Physics, Bristol 1977) p.107

9.21 J. Hafner: Phys. Rev. A **16**, 351 (1977)

9.22 J. Hafner: Phys. Rev. B **21**, 406 (1980)

9.23 E.H. Henninger, R.C. Buschert, L. Heaton: J. Chem. Phys. **44**, 1758 (1966)

9.24 J.L. Lebowitz, J.S. Rowlinson: J. Chem. Phys. **41**, 133 (1964);
J.L. Lebowitz, J.K. Percus: Phys. Rev. **144**, 251 (1966)

9.25 G.A. Mansoori, N.F. Carnahan, K.E. Starling, T.W. Leland: J. Chem. Phys. **54**, 1523 (1971)

9.26 I.H. Umar, A. Meyer, M. Watabe, W.H. Young: J. Phys. F **4**, 1691 (1974)

9.27 J.M. Harder, M. Silbert, I. Yokoyama, W.H. Young: J. Phys.F **10**, 1101 (1980)

9.28 I. Yokoyama, M.J. Stott, M. Watabe, W.H. Young, M. Hasegawa: J. Phys. F **9**, 207 (1979)

9.29   I.H.Umar: M. Watabe, W.H. Young: Phil. Mag. **30**, 957 (1974)
9.30   V.P. Kazimirov, G.I. Batolin: Fiz. Met. Metalloved. **47**, 689 (1979)
9.31   R.N. Singh: J. Phys. F **10**, 1411 (1980)
9.32   N.A. Batolin, B.R. Gelchinskij, V.A. Poluchin, V.F. Ukhov, O.A. Esin: Dokl. Acad. Nauk SSSR **222**, 1323 (1975)
9.33   Y. Waseda, K.T. Jacob, S. Tamaki: Z. Naturforsch. **34a**, 320 (1979)
9.34   S.M.M. Rahman: J. Phys. F **11**, 2301 (1981)
9.35   J.E. Enderby, P.A. Egelstaff, D.M. North: Philos. Mag. **14**, 961 (1966)
9.36   H.C. Longuet-Higgins: Proc. R. Soc. London **A205**, 247 (1951)
9.37   A.B. Bhatia, W.H. Hargrove, N.H. March: J. Phys. C **6**, 621 (1973)
       V.K. Ratti, A.B. Bhatia: Nuovo Cimento **43B**, 1 (1978)
9.38   W. van der Lugt, W. Geertsma: Proc. 5th Int'l. Conf. on Liquid and Amorphous Metals, ed. by C.N.J. Wagner, W.A. Johnson, J. Non-Crys. Solids **61&62**, 187 (1984)
9.39   S. Tamaki: Proc. 6th Intern. Conf. on Liquid and Amorphous Metals, ed. by E. Lüscher, F. Hensel, W. Gläser: Z. Phys. Chem., in press
9.40   A.B. Bhatia, W.H. Hargrove, D.E. Thornton: Phys. Rev. B **9**, 435 (1974)
9.41   B. Predel, G. Oehme: Z. Metallkd. **65**, 509 (1974); ibid **67**, 826 (1976)
9.42   F. Sommer: Z. Metallkd. **73**, 72 (1982)
9.43   K. Hoshino: J. Phys. F **13**, 1981 (1983)
9.44   A.P. Copestake, R. Evans: Proc. Intern. Conference on Ionic Liquids, Molten Salts and Polyelectrolytes, ed. by K.H. Bennemann, D. Quitmann, Lecture Notes Phys., Vol.172 (Springer, Berlin, Heidelberg 1982) p.86
9.45   A.P. Copestake, R. Evans, M.M. Telo da Gama: J. Physique **41**, C8-321 (1980)
9.46   M.S. Wertheim: Phys. Rev. Lett. **10**, 321 (1963)
9.47   E. Waisman: J. Chem. Phys. **59**, 495 (1973)
9.48   J. Hafner, A. Pasturel, P. Hicter: J. Phys. F **14**, 1137 (1984)
9.49   J. Hafner, A. Pasturel, P. Hicter: J. Phys. F **14**, 2279 (1984)
9.50   J. Hafner, A. Pasturel, P. Hicter: Z. Metallkd. **76**, 432 (1985)
9.51   S. Tamaki, T. Ishiguro, S. Takeda: J. Phys. F **12**, 1613 (1982)
9.52   S. Matsunaga, T. Ishiguro, S. Tamaki: J. Phys. F **13**, 587 (1983)
       B.P. Alblas, W. van der Lugt, J. Dijkstra, W. Geertsma, C. van Dijk: J. Phys. F **13**, 2465 (1983)
9.53   A. Pasturel, J. Hafner, P. Hicter, Phys. Rev. B **32**, 5009 (1985)
9.54   H. Ruppersberg, H. Reiter: J. Phys. F **12**, 1311 (1982)
9.55   C. Holzhey, F. Brouers, J.R. Franz, W. Schirmacher: J. Phys. F **12**, 2601 (1982)

9.56   C. Holzhey, F. Brouers, J.R. Franz, W. Schirmacher: Proc. 5th
        Intern. Conf. on Liquid and Amorphous Metals, ed. by C.N.J.
        Wagner, W.A Johnson, J. Non-Cryst. Solids **61&62**, 65 (1984)
9.57   L.M. Falicov, F. Yndurain: Phys. Rev. B **12**, 5664 (1975)
9.58   R.C. Kittler, L.M. Falicov: J. Phys. C **9**,4259 (1976)
        R. Evans, M.M. Telo da Gama, Philos. Mag. B **41**, 351 (1980)
9.59   J. Hafner, A. Pasturel: Proc. 3rd Intern. Conf. on the Structure
        of Non-Crystalline Materials, ed. by Chr. Janot (Les Editions
        de Physique 1985) p.387;
        A. Pasturel, J. Hafner: Phys. Rev. B **33**, 8357 (1986)
9.60   D. Stroud: Phys. Rev. B **8**, 1308 (1973)
9.61   D. Stroud: Phys. Rev. B **7**, 4405 (1973)
9.62   D.J. Stevenson: Phys. Rev. B **12**, 3999 (1975)
9.63   A. Kumar, H.R. Krishnamurthy, E.S. R. Gopal: Phys. Rep. **98**,
        57 (1983)
9.64   J. Hafner, W. Jank: unpublished
9.65   H.E. Stanley: *Introduction to Phase Transitions and Critical
        Phenomena* (Clarendon, Oxford 1971)
9.66   R.L. Henderson, N.W. Ashcroft: Phys. Rev. A **13**, 859 (1976)
9.67   G. Chabrier, J.P. Hansen: J. Phys. F **18**, L77 (1985);
        G. Chabrier, J.P. Hansen: Mol. Phys. **50**, 901 (1983)
9.68   J.D. Weeks, D. Chandler: Phys. Rev. Lett. **25**, 149 (1970)
9.69   J.D. Weeks, D. Chandler, H.C. Andersen: J. Chem. Phys. **54**,
        5237 (1971)
9.70   J.D. Weeks, D. Chandler, H.C. Andersen: J. Chem. Phys. **55**,
        5422 (1971)
9.71   S. Sung, D. Chandler: J. Chem. Phys. **56**, 4989 (1972)
9.72   L. Verlet, J.J. Weis: Phys. Rev. A **5**, 939 (1972)
9.73   B.N. Perry, M. Silbert: Mol. Phys. **37**, 1823 (1979)
9.74   S.I. Sandler: Chem. Phys. Lett. **33**, 351 (1975)
9.75   L.L. Lee, D. Levesque: Mol. Phys. **26**, 1351 (1973)
9.76   G. Kahl, J. Hafner: J.Phys. F **15**, 1627 (1985)
9.77   B.P. Alblas, W. van der Lugt, O. Mensies, C. van Dijk: Physica
        106B, 22 (1981)
9.78   R.E. Jacobs, H.C Andersen: Chem. Phys. **10**, 73 (1975)
9.79   M.M. Telo da Gama, R. Evans: Mol. Phys. **41**, 1091 (1980)
9.80   H.C. Andersen, D. Chandler: J. Chem. Phys. **53**, 547 (1970)
9.81   H.C. Andersen, D. Chandler: J. Chem. Phys. **57**, 1918 (1972)
9.82   H.C. Andersen, D. Chandler, J.D. Weeks: J. Chem. Phys. **56**,
        3812 (1972)
9.83   H.C. Andersen, D. Chandler: J. Chem. Phys. **55**, 1497 (1971)
9.84   S.H. Sung, D. Chandler, B.J. Alder: J. Chem. Phys. **61**, 932
        (1974)
9.85   G. Kahl, J. Hafner: Phys. Chem. Liquids (in press)
9.86   F. Cyrot-Lackmann, M. Cyrot: Solid State Commun. **22**, 517
        (1977)

9.87   M.J. Kelly, D. Bullett: J. Phys. C 12, 2531 (1979)

9.88   T. Fujiwara: J. Phys. F 12, 661 (1981)

9.89   T. Fujiwara: *Topological Disorder in Condensed Matter*, ed. by F. Yonezawa, T. Ninomiya, Solid-State Sci., Vol.46 (Springer, Berlin, Heidelberg 1983) p.111

9.90   A.B. Bhatia: *Cooperative Phenomena*, ed. by H. Haken, M. Wagner (Springer, Berlin, Heidelberg 1973) p.49

9.91   S.N. Khanna, F. Cyrot-Lackmann, P. Hicter: J. Chem. Phys. 73, 4636 (1980)

9.92   A. Pasturel, P. Hicter, F. Cyrot-Lackmann: J. Less Common Metals 86, 181 (1982)

9.93   A. Pasturel, P. Hicter, D. Mayou, F. Cyrot-Lackmann: Scr. Metall. 17, 841 (1983)

9.94   A. Meyer, M.J. Stott, W.H. Young: Phil. Mag. 33, 381 (1976)

9.95   A. Pasturel, C. Colinet, P. Hicter: J. Phys. F 15, L81 (1985)

9.96   M. Parrinello, M.P. Tosi: Riv. Nuovo Cimento 2, 1 (1979)

9.97   G. Jacucci, I.R. McDonald: *Liquid and Amorphous Metals*, ed. by E. Lüscher, H. Coufal (Sijthoff and Nordhoff, Alphen van Rijn, Netherlands 1980)

9.98   P.T. Cummings, G. Stell: J. Chem. Phys. 78, 1917 (1983)

9.99   J. Hafner, W. Jank: Physica B, submitted

9.100   G. Kahl, G. Pastore: to be submitted

9.101   J. Bosse, G. Jacucci, M. Ronchetti, W. Schirmacher: Phys. Rev. Lett. 37, 3277 (1986)

10.1   J. Hafner: Phys. Rev. B 21, 406 (1980)

10.2   J. Hafner: Phys. Rev. B 28, 1734 (1983)

10.3   J. Hafner, L. von Heimendahl: Phys. Rev. Lett. 42, 385 (1979)

10.4   M. Hasegawa, W.H. Young: J. Phys. F. 7, 2271 (1977); T. Soma, H. Matsuo, H. Funaki: Phys. Status Solidi B 108, 221 (1981)

10.5   I.R. Yukhnovskij, Z.A. Gurskij, W.I. Baranitskij: Ukr. Phys. J. 29, 1389 (1984); Z.A. Gurskij, W.I. Baranitskij: Fiz. Met. Metalloved. 57, 883 (1984)

10.6   A.I. Landa, V.E. Panin, M.F.Zhorovkov: Phys. Status Solidi B 108, 113 (1981)

10.7   Z.A. Gurskij, W.I. Baranitzkij: Preprint ITF-82-3P, Ukrainian Academy of Sciences

10.8   T.B. Massalski: Proc. 4th Int'l. Conf. on Rapidly Quenched Metals, Sendai 1981, ed. by T. Masumoto, K. Suzuki (The Japan Inst. of Metals, Sendai 1982) p.203

10.9   A.B. Bhatia, N.H. March: Phys. Lett. A 41, 397 (1972)

10.10   A.B. Bhatia, N.H. March: J. Phys. F 5, 1100 (1975)

10.11   A.B. Bhatia, N.H. March: Phys. Lett. A **51**, 401 (1975)
10.12   A.B. Bhatia, N.H. March, L. Rivaud: Phys. Lett. A **47**, 303 (1978)
10.13   W. Geertsma: Physica **132** B, 337 (1985)
10.14   F. Sommer: Proc. 5th Intern. Conf. on Rapidly Quenched Metals, Würzburg 1984, ed. by S. Steeb, H. Warlimont (North-Holland, Amsterdam 1985) p.153; Z. Metallkd. **72**, 219 (1981)
10.15   S. Tamaki, N.E. Cusack: J. Phys. F **9**, 403 (1979)
10.16   J. Hafner: unpublished
10.17   J.L. Barrat, M. Baus, J.P. Hansen. Phys. Rev. Lett. **56**, 1063 (1986);
        J.L. Barrat, M. Baus, J.P. Hansen: J. Phys. C **20**, 1413 (1987)
10.18   J.L. Barrat: J. Phys. C **20**, 1031 (1987)
10.19   M. Baus, J.L. Colot: J. Phys. C **18**, L365 (1985)
10.20   A.A. Mbaye, L.G. Ferreira, A. Zunger: Phys. Rev. Lett. **58**, 49 (1987)
10.21   R. Podloucky, H.J.F. Jansen, X.Q. Guo, A.J. Freeman: Phys. Rev. Lett., submitted

11.1   For a recent overview on metastable crystalline alloys, see, e.g., E. Hornbogen: *Proc. 5th Int'l. Conf. on Rapidly Quenched Metals*, ed. by S. Steeb, H. Warlimont (North-Holland, Amsterdam 1985) p.785;
        B.C. Giessen: *Advances in X-Ray Analysis*, Vol. 22, ed. by C.S. Barrett, G.R. Mallet, J.B. Newkirk (Plenum, New York 1969) p.23
11.2   D. Shechtmann, I. Blech, D. Gratias, J.W. Cahn: Phys. Rev. Lett. **53**, 1951 (1984)
11.3   For reviews on amorphous elements and alloys, see e.g. *Glassy Metals* I&II, ed. by H. J. Güntherodt, H. Beck, Topics Appl. Phys., Vols.46 and 53 (Springer, Berlin, Heidelberg 1981&82);
        Amorphous Metals and Semiconductors, ed. by P. Haasen, R.I. Jaffee (Acta-Scripta Metallurg. Proc. Series Vol.3 (Pergamon, Oxford 1986)
11.4   R.B. Schwarz, W.L. Johnson: Phys. Rev. Lett. **51**, 415 (1983);
        K. Samwer, A. Regenbrecht, H. Schröder: Proc. 5th Int'l. Conf. on Rapidly Quenched Metals, ed. by S. Steeb, H. Warlimont (North-Holland, Amsterdam 1985) p.1577
11.5   N. Saunders, A.P. Miodownik: J. Mater. Res. 1, 38 (1986)
11.6   B.G. Lewis, H.A. Davies: Proc. 3rd Int'l. Conf. on Liquid Metals, Bristol 1976, ed. by R. Evans, D.A. Greenwood (The Inst. of Phys., Bristol 1977) p.274;
        H.S. Chen, K.A. Jackson: *Metallic Glasses*, ed. by J.J. Gilman, H.J. Leamy (American Society for Metals, Metal Park 1978) p.74;

D.R. Uhlman: J. Non-Cryst. Solids 7, 337 (1972)

11.7   F. Sommer: Proc. 5th Int'l. Conf. on Rapidly Quenched Metals, Würzburg 1984, ed. by S. Steeb, H. Warlimont (North-Holland, Amsterdam 1985) p.153; Z. Metallkd. 72, 219 (1981)

11.8   J. Hafner: Phys. Rev. B 28, 1734 (1983)

11.9   T.B. Massalski: Proc. 4th Int'l. Conf. on Rapidly Quenched Metals, ed. by T. Masumoto, K. Suzuki (The Japan Institute of Metals, Sendai 1982) p.203

11.10   J. Jäckle: Phil. Mag. B44, 533 (1981); J. Chem. Phys. 79, 4463 (1983);
S.T. Chui, G.O. Williams, H.L. Frisch: Phys. Rev. B 26, 171 (1982)

11.11   W. Kauzmann: Chem. Rev. 43, 219 (1948);
See also J.W. Allen, A.C. Wright, G.A.N. Connell: J. Non-Cryst. Solids 42, 509 (1980)

11.12   C.A. Angell, W. Sichina, Ann. Acad. Sci. (N.Y.) 279, 53 (1976)

11.13   J.N. Cape, L.V. Woodstock: J. Chem. Phys. 72, 976 (1980)

11.14   G. Grest, M.H. Cohen: Adv. Chem. Phys. 48, 455 (1981)

11.15   H.A. Davies: Amorphous Metallic Materials, ed. by P. Duhaj, P. Mrafko (Slovak Acad. of Sci., Bratislava 1980) p.107

11.16   J.D. Bernal: Nature 185, 68 (1960)

11.17   See, e.g., P.H. Gaskell in [Ref.11.3b, Chap.2]

11.18   L. von Heimendahl: J. Phys. F 5, L141 (1975);
J.A. Baker, M. Hoare, J.L. Finney: Nature 257, 120 (1975)

11.19   See, e.g., M. Grabow, H.C. Andersen: Proc. Int'l. Conf. on the Theory of the Structure of Non-Cryst. Solids, ed. by D. Adler, J. Bicerano; J. Non-Cryst. Solids 75, 225 (1985)

11.20   F.N. Stillinger, T.A. Weber: Phys. Rev. A 25, 978 (1982);
T.A. Weber, F.H. Stillinger: Phys. Rev. B 31, 1954 (1985)

11.21   F.F. Abraham: J. Chem. Phys. 72, 359 (1980)

11.22   J. Hafner: Phys. Rev. B 27, 678 (1983)

11.23   J. Hafner, A. Pasturel: Proc. 3rd Int'l. Conf. on the Structure of Non-Crystalline Meterials, ed. by Chr. Janot (Les Editions de Physique, Paris 1985) p.367

11.24   E. Nassif, P. Lamparter, S. Steeb: Z. Naturforsch. 38a, 1206 (1983)

11.25   F. Spaepen, G.S. Cargill: Proc. 5th Int'l. Conf. on Metals, ed. by S. Steeb, H. Warlimont (North-Holland, Amsterdam 1985) p.581

11.26   H. Rudin, S. Jost, H.J. Güntherodt: J. Non-Cryst. Solids 61&62, 291 (1984)

11.27   E. Nassif, P. Lamparter, W. Sperl, S. Steeb: Z. Naturforsch. 38a, 142 (1983)

11.28   T. Fujiwara: J. Phys. F 12, 661 (1981)

11.29   T. Fujiwara, H.S. Chen, Y. Waseda: Z. Naturforsch. 37a, 611 (1982)

11.30 D. Pettifor, R. Podloucky: Phys. Rev. Lett **53**, 1080 (1984); J. Phys. C **19**, 315 (1986)

11.31 E. Nassif, P. Lamparter, B. Sedelmeyer, S. Steeb: Z. Naturforsch. **38a**, 1093 (1983)

11.32 J. Hafner: Phys. Rev. B **21**, 406 (1980)

11.33 B.C. Giessen: *Alloy Phase Diagrams*, ed. by L.H. Bennett, T.B. Massalski, B.C. Giessen (North-Holland, New York 1983) p.263

11.34 F.C. Frank: Proc. Roy. Soc. London **215**, 43 (1952)

11.35 H.S.M. Coxeter: *Regular Polytopes* (Dover, New York 1983)

11.36 M. Kléman, J.F. Sadoc: J. Physique. Lett. **40**, L569 (1979)

11.37 D.R. Nelson: Phys. Rev. Lett. **50**, 983 (1983); Phys. Rev. B **23**, 5515 (1983)

11.38 F.C. Frank, J.S. Kasper: Acta Cryst. **11**, 184 (1958); ibid. **12**, 483 (1959)

11.39 J. L. Finney: Proc. R. Soc. **A319**, 479 (1970)

11.40 J.F. Sadoc: J. Physique **41**, C8-36 (1980)

11.41 D.R. Nelson, M. Widom: Nucl. Phys. B **240**, 113 (1984)

11.42 S. Sachdev, D.R. Nelson: Phys. Rev. Lett. **53**, 1947 (1984)

11.43 J. Hafner: J. Physique **46**, C9-69 (1985)

11.44 S.R. Nagel, J. Tauc: Phys. Rev. Lett. **35**, 380 (1975)

11.45 P. Häussler, F. Baumann, J. Krieg, G. Indlekofer, P. Oelhafen, H.J. Güntherodt: J. Non-Cryst. Solids **61&62**, 1249 (1984)

11.46 J. Hafner, L. von Heimendahl: Phys. Rev. Lett. **42**, 385 (1979)

11.47 H. Beck, R. Oberle: Proc. 3rd Int'l. Conf. on Rapidly Quenched Metals, ed. by B. Cantor (The Metals Soc., London 1980)

11.48 H. Leitz, W. Buckel: Z. Phys. B **35**, 73 (1979)

11.49 P. Häussler, F. Baumann: Physica **108** B, 909 (1981)

11.50 U. Mizutani: Prog. Mater. Sci. **28**, 97 (1983)

11.51 R. Alben, M. Blume, H. Krakauer, L. Schwartz: Phys. Rev. B **12**, 4090 (1975)

11.52 R. Haydock, V. Heine, M.J. Kelly: J. Phys. C **5**, 2845 (1972); ibid **8**, 2591 (1975)

11.53 J. Hafner: J. Phys. C **16**, 5773 (1983)

11.54 W. Fembacher: Dissertation, University of Regensburg (1985) unpublished

11.55 S.S. Jaswal: J. Non-Cryst. Solids **75**, 373 (1985)

11.56 J. Hafner: J. Phys. F **14**, L287 (1981);
G.S. Grest, S.R. Nagel, A. Rahman: Phys. Rev. B **29**, 5968 (1984)

11.57 J.B. Suck, H. Rudin: In [Ref.11.3b, Chap.7]

11.58 D. Weaire, M.F. Ashby, J. Logan, M.H. Weins: Acta Metall. **19**, 779 (1971)

11.59 J. Hafner: J. Non-Cryst. Solids **75**, 253 (1985)

11.60 S.J. Poon, T.H. Geballe: Phys. Rev. B **18**, 233 (1978)

11.61 G.F. Weir, M.A. Howson, B.L. Gallagher, G.J. Morgan: Phil. Mag. **B47**, 136 (1983)

11.62 U. Krey, R. Jeschek, W. Fembacher: J. Non-Cryst. Solids **61&62**, 1161 (1984)

11.63 D. Levine, P.J. Steinhardt: Phys. Rev. Lett. **53**, 2477 (1984)

11.64 R. Penrose: Bull. Inst. Math. Appl. **10**, 266 (1974)

11.65 A.L. Mackay, Physica **114A**, 609 (1982)

11.66 V. Elser, C.L. Henley; Phys. Rev. Lett. **55**, 2883 (1985)

11.67 M. Duneau, A. Katz: Phys. Rev. Lett. **54**, 2688 (1985)

11.68 C.L. Henley, V. Elser: Philos. Mag. B **53**, L59 (1986)

11.69 D.R. Nelson, S. Sachdev: *Amorphous Metals and Semiconductors*, ed. by P. Haasen, R.I. Jaffee (Acta-Scripta Mtallurgica), Proc. Series Vol.3 (Pergamon, Oxford 1986) p.28

11.70 A.D.J. Haymet: Chem. Phys. Lett. **122**, 324 (1985)

11.71 M. Jaric: Phys. Rev. Lett. **55**, 607 (1985)

11.72 J.W. Cahn, D. Shechtman, D. Gratias: J. Mater. Res. **1**, 13 (1986)

11.73 P.J. Steinhardt, D.R. Nelson, M. Ronchetti: Phys. Rev. Lett. **47**, 1297 (1981); Phys. Rev. B **28**, 784 (1983)

11.74 E. Leutheusser: Phys. Rev. A **29**, 2765 (1984)

11.75 U. Bengtzelius, W. Götze, A. Sjölander: J. Phys. C **17**, 5915 (1984)

11.76 W. Götze: In *Amorphous and Liquid Materials*, ed. by E. Lüscher, G. Fritsch, G. Jacucci, NATO Adv. Study Inst. Ser. E, Vol.118 (Nijhoff, Dordrecht 1987)

11.77 W. Götze: Proc. 6th Int'l. Conf. on Liquid and Amorphous Metals, ed. by F. Hensel, E. Lüscher, Z. Phys. Chem., in press

11.78 J. Hafner: Phys. Rev. B, submitted

11.79 J. Hafner, S.S. Jaswal: Phys. Rev. Lett., submitted; S.S. Jaswal, J. Hafner: to be published

12.1 See e.g. the articles of R.M. Martin, O. Nielsen *Electronic Structure, Dynamics, and the Quantum Structural Properties of Condensed Matter*, ed. by J.R. Devreese, P. Van Camp, NATO-ASI Series B, Vol.121 (Plenum, New York 1985)

12.2 F.H. Stillinger, T.A. Weber: Phys. Rev. B **31**, 5262 (1985)

12.3 T. Takai, T. Haliciogliu, W.A. Tiller: Scr. Metall. **19**, 709 (1985)

12.4 R. Biswas, D. Hamann, Phys. Rev. Lett. **55**, 2001 (1985)

12.5 J. Tersoff: Phys. Rev. Lett. **56**, 632 (1986)

A.1   P. Hohenberg, W. Kohn: Phys. Rev. **136** B, 864 (1964)
A.2   W. Kohn, L.J. Sham: Phys. Rev. **140** A, 1133 (1965)
A.3   L. Hedin, S. Lundquist: Solid State Phys. **23**, 1 (1969)
A.4   J.A. Moriarty: Phys. Rev. B **16**, 2537 (1979)
A.5   J.A. Moriarty: Phys. Rev. B **26**, 1754 (1982)
A.6   I. Lindgren: Int. J. Quantum Chem. Symp. **5**, 411 (1971);
      I. Lindgren, K. Schwarz: Phys. Rev. A **5**, 2572 (1972)
A.7   J.A. Moriarty: Phys. Rev. B **1**, 1363 (1970)
A.8   J.A. Moriarty: Phys. Rev. B **8**, 1338 (1983)
A.9   J. Hafner: Z. Phys. B **22**, 351 (1975)
A.10 J. Hafner: Z. Phys. B **24**, 41 (1976)
A.11 L. Hedin, B.I. Lundquist: J. Phys. C **4**, 2064 (1971)
A.12 R. Taylor: J. Phys. F **8**, 1699 (1978)
A.13 J. Hafner, V. Heine: J.Phys. F **16**, 1429 (1986)
A.14 B.J. Austin, V. Heine, L.J. Sham: Phys. Rev. **127**, 276 (1962)
A.15 J.C. Phillips, L. Kleinman: Phys. Rev. **116**, 287 and 880(1959)
A.16 J.P. Perdew, S.H. Vosko: Phys. Status Solidi B **63**, K47 (1974)
A.17 P.O. Löwdin: J. Math. Phys. **3**, 969 (1962)
A.18 M.H. Cohen, V. Heine: Phys. Rev. **122**, 1821 (1961)
A.19 L.R. Kahn, W.R. Goddard III: Chem. Phys. Lett. **2**, 667 (1968)
A.20 A. Zunger, M.H. Cohen: Phys. Rev. B **18**, 5449 (1978)
A.21 A. Zunger, M.A. Ratner: Chem. Phys. **30**, 423 (1978)
A.22 W.A. Harrison: *Pseudopotentials in the Theory of Metals* (Benjamin, New York 1966)
A.23 J. Hafner: J. Phys. F **6**, 1243 (1976)
A.24 G.P. Kerker: J. Phys. C **13**, L189 (1980)
A.25 D.R. Hamann, M. Schlüter, C. Chiang: Phys. Rev. Lett. **43**, 1494 (1979)
A.26 D. Weaire, A.R. Williams: Phys. Rev. B **14**, 859 (1976)
A.27 R.S. Shaw, W.A. Harrison: Phys. Rev. **163**, 604 (1967)
A.28 W.C. Topp, J.J. Hopfield: Phys. Rev. B **7**, 1295 (1974)
A.29 J.T. Devreese, F. Brosens: *Electronic Structure Dynamics, and Quantum Structural Properties of Condensed matter*, ed. by J.T. Devreese, P. Van Champ, NATO-ASI Series B, Vol.121 (Plenum, New York 1985) p.9
A.30 A. Zunger: J. Vac. Sci. Technol. **16**, 1337 (1979)
A.31 P.A. Christiansen, Y.S. Lee, K.S. Pitzer: J. Chem. Phys. **71**, 4445 (1979)
A.32 G.B. Bachelet, H.S. Greenside, G.A. Baraff, M. Schlüter: Phys. Rev. B **24**, 4745 (1981)
A.33 G.B. Bachelet, D.R. Hamann, M. Schlüter: Phys. Rev. B **26**, 4199 (1982)
A.34 C.O. Almbladh, A.L. Morales, J. Phys. F **15**, 991 (1985)
A.35 M. Kolar, J. Masek: Proc. 13th Int'l. Symp. on Electronic Structure of Metals and Alloys, ed. by P. Ziesche (Technical Univ. Dresden 1983) p.42

B.1 J. Lindhard: K. Dan. Vidensk. Selsk. Mat. Fys. Medd. **28**, 28 (1954)

B.2 W. Kohn, L.J. Sham: Phs. Rev. **140 A**, 1133 (1965)

B.3 L. Hedin, S. Lundqvist: Solid State Phys. **23**, 1 (1969)

B.4 R. Taylor: J. Phys. F **8**, 1699 (1978)

B.5 P. Hohenberg, W.Kohn: Phys. Rev. **136 B**, 864 (1964)

B.6 D. Forster: *Hydrodynamic Fluctuations, Broken Symmetry, and Correlation Functions* (Benjamin, New York 1975)

B.7 K.S. Singwi, M.P. Tosi: Solid State Phys. **36**, 177 (1981)

B.8 J. Hafner, V.Heine: J. Phys. F **16**, 1429 (1986)

B.9 J. Hubbard: Proc. Roy. Soc. London A **243**, 336 (1958)

B.10 P. Vashishta, K.S. Singwi: Phys. Rev. B **6**, 875 (1972)

B.11 D.J.W. Geldart, R. Taylor: Can J. Phys. **48**, 167 (1970)

B.12 S. Ichimaru, K. Utsumi: Phys. Rev. B **24**, 7385 (1981)

B.13 S. Ichimaru: Rev. Mod. Phys. **54**, 1017 (1982)

B.14 D.M. Ceperley, B.J. Alder: Phys. Rev. Lett. **45**, 566 (1980)

B.15 R. Evans: *Electrons in Disordered Metals and at Metallic Surfaces*, ed. by P. Phariseau, B.L. Györffy, L. Scheire, NATO-Adv. Study Inst. Series Vol. B **42** (Plenum, New York 1979) p.417

C.1 K. Fuchs: Proc. Roy. Soc. London A **151**, 515 (1935)

C.2 C.A. Sholl: Proc. Phys. Soc. London **92**, 434 (1967)

C.3 W.J. Carr: Phys. Rev. **122**, 1437 (1961)

C.4 D. Weaire, A.R. Williams: Phil. Mag. **19**, 1105 (1969)

C.5 V. Heine, D.Weaire: Solid State Phys. **24**, 247 (1970)

C.6 K. Ebina, T. Nakamura: Phys. Rev. B **32**, 2614 (1985)

C.7 M.T. Yin, M.L. Cohen: Phys. Rev. B **26**, 5668 (1982)

C.8 J. Hafner: Phys. Rev. B **10**, 4151 (1974)

C.9 P. Villars, L.D. Calvert: *Pearson's Handbook of Crystallographic Data for Intermetallic Phases*, Vols.1-3 (American Soc. for Metals, Metals Park, Ohio 1985)

C.10 H. Olijnyk, W.B. Holzapfel: Phys. Lett **99A**, 381 (1983)

C.11 K. Takemura, S. Minomura, O. Shimomura: Phys. Rev. Lett. **49**, 1772 (1982); **51**, 1603 (1983) (Erratum)

C.12 G.L. Krasko: Sov. Phys. - JETP Lett. **13**, 155 (1971)

C.13 J.E. Inglesfield: J. Phys. C **2**, 1285 and 1293 (1969); Acta Met. **17**, 1395 (1969)

C.14 H. Jones: Phys. Rev. A **8**, 3215 (1973)

C.15 I.H. Umar, A. Meyer, M. Watabe, W.H. Young: J. Phys. F **4**, 1691 (1974)

C.16 M. Ross, D. Seale: Phys. Rev. A **9**, 396 (1974)

C.17 A. Pasturel, J. Hafner, P. Hicter: Phys. Rev. B **32**, 5009 (1985)

D.1 C.A. Croxton: *Liquid State Physics - A Statistical Mechanical Introduction* (Cambridge Univ. Press, Cambridge 1974)

D.2   J.G. Kirkwood: J. Chem. Phys. **3**, 300 (1935)

D.3   R. Abe: Prog. Theoret. Phys. **19**, 57 and 406 (1958)

D.4   G.S. Rushbrooke, H.I. Scoins: Proc. Roy. Soc. London A **216**, 204 (1953)

D.5   W.G. Madden, S.A. Rice: J. Chem. Phys. **72**, 4208 (1980)

D.6   W.G. Madden: J. Chem. Phys. **75**, 1984 (1981)

D.7   Y. Rosenfeld, N.W. Ashcroft: Phys. Rev. A **20**, 1208 (1979)

D.8   F. Lado, S.M. Foiles, N.W. Ashcroft: Phys. Rev. A **28**, 2374 (1983)

D.9   D.D. Carley: J. Chem. Phys. **67**, 1267 (1977)

D.10   F.J. Rogers, D.A. Young: Phys. Rev. A **30**, 999 (1984)

D.11   J.P. Hansen, G. Zerah: Phys. Lett. A **108**, 277 (1985)

D.12   G. Zerah, J.P. Hansen: J. Chem. Phys. **84**, 2336 (1986)

D.13   J. Hafner: Phys. Rev. A **16**, 351 (1977)

D.14   M. Hasegawa, W.H. Young: J. Physique **41**, C8-567 (1980)

D.15   W. Olivares, D.A. McQuarrie: J. Chem. Phys. **65**, 3604 (1976)

D.16   Y. Rosenfeld: J. Stat. Phys. **37**, 215 (1984); ibid **42**, 437 (1986)

D.17   R.J. Baxter: J. Chem. Phys. **52**, 4559 (1970)

D.18   F. Lado: J. Chem. Phys. **59**, 4830 (1973)

D.19   M. Wertheim: Phys. Rev. Lett. **8**, 321 (1963)

D.20   N.W. Ashcroft, J. Lekner: Phys. Rev. **165**, 83 (1966)

D.21   N.F. Carnahan, K.E. Starling: J. Chem. Phys. **51**, 635 (1969)

D.22   J.L. Lebowitz: Phys. Rev. **133 A**, 895 (1964)

D.23   J.L. Lebowitz, J.S. Rowlinson: J. Chem. Phys. **41**, 133 (1964)

D.24   G.A. Mansoori, N.F Carnahan, K.E. Starling,,T.W. Leland: J. Chem. Phys. **54**, 1523 (1971)

D.25   N.W. Ashcroft, D.C. Langreth: Phys. Rev. **156**, 685 (1967)

D.26   J.E. Enderby, D.M. North: J. Phys. Chem. Liq. **1**, 1 (1968)

D.27   M. Shimoji: *Liquid Metals* (Academic, New York 1977) Sect.307

D.28   E. Waisman: Mol. Phys. **25**, 45 (1973)

D.29   R.G. Palmer, J.D. Weeks: J. Chem. Phys. **58**, 4171 (1973)

D.30   J.S. Hoye, G.S. Stell: J. Chem. Phys. **67**, 439 (1977)

D.31   C. Hausleitner: Diplomarbeit, Tech. Univ. Vienna (1987) unpublished;
C. Hausleitner, J, Hafner: to be published

D.32   E. Waisman, J.L. Lebowitz: J. Chem. Phys. **56**, 3086 and 3093 (1972)

D.33   M. Gillan, B. Larsen, M.P. Tosi, N.H. March: J. Phys. C **9**, 889 (1976)

D.34   L. Blum: Mol. Phys. **30**, 1529 (1975);
L. Blum, J.S. Hoye: J. Phys. Chem. **81**, 1311 (1977)

D.35   M. Ginoza: J. Phys. Soc. Jpn. **54**, 2783 (1985)

D.36   E. Waisman: J. Chem. Phys. **59**, 495 (1973)

D.37   J. Hafner, A. Pasturel, P. Hicter: J. Phys. F **14**, 1137 (1984) ;136;1⌐

Subject Index

Born-Oppenheimer approximation,
see Adiabatic approximation
Bulk modulus, see Compressibility

Ca
– crystal structure 72, 88
– liquid structure 108
– pressure-induced phase transition 88
Ca-Li (liquid), concentration fluctuations
248
CaLi$_2$
– electron density 234
– interatomic potentials 280
– stability 218–220
– structural energy differences 223
Ca-Mg
– phase diagram 283
– glass transition 295–296
Ca-Mg (amorphous)
– computer simulation 302
– structure 299
Ca-Mg (liquid), entropy, heat and volume
of formation 249, 284
CaMg$_2$
– electron density 234
– heat of formation 284
– phonon density of states 236
– stability 218–220
– structural energy differences 223
Ca-Zn (crystalline)
– crystal structures 230
– interatomic potentials 231
Cancellation theorem 17–18, 35–36,
320–322
Canonical ensemble 30, 104, 341
Cd, crystal structure 80–83
Cd-Mg, phase diagram 284
Cd-Mg (crystalline)
– crystal structures 174–176
– entropy of formation 204
– heat of formation 174, 204
– order-disorder transition 193–194,
205–206
– volume of formation 204
Cd-Na (crystalline), heat of formation
158, 161
Cd-Zn, phase diagram 284
Charge transfer 59, 154–156, 261–262
Chemical compression 58–59, 212–213
Chemical coordinates 207–210
Chemical ordering 59–63, 242, 248, 251,
254–266, 300
Chemical potential 7, 154, 156
– model of alloy formation 154–156
Chemical scale 209–210
Clausius-Clapeyron relation 134–136
Closure relation 104–107, 255, 343–347
– hypernetted chain (HNC) 105, 345

– mean spherical approximation (MSA)
105–106, 255, 345
– mixed, see Closure relation, Rogers-
Young; Closure relation, Zerah-Hansen
– optimized random phase approximation (ORPA) 113, 272–273
– Percus-Yevick (PY) 105, 345
– Rogers-Young (RY) 346
– Zerah-Hansen (ZH) 346
Cluster-Bethe-lattice method (CBLM)
28, 261–262, 275–276
Cohen-Heine criterion 36, 320
Coherent-potential approximation (CPA)
26
– for phonons in alloys 179–201
Cohesion
– of simple metals 67–73
– of transition metals 72–76
Collective exitation 21, 129–131, 278–281
– in crystals, see Phonons
– in liquid alloys 278–281
– in liquids 129–131
– in metallic glasses 307–309
Common tangent construction 6, 262, 283
Compressibility 44
– pressure derivative of 73
– problem 44–45
– of electron gas 69, 328–330
– of liquid metals 108
– of simple metals 68–69, 73
– sum rule 47–48, 328
Computer simulation 103–104
– of liquid alloys 243–245
– of liquid metals 103–104
– of metallic glasses 298–306
– see also Molecular dynamics
Concentration
– fluctuations 55, 242, 248, 253, 255–256,
286–288, 300–301
– waves 190–192
Configurational coordinates 29
Configurational free energy,
see Free energy
Configurational entropy, see Entropy
Coordination number
– in crystals 81
– in liquids 117–118, 274–275
– in metallic glasses 303–305
Core radius 39, 53
– and crystal structure 80–83
– of simple metals 67–70
Correlation
– energy 12, 67, 155, 160, 316–319
– function, see Direct correlation
function, Pair correlation function
– potential 13, 315–319
Covalency 77, 194, 237–238
Critical cooling rate 297